Mycotoxins in Animal Products

Martin Weidenbörner

Mycotoxins in Animal Products

Milk and Milk Products, and Meat

Springer

Martin Weidenbörner
Bonn, Germany

ISBN 978-3-030-30918-3 ISBN 978-3-030-30919-0 (eBook)
https://doi.org/10.1007/978-3-030-30919-0

This Springer imprint is published by the registered company Springer Nature Switzerland AG
The registered company address is: Gewerbestrasse 11, 6330 Cham, Switzerland

Holger Hindorf

Preface

Mycotoxins in Animal Products—Milk and Milk Products, and Meat (Volume III) deals with mycotoxin contamination of foodstuffs of animal origin. Although not mentioned in the title of the book, mycotoxin contamination of, e.g., aquatic foodstuff, egg, fish, and seafood, as well as edible insects, is also listed. Nut butters and sesame butter are mentioned in *Mycotoxins in Animal Products—Milk and Milk Products, and Meat* as well as in *Mycotoxins in Plants and Plant Products—Cocoa, Coffee, Fruits and Fruit Products, Medicinal Plants, Nuts, Spices, Wine* because their ingredients are of both animal and plant origin. This is also true for a few other articles dealing with animal and plant products already mentioned in *Mycotoxins in Plants and Plant Products—Cereals and Cereal Products* or in Volume II. Human breast milk is a very important foodstuff but listed separately behind all other items in the chapter "Mycotoxin Contamination of Animal Products as well as Human Breast Milk" because it is of human origin but not an animal product.

Since mankind has existed, the need for foodstuff has been indispensable. But food supplies can be attacked by microorganisms. If suitable conditions prevail, microorganisms can grow and subsequently spoil the foodstuff. Of these organisms, filamentous fungi are of special interest. Once they grow, they not only spoil the foodstuff but can also contaminate it with mycotoxins. The presence of these invisible mycotoxins in foodstuff is undesirable and can cause serious problems. High levels of mycotoxins can even cause death shortly after exposure, whereas low levels of mycotoxins can cause disorders in various organs and/or impair immunity. In the end, mycotoxins are a cost factor in human health.

In contrast to bacterial contamination of foodstuff, the enhanced contamination by fungi becomes more apparent. Each consumer knows about the appearance of moldy foodstuff, and the awareness of the public concerning the health hazard of foodstuffs is great. Furthermore, consumers demand high-quality foodstuff. As a prerequisite for producing such foodstuff, the food industry as a whole must be aware of fungi and their corresponding mycotoxins. In this case, the third volume, *Mycotoxins in Animal Products—Milk and Milk Products, and Meat*, gives excellent information about the main mycotoxins occurring in foodstuff of animal origin.

Mycotoxin contamination in foodstuff of animal origin is often represented by metabolites of the *Aspergillus flavus* group, the aflatoxins, predominantly aflatoxin M_1 (AFM_1), which frequently occurs in different types of milk and dairy products. Next to AFM_1 ochratoxin A (OTA, *Aspergillus* spp. and *Penicillium* spp.) is often found in animal foodstuff, especially in pigs. Mycotoxins like patulin (PAT) and penicillic acid (PA, *Aspergillus* spp. and *Penicillium* spp.), mycophenolic acid (MA) and roquefortine C (ROQ C, *Penicillium* spp.), sterigmatocystin (STG, *Aspergillus* spp.), and zearalenone (ZEA, *Fusarium* spp.) appeared to a lesser extent. The occurrence of mycotoxins, such as fumonisin/s, depsipeptides (e.g., beauvericin (BEA) and enniatins), trichothecenes (*Fusarium* spp.), penitrem A (PEA, *Penicillium* spp.), and *Alternaria* toxins, is normally rare in foodstuff of animal origin. Specialists of these mycotoxin-producing fungi can also grow during foodstuff is preserved, e.g., in a refrigerator. All these mycotoxins can be very harmful to humans when ingested.

In the early 1960s, mycotoxin research started with the discovery of aflatoxins in peanut meal causing turkey X disease. In the same decade, monohydroxylated AFM_1 (a possible human carcinogen), a metabolization product of AFB_1 produced in the animal's rumen, in milk was first reported. In many kinds of milk and cheese, AFM_1 is the main mycotoxin. If feedstuff is contaminated by aflatoxin, there is a high probability that AFM_1 occurred in the milk and the corresponding dairy products. Depending on various factors like nutritional and physical status of the animal, up to ~6% of the AFB_1 ingested with the feed is transformed by the liver (cytochrome 450) to AFM_1. Because of the water-soluble hydroxyl group, AFM_1 can easily be excreted through, e.g., the milk (indirect contamination). AFM_1 is not completely inactivated by heat. Therefore, it occurs not only in raw milk but also in pasteurized and sterilized milk samples. Its presence in human breast milk is a result of aflatoxin/s contamination of foodstuffs. Together with AFM_1, OTA is the most common mycotoxin in human breast milk. But in contrast to AFM_1, this mycotoxin predominates in kidneys and blood of pigs due to the consumption of ochratoxin A-contaminated feedstuffs.

During the manufacture of cheese, maximum concentrations of AFM_1 arise due to the high affinity of AFM_1 to the casein fraction. AFB_1 and AFG_1 occur only to a minor extent. AFM_1 concentration in the corresponding cheese can be four times higher than that in cheese milk. However, in yogurt AFM_1 concentration is lowered by different factors such as low pH and formation of organic acids or lactic acid bacteria.

For the protection of consumers, the EU has set a limit of 0.050 μg AFM_1/kg milk.

During cheese manufacture, direct contamination (intentional or accidental) of cheeses such as Camembert and Roquefort with molds is typical. Although selected strains of *P. camemberti* (cyclopiazonic acid, CPA) and *P. roqueforti* (mycophenolic acid and roquefortine C) are used as ripening cultures, they may produce the aforementioned mycotoxins. However, these metabolites exhibit only lower toxicity and occur in low to very low levels. CPA is also one of the few mycotoxins which is considered to transfer into milk. In cheese citrinin (CIT) is often associated with OTA, but not frequently found in this kind of foodstuff. In comparison with cereals and their products, the intake of ochratoxin A from cheese seems to be negligible. If *A. versicolor* is growing on the cheese's surface, STG contamination may be a result of that postcontamination. Aflatoxin/s, citrinin, penitrem A, roquefortine C, and sterigmatocystin are stable in cheese. In contrast patulin, penicillic acid, and PR toxin which can be transformed to, e.g., PR imine or PR amide do not persist.

Whereas zearalenone could be detected in different animal products, other mycotoxins such as the fumonisins, enniatins, trichothecenes, or *Alternaria* toxins play only a minor role in the contamination of foodstuffs of animal origin.

This book provides a basic overview of the most well-known mycotoxins in animal products. Their degree of contamination and the range of concentration of mycotoxins, as well as the mean of the positive samples (as far as possible), are shown in each case. These basic information are intended for readers interested in this topic.

Mycotoxin-contaminated foodstuffs are listed in alphabetical order. Terms with brackets followed single items. All types of butter, cheese, milk, and sausage are listed together to get a quicker overview about mycotoxins involved in the corresponding foodstuffs. This was done as far as possible. Data of more than 160 new publications are added to Volume III. In the end, data of more than 560 articles have been considered in the book. Each item contains the relevant information as far as possible. The present book is the last of three volumes, dealing with the mycotoxin contamination of plants and plant products as well as of animal origin.

For detailed information, the more interested reader is referred to the points "Co-contamination" and "Further contamination." Co-contaminations of a sample with two or more mycotoxins are listed as described in the corresponding article. If "For detailed information see the article" occurs behind "Co-contamination," the corresponding publication dealt with this co-contamination mostly from a general point of view. All entries contain the point "Further contamination," which gives additional information about mycotoxin contamination of foodstuff of animal origin. Moreover, if "see also ..." is written at the end of "Further contamination," a look into Chapter 2: Further Mycotoxins and Microbial Metabolites (e.g. Chicken gizzard, zearalenone, literature[326]), is appropriate. Here, the remaining mycotoxins/microbial metabolites of an animal product are listed. Finally, the book contains plenty of information if totally utilized.

The book also shows whether a food of animal origin is predisposed to mycotoxin contamination, documented by the number of different mycotoxins and the number of citations for the corresponding food item. This was done as far as possible for each case of mycotoxin contamination. Documenting this, more than 560 publications have been used which are all available at German libraries.

The main part of the book represents Chapter 1: Mycotoxin Contamination of Animal Products as well as Human Breast Milk. It discusses in detail mycotoxin contamination of animal products and human breast milk. This is followed by Chapter 2: Further Mycotoxins and Microbial Metabolites, Chapter 3: Mycotoxins and Their Animal Product Spectrum as well as Human Breast Milk, Chapter 4: Animal Products as well as Human Breast Milk and Their Mycotoxins, and Chapter 5: Mycotoxin Contamination in Conventional and Organic Animal Products. This chapter lists articles with their index numbers which compare mycotoxin contamination in conventional and organic production. Chapter 6 documenting the Numerical and Alphabetical Bibliography follows.

The chief publications considered are in English language. Articles in other languages have been chosen only if comprehensive summaries or tables, listing detailed results, were provided in English. Articles that unequivocally describe mycotoxin contamination of animal products (already apparent in the title) have been selected. There are exceptions, however. Also, the most cited publications that satisfied the aforementioned requirements have been considered for the book. Determination methods of mycotoxins are quite beyond the scope of this book (see the title of the book). The focus is on natural mycotoxin contamination of animal products as well as human breast milk. Publications containing data about mycotoxins due to artificial contamination (infected by direct fungal inoculation) have not been included in this volume. Except for a few articles, the publications cited are mainly dealing with natural mycotoxin contamination of marketed animal products. *Mycotoxins in Animals and Animal Products—Milk and Milk Products, and Meat* will provide more information and contribute to the transparency of animal products consumed by humans in relation to mycotoxin contamination. It is a suitable reference text for mycotoxin contamination in this kind of foodstuff.

This fundamental book is especially written for scientists and researchers who are interested in mycotoxin contamination of foodstuff of animal products for human consumption and will be especially suitable for those working in food microbiology, food technology, and the food industry (e.g., food producers, supervisors of food, food traders), as well as ministries, offices, and departments of farming and environmental regulation at national and international levels, bureaus, associations, agricultural bodies, mycologists, mycotoxicologists, toxicologists, biologists, chemists, biochemists, plant pathologists, supervisors in food quality control, lawyers and experts in food law, students of the respective fields, and other interested groups.

Bonn, Germany Martin Weidenbörner

How to Handle the Book

The different mycotoxins in foodstuffs of animal origin are classified by the fungal genera which produce them. However, also other fungal genera may be able to produce a special mycotoxin, but only main producers or relevant species interesting for this book are chosen. For a quick overview, each mycotoxin can furthermore be looked up for its presence in animal products as well as human breast milk at the end of the book (Chapter 3: Mycotoxins and Their Animal Product Spectrum as well as Human Breast Milk) (Tables 1–5) and vice versa (Chapter 4: Animal Products as well as Human Breast Milk and Their Mycotoxins).

In Chapter 1: "Mycotoxin Contamination of Animal Products as well as Human Breast Milk (Volume III)," mycotoxins are arranged as follows:

Alternaria Toxins
Altenuene (ALT), alternariol methyl ether (AME), tenuazonic acid (TA)

Aspergillus Toxins
Aflatoxin/s: Aflatoxin B_1 (AFB_1), aflatoxin B_2 (AFB_2), aflatoxin G_1 (AFG_1), aflatoxin G_2 (AFG_2), aflatoxin M_1 (AFM_1), aflatoxin M_2 (AFM_2), aflatoxin/s (AF/AFS)
Other ***Aspergillus*** Toxins: Sterigmatocystin (STG)

Aspergillus and ***Penicillium*** Toxins
Citrinin (CIT), cyclopiazonic acid (CPA), ochratoxin A (OTA), ochratoxin B (OTB), patulin (PAT), penicillic acid (PA)

Fusarium Toxins
Depsipeptides: Beauvericin (BEA), enniatin A (ENA), enniatin A_1 (ENA_1), enniatin B (ENB), enniatin B_1 (ENB_1)
Fumonisin/s: Fumonisin B_1 (FB_1)
Type A Trichothecenes: HT-2 toxin (HT-2), neosolaniol (NEO), T-2 toxin (T-2)
Type B Trichothecenes: Deoxynivalenol (DON), 3-acetyldeoxynivalenol (3-AcDON), 15-acetyldeoxynivalenol (15-AcDON),
Fusarenon-X (FUS-X), nivalenol (NIV)
Other ***Fusarium*** Toxins: Zearalenone (ZEA)

Penicillium Toxins
Isofumigaclavine A (IFC A), isofumigaclavine B (IFC B), mycophenolic acid (MA), penitrem A (PEA), roquefortine C (ROQ C)

Each declaration of mycotoxin contamination of animal products as well as human breast milk comprises seven main information:

Incidence:	3/7 = 3 positives for aflatoxin contamination in relation to 7 investigated
Concentration range:	minimum and maximum residue values or only the maximum residue value
Ø Concentration:	Ø mycotoxin contamination for the positive samples (as far as possible)
Sample origin:	location, where the sample/s come/s from (as far as possible)
Sample year:	in which month/year the sample was collected/investigated (as far as possible)
Country:	origin of publication
Literature number:	Italy[43]

Usually, the highest mycotoxin value or the lowest and the highest value of mycotoxin contamination is given. In general, mean values of the mycotoxin-positive samples are stated. The presented concentrations (measuring units) occur always in µg/kg or µg/l. If a variant of a trial should not be listed, "no contamination" was determined. In some cases, a variant may be stated although mycotoxin concentration has not been detected. In this case "no contamination" is written.

Each entry is marked with an index number, which is located behind the name of the involved country/countries where the publication has been carried out. This index number stands for the article where the presented results can be checked. It occurs again in the numerical bibliography. The index number refers to the title of the corresponding article. In the alphabetical bibliography, the literature is additionally arranged according to the author's first name of publication. Behind each literature arranged alphabetically, there is a number in brackets giving information about the position of the literature in the numerical bibliography, e.g., (43) in the alphabetical bibliography denotes the 43rd position in the numerical bibliography.

To easily find further mycotoxin-contaminated animal or animal products as well as human breast milk in an entry, the reader has to keep in mind the following scheme:

"Further contamination": **Cheese (goat cheese), AFM_1, literature**[43]

- Title of item: **Cheese (goat cheese)**
- Mycotoxin: **AFM_1**
- Index number: **literature**[43]

With these facts, further contaminated/co-contaminated animal products presented in the original publication can be found in this book. Moreover, all mycotoxins, contaminating the same animal product (listed in "Further contamination"), are quoted in this book. Additionally, the corresponding data in "Further contamination" are marked with an index number. This number can be related to the article, which is presented in the numerical bibliography.

Acknowledgments

I would like to thank my family, especially my wife, for the patience, valuable advice, and support they have provided in writing this book. I also wish to thank Mrs. Renate Frohn and Mrs. Ursula Kleinheyer-Thomas as well as the whole staff from the Branch Library for Medicine, Science and Agricultural Science, University of Bonn, for their support in searching and finding the corresponding publications. I would also like to thank Dr. Ursula Monnerjahn-Karbach for the valuable advice and Mr. Santhamurthy Ramamoorthy/SPi Global for excellent work.

Contents

Abbreviations

15-AcDON	15-Acetyldeoxynivalenol
3-AcDON	3-Acetyldeoxynivalenol
AFB_1	Aflatoxin B_1
AFB_2	Aflatoxin B_2
AFG_1	Aflatoxin G_1
AFG_2	Aflatoxin G_2
AFL	Aflatoxicol
AFM_1	Aflatoxin M_1
AFM_2	Aflatoxin M_2
AF/S	Aflatoxin/s
ALT	Altenuene
AME	Alternariol methyl ether
BEA	Beauvericin
BEN	Balkan endemic nephropathy
CIT	Citrinin
CMA	Central Marketing Association of German Agricultural Producers mbH
CPA	Cyclopiazonic acid
DON	Deoxynivalenol
ENA	Enniatin A
ENA_1	Enniatin A_1
ENB	Enniatin B
ENB_1	Enniatin B_1
EU	European Union
FB_1	Fumonisin B_1
FUS-X	Fusarenon-X
GAU	Gujarat Agricultural University
GDR	German Democratic Republic
HPLC	High performance liquid chromatography
HT-2	HT-2 toxin
IFC A	Isofumigaclavine A (roquefortine A)
IFC B	Isofumigaclavine B (roquefortine B)
LOD	Limit of detection
LOQ	Limit of quantification
MA	Mycophenolic acid
MPN	Danish mycotoxic porcine nephropathy
MSRM	Migliarino, San Rossore, Massaciuccoli (natural park)
nc	No comment
nd	Not detected
NEO	Neosolaniol
NIV	Nivalenol
NWFP	North West Frontier Province
OTA	Ochratoxin A
OT-α	Ochratoxin-α

OTAME	Ochratoxin A methyl ester
OTB	Ochratoxin B
PA	Penicillic acid
PAT	Patulin
PAU	Punjab Agricultural University
PEA	Penitrem A
pos	Positive
pr	Present
RADIUS	Agronomy and Biotechnology of the Thomas More University of Applied Sciences, Belgium
ROQ C	Roquefortine C, roquefortine (isofumigaclavine C)
sa	Sample/s
STG	Sterigmatocystin
T-2	T-2 toxin
TA	Tenuazonic acid
TLC	Thin layer chromatography
tr	Traces
USP	University of São Paulo
ZEA	Zearalenone
α-ZEL	α-Zearalenol
ß-ZEL	ß-Zearalenol

Mycotoxin Contamination of Animal Products as well as Human Breast Milk

Aquatic food may contain the following mycotoxins:

Alternaria Toxins

Altenuene

incidence: 1/5*, conc.: 8.77 µg/kg, sample origin: vegetable markets, grain shops, farmer's markets, or rural household in 5 provinces (Sixth China Total Diet Study), China, sample year: unknown, country: China[536], *aquatic food and aquatic food products

- Co-contamination: not reported
- Further contamination of animal products reported in the present book/article[536]: Aquatic food, AME, BEA, ENB, ENB_1, and TA, literature[536]; Eggs, AME, BEA, ENA, ENA_1, ENB, ENB_1, and TA, literature[536]; Meat, ENB, literature[536]; Milk, BEA, ENB, and TA, literature[536]; see also Aquatic food, Eggs, chapter: Further Mycotoxins and Microbial Metabolites

Alternariol Methyl Ether

incidence: 1/5*, conc.: 0.38 µg/kg, sample origin: vegetable markets, grain shops, farmer's markets, or rural household in 5 provinces (Sixth China Total Diet Study), China, sample year: unknown, country: China[536], *aquatic food and aquatic food products

- Co-contamination: not reported
- Further contamination of animal products reported in the present book/article[536]: Aquatic food, ALT, BEA, ENB, ENB_1, and TA, literature[536]; Eggs, AME, BEA, ENA, ENA_1, ENB, ENB_1, and TA, literature[536]; Meat, ENB, literature[536]; Milk, BEA, ENB, and TA, literature[536]; see also Aquatic food, Eggs, chapter: Further Mycotoxins and Microbial Metabolites

Tenuazonic Acid

incidence: 1/5*, conc.: 0.68 µg/kg, sample origin: vegetable markets, grain shops, farmer's markets, or rural household in 5 provinces (Sixth China Total Diet Study), China, sample year: unknown, country: China[536], *aquatic food and aquatic food products

- Co-contamination: not reported
- Further contamination of animal products reported in the present book/article[536]: Aquatic food, AME, BEA, ENB, ENB_1, and TA, literature[536]; Eggs, AME, BEA, ENA, ENA_1, ENB, ENB_1, and TA, literature[536]; Meat, ENB, literature[536]; Milk, BEA, ENB, and TA, literature[536]; see also Aquatic food, Eggs, chapter: Further Mycotoxins and Microbial Metabolites

Fusarium Toxins

Depsipeptides

Beauvericin

incidence: 2/5*, conc. range: 0.32–1.02 µg/kg, ∅ conc.: 0.67 µg/kg, sample origin: vegetable markets, grain shops, farmer's markets, or rural household in 5 provinces (Sixth China Total Diet Study), China, sample year: unknown, country: China[536], *aquatic food and aquatic food products

- Co-contamination: not reported
- Further contamination of animal products reported in the present book/article[536]: Aquatic food, ALT, AME, ENB, ENB_1, and TA, literature[536]; Eggs, AME, BEA, ENA, ENA_1, ENB, ENB_1, and TA, literature[536]; Meat, ENB, literature[536]; Milk, BEA, ENB, and TA, literature[536]; see also Aquatic food, Eggs, chapter: Further Mycotoxins and Microbial Metabolites

Enniatin B

incidence: 3/5*, conc. range: 0.29–0.65 µg/kg, sample origin: vegetable markets, grain shops, farmer's markets, or rural household in 5 provinces (Sixth China Total Diet Study), China, sample year: unknown, country: China[536], *aquatic food and aquatic food products

- Co-contamination: not reported
- Further contamination of animal products reported in the present book/article[536]: Aquatic food, ALT, AME, BEA, ENB_1, and TA, literature[536]; Eggs, AME, BEA, ENA, ENA_1, ENB, ENB_1, and TA, literature[536]; Meat, ENB, literature[536]; Milk, BEA, ENB, and TA, literature[536]; see also Aquatic food, Eggs, chapter: Further Mycotoxins and Microbial Metabolites

Enniatin B_1

incidence: 1/5*, conc.: 0.22 µg/kg, sample origin: vegetable markets, grain shops, farmer's markets, or rural household in 5 provinces (Sixth China Total Diet Study), China, sample year: unknown, country: China[536], *aquatic food and aquatic food products

- Co-contamination: not reported
- Further contamination of animal products reported in the present book/article[536]: Aquatic food, ALT, AME, BEA, ENB, and TA, literature[536]; Eggs, AME, BEA, ENA, ENA_1, ENB, ENB_1, and TA, literature[536]; Meat, ENB, literature[536]; Milk, BEA, ENB, and TA, literature[536]; see also Aquatic food, Eggs, chapter: Further Mycotoxins and Microbial Metabolites

Ayran may contain the following mycotoxins:

Aspergillus Toxins

Aflatoxin M_1

incidence: 4/6, conc. range: 0.01589–0.02735 µg/l, ∅ conc.: 0.0213 µg/l, sample origin: rural regions in Hamadan, Ilam, Kermanshah, and Kurdistan (provinces), Iran, sample year: winter and summer seasons 2014, country: Iran[328]

- Co-contamination: not reported
- Further contamination of animal products reported in the present book/article[328]: Cheese, Kashk, Milk (cow milk), Milk (goat milk), Milk (sheep milk), Tarkhineh, Yogurt, AFM_1, literature[328]

incidence: 16/71*, conc. range: 0.013–0.053 µg/l, sample origin: dairy ranches, supermarkets, and retail outlets in Esfahan, Shiraz, and Tabriz (cities) as well as Tehran (capital), Iran, sample year: 2008, country: Iran[440], *industrial Ayran (doogh)

incidence: 9/65*, conc. range: 0.013–0.029 µg/l, sample origin: dairy ranches, supermarkets, and retail outlets in Esfahan, Shiraz, and Tabriz (cities) as well as Tehran (capital), Iran, sample year: 2008, country: Iran[440], *traditional Ayran (doogh)

- Co-contamination: not reported

M. Weidenbörner, *Mycotoxins in Animal Products*, https://doi.org/10.1007/978-3-030-30919-0_1

- Further contamination of animal products reported in the present book/article[440]: Cheese (Lighvan cheese), Kashk, Milk (cow milk), Milk (goat milk), Milk (sheep milk), Yogurt, AFM_1, literature[440]

incidence: 1/55, conc.: 0.023 µg/kg, sample origin: Corum (province), Turkey, sample year: October 2012–March 2015, country: Turkey[510]

- Co-contamination: not reported
- Further contamination of animal products reported in the present book/article[510]: Milk (cow milk), Yogurt, AFM_1, literature[510]

incidence: 110/110*, conc. range: <0.010 µg/l (41 sa.), 0.011–0.025 µg/l (49 sa.), 0.026–0.050 µg/l (7 sa.), 0.051–0.100 µg/l (9 sa.), >0.100 µg/l (3 sa., maximum: 0.152 µg/l), Ø conc.: 0.0464 µg/l, sample origin: dairy plants in Tehran (province), Iran, sample year: 2009, country: Iran[521], *pasteurized industrial Ayran (doogh)

incidence: 115/115*, conc. range: <0.010 µg/l (33 sa.), 0.011–0.025 µg/l (52 sa.), 0.026–0.050 µg/l (16 sa.), 0.051–0.100 µg/l (13 sa.), >0.100 µg/l (2 sa., maximum: 0.211 µg/l), Ø conc.: 0.0523 µg/l, sample origin: local people in Tehran (province), Iran, sample year: 2009, country: Iran[521], *pasteurized traditional Ayran (doogh)

- Co-contamination: not reported
- Further contamination of animal products reported in the present book/article[521]: not reported

Ayran (doogh) is a full-fat yogurt-based beverage enriched with salt and spices.
see also **Yogurt**

Baby food see **Milk (infant formula)**

Bacon may contain the following mycotoxins:

Aspergillus and ***Penicillium*** Toxins

Ochratoxin A

incidence: 2/34*, conc. range: 0.95–1.23 µg/kg, sample origin: markets, fairs, or directly from the producing household, Primorje, Zagorje, Medimurje, Slavonia, and Baranja (regions), Croatia, sample year: 2011–2014, country: Croatia/Italy[500], *Slavonian type

- Co-contamination: not reported
- Further contamination of animal products reported in the present book/article[500]: Ham, Sausage, Sausage (liver sausage), AFB_1 and OTA, literature[500]

Barfi see **Dairy products**

Beef meat may contain the following mycotoxins:

Aspergillus Toxins

Aflatoxin/s

incidence: 6/20*, conc. range: 1.10–8.32 µg/kg**, sample origin: local markets, Jordan, sample year: January–May 2007, country: Jordan[157], *imported beef meat, **AFB_1, AFB_2, AFG_1, AFG_2, and AFM_1

- Co-contamination: not reported
- Further contamination of animal products reported in the present book/article[157]: Eggs, Meat, Milk, AF/S, literature[157]

For detailed information see the article.

incidence: 25/25*, conc. range: <1 µg/kg** (12 sa.), 1–<2 µg/kg** (11 sa.), 2–<3 µg/kg** (2 sa., maximum: 2.1 µg/kg**), Ø conc.: 1.12 µg/kg**, sample origin: supermarkets in Mansoura (city), Egypt, sample year: unknown, country: Egypt[392], *ready-to-eat beef luncheon, **total AFS

- Co-contamination: 25 sa. co-contaminated with AFS and OTA
- Further contamination of animal products reported in the present book/article[392]: Beef meat, OTA, literature[392]; Beefburger, AF/S and OTA, literature[392]

Aspergillus and ***Penicillium*** Toxins

Ochratoxin A

incidence: 1/58, conc.: 0.03 µg/kg, sample origin: CMA, Germany, sample year: unknown, country: Germany[1]

- Co-contamination: not reported
- Further contamination of animal products in the present book/article[1]: Joints, Meat products, Pig kidney, Pig liver, Pork meat, Sausage (beef sausage), Sausage (blood sausage), Sausage (Bologna sausage), Sausage (liver sausage), Sausage (poultry sausage), Sausage (raw sausage), OTA, literature[1]

incidence: 25/25*, conc. range: <1 µg/kg (1 sa.), 3–<5 µg/kg (11 sa.), 5–<7 µg/kg (8 sa.), 7–9 µg/kg (5 sa., maximum: 8.5 µg/kg), Ø conc.: 5.23 µg/kg, sample origin: supermarkets in Mansoura (city), Egypt, sample year: unknown, country: Egypt[392], *ready-to-eat beef luncheon

- Co-contamination: 25 sa. co-contaminated with AFS and OTA
- Further contamination of animal products reported in the present book/article[392]: Beef meat, AF/S, literature[392]; Beefburger, AF/S and OTA, literature[392]

Fusarium Toxins

Zearalenone

incidence: 5/25*, conc. range: 3.2–11.8 µg/kg, Ø conc.: 8.7 µg/kg, sample origin: shops and supermarkets in Alexandria (province), Egypt, sample year: unknown, country: Egypt[141], *fresh beef meat

- Co-contamination: not reported
- Further contamination of animal products reported in the present book/article[141]: Beefburger, Cheese (Hard Roumy cheese), Cheese (Karish cheese), Meat, Milk, Milk powder, Sausage, ZEA, literature[141]

Beefburger may contain the following mycotoxins:

Aspergillus Toxins

Aflatoxin B_1

incidence: 5/25, conc. range: 8 µg/kg, sample origin: local companies in Cairo (capital), Egypt, sample year: unknown, country: Egypt[2]

- Co-contamination: not reported
- Further contamination of animal products in the present book/article[2]: Hot-dog, Kubeba, Meat, Sausage, AFB_1 and AFB_2, literature[2]

Aflatoxin/s

incidence: 25/25, conc. range: <1 µg/kg* (4 sa.), 1–<2 µg/kg* (7 sa.), 2–<3 µg/kg* (2 sa.), 3–<5 µg/kg* (7 sa.), 5–<7 µg/kg* (4 sa.), 7.5 µg/kg* (1 sa.), Ø conc.: 3.22 µg/kg*, sample origin: supermarkets in Mansoura (city), Egypt, sample year: unknown, country: Egypt[392], *total AFS

- Co-contamination: 15 sa. co-contaminated with AFS and OTA
- Further contamination of animal products reported in the present book/article[392]: Beef meat, AF/S and OTA, literature[392]; Beefburger, OTA, literature[392]

Aspergillus **and** ***Penicillium*** **Toxins**

Ochratoxin A

incidence: 1/1, conc.: 0.44 µg/kg, sample origin: supermarkets and fast-food chains in Quebec (city), Canada, sample year: September–October 2008, country: Canada[87]

incidence: 1/1, conc.: 0.22 µg/kg, sample origin: supermarkets and fast-food chains in Calgary (city), Canada, sample year: September–October 2009, country: Canada[87]

- Co-contamination: not reported
- Further contamination of animal products reported in the present book/article[87]: Butter (peanut butter), Chicken nuggets, Chickenburger, Hot-dog, Ice cream, Meat, Meat products, Pork meat, Sausage, OTA, literature[87]

incidence: 25/25, conc. range: 2–<3 µg/kg (5 sa.), 3–<5 µg/kg (11 sa.), 5–<7 µg/kg (7 sa.), 7–9 µg/kg (2 sa., maximum: 7.6 µg/kg), Ø conc.: 4.55 µg/kg, sample origin: supermarkets in Mansoura (city), Egypt, sample year: unknown, country: Egypt[392]

- Co-contamination: 15 sa. co-contaminated with AFS and OTA
- Further contamination of animal products reported in the present book/article[392]: Beef meat, AF/S and OTA, literature[392]; Beefburger, AF/S literature[392]

Fusarium **Toxins**

Zearalenone

incidence: 3/20, conc. range: 1.6–6.7 µg/kg, Ø conc.: 4.8 µg/kg, sample origin: shops and supermarkets in Alexandria (province), Egypt, sample year: unknown, country: Egypt[141]

- Co-contamination: not reported
- Further contamination of animal products reported in the present book/article[141]: Beef meat, Cheese (Hard Roumy cheese), Cheese (Karish cheese), Meat, Milk, Milk powder, Sausage, ZEA, literature[141]

Beverages see Dairy beverages

Black pudding see Sausage

Blue cheese dressing see Cheese (blue-veined cheese)

Boar kidney see Wild boar kidney

Boar liver see Wild boar liver

Boar muscle see Wild boar muscle

Bonga see Fish

Broiler see Chicken

Butter may contain the following mycotoxins:

Aspergillus **Toxins**

Aflatoxin M_1

incidence: 52/64, conc. range: 0.001–0.010 µg/kg (3 sa.), 0.011–0.050 µg/kg (29 sa.), 0.051–0.100 µg/kg (2 sa.), 0.101–0.250 µg/kg (7 sa.), >0.250 µg/kg (11 sa., maximum: 2.2 µg/kg), sample origin: Istanbul (city), Turkey, sample year: 2001, country: Turkey[10]

- Co-contamination: not reported
- Further contamination of animal products reported in the present book/article[10]: Cheese (white cheese), AFM_1, literature[10]

incidence: 25/27, conc. range: <0.001 µg/kg (8 sa.), 0.001–0.010 µg/kg (1 sa.), 0.011–0.050 µg/kg (15 sa.), 0.051–0.100 µg/kg (1 sa.), sample origin: markets in Ankara (capital), Turkey, sample year: September 2002–2003, country: Turkey[11]

- Co-contamination: not reported
- Further contamination of animal products reported in the present book/article[11]: Cheese (cream cheese), Cheese (Kashar cheese), Cheese (white cheese), AFM_1, literature[11]

incidence: 3/10, conc. range: 0.025–0.050 µg/kg (1 sa.), 0.051–0.100 µg/kg (2 sa., maximum: 0.070 µg/kg), Ø conc.: 0.057 µg/kg, sample origin: supermarkets in Adana (city), Turkey, sample year: unknown, country: Turkey[12]

- Co-contamination: not reported
- Further contamination of animal products reported in the present book/article[12]: Cheese (Kashar cheese), Milk (UHT milk), Cheese (white cheese), AFM_1, literature[12]

incidence: 92/92, conc. range: 0.001–0.050 µg/kg (66 sa.), 0.051–0.100 µg/kg (5 sa.), 0.101–0.250 µg/kg (6 sa.), >0.250 µg/kg (15 sa., maximum 7.0 µg/kg), Ø conc.: 0.236 µg/kg, sample origin: supermarkets in Izmir, Kayseri, Konya, and Tekirdag (cities) as well as Istanbul (capital), Turkey, sample year: March–June 2005, country: Turkey[13]

- Co-contamination: not reported
- Further contamination of animal products reported in the present book/article[13]: Cheese (cream cheese), AFM_1, literature[13]

incidence: 10/10, conc. range: 0.0047–0.0167 µg/kg, Ø conc.: 0.01428 µg/kg, sample origin: Sistan and Baluchestan (provinces), south-east of Iran, sample year: summer and winter 2015, country: Iran[417]

- Co-contamination: not reported
- Further contamination of animal products reported in the present book/article[417]: Cheese (white cheese), Milk, Yogurt, AFM_1, literature[417]

For detailed information see the article.

incidence: 39/45, conc. range: 0.011–0.025 µg/kg (1 sa.), 0.026–0.050 µg/kg (15 sa.), 0.051–0.100 µg/kg (9 sa.), 0.101–0.250 µg/kg (13 sa.), 0.320 µg/kg (1 sa.), sample origin: various districts in Burdur (city), Turkey, sample year: 2008, country: Turkey[418]

- Co-contamination: not reported
- Further contamination of animal products reported in the present book/article[418]: Cheese (white cheese), Ice cream, Milk (cow milk), Milk powder, Yogurt, AFM_1, literature[418]

incidence: 8/31, conc. range: 0.013–0.026 µg/kg, sample origin: retail stores and supermarkets in Esfahan, Shiraz, and Yazd (cities), as well as Tehran (capital), Iran, sample year: winter and summer 2009, country: Iran[435]

- Co-contamination: not reported
- Further contamination of animal products reported in the present book/article[435]: Cheese (white cheese), Ice cream, Milk, Yogurt, AFM_1, literature[435]

For detailed information see the article.

incidence: 33/74, conc. range: <0.050 µg/kg (16 sa.), >0.050–0.150 µg/kg (11 sa.), >0.151 µg/kg (6 sa., maximum: 0.4134 µg/kg), Ø conc.: 0.15629 µg/kg, sample origin: main districts of Punjab (province), Pakistan, sample year: November 2010–April 2011, country: Pakistan[451]

- Co-contamination: not reported

- Further contamination of animal products reported in the present book/article[451]: Cheese (cream cheese), Cheese (white cheese), Milk, Yogurt, AFM_1, literature[451]

incidence: 4/10, conc. range: 0.007–0.0074 µg/l, Ø conc.: 0.0072 µg/l, sample origin: dairy farms, sale points, bazaars, and markets in Faisalabad (city), Pakistan, sample year: unknown, country: Pakistan[482]

- Co-contamination: not reported
- Further contamination of animal products reported in the present book/article[482]: Milk (cow milk), Yogurt, AFM_1, literature[482]

For detailed information see the article.

Butter (cocoa butter) may contain the following mycotoxins:

Aspergillus Toxins

Aflatoxin B_1

incidence: 7/25, conc. range: ≤0.38 µg/kg, sample origin: different stages of manufacture at processing plants, Brazil, sample year: unknown, country: Brazil[14]

- Co-contamination: not reported
- Further contamination of animal products reported in the present book/article[14]: Butter (cocoa butter), AFB_2, literature[14]

Aflatoxin B_2

incidence: 1/25, conc.: 0.04 µg/kg, sample origin: different stages of manufacture at processing plants, Brazil, sample year: unknown, country: Brazil[14]

- Co-contamination: not reported
- Further contamination of animal products reported in the present book/article[14]: Butter (cocoa butter), AFB_1, literature[14]

Aspergillus and *Penicillium* Toxins

Ochratoxin A

incidence: 20/25, conc. range: ≤0.06 µg/kg, sample origin: processing plants in the region of Ilhéus (city) in Bahia (state) and São Paulo (city), São Paulo (state), Brazil, sample year: unknown, country: Brazil[15]

- Co-contamination: not reported
- Further contamination of animal products reported in the present book/article[15]: not reported

incidence: 1?/5, conc.: 0.08 µg/kg, sample origin: Canada, sample year: unknown, country: Canada[16]

- Co-contamination: For detailed information see the article.
- Further contamination of animal products reported in the present book/article[16]: not reported

incidence: 24/36, conc. range: ≤1.88 µg/kg, sample origin: Kumba? (city), Cameroon, sample year: 2005–2007, country: Cameroon/France[17]

- Co-contamination: not reported
- Further contamination of animal products reported in the present book/article[17]: not reported

Butter (nut butter) may contain the following mycotoxins:

Aspergillus Toxins

Aflatoxin B_1

incidence: 2/6*, conc. range: 0.2 µg/kg (1 sa.), 2.5 µg/kg** (1 sa.), Ø conc.: 1.35 µg/kg, sample origin: retail outlets, UK, sample year: September 2001, country: UK[9], *included almond, cashew, hazelnut, and 3 nut butter**

- Co-contamination: 1 sa. co-contaminated with AFB_1, AFB_2, AFG_1, AFG_2, and OTA; 1 sa. contaminated solely with AFB_1
- Further contamination of animal products reported in the present book/article[9]: Butter (nut butter), AFB_2, AFG_1, AFG_2, and OTA, literature[9]; Butter (peanut butter), AFB_1, AFB_2, AFG_1, AFG_2, and OTA, literature[9]

incidence: 1/1*, conc.: 2.9 µg/kg, sample origin: retail outlets, UK, sample year: September 2001, country: UK[9], *crunchy 3 nut butter

- Co-contamination: 1 sa. co-contaminated with AFB_1, AFB_2, AFG_1, and AFG_2
- Further contamination of animal products reported in the present book/article[9]: Butter (nut butter), AFB_2, AFG_1, AFG_2, and OTA, literature[9]; Butter (peanut butter), AFB_1, AFB_2, AFG_1, AFG_2, and OTA, literature[9]

Aflatoxin B_2

incidence: 1/6*, conc.: 0.6 µg/kg, sample origin: retail outlets, UK, sample year: September 2001, country: UK[9], *included almond, cashew, hazelnut, and 3 nut butter

- Co-contamination: 1 sa. co-contaminated with AFB_1, AFB_2, AFG_1, AFG_2, and OTA
- Further contamination of animal products reported in the present book/article[9]: Butter (nut butter), AFB_1, AFG_1, AFG_2, and OTA, literature[9]; Butter (peanut butter), AFB_1, AFB_2, AFG_1, AFG_2, and OTA, literature[9]

incidence: 1/1*, conc.: 0.6 µg/kg, sample origin: retail outlets, UK, sample year: September 2001, country: UK[9], *crunchy 3 nut butter

- Co-contamination: 1 sa. co-contaminated with AFB_1, AFB_2, AFG_1, and AFG_2
- Further contamination of animal products reported in the present book/article[9]: Butter (nut butter), AFB_1, AFG_1, AFG_2, and OTA, literature[9]; Butter (peanut butter), AFB_1, AFB_2, AFG_1, AFG_2, and OTA, literature[9]

Aflatoxin G_1

incidence: 1/6*, conc.: 0.8 µg/kg, sample origin: retail outlets, UK, sample year: September 2001, country: UK[9], *included almond, cashew, hazelnut, and 3 nut butter

- Co-contamination: 1 sa. co-contaminated with AFB_1, AFB_2, AFG_1, AFG_2, and OTA
- Further contamination of animal products reported in the present book/article[9]: Butter (nut butter), AFB_1, AFB_2, AFG_2, and OTA, literature[9]; Butter (peanut butter), AFB_1, AFB_2, AFG_1, AFG_2, and OTA, literature[9]

incidence: 1/1*, conc.: 1 µg/kg, sample origin: retail outlets, UK, sample year: September 2001, country: UK[9], *crunchy 3 nut butter

- Co-contamination: 1 sa. co-contaminated with AFB_1, AFB_2, AFG_1, and AFG_2
- Further contamination of animal products reported in the present book/article[9]: Butter (nut butter), AFB_1, AFB_2, AFG_2, and OTA, literature[9]; Butter (peanut butter), AFB_1, AFB_2, AFG_1, AFG_2, and OTA, literature[9]

Aflatoxin G_2

incidence: 1/6*, conc.: 0.3 µg/kg, sample origin: retail outlets, UK, sample year: September 2001, country: UK[9], *included almond, cashew, hazelnut, and 3 nut butter

- Co-contamination: 1 sa. co-contaminated with AFB_1, AFB_2, AFG_1, AFG_2, and OTA
- Further contamination of animal products reported in the present book/article[9]: Butter (nut butter), AFB_1, AFB_2, AFG_1, and OTA, literature[9]; Butter (peanut butter), AFB_1, AFB_2, AFG_1, AFG_2, and OTA, literature[9]

incidence: 1/1*, conc.: 0.3 μg/kg, sample origin: retail outlets, UK, sample year: September 2001, country: UK[9], *crunchy 3 nut butter

- Co-contamination: 1 sa. co-contaminated with AFB_1, AFB_2, AFG_1, and AFG_2
- Further contamination of animal products reported in the present book/article[9]: Butter (nut butter), AFB_1, AFB_2, AFG_1, and OTA, literature[9]; Butter (peanut butter), AFB_1, AFB_2, AFG_1, AFG_2, and OTA, literature[9]

Aspergillus and *Penicillium* Toxins

Ochratoxin A

incidence: 2/6*, conc. range: 0.7–2.0 μg/kg, ∅ conc.: 1.35 μg/kg, sample origin: retail outlets, UK, sample year: September 2001, country: UK[9], *included almond, cashew, hazelnut, and 3 nut butter

- Co-contamination: 1 sa. co-contaminated with AFB_1, AFB_2, AFG_1, AFG_2, and OTA; 1 sa. contaminated solely with OTA
- Further contamination of animal products reported in the present book/article[9]: Butter (nut butter), AFB_1, AFB_2, AFG_1, and AFG_2, literature[9]; Butter (peanut butter), AFB_1, AFB_2, AFG_1, AFG_2, and OTA, literature[9]

incidence: 0/1*, conc.: no contamination, sample origin: retail outlets, UK, sample year: September 2001, country: UK[9], *crunchy 3 nut butter

- Co-contamination: For detailed information see the article.
- Further contamination of animal products reported in the present book/article[9]: Butter (nut butter), AFB_1, AFB_2, AFG_1, and AFG_2, literature[9]; Butter (peanut butter), AFB_1, AFB_2, AFG_1, AFG_2, and OTA, literature[9]

Butter (peanut butter) may contain the following mycotoxins:

Aspergillus Toxins

Aflatoxin B_1

incidence: 19/29, conc. range: LOQ-2 μg/kg (13 sa.), >2–5 μg/kg (5 sa.), 7.5 μg/kg (1 sa.), ∅ conc.: 1.82 μg/kg, sample origin: retail outlets, UK, sample year: September 2001, country: UK[9]

- Co-contamination: 2 sa. co-contaminated with AFB_1, AFB_2, AFG_1, AFG_2, and OTA; 3 sa. co-contaminated with AFB_1, AFB_2, AFG_1, and AFG_2; 3 sa. co-contaminated with AFB_1, AFB_2, and AFG_1; 2 sa. co-contaminated with AFB_1 and AFB_2; 4 sa. co-contaminated with AFB_1 and AFG_1; 1 sa. co-contaminated with AFB_1 and OTA; 4 sa. contaminated solely with AFB_1
- Further contamination of animal products reported in the present book/article[9]: Butter (nut butter), AFB_1, AFB_2, AFG_1, AFG_2, and OTA, literature[9]; Butter (peanut butter), AFB_2, AFG_1, AFG_2, and OTA, literature[9]

incidence: 16/27, conc. range: LOQ-2 μg/kg (14 sa.), 2.5 μg/kg (1 sa.), 2.9 μg/kg (1 sa.), ∅ conc.: 1.14 μg/kg, sample origin: retail outlets, UK, sample year: September 2001, country: UK[9]

- Co-contamination: 2 sa. co-contaminated with AFB_1, AFB_2, AFG_1, and AFG_2; 1 sa. co-contaminated with AFB_1, AFB_2, and AFG_1; 2 sa. co-contaminated with AFB_1, AFB_2, AFG_1, and OTA; 1 sa. co-contaminated with AFB_1 and AFB_2; 3 sa. co-contaminated with AFB_1 and AFG_1; 7 sa. contaminated solely with AFB_1
- Further contamination of animal products reported in the present book/article[9]: Butter (nut butter), AFB_1, AFB_2, AFG_1, AFG_2, and OTA, literature[9]; Butter (peanut butter), AFB_2, AFG_1, AFG_2, and OTA, literature[9]

incidence: 7/16* **, conc. range: 2–5 μg/kg (6 sa.), 7 μg/kg (1 sa.), sample origin: retail outlets in 20 different regions, England, sample year: January 1982, country: UK[18], *regular food, **peanut butter, smooth

incidence: 5/16* **, conc. range: 2–5 μg/kg (3 sa.), 6–10 μg/kg (1 sa.), 12 μg/kg (1 sa.), sample origin: retail outlets in 20 different regions, England, sample year: January 1982, country: UK[18], *regular food, **peanut butter, crunchy

incidence: 4/11* **, conc. range: 6–10 μg/kg (2 sa.), 31–100 μg/kg (2 sa., maximum: 49 μg/kg), sample origin: retail outlets in 20 different regions, England, sample year: January 1982, country: UK[18], *health food, **peanut butter, smooth

incidence: 10/14* **, conc. range: 2–5 μg/kg (2 sa.), 6–10 μg/kg (1 sa.), 11–30 μg/kg (1 sa.), 31–100 μg/kg (1 sa.), >100 μg/kg (5 sa., maximum: 318 μg/kg), sample origin: retail outlets in 20 different regions, England, sample year: January 1982, country: UK[18], *health food, **peanut butter, crunchy

incidence: 5/6* **, conc. range: 11–30 μg/kg (3 sa.), 31–100 μg/kg (2 sa., maximum: 76 μg/kg), sample origin: retail outlets in 20 different regions, sample year: June 1983, country: UK[18], *health food, **peanut butter, smooth

incidence: 7/9* **, conc. range: 2–5 μg/kg (5 sa.), 6–10 μg/kg (1 sa.), 58 μg/kg (1 sa.), sample origin: retail outlets in 20 different regions, England, sample year: June 1983, country: UK[18], *health food, **peanut butter, crunchy

incidence: 1/4* **, conc.: 13 μg/kg, sample origin: retail outlets in 20 different regions, England, sample year: December 1984, country: UK[18], *health food, **peanut butter, smooth

incidence: 7/15* **, conc. range: 6–10 μg/kg (1 sa.), 11–30 μg/kg (3 sa.), 31–100 μg/kg (3 sa., maximum: 73 μg/kg), sample origin: retail outlets in 20 different regions, England, sample year: December 1984, country: UK[18], *health food, **peanut butter, crunchy

- Co-contamination: not reported
- Further contamination of animal products reported in the present book/article[18]: Butter (peanut butter), AF/S literature[18]

incidence: 8/8, ∅ conc.: 1.3 μg/kg, sample origin: shops, supermarkets and wholesalers, France, sample year: 1989–1990, country: France[19]

- Co-contamination: not reported
- Further contamination of animal products reported in the present book/article[19]: not reported

incidence: 22?/26, conc. range: 0.09–2.65 μg/kg, sample origin: unknown, sample year: unknown, country: Korea[20]

- Co-contamination: not reported
- Further contamination of animal products reported in the present book/article[20]: Butter (peanut butter), AFB_2, AFG_1, and AFG_2, literature[20]

incidence: 5/5, conc. range: 1.30–6.44 μg/kg, ∅ conc.: 3.60 μg/kg, sample origin: grocery markets in Busan, Daegu, Daejeon, Gangneung, and Gwangju (cities) as well as Seoul (capital), Korea, sample year: May 2004–June 2005, country: Korea[21]

- Co-contamination: not reported
- Further contamination of animal products reported in the present book/article[21]: not reported

incidence: 2/2, conc. range: 3.5–5.2 μg/kg, ∅ conc.: 4.4 μg/kg, sample origin: local marketing institutions, GDR, sample year: unknown, country: GDR[22]

- Co-contamination: 2 sa. co-contaminated with AFB_1, AFB_2, AFG_1, and AFG_2
- Further contamination of animal products reported in the present book/article[22]: Butter (peanut butter), AFB_2, AFG_1, and AFG_2, literature[22]; Dairy products, AFB_1, literature[22]

incidence: 20*/101**, conc. range: 64–1736 μg/kg, sample origin: 16 districts of eastern Nepal, sample year: 1995–2003, country: Nepal[23], *>30 μg/kg, **peanut butter/vegetable oil

- Co-contamination: not reported
- Further contamination of animal products reported in the present book/article[23]: not reported

incidence: 3?/6, conc. range: 0.6–2.4 μg/kg, sample origin: unknown, sample year: unknown, country: Japan[24]

- Co-contamination: not reported
- Further contamination of animal products reported in the present book/article[24]: Butter (peanut butter), AFB_2 and AFG_1, literature[24]; Cheese, AFM_1, literature[24]

incidence: 7/8, conc. range: 2.0–4.0 μg/kg (1 sa.), >4.0–≤249.0 μg/kg (6 sa.), sample origin: supermarkets and/or traditional markets in Yogyakarta (city), Indonesia, sample year: 1998–1999, country: Indonesia/Austria/Canada[25]

- Co-contamination: not reported
- Further contamination of animal products reported in the present book/article[25]: not reported

incidence: 1/13, conc.: 20 μg/kg, sample origin: main kitchen store, California (state), USA, sample year: August 1985, country: USA[45]

- Co-contamination: not reported
- Further contamination of animal products reported in the present book/article[45]: not reported

incidence: 40/62, conc. range: 0.5–54.6 μg/kg, ∅ conc.: 7.10 μg/kg, sample origin: small markets in several urban Shanghai (city) areas, China, sample year: April 1988–1989, country: China/USA[46]

- Co-contamination: not reported
- Further contamination of animal products reported in the present book/article[46]: Pig liver, AFB_1, literature[46]; Milk, Milk powder, AFM_1, literature[46]

incidence: 10*/21, conc. range: 0.17–2.59 μg/kg, ∅ conc.: 1.07 μg/kg, sample origin: local supermarkets and small retail shops across Japan, sample year: summer 2004–winter 2005, country: Japan[47], *domestic and imported products

- Co-contamination: 4 sa. co-contaminated with AFB_1, AFB_2, AFG_1, and AFG_2; 3 sa. co-contaminated with AFB_1 and AFB_2; 3 sa. contaminated solely with AFB_1
- Further contamination of animal products reported in the present book/article[47]: Butter (peanut butter), AFB_2, AFG_1, and AFG_2, literature[47]

incidence: 3/4, conc. range: 0.6–1.4 μg/kg, ∅ conc.: 1.0 μg/kg, sample origin: retail stores in Osaka (prefecture), Japan, sample year: 1988–1992, country: Japan[48]

- Co-contamination: 2 sa. co-contaminated with AFB_1, AFB_2, and AFG_1; 1 sa. co-contaminated with AFB_1 and AFB_2
- Further contamination of animal products reported in the present book/article[48]: Butter (peanut butter), AFB_2 and AFG_1, literature[48]

incidence: 2/2, conc. range: 0.76–5.90 μg/kg, ∅ conc.: 3.33 μg/kg, sample origin: local stores in Munich (city), Germany, sample year: unknown, country: Germany[49]

- Co-contamination: 1 sa. co-contaminated with AFB_1, AFB_2, and AFG_1; 1 sa. co-contaminated with AFB_1 and AFB_2
- Further contamination of animal products reported in the present book/article[49]: Butter (peanut butter), AFB_2, and AFG_1, literature[49]

incidence: 21?/74, conc. range: 1.2–73 μg/kg, ∅ conc.: 40.6 μg/kg, sample origin: sites of primary storage or processing (domestic products) and sites of importation, Cyprus, sample year: 1992–1996, country: Cyprus[50]

- Co-contamination: not reported
- Further contamination of animal products reported in the present book/article[50]: Butter (peanut butter), AFB_2, AFG_1, and AFG_2, literature[50]; Milk, Milk (cow milk), AFM_1, literature[50]

incidence: 14/30, conc. range: 5–117 μg/kg, ∅ conc.: 25 μg/kg, sample origin: markets in 16 areas (cities and counties), Taiwan, sample year: 1977–1978, country: Taiwan[51]

- Co-contamination: not reported
- Further contamination of animal products reported in the present book/article[51]: Butter (peanut butter), AFB_2, AFG_1, and AFG_2, literature[51]

incidence: 6/50* **, conc. range: 2.5–5 μg/kg (2 sa.), 11–≤28.6 μg/kg (4 sa.), sample origin: retail Health Food outlets in 25 different geographical locations, UK, sample year: 1986, country: UK[52], *health food, **peanut butter, smooth

incidence: 9/79* **, conc. range: 2.5–5 μg/kg (2 sa.), 5.1–10 μg/kg (3 sa.), 11–30 μg/kg (3 sa.), 41.0 μg/kg (1 sa.), sample origin: retail Health Food outlets in 25 different geographical locations, UK, sample year: 1986, country: UK[52], *health food, **peanut butter, crunchy

- Co-contamination: not reported
- Further contamination of animal products reported in the present book/article[52]: Butter (peanut butter), AF/S, literature[52]

incidence: 10/63, conc. range: 2–20 μg/kg, ∅ conc.: 7 μg/kg, sample origin: supermarkets in East Lansing (city), USA, sample year: unknown, country: USA[53]

- Co-contamination: not reported
- Further contamination of animal products reported in the present book/article[53]: not reported

incidence: 9/16, conc. range: 20–730 μg/kg, ∅ conc.: 218 μg/kg, sample origin: USA and Taiwan, sample year: unknown, country: Taiwan[54]

- Co-contamination: not reported
- Further contamination of animal products reported in the present book/article[54]: not reported

incidence: 2/2, conc. range: 1.8–5.5 μg/kg, ∅ conc.: 3.7 μg/kg, sample origin: unknown, sample year: unknown, country: Denmark[55]

- Co-contamination: 2 sa. co-contaminated with AFB_1, AFB_2, AFG_1, and AFG_2

- Further contamination of animal products reported in the present book/article[55]: Butter (peanut butter), AFB_2, AFG_1, and AFG_2, literature[55]

incidence: 15?/21, conc. range: 3.2–16.0 μg/kg, sample origin: storage depots and retail outlets, Botswana, sample year: January 1996–December 1997, country: Botswana[56]

- Co-contamination: not reported
- Further contamination of animal products reported in the present book/article[56]: Butter (peanut butter), AFB_2, AFG_1, and AFG_2, literature[56]

incidence: 14/14*, conc. range: 29–128 μg/kg, ∅ conc.: 63.9 μg/kg, sample origin: street sellers, Khartoum (capital, state) area, Sudan, sample year: 2005–2006, country: Sudan[57], *traditionally processed peanut butter (homemade)

incidence: 10/10*, conc. range: 21–131 μg/kg, ∅ conc.: 54.5 μg/kg, sample origin: street sellers, Khartoum (capital, state) north area, Sudan, sample year: 2005–2006, country: Sudan[57], *traditionally processed peanut butter (homemade)

incidence: 12/12*, conc. range: 17–170 μg/kg, ∅ conc.: 101 μg/kg, sample origin: street sellers, Omdurman (city) area, Sudan, sample year: 2005–2006, country: Sudan[57], *traditionally processed peanut butter (homemade)

incidence: 10/14, conc. range: 1–57 μg/kg, ∅ conc.: 14.5 μg/kg, sample origin: retail stores, Khartoum (state), Sudan, sample year: 2005–2006, country: Sudan[57]

- Co-contamination: not reported
- Further contamination of animal products reported in the present book/article[57]: not reported

For detailed information see the article.

incidence: 2/2, conc. range: 5.82–6.44 μg/kg, ∅ conc.: 6.13 μg/kg, sample origin: supermarkets, local stores, and markets in Busan, Daejeon, Gwangju (cities), and Seoul (capital), Korea, sample year: June–July 2004, country: Korea[58]

- Co-contamination: not reported
- Further contamination of animal products reported in the present book/article[58]: Butter (peanut butter), AFB_2, literature[58]

incidence: 19/20, conc. range: 2.06–63.72 μg/kg, ∅ conc.: 16.6 μg/kg, sample origin: markets in Ankara (capital), Turkey, sample year: unknown, country: Turkey[59]

- Co-contamination: 19 sa. co-contaminated with AFB_1, AFB_2, and AFG_1
- Further contamination of animal products reported in the present book/article[59]: Butter (peanut butter), AFB_2 and AFG_1, literature[59]

incidence: 5/5, conc. range: 0.62–14.3 μg/kg, ∅ conc.: 4.6 μg/kg, sample origin: unknown, sample year: unknown, country: USA[60]

- Co-contamination: 4 sa. co-contaminated with AFB_1, AFB_2, AFG_1, and AFG_2; 1 sa. co-contaminated with AFB_1 and AFB_2
- Further contamination of animal products reported in the present book/article[60]: Butter (peanut butter), AFB_2, AFG_1, and AFG_2, literature[60]

incidence: 41/50*, conc. range: 0.39–68.51 μg/kg, sample origin: retail markets in Changchun, Chengdu, Shanghai, Shijiazhuang, Zhengzhou (cities), and Beijing (capital), China, sample year: 2007, country: China[61], *manufactured domestically

- Co-contamination: 41 sa. co-contaminated with AFB_1 and AFB_2; no further information available
- Further contamination of animal products reported in the present book/article[61]: Butter (peanut butter), AFB_2, AFG_1, and AFG_2, literature[61]

incidence: 1/1, conc.: 26.3 μg/kg, sample origin: market in Khartoum (capital), Sudan, sample year: February 1981–1984?, country: UK/Sudan[62]

- Co-contamination: 1 sa. co-contaminated with AFB_1, AFB_2, and AFG_1
- Further contamination of animal products reported in the present book/article[62]: Butter (peanut butter), AFB_2 and AFG_1, literature[62]

incidence: 7?/12, conc. range: 13.3–56.6 μg/kg, sample origin: supermarkets and village stores in 5 districts of Penang (state) including Penang Island and Wellesley (province), Malaysia, sample year: unknown, country: Malaysia[63]

- Co-contamination: not reported
- Further contamination of animal products reported in the present book/article[63]: Butter (peanut butter), AFB_2, literature[63]

incidence: 4/10, conc. range: 5–61 μg/kg, sample origin: traditional markets and supermarkets, Indonesia, sample year: 2001–2002, country: Austria/Indonesia/Thailand[64]

- Co-contamination: not reported
- Further contamination of animal products reported in the present book/article[64]: Butter (peanut butter), AF/S, literature[64]

incidence: 31?/33, conc. range: ≤51.4 μg/kg, sample origin: local stores in Hangzhou, Ningbo, and Quzhou (cities), Zhejiang (province), China, sample year: unknown, country: China[65]

- Co-contamination: not reported
- Further contamination of animal products reported in the present book/article[65]: Butter (peanut butter), AFB_2, AFG_1, AFG_2, AFM_1, and AFM_2, literature[65]

incidence: 31/75, conc. range: 0.1–2.1 μg/kg, ∅ conc.: 0.5 μg/kg, sample origin: available in New Zealand, sample year: 2010, country: New Zealand[66]

- Co-contamination: not reported
- Further contamination of animal products reported in the present book/article[66]: Butter (peanut butter), AF/S, literature[66]

incidence: 28/43, conc. range: 73.9–534 μg/kg, ∅ conc.: 223 μg/kg, sample origin: households, small factories, supermarkets, and grocery shops in Khartoum State, Sudan, sample year: January 2010, country: Sudan[67]

- Co-contamination: not reported
- Further contamination of animal products reported in the present book/article[67]: Butter (peanut butter), AFB_2, AFG_1, and AFG_2, literature[67]

incidence: 14/14*, conc. range: 13.2–40.6 μg/kg, sample origin: market in Lilongwe City, Malawi, sample year: December 2012, country: Malawi[68], *peanut butter, local

- Co-contamination: 14 sa. co-contaminated with AFB_1, AFB_2, AFG_1, and AFG_2
- Further contamination of animal products reported in the present book/article[68]: Butter (peanut butter), AFB_2, AFG_1, and AFG_2, literature[68]

incidence: 8/11*, conc. range: 0.5–1.4 μg/kg, sample origin: market in Lilongwe City, Malawi, sample year: December 2012, country: Malawi[68], *peanut butter, imported

- Co-contamination: not reported
- Further contamination of animal products reported in the present book/article[68]: Butter (peanut butter), AFB_2, AFG_1, and AFG_2, literature[68]

incidence: 10/11, conc. range: 6.1–191 μg/kg, ∅ conc.: 55.7 μg/kg, sample origin: retail shops and vendors at informal markets in Bulawayo metropolitan area, southern Zimbabwe, sample year: October–November 2011, country: Botswana/South Africa[69]

- Co-contamination: 1 sa. co-contaminated with AFB_1, AFB_2, AFG_1, and AFG_2; 3 sa. co-contaminated with AFB_1, AFB_2, and AFG_1; 1 sa. co-contaminated with AFB_1 and AFG_1; 5 sa. contaminated solely with AFB_1
- Further contamination of animal products reported in the present book/article[69]: Butter (peanut butter), AFB_2, AFG_1, and AFG_2, literature[69]

Aflatoxin B_2

incidence: 10/29, conc. range: LOQ-2 μg/kg (10 sa., maximum: 1.8 μg/kg), ∅ conc.: 0.72 μg/kg, sample origin: retail outlets, UK, sample year: September 2001, country: UK[9]

- Co-contamination: 2 sa. co-contaminated with AFB_1, AFB_2, AFG_1, AFG_2, and OTA; 3 sa. co-contaminated with AFB_1, AFB_2, AFG_1, and AFG_2; 3 sa. co-contaminated with AFB_1, AFB_2, and AFG_1; 2 sa. co-contaminated with AFB_1 and AFB_2
- Further contamination of animal products reported in the present book/article[9]: Butter (nut butter), AFB_1, AFB_2, AFG_1, AFG_2, and OTA, literature[9]; Butter (peanut butter), AFB_1, AFG_1, AFG_2, and OTA, literature[9]

incidence: 6/27, conc. range: LOQ-2 μg/kg (6 sa., maximum: 0.6 μg/kg), ∅ conc.: 0.3 μg/kg, sample origin: retail outlets, UK, sample year: September 2001, country: UK[9]

- Co-contamination: 2 sa. co-contaminated with AFB_1, AFB_2, AFG_1, and AFG_2; 1 sa. co-contaminated with AFB_1, AFB_2, and AFG_1; 2 sa. co-contaminated with AFB_1, AFB_2, AFG_1, and OTA; 1 sa. co-contaminated with AFB_1 and AFB_2
- Further contamination of animal products reported in the present book/article[9]: Butter (nut butter), AFB_1, AFB_2, AFG_1, AFG_2, and OTA, literature[9]; Butter (peanut butter), AFB_1, AFG_1, AFG_2, and OTA, literature[9]

incidence: 22?/26, conc. range: 0.04–0.65 μg/kg, sample origin: unknown, sample year: unknown, country: Korea[20]

- Co-contamination: not reported
- Further contamination of animal products reported in the present book/article[20]: Butter (peanut butter), AFB_1, AFG_1, and AFG_2, literature[20]

incidence: 2/2, conc. range: 0.5–0.6 μg/kg, ∅ conc.: 0.55 μg/kg, sample origin: local marketing institutions, GDR, sample year: unknown, country: GDR[22]

- Co-contamination: 2 sa. co-contaminated with AFB_1, AFB_2, AFG_1, and AFG_2
- Further contamination of animal products reported in the present book/article[22]: Butter (peanut butter), AFB_1, AFG_1, and AFG_2, literature[22]; Dairy products, AFB_1, literature[22]

incidence: 1/6, conc.: 0. 4 μg/kg, sample origin: unknown, sample year: unknown, country: Japan[24]

- Co-contamination: not reported
- Further contamination of animal products reported in the present book/article[24]: Butter (peanut butter), AFB_1 and AFG_1, literature[24]; Cheese, AFM_1, literature[24]

incidence: 7*/21, conc. range: 0.16–0.52 μg/kg, ∅ conc.: 0.27 μg/kg, sample origin: local supermarkets and small retail shops across Japan, sample year: summer 2004–winter 2005, country: Japan[47], *domestic and imported products

- Co-contamination: 4 sa. co-contaminated with AFB_1, AFB_2, AFG_1, and AFG_2; 3 sa. co-contaminated with AFB_1 and AFB_2
- Further contamination of animal products reported in the present book/article[47]: Butter (peanut butter), AFB_1, AFG_1, and AFG_2, literature[47]

incidence: 3/4, conc. range: 0.1–0.3 μg/kg, ∅ conc.: 0.2 μg/kg, sample origin: retail stores in Osaka (prefecture), Japan, sample year: 1988–1992, country: Japan[48]

- Co-contamination: 2 sa. co-contaminated with AFB_1, AFB_2, and AFG_1; 1 sa. co-contaminated with AFB_1 and AFB_2
- Further contamination of animal products reported in the present book/article[48]: Butter (peanut butter), AFB_1 and AFG_1, literature[48]

incidence: 2/2, conc. range: 0.12–1.38 μg/kg, ∅ conc.: 0.75 μg/kg, sample origin: local stores in Munich (city), Germany, sample year: unknown, country: Germany[49]

- Co-contamination: 1 sa. co-contaminated with AFB_1, AFB_2, and AFG_1; 1 sa. co-contaminated with AFB_1 and AFB_2
- Further contamination of animal products reported in the present book/article[49]: Butter (peanut butter), AFB_1 and AFG_1, literature[49]

incidence: 21?/74, conc. range: 0.3–9 μg/kg, ∅ conc.: 6.4 μg/kg, sample origin: sites of primary storage or processing (domestic products) and sites of importation, Cyprus, sample year: 1992–1996, country: Cyprus[50]

- Co-contamination: not reported
- Further contamination of animal products reported in the present book/article[50]: Butter (peanut butter), AFB_1, AFG_1, and AFG_2, literature[50]; Milk, Milk (cow milk), AFM_1, literature[50]

incidence: 5/30, conc. range: 5–50 μg/kg, ∅ conc.: 18 μg/kg, sample origin: markets in 16 areas (cities and counties), Taiwan, sample year: 1977–1978, country: Taiwan[51]

- Co-contamination: not reported
- Further contamination of animal products reported in the present book/article[51]: Butter (peanut butter), AFB_1, AFG_1, and AFG_2, literature[51]

incidence: 2/2, conc. range: 0.4–1.0 μg/kg, ∅ conc.: 0.7 μg/kg, sample origin: unknown, sample year: unknown, country: Denmark[55]

- Co-contamination: 2 sa. co-contaminated with AFB_1, AFB_2, AFG_1, and AFG_2
- Further contamination of animal products reported in the present book/article[55]: Butter (peanut butter), AFB_1, AFG_1, and AFG_2, literature[55]

incidence: 15?/21, conc. range: 1.6–20.0 μg/kg, sample origin: storage depots and retail outlets, Botswana, sample year: January 1996–December 1997, country: Botswana[56]

- Co contamination: not reported
- Further contamination of animal products reported in the present book/article[56]: Butter (peanut butter), AFB_1, AFG_1, and AFG_2, literature[56]

incidence: 1/2, conc.: 1.15 μg/kg, sample origin: supermarkets, local stores, and markets in Busan, Daejeon, Gwangju (cities), and Seoul (capital), Korea, sample year: June–July 2004, country: Korea[58]

- Co-contamination: not reported
- Further contamination of animal products reported in the present book/article[58]: Butter (peanut butter), AFB_1, literature[58]

incidence: 20/20, conc. range: 0.06–4.68 µg/kg, Ø conc.: 1.23 µg/kg, sample origin: markets in Ankara (capital), Turkey, sample year: unknown, country: Turkey[59]

- Co-contamination: 19 sa. co-contaminated with AFB_1, AFB_2, and AFG_1; 1 sa. co-contaminated with AFB_2 and AFG_1
- Further contamination of animal products reported in the present book/article[59]: Butter (peanut butter), AFB_1 and AFG_1, literature[59]

incidence: 5/5, conc. range: 0.14–3.3 µg/kg, Ø conc.: 1.0 µg/kg, sample origin: unknown, sample year: unknown, country: USA[60]

- Co-contamination: 4 sa. co-contaminated with AFB_1, AFB_2, AFG_1, and AFG_2; 1 sa. co-contaminated with AFB_1 and AFB_2
- Further contamination of animal products reported in the present book/article[60]: Butter (peanut butter), AFB_1, AFG_1, and AFG_2, literature[60]

incidence: 41/50*, conc. range: ≤5.52 µg/kg, sample origin: retail markets in Changchun, Chengdu, Shanghai, Shijiazhuang, Zhengzhou (cities), and Beijing (capital), China, sample year: 2007, country: China[61], *manufactured domestically

- Co-contamination: 41 sa. co-contaminated with AFB_1 and AFB_2; no further information available
- Further contamination of animal products reported in the present book/article[61]: Butter (peanut butter), AFB_1, AFG_1, and AFG_2, literature[61]

incidence: 1/1, conc.: 9.72 µg/kg, sample origin: market in Khartoum (capital), Sudan, sample year: February 1981–1984?, country: UK/Sudan[62]

- Co-contamination: 1 sa. co-contaminated with AFB_1, AFB_2, and AFG_1
- Further contamination of animal products reported in the present book/article[62]: Butter (peanut butter), AFB_1 and AFG_1, literature[62]

incidence: 7?/12, conc. range: 3.31–10.8 µg/kg, sample origin: supermarkets and village stores in 5 districts of Penang (state) including Penang Island and Wellesley (province), Malaysia, sample year: unknown, country: Malaysia[63]

- Co-contamination: not reported
- Further contamination of animal products reported in the present book/article[63]: Butter (peanut butter), AFB_1, literature[63]

incidence: 31?/33, conc. range: ≤14.0 µg/kg, sample origin: local stores in Hangzhou, Ningbo, and Quzhou (cities), Zhejiang (province), China, sample year: unknown, country: China[65]

- Co-contamination: not reported
- Further contamination of animal products reported in the present book/article[65]: Butter (peanut butter), AFB_1, AFG_1, AFG_2, AFM_1, and AFM_2, literature[65]

incidence: 42/43, conc. range: 0.18–23.9 µg/kg, Ø conc.: 3.20 µg/kg, sample origin: households, small factories, supermarkets, and grocery shops in Khartoum State, Sudan, sample year: January 2010, country: Sudan[67]

- Co-contamination: not reported
- Further contamination of animal products reported in the present book/article[678]: Butter (peanut butter), AFB_1, AFG_1, and AFG_2, literature[67]

incidence: 14/14*, conc. range: 1.7–7.2 µg/kg, sample origin: market in Lilongwe City, Malawi, sample year: December 2012, country: Malawi[68], *peanut butter, local

- Co-contamination: 14 sa. co-contaminated with AFB_1, AFB_2, AFG_1, and AFG_2
- Further contamination of animal products reported in the present book/article[68]: Butter (peanut butter), AFB_1, AFG_1, and AFG_2, literature[68]

incidence: 8/11*, conc. range: 0.2–0.4 µg/kg, sample origin: market in Lilongwe City, Malawi, sample year: December 2012, country: Malawi[68], *peanut butter, imported

- Co-contamination: not reported
- Further contamination of animal products reported in the present book/article[68]: Butter (peanut butter), AFB_1, AFG_1, and AFG_2, literature[68]

incidence: 4/11, conc. range: 6.2–25.7 µg/kg, Ø conc.: 15.8 µg/kg, sample origin: retail shops and vendors at informal markets in Bulawayo metropolitan area, southern Zimbabwe, sample year: October–November 2011, country: Botswana/South Africa[69]

- Co-contamination: 1 sa. co-contaminated with AFB_1, AFB_2, AFG_1, and AFG_2; 3 sa. co-contaminated with AFB_1, AFB_2, and AFG_1
- Further contamination of animal products reported in the present book/article[69]: Butter (peanut butter), AFB_1, AFG_1, and AFG_2, literature[69]

Aflatoxin G_1

incidence: 13/29, conc. range: LOQ-2 µg/kg (13 sa., maximum: 1.4 µg/kg), Ø conc.: 0.79 µg/kg, sample origin: retail outlets, UK, sample year: September 2001, country: UK[9]

- Co-contamination: 2 sa. co-contaminated with AFB_1, AFB_2, AFG_1, AFG_2, and OTA; 3 sa. co-contaminated with AFB_1, AFB_2, AFG_1, and AFG_2; 3 sa. co-contaminated with AFB_1, AFB_2, and AFG_1; 4 sa. co-contaminated with AFB_1 and AFG_1; 1 sa. contaminated solely with AFG_1
- Further contamination of animal products reported in the present book/article[9]: Butter (nut butter), AFB_1, AFB_2, AFG_1, AFG_2, and OTA, literature[9]; Butter (peanut butter), AFB_1, AFB_2, AFG_2, and OTA, literature[9]

incidence: 8/27, conc. range: LOQ-2 µg/kg (8 sa., maximum: 1 µg/kg), Ø conc.: 0.44 µg/kg, sample origin: retail outlets, UK, sample year: September 2001, country: UK[9]

- Co-contamination: 2 sa. co-contaminated with AFB_1, AFB_2, AFG_1, and AFG_2; 1 sa. co-contaminated with AFB_1, AFB_2, and AFG_1; 2 sa. co-contaminated with AFB_1, AFB_2, AFG_1, and OTA; 3 sa. co-contaminated with AFB_1 and AFG_1
- Further contamination of animal products reported in the present book/article[9]: Butter (nut butter), AFB_1, AFB_2, AFG_1, AFG_2, and OTA, literature[9]; Butter (peanut butter), AFB_1, AFB_2, AFG_2, and OTA, literature[9]

incidence: 22?/26, conc. range: 0.13–2.21 µg/kg, sample origin: unknown, sample year: unknown, country: Korea[20]

- Co-contamination: not reported
- Further contamination of animal products reported in the present book/article[20]: Butter (peanut butter), AFB_1, AFB_2, and AFG_2, literature[20]

incidence: 2/2, conc. range: 3.5–5.2 µg/kg, Ø conc.: 4.4 µg/kg, sample origin: local marketing institutions, GDR, sample year: unknown, country: GDR[22]

- Co-contamination: 2 sa. co-contaminated with AFB_1, AFB_2, AFG_1, and AFG_2
- Further contamination of animal products reported in the present book/article[22]: Butter (peanut butter), AFB_1, AFB_2, and AFG_2, literature[22]; Dairy products, AFB_1, literature[22]

incidence: 3?/6, conc. range: 0.1–0.4 μg/kg, sample origin: unknown, sample year: unknown, country: Japan[24]
- Co-contamination: not reported
- Further contamination of animal products reported in the present book/article[24]: Butter (peanut butter), AFB_1 and AFB_2, literature[24]; Cheese, AFM_1, literature[24]

incidence: 4*/21, conc. range: 0.17–0.81 μg/kg, ∅ conc.: 0.4 μg/kg, sample origin: local supermarkets and small retail shops across Japan, sample year: summer 2004–winter 2005, country: Japan[47], *domestic and imported products
- Co-contamination: 4 sa. co-contaminated with AFB_1, AFB_2, AFG_1, and AFG_2
- Further contamination of animal products reported in the present book/article[47]: Butter (peanut butter), AFB_1, AFB_2, and AFG_2, literature[47]

incidence: 2/4, conc. range: 0.3 μg/kg, ∅ conc.: 0.3 μg/kg, sample origin: retail stores in Osaka (prefecture), Japan, sample year: 1988–1992, country: Japan[48]
- Co-contamination: 2 sa. co-contaminated with AFB_1, AFB_2, and AFG_1
- Further contamination of animal products reported in the present book/article[48]: Butter (peanut butter), AFB_1 and AFB_2, literature[48]

incidence: 1/2, conc.: 0.52 μg/kg, sample origin: local stores in Munich (city), Germany, sample year: unknown, country: Germany[49]
- Co-contamination: 1 sa. co-contaminated with AFB_1, AFB_2, and AFG_1
- Further contamination of animal products reported in the present book/article[49]: Butter (peanut butter), AFB_1 and AFB_2, literature[49]

incidence: 21?/74, conc. range: <0.4–0.9 μg/kg, ∅ conc.: 0.6 μg/kg, sample origin: sites of primary storage or processing (domestic products) and sites of importation, Cyprus, sample year: 1992–1996, country: Cyprus[50]
- Co-contamination: not reported
- Further contamination of animal products reported in the present book/article[50]: Butter (peanut butter), AFB_1, AFB_2, and AFG_2, literature[50]; Milk, Milk (cow milk), AFM_1, literature[50]

incidence: 5/30, conc. range: 6–135 μg/kg, ∅ conc.: 54 μg/kg, sample origin: markets in 16 areas (cities and counties), Taiwan, sample year: 1977–1978, country: Taiwan[51]
- Co-contamination: not reported
- Further contamination of animal products reported in the present book/article[51]: Butter (peanut butter), AFB_1, AFB_2, and AFG_2, literature[51]

incidence: 2/2, conc. range: 0.6–1.2 μg/kg, ∅ conc.: 0.9 μg/kg, sample origin: unknown, sample year: unknown, country: Denmark[55]
- Co-contamination: 2 sa. co-contaminated with AFB_1, AFB_2, AFG_1, and AFG_2
- Further contamination of animal products reported in the present book/article[55]: Butter (peanut butter), AFB_1, AFB_2, and AFG_2, literature[55]

incidence: 15?/21, conc. range: 3.2–20.0 μg/kg, sample origin: storage depots and retail outlets, Botswana, sample year: January 1996–December 1997, country: Botswana[56]
- Co-contamination: not reported
- Further contamination of animal products reported in the present book/article[56]: Butter (peanut butter), AFB_1, AFB_2, and AFG_2, literature[56]

incidence: 20/20, conc. range: 1.66–32.78 μg/kg, ∅ conc.: 10.6 μg/kg, sample origin: markets in Ankara (capital), Turkey, sample year: unknown, country: Turkey[59]
- Co-contamination: 19 sa. co-contaminated with AFB_1, AFB_2, and AFG_1; 1 sa. co-contaminated with AFB_2 and AFG_1
- Further contamination of animal products reported in the present book/article[59]: Butter (peanut butter), AFB_1 and AFB_2, literature[59]

incidence: 4/5, conc. range: 0.11–2.01 μg/kg, ∅ conc.: 0.7 μg/kg, sample origin: unknown, sample year: unknown, country: USA[60]
- Co-contamination: 4 sa. co-contaminated with AFB_1, AFB_2, AFG_1, and AFG_2
- Further contamination of animal products reported in the present book/article[60]: Butter (peanut butter), AFB_1, AFB_2, and AFG_2, literature[60]

incidence: 40/50*, conc. range: ≤21.22 μg/kg, sample origin: retail markets in Changchun, Chengdu, Shanghai, Shijiazhuang, Zhengzhou (cities), and Beijing (capital), China, sample year: 2007, country: China[61], *manufactured domestically
- Co-contamination: not reported
- Further contamination of animal products reported in the present book/article[61]: Butter (peanut butter), AFB_1, AFB_2, and AFG_2, literature[61]

incidence: 1/1, conc.: 84.5 μg/kg, sample origin: market in Khartoum (capital), Sudan, sample year: February 1981–1984?, country: UK/Sudan[62]
- Co-contamination: 1 sa. co-contaminated with AFB_1, AFB_2, and AFG_1
- Further contamination of animal products reported in the present book/article[62]: Butter (peanut butter), AFB_1 and AFB_2, literature[62]

incidence: 31?/33, conc. range: ≤20.8 μg/kg, sample origin: local stores in Hangzhou, Ningbo, and Quzhou (cities), Zhejiang (province), China, sample year: unknown, country: China[65]
- Co-contamination: not reported
- Further contamination of animal products reported in the present book/article[65]: Butter (peanut butter), AFB_1, AFB_2, AFG_2, AFM_1, and AFM_2, literature[65]

incidence: 43/43, conc. range: 26.5–401 μg/kg, ∅ conc.: 137 μg/kg, sample origin: households, small factories, supermarkets, and grocery shops in Khartoum State, Sudan, sample year: January 2010, country: Sudan[67]
- Co-contamination: not reported
- Further contamination of animal products reported in the present book/article[67]: Butter (peanut butter), AFB_1, AFB_2, and AFG_2, literature[67]

incidence: 14/14*, conc. range: 14.8–65.0 μg/kg, sample origin: market in Lilongwe City, Malawi, sample year: December 2012, country: Malawi[68], *peanut butter, local
- Co-contamination: 14 sa. co-contaminated with AFB_1, AFB_2, AFG_1, and AFG_2

- Further contamination of animal products reported in the present book/article[68]: Butter (peanut butter), AFB_1, AFB_2, and AFG_2, literature[68]

incidence: 7/11*, conc. range: 0.5–1.8 µg/kg, sample origin: market in Lilongwe City, Malawi, sample year: December 2012, country: Malawi[68], *peanut butter, imported

- Co-contamination: not reported
- Further contamination of animal products reported in the present book/article[68]: Butter (peanut butter), AFB_1, AFB_2, and AFG_2, literature[68]

incidence: 5/11, conc. range: 9.3–47.1 µg/kg, ∅ conc.: 25.5 µg/kg, sample origin: retail shops and vendors at informal markets in Bulawayo metropolitan area, southern Zimbabwe, sample year: October–November 2011, country: Botswana/South Africa[69]

- Co-contamination: 1 sa. co-contaminated with AFB_1, AFB_2, AFG_1, and AFG_2; 3 sa. co-contaminated with AFB_1, AFB_2, and AFG_1; 1 sa. co-contaminated with AFB_1 and AFG_1
- Further contamination of animal products reported in the present book/article[69]: Butter (peanut butter), AFB_1, AFB_2, and AFG_2, literature[69]

Aflatoxin G_2

incidence: 5/29, conc. range: LOQ-2 µg/kg (5 sa., maximum: 0.5 µg/kg), ∅ conc.: 0.38 µg/kg, sample origin: retail outlets, UK, sample year: September 2001, country: UK[9]

- Co-contamination: 2 sa. co-contaminated with AFB_1, AFB_2, AFG_1, AFG_2, and OTA; 3 sa. co-contaminated with AFB_1, AFB_2, AFG_1, and AFG_2
- Further contamination of animal products reported in the present book/article[9]: Butter (nut butter), AFB_1, AFB_2, AFG_1, AFG_2, and OTA, literature[9]; Butter (peanut butter), AFB_1, AFB_2, AFG_1, and OTA, literature[9]

incidence: 2/27, conc. range: LOQ-2 µg/kg (2 sa., maximum: 0.3 µg/kg), ∅ conc.: 0.25 µg/kg, sample origin: retail outlets, UK, sample year: September 2001, country: UK[9]

- Co-contamination: 2 sa. co-contaminated with AFB_1, AFB_2, AFG_1, and AFG_2
- Further contamination of animal products reported in the present book/article[9]: Butter (nut butter), AFB_1, AFB_2, AFG_1, AFG_2, and OTA, literature[9]; Butter (peanut butter), AFB_1, AFB_2, AFG_1, and OTA, literature[9]

incidence: 2/26, conc. range: 0.21–0.30 µg/kg, ∅ conc.: 0.255 µg/kg, sample origin: unknown, sample year: unknown, country: Korea[20]

- Co-contamination: not reported
- Further contamination of animal products reported in the present book/article[20]: Butter (peanut butter), AFB_1, AFB_2, and AFG_1, literature[20]

incidence: 2/2, conc. range: 1.3–1.7 µg/kg, ∅ conc.: 1.5 µg/kg, sample origin: local marketing institutions, GDR, sample year: unknown, country: GDR[22]

- Co-contamination: 2 sa. co-contaminated with AFB_1, AFB_2, AFG_1, and AFG_2
- Further contamination of animal products reported in the present book/article[22]: Butter (peanut butter), AFB_1, AFB_2, and AFG_1, literature[22]; Dairy products, AFB_1, literature[22]

incidence: 4*/21, conc. range: 0.12–0.46 µg/kg, ∅ conc.: 0.21 µg/kg, sample origin: local supermarkets and small retail shops across Japan, sample year: summer 2004–winter 2005, country: Japan[47], *domestic and imported products

- Co-contamination: 4 sa. co-contaminated with AFB_1, AFB_2, AFG_1, and AFG_2
- Further contamination of animal products reported in the present book/article[47]: Butter (peanut butter), AFB_1, AFB_2, and AFG_1, literature[47]

incidence: 1/74, conc.: 0.3 µg/kg, sample origin: sites of primary storage or processing (domestic products) and sites of importation, Cyprus, sample year: 1992–1996, country: Cyprus[50]

- Co-contamination: not reported
- Further contamination of animal products reported in the present book/article[50]: Butter (peanut butter), AFB_1, AFB_2, and AFG_1, literature[50]; Milk, Milk (cow milk), AFM_1, literature[50]

incidence: 3/30, conc. range: 33–124 µg/kg, ∅ conc.: 68 µg/kg, sample origin: markets in 16 areas (cities and counties), Taiwan, sample year: 1977–1978, country: Taiwan[51]

- Co-contamination: not reported
- Further contamination of animal products reported in the present book/article[51]: Butter (peanut butter), AFB_1, AFB_2, and AFG_1, literature[51]

incidence: 2/2, conc. range: 0.2 µg/kg, ∅ conc.: 0.2 µg/kg, sample origin: unknown, sample year: unknown, country: Denmark[55]

- Co-contamination: 2 sa. co-contaminated with AFB_1, AFB_2, AFG_1, and AFG_2
- Further contamination of animal products reported in the present book/article[55]: Butter (peanut butter), AFB_1, AFB_2, and AFG_1, literature[55]

incidence: 15?/21, conc. range: 1.6–20.0 µg/kg, sample origin: storage depots and retail outlets, Botswana, sample year: January 1996–December 1997, country: Botswana[56]

- Co-contamination: not reported
- Further contamination of animal products reported in the present book/article[56]: Butter (peanut butter), AFB_1, AFB_2, and AFG_1, literature[56]

incidence: 4/5, conc. range: 0.11–0.44 µg/kg, ∅ conc.: 0.2 µg/kg, sample origin: unknown, sample year: unknown, country: USA[60]

- Co-contamination: 4 sa. co-contaminated with AFB_1, AFB_2, AFG_1, and AFG_2
- Further contamination of animal products reported in the present book/article[60]: Butter (peanut butter), AFB_1, AFB_2, and AFG_1, literature[60]

incidence: 37/50*, conc. range: ≤6.36 µg/kg, sample origin: retail markets in Changchun, Chengdu, Shanghai, Shijiazhuang, Zhengzhou (cities), and Beijing (capital), China, sample year: 2007, country: China[61], *manufactured domestically

- Co-contamination: not reported
- Further contamination of animal products reported in the present book/article[61]: Butter (peanut butter), AFB_1, AFB_2, and AFG_1, literature[61]

incidence: 31?/33, conc. range: ≤4.5 µg/kg, sample origin: local stores in Hangzhou, Ningbo, and Quzhou (cities), Zhejiang (province), China, sample year: unknown, country: China[65]

- Co-contamination: not reported
- Further contamination of animal products reported in the present book/article[65]: Butter (peanut butter), AFB_1, AFB_2, AFG_1, AFM_1, and AFM_2, literature[65]

incidence: 4/43, conc. range: 8.60–30.1 µg/kg, ∅ conc.: 18.5 µg/kg, sample origin: households, small factories, supermarkets, and grocery shops in Khartoum State, Sudan, sample year: January 2010, country: Sudan[67]

- Co-contamination: not reported
- Further contamination of animal products reported in the present book/article[67]: Butter (peanut butter), AFB_1, AFB_2, and AFG_1, literature[67]

incidence: 14/14*, conc. range: 1.7–6.4 µg/kg, sample origin: market in Lilongwe City, Malawi, sample year: December 2012, country: Malawi[68], *peanut butter, local
- Co-contamination: 14 sa. co-contaminated with AFB_1, AFB_2, AFG_1, and AFG_2
- Further contamination of animal products reported in the present book/article[68]: Butter (peanut butter), AFB_1, AFB_2, and AFG_1, literature[68]

incidence: 6/11*, conc. range: 0.2–0.7 µg/kg, sample origin: market in Lilongwe City, Malawi, sample year: December 2012, country: Malawi[68], *peanut butter, imported
- Co-contamination: not reported
- Further contamination of animal products reported in the present book/article[68]: Butter (peanut butter), AFB_1, AFB_2, and AFG_1, literature[68]

incidence: 1/11, conc.: 8.8 µg/kg, sample origin: retail shops and vendors at informal markets in Bulawayo metropolitan area, Zimbabwe, sample year: October–November 2011, country: Botswana/South Africa[69]
- Co-contamination: 1 sa. co-contaminated with AFB_1, AFB_2, AFG_1, and AFG_2
- Further contamination of animal products reported in the present book/article[69]: Butter (peanut butter), AFB_1, AFB_2, and AFG_1, literature[69]

Aflatoxin M_1
incidence: 31?/33, conc. range: ≤4.2 µg/kg, sample origin: local stores in Hangzhou, Ningbo, and Quzhou (cities), Zhejiang (province), China, sample year: unknown, country: China[65]
- Co-contamination: not reported
- Further contamination of animal products reported in the present book/article[65]: Butter (peanut butter), AFB_1, AFB_2, AFG_1, AFG_2, and AFM_2, literature[65]

Aflatoxin M_2
incidence: 31?/33, conc. range: ≤1.8 µg/kg, sample origin: local stores in Hangzhou, Ningbo, and Quzhou (cities), Zhejiang (province), China, sample year: unknown, country: China[65]
- Co-contamination: not reported
- Further contamination of animal products reported in the present book/article[65]: Butter (peanut butter), AFB_1, AFB_2, AFG_1, AFG_2, and AFM_1, literature[65]

Aflatoxin/s
incidence: 7/16* **, conc. range: 2–5 µg/kg*** (6 sa.), 8 µg/kg*** (1 sa.), sample origin: retail outlets in 20 different regions, England, sample year: January 1982, country: UK[18], *regular food, **peanut butter, smooth, ***AFB_1, AFB_2, AFG_1, and AFG_2

incidence: 6/16* **, conc. range: 2–5 µg/kg*** (4 sa.), 6–10 µg/kg*** (1 sa.), 14 µg/kg*** (1 sa.), sample origin: retail outlets in 20 different regions, England, sample year: January 1982, country: UK[18], *regular food, **peanut butter, crunchy, ***AFB_1, AFB_2, AFG_1, and AFG_2

incidence: 6/11* **, conc. range: 2–5 µg/kg*** (1 sa.), 6–10 µg/kg*** (2 sa.), 11–30 µg/kg*** (1 sa.), 31–100 µg/kg*** (2 sa., maximum: 85 µg/kg***), sample origin: retail outlets in 20 different regions, England, sample year: January 1982, country: UK[18], *health food, **peanut butter, smooth, ***AFB_1, AFB_2, AFG_1, and AFG_2

incidence: 10/14* **, conc. range: 2–5 µg/kg*** (1 sa.), 6–10 µg/kg*** (2 sa.), 11–30 µg/kg*** (1 sa.), 31–100 µg/kg*** (1 sa.), >100 µg/kg*** (5 sa., maximum: 345 µg/kg***), sample origin: retail outlets in 20 different regions, England, sample year: January 1982, country: UK[18], *health food, **peanut butter, crunchy, ***AFB_1, AFB_2, AFG_1, and AFG_2

incidence: 6/6* **, conc. range: 6–10 µg/kg*** (1 sa.), 11–30 µg/kg*** (1 sa.), 31–100 µg/kg*** (3 sa.), 175 µg/kg*** (1 sa.), sample origin: retail outlets in 20 different regions, sample year: June 1983, country: UK[18], *health food, **peanut butter, smooth, ***AFB_1, AFB_2, AFG_1, and AFG_2

incidence: 7/9* **, conc. range: 2–5 µg/kg*** (3 sa.), 6–10 µg/kg*** (2 sa.), 11–30 µg/kg*** (1 sa.), 211 µg/kg*** (1 sa.), sample origin: retail outlets in 20 different regions, England, sample year: June 1983, country: UK[18], *health food, **peanut butter, crunchy, ***AFB_1, AFB_2, AFG_1, and AFG_2

incidence: 1/4* **, conc.: 27 µg/kg***, sample origin: retail outlets in 20 different regions, England, sample year: December 1984, country: UK[18], *health food, **peanut butter, smooth, ***AFB_1, AFB_2, AFG_1, and AFG_2

incidence: 7/15* **, conc. range: 6–10 µg/kg*** (1 sa.), 11–30 µg/kg*** (1 sa.), 31–100 µg/kg*** (2 sa.), >100 µg/kg*** (3 sa., maximum: 147 µg/kg***), sample origin: retail outlets in 20 different regions, England, sample year: December 1984, country: UK[18], *health food, **peanut butter, crunchy, ***AFB_1, AFB_2, AFG_1, and AFG_2
- Co-contamination: not reported
- Further contamination of animal products reported in the present book/article[18]: Butter (peanut butter), AFB_1, literature[18]

incidence: 6/50* **, conc. range: 2.5–5 µg/kg*** (2 sa.), 11–30 µg/kg*** (3 sa.), 40.1 µg/kg*** (1 sa.), sample origin: retail Health Food outlets in 25 different geographical locations, UK, sample year: 1986, country: UK[52], *health food, **peanut butter, smooth, *** AFB_1, AFB_2, AFG_1, and AFG_2

incidence: 12/79* **, conc. range: 2.5–5 µg/kg*** (4 sa.), 5.1–10 µg/kg*** (4 sa.), 11–30 µg/kg*** (2 sa.), 31–≤53.3 µg/kg*** (2 sa.), sample origin: retail Health Food outlets in 25 different geographical locations, UK, sample year: 1986, country: UK[52], *health food, **peanut butter, crunchy, *** AFB_1, AFB_2, AFG_1, and AFG_2
- Co-contamination: not reported
- Further contamination of animal products reported in the present book/article[52]: Butter (peanut butter), AFB_1, literature[52]

incidence: 4/10, conc. range: 7–228 µg/kg*, sample origin: traditional markets and supermarkets, Indonesia, sample year: 2001–2002, country: Austria/Indonesia/Thailand[64], *AFB_1, AFB_2, AFG_1, and AFG_2
- Co-contamination: not reported
- Further contamination of animal products reported in the present book/article[64]: Butter (peanut butter), AFB_1, literature[64]

incidence: 31/75, conc. range: 0.1–3.4 µg/kg*, Ø conc.: 0.7 µg/kg*, sample origin: available in New Zealand, sample year: 2010, country: New Zealand[66], *AFS total
- Co-contamination: not reported
- Further contamination of animal products reported in the present book/article[66]: Butter (peanut butter), AFB_1, literature[66]

incidence: 478/483, Ø conc.: 143.6 µg/kg*, sample origin: greater Manila (capital) area, Philippines, sample year: unknown, country: Philippines[70], *AF of pos. sa.?

- Co-contamination: not reported
- Further contamination of animal products reported in the present book/article[70]: Fish products, Meat products, AF/S, literature[70]

incidence: 98/2092, conc. range: 5.0–9.9 μg/kg* (47 sa.), 10.0–14.9 μg/kg* (36 sa.), 15.0–19.9 μg/kg* (12 sa.), >25 μg/kg* (3 sa.), sample origin: brokers, distributors, growers, processors, and retail trade, Canada, sample year: 1970–1975, country: Canada[71], *AFB_1, AFB_2, AFG_1, and AFG_2

- Co-contamination: not reported
- Further contamination of animal products reported in the present book/article[71]: not reported

incidence: 17/104, conc. range: ≤20 μg/kg* (13 sa.), >20 μg/kg* (4 sa., maximum: 27 μg/kg*), ∅ conc.: 14 μg/kg*, sample origin: FDA districts, USA, sample year: 1986, country: USA[72], *AFB_1, AFB_2, AFG_1, and AFG_2

incidence: 1/3, conc.: 43 μg/kg*, sample origin: imported to USA, sample year: 1986, country: USA[72], *AFB_1, AFB_2, AFG_1, and AFG_2

- Co-contamination: not reported
- Further contamination of animal products reported in the present book/article[72]: not reported

incidence: 13/42, conc.: 1.0–3.9 μg/kg* (8 sa.), 4.0–10.0 μg/kg* (4 sa.), 21 μg/kg* (1 sa.), sample origin: retail sa. from Norwich area and health food stores, market stalls, and supermarkets across UK, sample year: unknown, country: UK[74], *AFB_1, AFB_2, AFG_1, and AFG_2

- Co-contamination: not reported
- Further contamination of animal products reported in the present book/article[74]: not reported

incidence: 10/18, conc. range: 1.0–<5.0 μg/kg* (1 sa.), 5.0–<10.0 μg/kg* (3 sa.), 10.0–<50.0 μg/kg* (4 sa.), 50.0–<100.0 μg/kg* (1 sa.), 775 μg/kg* (1 sa.), sample origin: commercially available, sample year: unknown, country: UK[75], *AFB_1, AFB_2, AFG_1, and AFG_2

- Co-contamination: not reported
- Further contamination of animal products reported in the present book/article[75]: not reported

incidence: 7/7, conc. range: 0.32–13.26 μg/kg*, sample origin: sales outlets, Qatar, sample year: October 2002, country: Qatar[76], *AFB_1, AFB_2, AFG_1, and AFG_2

- Co-contamination: not reported
- Further contamination of animal products reported in the present book/article[76]: Custard powder, AF/S and DON, literature[76]

incidence: 256/400, conc. range: 32–54 μg/kg*, ∅ conc.: 43.5 μg/kg*, sample origin: local market and processing centers, central region of Sudan, sample year: unknown, country: Sudan/Saudi Arabia[77], *AFB_1, AFB_2, AFG_1, and AFG_2

- Co-contamination: not reported
- Further contamination of animal products reported in the present book/article[77]: not reported

For detailed information see the article.

incidence: 4/4, conc. range: 6.50–14.90 μg/kg*, ∅ conc.: 10.7 μg/kg*, sample origin: unknown, sample year: unknown, country: USA[78], *AFB_1, AFB_2, AFG_1, and AFG_2

- Co-contamination: not reported
- Further contamination of animal products reported in the present book/article[78]: not reported

incidence: 27/95, conc. range: <20 μg/kg*, ∅ conc.: 3.5 μg/kg*, sample origin: market basket sa. commercially available in Virginia (state), USA, sample year: 1982, country: USA[79], *AFB_1, AFB_2, AFG_1, and AFG_2

incidence: 273/355, conc. range: <20 μg/kg* (242 sa.), ≥20 μg/kg* (31 sa.), ∅ conc.: 7.7 μg/kg*, sample origin: market basket sa. commercially available in Virginia (state), USA, sample year: 1983, country: USA[79], *AFB_1, AFB_2, AFG_1, and AFG_2

incidence: 196/258, conc. range: <20 μg/kg* (189 sa.), ≥20 μg/kg* (7 sa.), ∅ conc.: 6.6 μg/kg*, sample origin: market basket sa. commercially available in Virginia (state), USA, sample year: 1984, country: USA[79], *AFB_1, AFB_2, AFG_1, and AFG_2

incidence: 184/235, conc. range: <20 μg/kg* (179 sa.), ≥20 μg/kg* (5 sa.), ∅ conc.: 5.7 μg/kg*, sample origin: market basket sa. commercially available in Virginia (state), USA, sample year: 1985, country: USA[79]

incidence: 435/465, conc. range: <20 μg/kg* (375 sa.), ≥20 μg/kg* (60 sa., maximum: 76.6 μg/kg*), ∅ conc.: 9.6 μg/kg*, sample origin: market basket sa. commercially available in Virginia (state), USA, sample year: 1986, country: USA[79], *AFB_1, AFB_2, AFG_1, and AFG_2

incidence: 404/449, conc. range: <20 μg/kg* (273 sa.), 20–50 μg/kg* (24 sa.), 50–100 μg/kg* (70 sa.), >100 μg/kg* (37 sa., maximum: 215 μg/kg*), ∅ conc.: 33.0 μg/kg*, sample origin: market basket sa. commercially available in Virginia (state), USA, sample year: 1987, country: USA[79], *AFB_1, AFB_2, AFG_1, and AFG_2

incidence: 187/322, conc. range: <20 μg/kg* (175 sa.), ≥20 μg/kg* (12 sa.), ∅ conc.: 8.9 μg/kg*, sample origin: market basket sa. commercially available in Virginia (state), USA, sample year: 1988, country: USA[79], *AFB_1, AFB_2, AFG_1, and AFG_2

incidence: 144/331, conc. range: <20 μg/kg*, ∅ conc.: 2.5 μg/kg*, sample origin: market basket sa. commercially available in Virginia (state), USA, sample year: 1989, country: USA[79], *AFB_1, AFB_2, AFG_1, and AFG_2

- Co-contamination: not reported
- Further contamination of animal products reported in the present book/article[79]: not reported

incidence: 16/35, conc. range: 0.1–5 μg/kg* (5 sa.), 6–10 μg/kg* (6 sa.), 11–20 μg/kg* (3 sa.), 21–35 μg/kg* (2 sa.), sample origin: commercial and imported food from retail outlets and points of entries, Malaysia, sample year: 1995–1999, country: Malaysia[80], *AF not specified

- Co-contamination: not reported
- Further contamination of animal products reported in the present book/article[80]: not reported

incidence: 10/11, conc. range: LOD-20 μg/kg* (1 sa.), 20–100 μg/kg* (3 sa.), 100–1000 μg/kg* (4 sa.), ≤2720 μg/kg* (2 sa.), sample origin: open-air markets in Port-au-Prince (capital), Haiti, sample year: July 2012, country: USA[81], *AFB_1, AFB_2, AFG_1, and AFG_2

incidence: 21/21, conc. range: 20–100 μg/kg* (2 sa.), 100–1000 μg/kg* (17 sa.), ≤1850 μg/kg* (2 sa.), sample origin: open-air markets in Cap Haitien, Haiti, sample year: July 2013, country: USA[81], *AFB_1, AFB_2, AFG_1, and AFG_2

- Co-contamination: not reported
- Further contamination of animal products reported in the present book/article[81]: not reported

incidence: 5/39, conc. range: 0.2–9.8 µg/kg*, ∅ conc.: 2.45 µg/kg*, sample origin: traditional markets, supermarkets, grocery stores, wholesale stores, and convenience stores in 25 counties, Taiwan, sample year: 2006, country: Taiwan[82], *AFB_1, AFB_2, AFG_1, and AFG_2

incidence: 5/5, conc. range: 1.2–2.8 µg/kg*, ∅ conc.: 1.96 µg/kg*, sample origin: traditional markets, supermarkets, grocery stores, wholesale stores, and convenience stores in 25 counties, Taiwan, sample year: 2009, country: Taiwan[82], *AFB_1, AFB_2, AFG_1, and AFG_2

incidence: 11/14, conc. range: 0.2–4.12 µg/kg*, ∅ conc.: 1.28 µg/kg*, sample origin: traditional markets, supermarkets, grocery stores, wholesale stores, and convenience stores in 25 counties, Taiwan, sample year: 2010, country: Taiwan[82], *AFB_1, AFB_2, AFG_1, and AFG_2

incidence: 8/10, conc. range: 0.2–32.5 µg/kg*, ∅ conc.: 5.5 µg/kg*, sample origin: traditional markets, supermarkets, grocery stores, wholesale stores, and convenience stores in 25 counties, Taiwan, sample year: 2011, country: Taiwan[82], *AFB_1, AFB_2, AFG_1, and AFG_2

- Co-contamination: For detailed information see the article.
- Further contamination of animal products reported in the present book/article[82]: not reported

incidence: 16/32, conc. range: ≤32.2 µg/kg*, sample origin: markets, superstores, malls, and venders of major districts of Punjab (province), Pakistan, sample year: unknown, country: Pakistan[83], *AFB_1, AFB_2, AFG_1, and AFG_2

- Co-contamination: not reported
- Further contamination of animal products reported in the present book/article[83]: not reported

incidence: 5/160, conc. range: 2–20 µg/kg*, sample origin: imported to New Zealand, sample year: 1999–2004, country: New Zealand[84], *AFB_1, AFB_2, AFG_1, and AFG_2

- Co-contamination: not reported
- Further contamination of animal products reported in the present book/article[84]: not reported

For detailed information see the article.

incidence: 31/75, conc. range: ≤3.4 µg/kg*, sample origin: New Zealand, sample year: 2010, country: New Zealand[85], *AFB_1, AFB_2, AFG_1, and AFG_2

- Co-contamination: not reported
- Further contamination of animal products reported in the present book/article[85]: not reported

incidence: 14/14*, conc. range: 34.2–115.6 µg/kg**, sample origin: supermarkets in Lilongwe (capital), Malawi, sample year: December 2012, country: Belgium/Malawi[86], *manufactured in Malawi, **AFB_1, AFB_2, AFG_1, and AFG_2

incidence: 8/11*, conc. range: ≤4.3 µg/kg**, sample origin: supermarkets in Lilongwe (capital), Malawi, sample year: December 2012, country: Belgium/Malawi[86], *imported from South Africa, **AFB_1, AFB_2, AFG_1, and AFG_2

- Co-contamination: not reported
- Further contamination of animal products reported in the present book/article[86]: not reported

Aspergillus **and *Penicillium*** Toxins

Ochratoxin A

incidence: 3/29, conc. range: 0.4–5.4 µg/kg, ∅ conc.: 2.6 µg/kg, sample origin: retail outlets, UK, sample year: September 2001, country: UK[9]

- Co-contamination: 2 sa. co-contaminated with AFB_1, AFB_2, AFG_1, AFG_2, and OTA; 1 sa. co-contaminated with AFB_1 and OTA
- Further contamination of animal products reported in the present book/article[9]: Butter (nut butter), AFB_1, AFB_2, AFG_1, AFG_2, and OTA, literature[9]; Butter (peanut butter), AFB_1, AFB_2, AFG_1, and AFG_2, literature[9]

incidence: 2/27, conc. range: LOQ-2 µg/kg (2 sa., maximum: 0.4 µg/kg), ∅ conc.: 0.35 µg/kg, sample origin: retail outlets, UK, sample year: September 2001, country: UK[9]

- Co-contamination: 2 sa. co-contaminated with AFB_1, AFB_2, AFG_1, and OTA
- Further contamination of animal products reported in the present book/article[9]: Butter (nut butter), AFB_1, AFB_2, AFG_1, AFG_2, and OTA, literature[9]; Butter (peanut butter), AFB_1, AFB_2, AFG_1, and AFG_2, literature[9]

incidence: 1/1, conc.: 0.04 µg/kg, sample origin: supermarkets and fast-food chains in Quebec (city), Canada, sample year: September–October 2008, country: Canada[87]

incidence: 0/1, conc.: no contamination, sample origin: supermarkets and fast-food chains in Calgary (city), Canada, sample year: September–October 2009, country: Canada[87]

- Co-contamination: not reported
- Further contamination of animal products reported in the present book/article[87]: Beefburger, Chicken nuggets, Chickenburger, Hot-dog, Ice cream, Meat, Meat products, Pork meat, Sausage, OTA, literature[87]

Butter (sesame butter) may contain the following mycotoxins:

Aspergillus Toxins

Aflatoxin/s

incidence: 1/14*, conc.: >176 µg/kg**, sample origin: retail markets, Turkey, sample year: unknown, country: Turkey[88], *Tahini, **AFB_1, AFB_2, AFG_1, and AFG_2

- Co-contamination: not reported
- Further contamination of animal products reported in the present book/article[88]: not reported

Cheese may contain the following mycotoxins:

Aspergillus Toxins

Aflatoxin B_1

incidence: 2/248, conc. range: 1 µg/kg, sample origin: Tunisia, sample year: unknown, country: Tunisia/USA[26]

- Co-contamination: 2 sa. co-contaminated with AFB_1 and AFM_1
- Further contamination of animal products reported in the present book/article[26]: Cheese, AFM_1, literature[26]

incidence: 6/26, conc. range: 7.9–15.0 µg/kg, sample origin: Division of Dairy Technology, National Dairy Research Institute, Karnal (city), India, sample year: September 1974–February 1975, country: India[27]

- Co-contamination: not reported
- Further contamination of animal products reported in the present book/article[27]: Cheese, AFB_2 and AFG_1, literature[27]; Dairy products, AFB_1 and AFG_1, literature[27]; Milk, Milk (buffalo milk), Milk (cow milk), Milk powder, AFM_1, literature[27]

Aflatoxin B_2

incidence: 1/26, conc.: tr, sample origin: Division of Dairy Technology, National Dairy Research Institute, Karnal (city), India, sample year: September 1974–February 1975, country: India[27]

- Co-contamination: not reported
- Further contamination of animal products reported in the present book/article[27]: Cheese, AFB_1 and AFG_1, literature[27]; Dairy products, AFB_1 and AFG_1, literature[27]; Milk, Milk (buffalo milk), Milk (cow milk), Milk powder, AFM_1, literature[27]

Aflatoxin G_1

incidence: 1/26, conc.: pr, sample origin: Division of Dairy Technology, National Dairy Research Institute, Karnal (city), India, sample year: September 1974–February 1975, country: India[27]

- Co-contamination: not reported
- Further contamination of animal products reported in the present book/article[27]: Cheese, AFB_1 and AFB_2, literature[27]; Dairy products, AFB_1 and AFG_1, literature[27]; Milk, Milk (buffalo milk), Milk (cow milk), Milk powder, AFM_1, literature[27]

Aflatoxin M_1

incidence: 11/45*, conc. range: 0.1–0.4 µg/kg, sample origin: Denmark, imported from Japan, sample year: unknown, country: Japan[24], *natural cheese

incidence: 8/40*, conc. range: 0.1–0.2 µg/kg, sample origin: Netherlands, imported from Japan, sample year: unknown, country: Japan[24], *natural cheese

incidence: 5/22*, conc. range: 0.1–0.4 µg/kg, sample origin: West Germany, imported from Japan, sample year: unknown, country: Japan[24], *natural cheese

incidence: 1/4*, conc.: 0.2 µg/kg, sample origin: UK, imported from Japan, sample year: unknown, country: Japan[24], *natural cheese

incidence: 19/80*, conc. range: 0.1–1.2 µg/kg, sample origin: different countries, imported from Japan, sample year: unknown, country: Japan[24], *natural cheese

- Co-contamination: not reported
- Further contamination of animal products reported in the present book/article[24]: Butter (peanut butter), AFB_1, AFB_2, and AFG_1, literature[24]

incidence: 2/248, conc. range: 6.2–10.6 µg/kg, ∅ conc.: 8.4 µg/kg, sample origin: Tunisia, sample year: unknown, country: Tunisia/USA[26]

- Co-contamination: 2 sa. co-contaminated with AFB_1 and AFM_1
- Further contamination of animal products reported in the present book/article[26]: Cheese, AFB_1, literature[26]

incidence: 9*/279, conc. range: 0.4–1.1 µg/kg, ∅ conc.: 0.8 µg/kg, sample origin: imported from Japan, sample year: unknown, country: Japan[29], *included Cheddar, Comte, Emmental, Gouda, Maribo, and Mozzarella cheese

- Co-contamination: not reported
- Further contamination of animal products reported in the present book/article[29]: not reported

incidence: 16/82*, conc. range: 0.005–0.050 µg/kg (9 sa.), 0.051–0.100 µg/kg (1 sa.), 0.101–0.250 µg/kg (4 sa.), 0.251–0.400 µg/kg (2 sa.), sample origin: France, imported from Italy, sample year: 1984, country: Italy[32], *included Brie, Camembert, Caprice des Dieux, Chamois d'Or, Le Roitelet, and Roquefort cheese

incidence: 9/34*, conc. range: 0.005–0.050 µg/kg (8 sa.), 0.051–0.100 µg/kg (1 sa.), sample origin: West Germany, imported from Italy, sample year: 1984, country: Italy[32], *Emmental cheese

incidence: 23/43*, conc. range: 0.005–0.050 µg/kg (4 sa.), 0.051–0.100 µg/kg (8 sa.), 0.101–0.250 µg/kg (11 sa.), sample origin: Netherlands, imported from Italy, sample year: 1984, country: Italy[32], *included Edam, Gouda, and Maasdam cheese

incidence: 124/338*, conc. range: 0.005–0.050 µg/kg (62 sa.), 0.051–0.100 µg/kg (26 sa.), 0.101–0.250 µg/kg (27 sa.), 0.251–0.400 µg/kg (7 sa.), >0.400 µg/kg (2 sa.), sample origin: northern Italy, sample year: spring–autumn 1985, country: Italy[32], included Grana Padano and Parmesan cheese

incidence: 6/78*, conc. range: 0.005–0.050 µg/kg (6 sa.), sample origin: central-southern Italy, sample year: spring–autumn 1985, country: Italy[32], *standard local cheeses

- Co-contamination: not reported
- Further contamination of animal products reported in the present book/article[32]: Milk (cow milk), AFM_1, literature[32]

incidence: 51/57*, conc. range: ≤0.810 µg/kg, sample origin: supermarkets and street milkmen in Bursa Province, Turkey, sample year: unknown, country: Turkey[33], *included white cheese (different kinds), Kashar cheese (fresh and old), Mihalic, skin bag, Urfa, and Van otlu cheese

- Co-contamination: not reported
- Further contamination of animal products reported in the present book/article[33]: Milk, AFM_1, literature[33]

incidence: 5/9*, conc. range: 20–72 µg/kg, ∅ conc.: 42.60 µg/kg, sample origin: southern Spain, sample year: unknown, country: Spain[34], *fresh cheese made of different types of milk (cow, ewe, goat, and mixed)

incidence: 5/9*, conc. range: 20–130 µg/kg, ∅ conc.: 73.80 µg/kg, sample origin: southern Spain, sample year: unknown, country: Spain[34], *semi-ripened cheese made of different types of milk (cow, ewe, goat, and mixed)

incidence: 6/17*, conc. range: 20–200 µg/kg, ∅ conc.: 105.33 µg/kg, sample origin: southern Spain, sample year: unknown, country: Spain[34], *ripened cheese made of different types of milk (cow, ewe, goat, and mixed)

- Co-contamination: not reported
- Further contamination of animal products reported in the present book/article[34]: not reported

incidence: 11/39, conc. range: 0.051–0.080 µg/kg (1 sa.), 0.081–0.110 µg/kg (2 sa.), >0.110 µg/kg (8 sa., maximum: 0.18844 µg/kg), sample origin: Ankara (capital), Turkey, sample year: 2004, country: Turkey[38]

- Co-contamination: not reported
- Further contamination of animal products reported in the present book/article[38]: Yogurt, AFM_1, literature[38]

incidence: 10/22*, conc. range: 0.089–0.902 µg/kg, ∅ conc.: 0.280 µg/kg, sample origin: retail stores, Iran, sample year: spring, country: Iran[39], *traditional cheese produced from ewe's and cow's milk

incidence: 5/22*, conc. range: 0.118–0.845 µg/kg, ∅ conc.: 0.317 µg/kg, sample origin: retail stores, Iran, sample year: summer, country: Iran[39], *traditional cheese produced from ewe's and cow's milk

incidence: 15/22*, conc. range: 0.082–1.157 µg/kg, ∅ conc.: 0.425 µg/kg, sample origin: retail stores, Iran, sample year: autumn, country: Iran[39], *traditional cheese produced from ewe's and cow's milk incidence: 17/22*, conc. range: 0.142–1.254 μg/kg, Ø conc.: 0.505 μg/kg, sample origin: retail stores, Iran, sample year: winter, country: Iran[39], *traditional cheese produced from ewe's and cow's milk

- Co-contamination: not reported
- Further contamination of animal products reported in the present book/article[39]: not reported

incidence: 2/13*, conc. range: 0.039**–0.223** μg/kg, Ø conc.: 0.131** μg/kg, sample origin: small food-processing plants in central part of Slovenia, sample year: November 2004–January 2005, country: Slovenia[40], *fresh salted and non-salted cheese, **fresh cheese

incidence: 2/13*, conc. range: 0.025–0.068 μg/kg, Ø conc.: 0.0465 μg/kg, sample origin: small food-processing plants in central part of Slovenia, sample year: November 2004–January 2005, country: Slovenia[40], *semihard cheese

- Co-contamination: not reported
- Further contamination of animal products reported in the present book/article[40]: Cheese curd, AFM_1, literature[40]

incidence: 8*/10, conc. range: 0.0250–0.0784 μg/kg, sample origin: Egypt, imported and purchased in supermarkets in Kuwait City (capital), Kuwait, sample year: January 2005–March 2007, country: Kuwait[41], *see the article for kind of contaminated cheese

incidence: 4*/4, conc. range: 0.0290–0.4516 μg/kg, sample origin: Greece, imported and purchased in supermarkets in Kuwait City (capital), Kuwait, sample year: January 2005–March 2007, country: Kuwait[41], *see the article for kind of contaminated cheese

incidence: 2*/4, conc. range: 0.0248–0.0312 μg/kg, Ø conc.: 0.028 μg/kg, sample origin: France, imported and purchased in supermarkets in Kuwait City (capital), Kuwait, sample year: January 2005–March 2007, country: Kuwait[41], *see the article for kind of contaminated cheese

incidence: 2*/4, conc. range: 0.0369–0.0572 μg/kg, Ø conc.: 0.0471 μg/kg, sample origin: Denmark, imported and purchased in supermarkets in Kuwait City (capital), Kuwait, sample year: January 2005–March 2007, country: Kuwait[41], *see the article for kind of contaminated cheese

incidence: 3*/3, conc. range: 0.0341–0.148 μg/kg, sample origin: Lebanon, imported and purchased in supermarkets in Kuwait City (capital), Kuwait, sample year: January 2005–March 2007, country: Kuwait[41], *see the article for kind of contaminated cheese

incidence: 2*/3, conc. range: 0.0398–0.125 μg/kg, Ø conc.: 0.0824 μg/kg, sample origin: UAE, imported and purchased in supermarkets in Kuwait City (capital), Kuwait, sample year: January 2005–March 2007, country: Kuwait[41], *see the article for kind of contaminated cheese

incidence: 2*/2, conc. range: 0.0395–0.186 μg/kg, Ø conc.: 0.1128 μg/kg, sample origin: Syria, imported and purchased in supermarkets in Kuwait City (capital), Kuwait, sample year: January 2005–March 2007, country: Kuwait[41], *see the article for kind of contaminated cheese

incidence: 1*/2, conc.: 0.0428 μg/kg, sample origin: Cyprus, imported and purchased in supermarkets in Kuwait City (capital), Kuwait, sample year: January 2005–March 2007, country: Kuwait[41], *see the article for kind of contaminated cheese

incidence: 2*/2, conc. range: 0.0238–0.0250 μg/kg, Ø conc.: 0.0244 μg/kg, sample origin: Saudi Arabia, imported and purchased in supermarkets in Kuwait City (capital), Kuwait, sample year: January 2005–March 2007, country: Kuwait[41], *see the article for kind of contaminated cheese

incidence: 1*/1, conc.: 0.0373 μg/kg, sample origin: Czech Republic, imported and purchased in supermarkets in Kuwait City (capital), Kuwait, sample year: January 2005–March 2007, country: Kuwait[41], *see the article for kind of contaminated cheese

incidence: 1*/1, conc.: 0.0272 μg/kg, sample origin: Jordan, imported and purchased in supermarkets in Kuwait City (capital), Kuwait, sample year: January 2005–March 2007, country: Kuwait[41], *see the article for kind of contaminated cheese

incidence: 1*/1, conc.: 0.0241 μg/kg, sample origin: Germany, imported and purchased in supermarkets in Kuwait City (capital), Kuwait, sample year: January 2005–March 2007, country: Kuwait[41], *see the article for kind of contaminated cheese

incidence: 1*/1, conc.: 0.0387 μg/kg, sample origin: Hungary, imported and purchased in supermarkets in Kuwait City (capital), Kuwait, sample year: January 2005–March 2007, country: Kuwait[41], *see the article for kind of contaminated cheese

incidence: 1*/1, conc.: 0.249 μg/kg, sample origin: Turkey, imported and purchased in supermarkets in Kuwait City (capital), Kuwait, sample year: January 2005–March 2007, country: Kuwait[41], *see the article for kind of contaminated cheese

incidence: 2*/2, conc. range: 0.0346–0.0784 μg/kg, Ø conc.: 0.0565 μg/kg, sample origin: Kuwait, purchased from supermarkets in Kuwait City (capital), Kuwait, sample year: January 2005–March 2007, country: Kuwait[41], *see the article for kind of contaminated cheese

- Co-contamination: not reported
- Further contamination of human breast milk and animal products reported in the present book/article[41]: Milk (cow milk), Milk (human breast milk), Milk (UHT milk), Milk powder, AFM_1, literature[41]

incidence: 2*/5, conc. range: 0.1–0.9 μg/kg*, Ø conc.: 0.5 μg/kg*, sample origin: imported and domestic sa., sample year: unknown, country: Japan[42], *imported sa. contaminated

- Co-contamination: 1 sa. co-contaminated with AFM_1 and AFM_2; 1 sa. contaminated solely with AFM_1
- Further contamination of animal products reported in the present book/article[42]: Cheese, AFM_2, literature[42]

incidence: 15/19*, conc. range: 10–50 μg/kg (7 sa.), 50–100 μg/kg (4 sa.), 100–150 μg/kg (4 sa.), sample origin: Belgian market, Belgium, sample year: unknown, country: Belgium[110], *Belgian cheeses

- Co-contamination: not reported
- Further contamination of animal products reported in the present book/article[110]: not reported

incidence: 4/16, conc. range: 0.010–0.068 μg/kg, Ø conc.: 0.033 μg/kg, sample origin: domestic, sample year: 1983, country: Japan[120]

- Co-contamination: not reported
- Further contamination of animal products reported in the present book/article[120]: Cheese (blue-veined cheese), Cheese (Brie cheese), Cheese (butter cheese), Cheese (Camembert cheese), Cheese (Cheddar cheese), Cheese (cream cheese), Cheese (Edam cheese), Cheese (Gouda cheese), Cheese (Havarti cheese), Cheese (Maribo cheese), Cheese (Mozzarella cheese), Cheese (Samsoe cheese), AFM_1, literature[120]

incidence: 2/2, conc. range: 0.0046–0.0222 µg/kg, ∅ conc.: 0.0134 µg/kg, sample origin: retail shops in Marang and Kuala Terengganu (town and city), Malaysia, sample year: July 2013?, country: Malaysia[163]

- Co-contamination: not reported
- Further contamination of animal products reported in the present book/article[163]: Dairy beverages, Milk, Milk (cow milk), Milk powder, Yogurt, AFM_1, literature[163]

incidence: 4/4, conc. range: 0.16–0.32 µg/l (4 sa.), sample origin: supermarkets in Heilongjiang (province), northeast China, sample year: March–May 2008, country: China/Russia/Korea[283]

- Co-contamination: not reported
- Further contamination of animal products reported in the present book/article[283]: Dairy products, Milk (cow milk), Milk powder, AFM_1, literature[283]

incidence: 13/25, conc. range: 0.0773–0.298 µg/kg, ∅ conc.: 0.18197 µg/kg, sample origin: rural regions in Hamadan, Ilam, Kermanshah, and Kurdistan (provinces), Iran, sample year: winter and summer seasons 2014, country: Iran[328]

- Co-contamination: not reported
- Further contamination of animal products reported in the present book/article[328]: Ayran, Kashk, Milk (cow milk), Milk (goat milk), Milk (sheep milk), Tarkhineh, Yogurt, AFM_1, literature[328]

incidence: 85/102*, conc. range: 0.025–<0.050 µg/kg (27 sa.), 0.050–0.150 µg/kg (54 sa.), 0.150–0.250 µg/kg (3 sa.), ~0.300 µg/kg (1 sa.), sample origin: local supermarkets and small-scale producers in Cuneo (city, commune), Italy, sample year: unknown, country: Italy[400], *included creamy, soft, semihard, hard, elastic, and blue cheese

- Co-contamination: not reported
- Further contamination of animal products reported in the present book/article[400]: not reported

For detailed information see the article.

incidence: 69/80, conc. range: 0.0143–0.5721 µg/l, sample origin: retail markets in Isfahan (province), Iran, sample year: February 2011–2012, country: Iran[424]

- Co-contamination: not reported
- Further contamination of animal products reported in the present book/article[424]: Ice cream, Yogurt, AFM_1, literature[424]

For detailed information see the article.

incidence: 3/10, conc. range: 0.091–0.30 µg/kg, ∅ conc.: 0.16 µg/kg, sample origin: households from employees of the University of São Paulo at Pirassununga (municipality), Brazil, sample year: June 2011–March 2012, country: Brazil[458]

- Co-contamination: not reported
- Further contamination of animal products reported in the present book/article[458]: Milk (cow milk), Milk powder, AFM_1, literature[458]

incidence: 136/197, conc. range: tr-0.23 µg/kg, ∅ conc.: 0.09 µg/kg, sample origin: Germany, sample year: May–August 1976, country: Germany[466]

- Co-contamination: not reported
- Further contamination of animal products reported in the present book/article[466]: not reported

For detailed information see the article.

incidence: 7/17, conc. range: ≤0.003–0.005 µg/kg (3 sa.), <0.010–0.020 µg/kg (4 sa., maximum: 0.018 µg/kg), sample origin: dairies in different areas on Sicily (island, autonomous region), Italy, sample year: January–June 2012, country: Italy[508]

- Co-contamination: not reported
- Further contamination of animal products reported in the present book/article[508]: Cream, Milk (cow milk), Milk (goat milk), Milk (sheep milk), AFM_1, literature[508]

incidence: 12/42, conc. range: 0.02–0.132 µg/kg, ∅ conc.: 0.061 µg/kg, sample origin: control points such as import, primary storage, and market, Cyprus, sample year: 2004–2013, country: Cyprus[542]

- Co-contamination: not reported
- Further contamination of animal products reported in the present book/article[542]: Cheese (Anari cheese), Cheese (white cheese), Ice cream, Milk, Milk powder, Yogurt, AFM_1, literature[542]

incidence: 7/8*, conc. range: 0.015–0.05 µg/kg (5 sa.), 0.05–0.15 µg/kg (2 sa.), ∅ conc.: 0.047 µg/kg, sample origin: supermarkets, South Korea, sample year: October 2013–March 2014, country: South Korea[561], *local cheese sa.

incidence: 9/53*, conc. range: 0.015–0.05 µg/kg (8 sa.), 0.05–0.15 µg/kg (1 sa.), ∅ conc.: 0.026 µg/kg, sample origin: supermarkets, South Korea, sample year: October 2013–March 2014, country: South Korea[561], *imported cheese sa.

- Co-contamination: not reported
- Further contamination of animal products reported in the present book/article[561]: Milk (cow milk), Yogurt, AFM_1, literature[561]

Aflatoxin M_2

incidence: 1*/5, conc.: <0.1 µg/kg*, sample origin: imported and domestic sa., sample year: unknown, country: Japan[42], *imported sa. contaminated

- Co-contamination: 1 sa. co-contaminated with AFM_1 and AFM_2
- Further contamination of animal products reported in the present book/article[42]: Cheese, AFM_1, literature[42]

Sterigmatocystin

incidence: 7/21*, conc. range: 0.03–1.23 µg/kg, ∅ conc.: 0.28 µg/kg, sample origin: supermarkets in Latvia and Belgium, sample year: 2008, country: Belgium/Latvia[111], *included hard, semihard, soft, and semisoft cheese

- Co-contamination: not reported
- Further contamination of animal products reported in the present book/article[111]: not reported

For detailed information see the article.

Aspergillus and ***Penicillium*** Toxins

Citrinin

incidence: 2/25*, conc. range: ≤50 µg/kg, sample origin: Environmental Health Officers, Leatherhead Food Research Association, and individual members of staff, UK, sample year: 1980, country: UK[115], *mold-spoiled cheese

incidence: 15/19*, conc. range: ≤50 µg/kg, sample origin: Environmental Health Officers, Leatherhead Food Research Association, and individual members of staff, UK, sample year: 1981–1982, country: UK[115], *mold-spoiled cheese

- Co-contamination: not reported
- Further contamination of animal products reported in the present book/article[115]: Cheese, OTA, literature[115]

Ochratoxin A

incidence: 2/33, conc. range: 0.060–0.110 μg/kg*, sample origin: regions around Detmold, Giessen, Hamburg, Jena, Karlsruhe, Kiel, Kulmbach, Trier (cities), Germany, sample year: unknown, country: Germany[100], *Harzer cheese

incidence: 8/38*, conc. range: ≤0.860 μg/kg*, sample origin: regions around Detmold, Giessen, Hamburg, Jena, Karlsruhe, Kiel, Kulmbach, Trier (cities), Germany, sample year: unknown, country: Germany[100], *cheese with ingredients (spices, etc.)
- Co-contamination: not reported
- Further contamination of animal products reported in the present book/article[100]: Cheese (fresh cheese), OTA, literature[100]

incidence: 15/25*, conc. range: ≤260 μg/kg, sample origin: Environmental Health Officers, Leatherhead Food Research Association, and individual members of staff, UK, sample year: 1980, country: UK[115], *mold-spoiled cheese

incidence: 3/19*, conc. range: ≤7 μg/kg, sample origin: Environmental Health Officers, Leatherhead Food Research Association, and individual members of staff, UK, sample year: 1981–1982, country: UK[115], *mold-spoiled cheese
- Co-contamination: not reported
- Further contamination of animal products reported in the present book/article[115]: Cheese, CIT, literature[115]

Penicillium Toxins

Mycophenolic Acid

incidence: 6/15*, conc. range: 10–100 μg/kg (1 sa.), 100–1000 μg/kg (3 sa.), 1000–5000 μg/kg (2 sa.), sample origin: commercial, sample year: unknown, country: France[118], *industrial French cheeses
- Co-contamination: not reported
- Further contamination of animal products reported in the present book/article[118]: Cheese (Bleu des Causses cheese), Cheese (blue-veined cheese), Cheese (Gorgonzola cheese), Cheese (Roquefort cheese), MA, literature[118]

For detailed information see the article.

Cheese (Akkawi cheese) see **Cheese (white cheese)**

Cheese (Anari cheese) may contain the following mycotoxins:

Aspergillus Toxins

Aflatoxin M_1

incidence: 14/39, conc. range: 0.007–0.057 μg/kg, ∅ conc.: 0.031 μg/kg, sample origin: control points such as import, primary storage, and market, Cyprus, sample year: 2004–2013, country: Cyprus[542]
- Co-contamination: not reported
- Further contamination of animal products reported in the present book/article[542]: Cheese, Cheese (white cheese), Ice cream, Milk, Milk powder, Yogurt, AFM_1, literature[542]

Cheese (Baladi cheese) see **Cheese (white cheese)**

Cheese (Bavaria blue cheese) see **Cheese (blue-veined cheese)**

Cheese (Bhutanese cheese) may contain the following mycotoxins:

Aspergillus Toxins

Aflatoxin B_1

incidence: 3/19*, conc. range: 68–212 μg/kg, ∅ conc.: 156 μg/kg, sample origin: shops in Thimphu (capital), Bhutan, sample year: September–October, country: India[119], *moldy brown and green growth on cheese surface (chuku = Bhutanese term for the cheese, chhurpi = Nepalese term for the cheese)
- Co-contamination: 1 sa. co-contaminated with AFB_1 and AFB_2; 2 sa. contaminated solely with AFB_1
- Further contamination of animal products reported in the present book/article[119]: Cheese, AFB_2, CIT, and OTA, literature[119]

Aflatoxin B_2

incidence: 1/19, conc.: 56 μg/kg, sample origin: shops in Thimphu (capital), Bhutan, sample year: September–October, country: India[119], *moldy brown and green growth on cheese surface (chuku = Bhutanese term for the cheese, chhurpi = Nepalese term for the cheese)
- Co-contamination: 1 sa. co-contaminated with AFB_1 and AFB_2
- Further contamination of animal products reported in the present book/article[119]: Cheese, AFB_1, CIT, and OTA, literature[119]

Aspergillus and ***Penicillium*** Toxins

Citrinin

incidence: 6/19, conc. range: 74–224 μg/kg, ∅ conc.: 125 μg/kg, sample origin: shops in Thimphu (capital), Bhutan, sample year: September–October, country: India[119], *moldy brown and green growth on cheese surface (chuku = Bhutanese term for the cheese, chhurpi = Nepalese term for the cheese)
- Co-contamination: 1 sa. co-contaminated with AFB_1, CIT, and OTA; 1 sa. co-contaminated with CIT and OTA; 4 sa. contaminated solely with CIT
- Further contamination of animal products reported in the present book/article[119]: Cheese, AFB_1, AFB_2, and OTA, literature[119]

Ochratoxin A

incidence: 5/19, conc. range: 42–116 μg/kg, ∅ conc.: 84.2 μg/kg, sample origin: shops in Thimphu (capital), Bhutan, sample year: September–October, country: India[119], *moldy brown and green growth on cheese surface (chuku = Bhutanese term for the cheese, chhurpi = Nepalese term for the cheese)
- Co-contamination: 1 sa. co-contaminated with AFB_1, CIT, and OTA; 1 sa. co-contaminated with CIT and OTA; 3 sa. contaminated solely with OTA
- Further contamination of animal products reported in the present book/article[119]: Cheese, AFB_1, AFB_2, and CIT, literature[119]

Cheese (Bleu des Causses cheese) may contain the following mycotoxins:

Penicillium Toxins

Mycophenolic Acid

incidence: 3/6, conc. range: 10–100 μg/kg (1 sa.), 100–1000 μg/kg (2 sa.), sample origin: commercial, sample year: unknown, country: France[118]

- Co-contamination: not reported
- Further contamination of animal products reported in the present book/article[118]: Cheese, Cheese (blue-veined cheese), Cheese (Gorgonzola cheese), Cheese (Roquefort cheese), MA, literature[118]

For detailed information see the article.

Cheese (Blue Castello cheese) see **Cheese (blue-veined cheese)**

Cheese (blue cheese) see **Cheese** and **Cheese (blue-veined cheese)**

Cheese (Blue Stilton cheese) see **Cheese (blue-veined cheese)**

Cheese (blue-veined cheese) may contain the following mycotoxins:

Aspergillus Toxins

Aflatoxin M_1

incidence: 7/7, conc. range: 0.084–0.556 µg/kg, ∅ conc.: 0.233 µg/kg, sample origin: Denmark, imported and purchased in Japan, sample year: 1981–1983, country: Japan[120]

- Co-contamination: not reported
- Further contamination of animal products reported in the present book/article[120]: Cheese, Cheese (Brie cheese), Cheese (butter cheese), Cheese (Camembert cheese), Cheese (Cheddar cheese), Cheese (cream cheese), Cheese (Edam cheese), Cheese (Gouda cheese), Cheese (Havarti cheese), Cheese (Maribo cheese), Cheese (Mozzarella cheese), Cheese (Samsoe cheese), AFM_1, literature[120]

Aspergillus and *Penicillium* Toxins

Penicillic Acid

incidence: 1/110*, conc.: 820 µg/kg, sample origin: local retail dealers and manufacturers, France/Netherland?, sample year: unknown, country: France/Netherlands[114], *in whole cheese

- Co-contamination: not reported
- Further contamination of animal products reported in the present book/article[114]: Cheese (blue-veined cheese), MA, literature[114]; Cheese (goat cheese), PA and PAT, literature[114]; Cheese (hard cheese), PA, PAT, MA, and STG, literature[114]; Milk powder, AFB_1 and AFM_1, literature[114]

Penicillium Toxins

Isofumigaclavine A

incidence: 7/7*, conc. range: 150–4700 µg/kg, ∅ conc.: 1668 µg/kg, sample origin: Denmark, imported and purchased in Canada, sample year: unknown, country: Canada[122], *included "mold-free" and "high-mold" sa.

incidence: 1/1, conc.: 1833 µg/kg, sample origin: Finland, imported and purchased in Canada, sample year: unknown, country: Canada[122]

incidence: 4/6*, conc. range: tr-170 µg/kg, sample origin: West Germany, imported and purchased in Canada, sample year: unknown, country: Canada[122], *included "mold-free" and "high-mold" sa.

incidence: 1/1, conc.: 785 µg/kg, sample origin: Quebec (province), Canada, sample year: unknown, country: Canada[122]

incidence: 2/3, conc. range: 100–150 µg/kg, ∅ conc.: 120 µg/kg, sample origin: Roquefort (village) in France, imported and purchased in Canada, sample year: unknown, country: Canada[122]

- Co-contamination: For detailed information see the article.
- Further contamination of animal products reported in the present book/article[122]: Cheese (blue-veined cheese), IFC B and ROQ C, literature[122]; Cheese (Gorgonzola cheese) ROQ C, literature[122]; Cheese (Stilton cheese), IFC A and ROQ C, literature[122]

For detailed information see the article.

Isofumigaclavine B

incidence: 4/7*, conc. range: tr, sample origin: Denmark, imported and purchased in Canada, sample year: unknown, country: Canada[122], *included "mold-free" and "high-mold" sa.

incidence: 1/1, conc.: tr, sample origin: Finland, imported and purchased in Canada, sample year: unknown, country: Canada[122]

incidence: 0/6*, conc.: no contamination, sample origin: West Germany, imported and purchased in Canada, sample year: unknown, country: Canada[122], *included "mold-free" and "high-mold" sa.

incidence: 1/1, conc.: tr, sample origin: Quebec (province), Canada, sample year: unknown, country: Canada[122]

incidence: 0/3, conc.: no contamination, sample origin: Roquefort (village) in France, imported and purchased in Canada, sample year: unknown, country: Canada[122]

- Co-contamination: For detailed information see the article.
- Further contamination of animal products reported in the present book/article[122]: Cheese (blue-veined cheese), IFC A and ROQ C, literature[122]; Cheese (Gorgonzola cheese) ROQ C, literature[122]; Cheese (Stilton cheese), IFC A and ROQ C, literature[122]

For detailed information see the article.

Mycophenolic Acid

incidence: 41/110, conc. range: 10–100 µg/kg (9 sa.), 100–1000 µg/kg (8 sa.), 1000–5000 µg/kg (15 sa.), 5000–10,000 µg/kg (5 sa.), 10,000–15,000 µg/kg (4 sa.), sample origin: local retail dealers and manufacturers, France/Netherland?, sample year: unknown, country: France/Netherlands[114]

- Co-contamination: not reported
- Further contamination of animal products reported in the present book/article[114]: Cheese (blue-veined cheese) PA, literature[114]; Cheese (goat cheese), PA and PAT, literature[114]; Cheese (hard cheese), PA, PAT, MA, and STG, literature[114]; Milk powder, AFB_1 and AFM_1, literature[114]

incidence: 3/12*, conc. range: 10–100 µg/kg (2 sa.), 100–1000 µg/kg (1 sa.), sample origin: commercial, sample year: unknown, country: France[118], *German blue cheese

- Co-contamination: not reported
- Further contamination of animal products reported in the present book/article[118]: Cheese, Cheese (Bleu des Causses cheese), Cheese (Gorgonzola cheese), Cheese (Roquefort cheese), MA, literature[118]

For detailed information see the article.

incidence: 1/10, conc.: 300 µg/kg, sample origin: imported from Denmark, France, and Germany as well as Finland (domestic) and purchased in supermarkets, Finland, sample year: 2003, country: Finland[123]

- Co-contamination: 1 sa. co-contaminated with MA and ROQ C
- Further contamination of animal products reported in the present book/article[123]: Cheese (blue-veined cheese), ROQ C, literature[123]

incidence: 4/32*, conc. range: 250–5000 μg/kg**, Ø conc.: 3375 μg/kg**, sample origin: German market, Germany, sample year: 1980–1981, country: Germany[434], *included Bavaria blue, Blauer Kommandör, Blue Castello, Blue Stilton, Bresse bleu, Danablue, Edelpilz, Gorgonzola, processed, and Roquefort** cheese (made from sheep milk)

- Co-contamination: not reported
- Further contamination of animal products reported in the present book/article[434]: not reported

incidence: 27/56*, conc. range: 15–3120 μg/kg, Ø conc.: 474 μg/kg, sample origin: different countries, purchased from supermarkets and retail markets, France/Italy, sample year: October–December 2011, country: France/Italy[439], *blue-veined cheese made from cow milk

incidence: 14/18*, conc. range: 176–6190 μg/kg, Ø conc.: 1376 μg/kg, sample origin: produced in France, Ireland, and Scotland and purchased from supermarkets and retail markets, France/Italy, sample year: October–December 2011, country: France/Italy[439], *blue-veined cheese made from ewe milk

incidence: 1/3*, conc.: 377 μg/kg, sample origin: produced in France and Netherlands and purchased from supermarkets and retail markets, France/Italy, sample year: October–December 2011, country: France/Italy[439], *blue-veined cheese made from goat milk

incidence: 2/4*, conc. range: 17–144 μg/kg, Ø conc.: 81 μg/kg, sample origin: produced in Spain and purchased from supermarkets and retail markets, France/Italy, sample year: October–December 2011, country: France/Italy[439], *blue-veined cheese made from cow and ewe, cow and goat, or cow, ewe, and goat milk

incidence: 0/5*, conc.: no contamination, sample origin: produced in Canada, Germany, New Zealand, as well as Poland and purchased from supermarkets and retail markets, France/Italy, sample year: October–December 2011, country: France/Italy[439], *blue-veined cheese made from unknown milk sources

- Co-contamination: 44 sa. co-contaminated with MA and ROQ C (in total)
- Further contamination of animal products reported in the present book/article[439]: Cheese (blue-veined cheese), ROQ C, literature[439]

For detailed information see the article.

incidence: 51/53*, conc. range: ≤11,000 μg/kg**, sample origin: imported and domestic sa., and purchased in retail shops in the area of Munich (city), Germany, sample year: unknown, country: Germany[537], *included blue, blue-red, blue-white**, Blue Stilton, Cabrales, Danablue, Fourme d'Ambert, Gorgonzola, and Roquefort cheese

- Co-contamination: not reported
- Further contamination of animal products reported in the present book/article[537]: not reported

Roquefortine C

incidence: 7/7*, conc. range: 60–2500 μg/kg, Ø conc.: 1136 μg/kg, sample origin: Denmark, imported and purchased in Canada, sample year: unknown, country: Canada[122], *included "mold-free" and "high-mold" sa.

incidence: 1/1, conc.: 970 μg/kg, sample origin: Finland, imported and purchased in Canada, sample year: unknown, country: Canada[122]

incidence: 4/6*, conc. range: 50–6800 μg/kg, Ø conc.: 966 μg/kg, sample origin: West Germany, imported and purchased in Canada, sample year: unknown, country: Canada[122], *included "mold-free" and "high-mold" sa.

incidence: 1/1, conc.: 1085 μg/kg, sample origin: Quebec (province), Canada, sample year: unknown, country: Canada[122]

incidence: 3/3, conc. range: 60–1500 μg/kg, Ø conc.: 487 μg/kg, sample origin: Roquefort (village) in France, imported and purchased in Canada, sample year: unknown, country: Canada[122]

- Co-contamination: For detailed information see the article.
- Further contamination of animal products reported in the present book/article[122]: Cheese (blue-veined cheese), IFC A and IFC B, literature[122]; Cheese (Gorgonzola cheese) ROQ C, literature[122]; Cheese (Stilton cheese), IFC A and ROQ C, literature[122]

For detailed information see the article.

incidence: 10/10, conc. range: 900–12,000 μg/kg, Ø conc.: 3590 μg/kg, sample origin: imported from Denmark, France, and Germany as well as Finland (domestic) and purchased in supermarkets, Finland, sample year: 2003, country: Finland[123]

- Co-contamination: 1 sa. co-contaminated with MA and ROQ C; 9 sa. contaminated solely with ROQ C
- Further contamination of animal products reported in the present book/article[123]: Cheese (blue cheese), MA, literature[123]

incidence: 1/1*, conc.: 800 μg/kg, sample origin: imported from Denmark and purchased in supermarkets, Finland, sample year: 2003, country: Finland[123], *blue-white cheese

- Co-contamination: not reported
- Further contamination of animal products reported in the present book/article[123]: Cheese (blue-veined cheese), MA, literature[123]

incidence: 12/12, conc. range: 162–651 μg/kg, Ø conc.: 424 μg/kg, sample origin: USA, sample year: unknown, country: USA[125]

incidence: 2/2*, conc. range: 18–72 μg/kg, Ø conc.: 45 μg/kg, sample origin: USA, sample year: unknown, country: USA[125], *blue cheese dressing

- Co-contamination: not reported
- Further contamination of animal products reported in the present book/article[125]: not reported

incidence: 56/56*, conc. range: 13–5140 μg/kg, Ø conc.: 765 μg/kg, sample origin: different countries, purchased from supermarkets and retail markets, France/Italy, sample year: October–December 2011, country: France/Italy[439], *blue-veined cheese made from cow milk

incidence: 18/18*, conc. range: 130–1720 μg/kg, Ø conc.: 617 μg/kg, sample origin: produced in France, Ireland, and Scotland and purchased from supermarkets and retail markets, France/Italy, sample year: October–December 2011, country: France/Italy[439], *blue-veined cheese made from ewe milk

incidence: 3/3*, conc.: range: 168–14,125 μg/kg, Ø conc.: 4990 μg/kg, sample origin: produced in France and Netherlands and purchased from supermarkets and retail markets, France/Italy, sample year: October–December 2011, country: France/Italy[439], *blue-veined cheese made from goat milk incidence: 4/4*, conc. range: 77–505 µg/kg, ∅ conc.: 327 µg/kg, sample origin: produced in Spain and purchased from supermarkets and retail markets, France/Italy, sample year: October–December 2011, country: France/Italy[439], *blue-veined cheese made from cow and ewe, cow and goat, or cow, ewe, and goat milk

incidence: 5/5*, conc. range: 11–719 µg/kg, ∅ conc.: 490 µg/kg, sample origin: produced in Canada, Germany, New Zealand, as well as Poland and purchased from supermarkets and retail markets, France/Italy, sample year: October–December 2011, country: France/Italy[439], *blue-veined cheese made from unknown milk sources

- Co-contamination: 44 sa. co-contaminated with MPA and ROQ C; 42 sa. contaminated solely with ROQ C (in total)
- Further contamination of animal products reported in the present book/article[439]: Cheese (blue-veined cheese), MA, literature[439]

For detailed information see the article.
see also **Cheese**

Cheese (Bresse bleu cheese) see **Cheese (blue-veined cheese)**

Cheese (Brie cheese) may contain the following mycotoxins:

Aspergillus Toxins

Aflatoxin M_1
incidence: 1/2, conc.: 0.4 µg/kg, sample origin: France, imported and purchased in grocery stores in Washington, DC (capital) area, USA, sample year: before 1985, country: USA[30]

- Co-contamination: not reported
- Further contamination of animal products reported in the present book/article[30]: Cheese (Camembert cheese), Cheese (Cheshire cheese), Cheese (Edam cheese), Cheese (fondue cheese), Cheese (Gouda cheese), Cheese (Gruyere cheese), AFM_1, literature[30]

incidence: 6/6, conc. range: 0.058–0.414 µg/kg, ∅ conc.: 0.195 µg/kg, sample origin: Denmark, imported and purchased in Japan, sample year: 1981–1983, country: Japan[120]

incidence: 5/5, conc. range: 0.100–0.714 µg/kg, ∅ conc.: 0.299 µg/kg, sample origin: France, imported and purchased in Japan, sample year: 1981, country: Japan[120]

incidence: 1/9, conc.: 0.055 µg/kg, sample origin: France, imported and purchased in Japan, sample year: 1983, country: Japan[120]

incidence: 2/2, conc. range: 0.024–0.029 µg/kg, ∅ conc.: 0.027 µg/kg, sample origin: West Germany, imported and purchased in Japan, sample year: 1981–1983, country: Japan[120]

- Co-contamination: not reported
- Further contamination of animal products reported in the present book/article[120]: Cheese, Cheese (blue-veined cheese), Cheese (butter cheese), Cheese (Camembert cheese), Cheese (Cheddar cheese), Cheese (cream cheese), Cheese (Edam cheese), Cheese (Gouda cheese), Cheese (Havarti cheese), Cheese (Maribo cheese), Cheese (Mozzarella cheese), Cheese (Samsoe cheese), AFM_1, literature[120]

see also **Cheese**

Cheese (butter cheese) may contain the following mycotoxins:

Aspergillus Toxins

Aflatoxin M_1
incidence: 5/5, conc. range: 0.025–0.041 µg/kg, ∅ conc.: 0.037 µg/kg, sample origin: West Germany, imported and purchased in Japan, sample year: 1981–1983, country: Japan[120]

- Co-contamination: not reported
- Further contamination of animal products reported in the present book/article[120]: Cheese, Cheese (blue-veined cheese), Cheese (Brie cheese), Cheese (Camembert cheese), Cheese (Cheddar cheese), Cheese (cream cheese), Cheese (Edam cheese), Cheese (Gouda cheese), Cheese (Havarti cheese), Cheese (Maribo cheese), Cheese (Mozzarella cheese), Cheese (Samsoe cheese), AFM_1, literature[120]

Cheese (Camembert cheese) may contain the following mycotoxins:

Aspergillus Toxins

Aflatoxin M_1
incidence: 33/65*, conc. range: 0.1–0.50 µg/kg (27 sa.), 0.51–1.0 µg/kg (6 sa., maximum: 0.73 µg/kg), ∅ conc.: 0.31 µg/kg, sample origin: unknown, sample year: September 1972–December 1974, country: Germany[28], *Camembert and Brie cheese

- Co-contamination: not reported
- Further contamination of animal products reported in the present book/article[28]: Cheese (fresh cheese), Cheese (hard cheese), Cheese (processed cheese), Milk (cow milk), Milk powder, Yogurt, AFM_1, literature[28]

incidence: 1/4, conc.: 0.8 µg/kg, sample origin: France, imported and purchased in grocery stores in Washington, DC (capital) area, USA, sample year: before 1985, country: USA[30]

- Co-contamination: not reported
- Further contamination of animal products reported in the present book/article[30]: Cheese (Brie cheese), Cheese (Cheshire cheese), Cheese (Edam cheese), Cheese (fondue cheese), Cheese (Gouda cheese), Cheese (Gruyere cheese), AFM_1, literature[30]

incidence: 7/7, conc. range: 0.055–0.479 µg/kg, ∅ conc.: 0.207 µg/kg, sample origin: Denmark, imported and purchased in Japan, sample year: 1981–1983, country: Japan[120]

incidence: 12/13, conc. range: 0.150–0.563 µg/kg, ∅ conc.: 0.327 µg/kg, sample origin: France, imported and purchased in Japan, sample year: 1981, country: Japan[120]

incidence: 6/12, conc. range: 0.013–0.023 µg/kg, ∅ conc.: 0.019 µg/kg, sample origin: France, imported and purchased in Japan, sample year: 1983, country: Japan[120]

incidence: 1/1, conc.: 0.018 µg/kg, sample origin: West Germany, imported and purchased in Japan, sample year: 1981–1983, country: Japan[120]

- Co-contamination: not reported
- Further contamination of animal products reported in the present book/article[120]: Cheese, Cheese (blue-veined cheese), Cheese (Brie cheese), Cheese (butter cheese), Cheese (Cheddar cheese), Cheese (cream cheese), Cheese (Edam cheese), Cheese (Gouda cheese), Cheese (Havarti cheese), Cheese (Maribo cheese), Cheese (Mozzarella cheese), Cheese (Samsoe cheese), AFM_1, literature[120]

incidence: 1/100, conc.: tr, sample origin: retail milk shops, France, sample year: winter 1976, country: France[127]

- Co-contamination: not reported
- Further contamination of animal products reported in the present book/article[127]: not reported

incidence: 1/1, conc.: 0.312 µg/kg, sample origin: unknown, sample year: unknown, country: Japan[128]

- Co-contamination: not reported
- Further contamination of animal products reported in the present book/article[128]: Cheese (cream cheese), Cheese (Gouda cheese), Cheese (Münster cheese), AFM_1, literature[128]

incidence: 2/2*, conc. range: 0.68**–0.95*** µg/kg, sample origin: local dairies, sample year: unknown, country: France/UK[129], *fate of AFM_1 during the manufacture of Camembert cheese, **content of AFM_1 after 15 days of ripening and ***initial content

- Co-contamination: not reported
- Further contamination of animal products reported in the present book/article[129]: Cheese curd, Cheese whey, Milk (cow milk), AFM_1, literature[129]

Aspergillus **and** ***Penicillium*** Toxins

Cyclopiazonic Acid

incidence: 11/20*, conc. range: 0.05–0.1 µg/kg (3 sa.), 0.1–0.2 µg/kg (5 sa.), 0.2–1.5 µg/kg (3 sa.), sample origin: local supermarket in Toulouse (city), France, sample year: unknown, country: France[116], *cheese crust of Camembert cheese

- Co-contamination: not reported
- Further contamination of animal products reported in the present book/article[116]: not reported

incidence: 6/6*, conc. range: 20–80 µg/kg, sample origin: different local markets in southern Italy, sample year: unknown, country: Italy[117], *white surface mold-cheeses

- Co-contamination: not reported
- Further contamination of animal products reported in the present book/article[117]: not reported

see also **Cheese**

Cheese (Caprice des Dieux cheese) see Cheese

Cheese (Cecil cheese) may contain the following mycotoxins:

Aspergillus Toxins

Aflatoxin M_1

incidence: 6/30, conc. range: 0.0531–0.115 µg/kg, ∅ conc.: 0.08258 µg/kg, sample origin: local markets in Kars (district), Turkey, sample year: winter months 2003, country: Turkey[130]

- Co-contamination: not reported
- Further contamination of animal products reported in the present book/article[130]: Cheese (Kashar cheese), AFM_1, literature[130]

Cheese (Chamois d'Or cheese) see Cheese

Cheese (Cheddar cheese) may contain the following mycotoxins:

Aspergillus Toxins

Aflatoxin M_1

incidence: 4/4, conc. range: 0.015–0.030 µg/kg, ∅ conc.: 0.020 µg/kg, sample origin: UK, imported and purchased in Japan, sample year: 1981–1983, country: Japan[120]

- Co-contamination: not reported
- Further contamination of animal products reported in the present book/article[120]: Cheese, Cheese (blue-veined cheese), Cheese (Brie cheese), Cheese (butter cheese), Cheese (Camembert cheese), Cheese (cream cheese), Cheese (Edam cheese), Cheese (Gouda cheese), Cheese (Havarti cheese), Cheese (Maribo cheese), Cheese (Mozzarella cheese), Cheese (Samsoe cheese), AFM_1, literature[120]

incidence: 12/12, conc. range: 0.02–0.05 µg/kg (8 sa.), 0.06–0.1 µg/kg (4 sa., maximum: 0.09 µg/kg), sample origin: retail outlets, England and Wales, sample year: unknown, country: UK[131]

- Co-contamination: not reported
- Further contamination of animal products reported in the present book/article[131]: Cheese (Cheshire cheese), Cheese (Double Gloucester cheese), Cheese (Lancashire cheese), Cheese (Leicester cheese), Cheese (Wensleydale cheese), Milk (cow milk), Milk (infant formula), Yogurt, AFM_1, literature[131]

see also **Cheese**

Cheese (Cheshire cheese) may contain the following mycotoxins:

Aspergillus Toxins

Aflatoxin M_1

incidence: 2/2, conc. range: 0.2–1.0 µg/kg, ∅ conc.: 0.6 µg/kg, sample origin: England, imported and purchased in grocery stores in Washington, DC (capital) area, USA, sample year: before 1985, country: USA[30]

- Co-contamination: not reported
- Further contamination of animal products reported in the present book/article[30]: Cheese (Brie cheese), Cheese (Camembert cheese), Cheese (Edam cheese), Cheese (fondue cheese), Cheese (Gouda cheese), Cheese (Gruyere cheese), AFM_1, literature[30]

incidence: 13/13, conc. range: 0.02–0.05 µg/kg (3 sa.), 0.06–0.1 µg/kg (5 sa.), 0.11–0.2 µg/kg (5 sa., maximum: 0.17 µg/kg), sample origin: retail outlets, England and Wales, sample year: unknown, country: UK[131]

- Co-contamination: not reported
- Further contamination of animal products reported in the present book/article[131]: Cheese (Cheddar cheese), Cheese (Double Gloucester cheese), Cheese (Lancashire cheese), Cheese (Leicester cheese), Cheese (Wensleydale cheese), Milk (cow milk), Milk (infant formula), Yogurt, AFM_1, literature[131]

Cheese (Chhurpi cheese) see Cheese (Bhutanese cheese)

Cheese (Chuku cheese) see Cheese (Bhutanese cheese)

Cheese (Civil cheese) may contain the following mycotoxins:

Aspergillus Toxins

Aflatoxin M_1

incidence: 21/25, conc. range: 0.51–0.250 µg/kg (1 sa.), 0.251–0.400 µg/kg (20 sa.), sample origin: local stores in Ardahan, Igdir, and Kars (cities), Turkey, sample year: unknown, country: Turkey[132]

- Co-contamination: not reported

- Further contamination of animal products reported in the present book/article[132]: Cheese (Graviera cheese), Cheese (Kashar cheese), Cheese (white cheese), AFM_1, literature[132]

incidence: 4/9, conc. range: 0.012–0.018 µg/kg, ∅ conc.: 0.01232 µg/kg, sample origin: retail markets in Erzurum (city), Turkey, sample year: unknown, country: Turkey[133]

- Co-contamination: not reported
- Further contamination of animal products reported in the present book/article[133]: Cheese (Kashar cheese), Cheese (Lor cheese), Cheese (Tulum cheese), Cheese (white cheese), AFM_1, literature[133]

Cheese (Comte cheese) see Cheese

Cheese (cottage cheese) may contain the following mycotoxins:

Aspergillus Toxins

Aflatoxin B_1

incidence: 1/20*, conc.: 104 µg/kg, sample origin: markets and homes in Sohag (governorate), Egypt, sample year: unknown, country: Egypt[134], *old cottage cheese (made from buffalo milk with low fat content, stored 3–12 months)

- Co-contamination: not reported
- Further contamination of animal products reported in the present book/article[134]: Cheese (cottage cheese), AFB_2 and OTA/OTB, literature[134]

Aflatoxin B_2

incidence: 1/20*, conc.: 89 µg/kg, sample origin: markets and homes in Sohag (governorate), Egypt, sample year: unknown, country: Egypt[134], *old cottage cheese (made from buffalo milk with low fat content, stored 3–12 months)

- Co-contamination: not reported
- Further contamination of animal products reported in the present book/article[134]: Cheese (cottage cheese), AFB_1 and OTA/OTB, literature[134]

Aspergillus and ***Penicillium*** Toxins

Ochratoxins (A, B)

incidence: 1/20*, conc.: 112 µg/kg, sample origin: markets and homes in Sohag (governorate), Egypt, sample year: unknown, country: Egypt[134], *old cottage cheese (made from buffalo milk with low fat content, stored 3–12 months)

- Co-contamination: not reported
- Further contamination of animal products reported in the present book/article[134]: Cheese (cottage cheese), AFB_1 and AFB_2, literature[134]

Cheese (cow cheese) may contain the following mycotoxins:

Aspergillus Toxins

Aflatoxin M_1

incidence: 5/31*, conc. range: 0.050–0.158 µg/kg, ∅ conc.: 0.0814 µg/kg, sample origin: sales centers in Apulia (region), Italy, sample year: unknown, country: Italy[43], *unripened/short-term ripened (<1 month)

incidence: 10/30*, conc. range: 0.050–0.114 µg/kg, ∅ conc.: 0.0707 µg/kg, sample origin: sales centers in Apulia (region), Italy, sample year: unknown, country: Italy[43], *medium-term ripened (1–3 months)

incidence: 10/31*, conc. range: 0.050–0.160 µg/kg, ∅ conc.: 0.073 µg/kg, sample origin: sales centers in Apulia (region), Italy, sample year: unknown, country: Italy[43], *long-term ripened (3–12 months)

- Co-contamination: not reported
- Further contamination of animal products reported in the present book/article[43]: Cheese (goat cheese), Cheese (goat-sheep cheese), Cheese (sheep cheese), AFM_1, literature[43]

Cheese (cream cheese) may contain the following mycotoxins:

Aspergillus Toxins

Aflatoxin M_1

incidence: 44/49, conc. range: <0.001 µg/kg (12 sa.), 0.011–0.050 µg/kg (10 sa.), 0.051–0.100 µg/kg (15 sa.), 0.101–0.250 µg/kg (7 sa.), sample origin: markets in Ankara (capital), Turkey, sample year: September 2002–2003, country: Turkey[11]

- Co-contamination: not reported
- Further contamination of animal products reported in the present book/article[11]: Butter, Cheese (Kashar cheese), Cheese (white cheese), AFM_1, literature[11]

incidence: 98/99, conc. range: 0.001–0.050 µg/kg (6 sa.), 0.051–0.100 µg/kg (17 sa.), 0.101–0.250 µg/kg (58 sa.), >0.250 µg/kg (18 sa., maximum 4.1 µg/kg), ∅ conc.: 0.330 µg/kg*, sample origin: supermarkets in Izmir, Kayseri, Konya, and Tekirdag (cities) as well as Istanbul (capital), Turkey, sample year: March–June 2005, country: Turkey[13], *of pos. sa.?

- Co-contamination: not reported
- Further contamination of animal products reported in the present book/article[13]: Butter, AFM_1, literature[13]

incidence: 9/9, conc. range: 0.037–0.134 µg/kg, ∅ conc.: 0.079 µg/kg, sample origin: Denmark, imported and purchased in Japan, sample year: 1981–1983, country: Japan[120]

- Co-contamination: not reported
- Further contamination of animal products reported in the present book/article[120]: Cheese, Cheese (blue-veined cheese), Cheese (Brie cheese), Cheese (butter cheese), Cheese (Camembert cheese), Cheese (Cheddar cheese), Cheese (Edam cheese), Cheese (Gouda cheese), Cheese (Havarti cheese), Cheese (Maribo cheese), Cheese (Mozzarella cheese), Cheese (Samsoe cheese), AFM_1, literature[120]

incidence: 2/2, conc. range: 0.064–0.087 µg/kg, ∅ conc.: 0.076 µg/kg, sample origin: unknown, sample year: unknown, country: Japan[128]

- Co-contamination: not reported
- Further contamination of animal products reported in the present book/article[128]: Cheese (Camembert cheese), Cheese (Gouda cheese), Cheese (Münster cheese), AFM_1, literature[128]

incidence: 8/200, conc. range: 0.100–0.700 µg/kg, ∅ conc.: 0.285 µg/kg, sample origin: different Turkish provinces, Turkey, sample year: January 2001–February 2002, country: Turkey[135]

- Co-contamination: not reported
- Further contamination of animal products reported in the present book/article[135]: Cheese (Kashar cheese), Cheese (white cheese), AFM_1, literature[135]

incidence: 68/94, conc. range: 0.050–0.150 µg/kg (23 sa.), 0.151–0.250 µg/kg (27 sa.), 0.251–0.450 µg/kg (10 sa.), 0.451–0.650 µg/kg (5 sa.), >0.651 µg/kg (3 sa., maximum: 0.7854 µg/kg), ∅ conc.: 0.2301 µg/kg, sample origin: supermarkets and retail outlets in Esfahan and Yazd (provinces), central part of Iran, sample year: October 2007–May 2008, country: Iran[136]

- Co-contamination: not reported
- Further contamination of animal products reported in the present book/article[136]: Cheese (white cheese), AFM_1, literature[136]

incidence: 89/150, conc. range: <0.050 µg/kg (25 sa.), >0.050–0.150 µg/kg (46 sa.), >0.151 µg/kg (18 sa., maximum: 0.4563 µg/kg), Ø conc.: 0.1729 µg/kg, sample origin: main districts of Punjab (province), Pakistan, sample year: November 2010–April 2011, country: Pakistan[451]

- Co-contamination: not reported
- Further contamination of animal products reported in the present book/article[451]: Butter, Cheese (white cheese), Milk, Yogurt, AFM_1, literature[451]

Penicillium Toxins

Penitrem A

incidence: 1/1*, conc.: nc, sample origin: unknown, sample year: unknown, country: USA[137], *visible moldy

- Co-contamination: not reported
- Further contamination of animal products reported in the present book/article[137]: not reported

see also **Cheese**

Cheese (Crescenza cheese) see **Cheese (soft cheese)**

Cheese (Danablu cheese) may contain the following mycotoxins:

Penicillium Toxins

Roquefortine C

incidence: 3/3, conc. range: 970–1470 µg/kg, Ø conc.: 1257 µg/kg, sample origin: purchased in Denmark and Iceland, sample year: unknown, country: Italy[124]

- Co-contamination: not reported
- Further contamination of animal products reported in the present book/article[124]: Cheese (Edelpilzkäse), Cheese (Gorgonzola cheese), Cheese (Roquefort cheese), Cheese (Stilton cheese), ROQ C, literature[124]

incidence: 3/3*, conc. range: 1170–2290 µg/kg, Ø conc.: 1883 µg/kg, sample origin: unknown, sample year: unknown, country: Netherlands[126], *3 batches of Danish Blue cheese

- Co-contamination: not reported
- Further contamination of animal products reported in the present book/article[126]: Cheese (Gorgonzola cheese), Cheese (Roquefort cheese), ROQ C, literature[126]

see also **Cheese (blue-veined cheese)**

Cheese (Domiati cheese) may contain the following mycotoxins:

Aspergillus Toxins

Aflatoxin M_1

incidence: 1/10, conc.: 0.5 µg/kg, sample origin: Egyptians markets, Egypt, sample year: 1999–2000, country: Egypt[138]

- Co-contamination: not reported
- Further contamination of human breast milk and animal products reported in the present book/article[138]: Cheese (Ras cheese), AFB_1, AFG_1, and AFM_1, literature[138]; Milk (cow milk), Milk powder, AFM_1, literature[138]; Milk (human breast milk), AFM_1 and OTA, literature[138]

Cheese (Double Gloucester cheese) may contain the following mycotoxins:

Aspergillus Toxins

Aflatoxin M_1

incidence: 13/13, conc. range: 0.02–0.05 µg/kg (4 sa.), 0.06–0.1 µg/kg (8 sa.), 0.13 µg/kg (1 sa.), sample origin: retail outlets, England and Wales, sample year: unknown, country: UK[131]

- Co-contamination: not reported
- Further contamination of animal products reported in the present book/article[131]: Cheese (Cheddar cheese), Cheese (Cheshire cheese), Cheese (Lancashire cheese), Cheese (Leicester cheese), Cheese (Wensleydale cheese), Milk (cow milk), Milk (infant formula), Yogurt, AFM_1, literature[131]

Cheese (Edam cheese) may contain the following mycotoxins:

Aspergillus Toxins

Aflatoxin M_1

incidence: 1/8, conc.: 0.1 µg/kg, sample origin: Holland, imported and purchased in grocery stores in Washington, DC (capital) area, USA, sample year: before 1985, country: USA[30]

- Co-contamination: not reported
- Further contamination of animal products reported in the present book/article[30]: Cheese (Brie cheese), Cheese (Camembert cheese), Cheese (Cheshire cheese), Cheese (fondue cheese), Cheese (Gouda cheese), Cheese (Gruyere cheese), AFM_1, literature[30]

incidence: 5/5, conc. range: 0.073–0.117 µg/kg, Ø conc.: 0.099 µg/kg, sample origin: Netherlands, imported and purchased in Japan, sample year: 1981–1983, country: Japan[120]

- Co-contamination: not reported
- Further contamination of animal products reported in the present book/article[120]: Cheese, Cheese (blue-veined cheese), Cheese (Brie cheese), Cheese (butter cheese), Cheese (Camembert cheese), Cheese (Cheddar cheese), Cheese (cream cheese), Cheese (Gouda cheese), Cheese (Havarti cheese), Cheese (Maribo cheese), Cheese (Mozzarella cheese), Cheese (Samsoe cheese), AFM_1, literature[120]

see also **Cheese**

Cheese (Edelpilzkäse) may contain the following mycotoxins:

Penicillium Toxins

Roquefortine C

incidence: 2/2, conc. range: 300–520 µg/kg, Ø conc.: 410 µg/kg, sample origin: purchased in Austria and Germany, sample year: unknown, country: Italy[124]

- Co-contamination: not reported
- Further contamination of animal products reported in the present book/article[124]: Cheese (Danablu cheese), Cheese (Gorgonzola cheese), Cheese (Roquefort cheese), Cheese (Stilton cheese), ROQ C, literature[124]

see also **Cheese (blue-veined cheese)**

Cheese (Emmental cheese) see **Cheese**

Cheese (Feta cheese) may contain the following mycotoxins:

Aspergillus **Toxins**

Aflatoxin M_1

incidence: 17/20, conc. range: 0.17–1.3 µg/kg, ∅ conc.: 0.41 µg/kg, sample origin: different plants of manufacture of Feta cheese in Tehran (province), Iran, sample year: May 2003, country: Iran[139]

incidence: 14/20, conc. range: 0.15–2.41 µg/kg, ∅ conc.: 0.35 µg/kg, sample origin: different plants of manufacture of Feta cheese in Tehran (province), Iran, sample year: August 2003, country: Iran[139]

incidence: 16/20, conc. range: 0.16–1.11 µg/kg, ∅ conc.: 0.36 µg/kg, sample origin: different plants of manufacture of Feta cheese in Tehran (province), Iran, sample year: November 2003, country: Iran[139]

incidence: 19/20, conc. range: 0.19–2.05 µg/kg, ∅ conc.: 0.52 µg/kg, sample origin: different plants of manufacture of Feta cheese in Tehran (province), Iran, sample year: February 2004, country: Iran[139]

- Co-contamination: not reported
- Further contamination of animal products reported in the present book/article[139]: not reported

see also **Cheese**

Cheese (fondue cheese) may contain the following mycotoxins:

Aspergillus **Toxins**

Aflatoxin M_1

incidence: 1/1, conc.: 0.4 µg/kg, sample origin: Switzerland, imported and purchased in grocery stores in Washington, DC (capital) area, USA, sample year: before 1985, country: USA[30]

- Co-contamination: not reported
- Further contamination of animal products reported in the present book/article[30]: Cheese (Brie cheese), Cheese (Camembert cheese), Cheese (Cheshire cheese), Cheese (Edam cheese), Cheese (Gouda cheese), Cheese (Gruyere cheese), AFM_1, literature[30]

Cheese (fresh cheese) may contain the following mycotoxins:

Aspergillus **Toxins**

Aflatoxin M_1

incidence: 27/80, conc. range: 0.1–0.20 µg/kg (13 sa.), 0.21–0.50 µg/kg (14 sa., maximum: 0.51 µg/kg), ∅ conc.: 0.23 µg/kg, sample origin: unknown, sample year: September 1972–December 1974, country: Germany[28]

- Co-contamination: not reported
- Further contamination of animal products reported in the present book/article[28]: Cheese (Camembert cheese), Cheese (hard cheese), Cheese (processed cheese), Milk (cow milk), Milk powder, Yogurt, AFM_1, literature[28]

Aspergillus **and** ***Penicillium*** **Toxins**

Ochratoxin A

incidence: 12/92*, conc. range: ≤0.060 µg/kg, sample origin: regions around Detmold, Giessen, Hamburg, Jena, Karlsruhe, Kiel, Kulmbach, Trier (cities), Germany, sample year: unknown, country: Germany[100], *fresh cheese and yogurt with ingredients

- Co-contamination: not reported
- Further contamination of animal products reported in the present book/article[100]: Cheese, OTA, literature[100]

see also **Cheese**

Cheese (goat cheese) may contain the following mycotoxins:

Aspergillus **Toxins**

Aflatoxin M_1

incidence: 11/20, conc. range: 0.015–0.050 µg/kg (5 sa.), 0.051–0.100 µg/kg (3 sa.), 0.101–0.200 µg/kg (3 sa, maximum: 0.200 µg/kg), sample origin: France, Greece, Italy, and Netherlands (domestic and imported from retail market), Italy, sample year: March–October 1996, country: Italy[36]

- Co-contamination: not reported
- Further contamination of animal products reported in the present book/article[36]: Cheese (soft cheese), Milk (goat milk), AFM_1, literature[36]

incidence: 1/4*, conc.: 0.090 µg/kg, sample origin: sales centers in Apulia (region), Italy, sample year: unknown, country: Italy[43], *unripened/short-term ripened (<1 month)

incidence: 1/4*, conc.: 0.250 µg/kg, sample origin: sales centers in Apulia (region), Italy, sample year: unknown, country: Italy[43], *medium-term ripened (1–3 months)

incidence: 0/4*, conc.: no contamination, sample origin: sales centers in Apulia (region), Italy, sample year: unknown, country: Italy[43], *long-term ripened (3–12 months)

- Co-contamination: not reported
- Further contamination of animal products reported in the present book/article[43]: Cheese (cow cheese), Cheese (goat-sheep cheese), Cheese (sheep cheese), AFM_1, literature[43]

Aspergillus **and** ***Penicillium*** **Toxins**

Patulin

incidence: 1/18*, conc.: 30 µg/kg, sample origin: local retail dealers and manufacturers, France/Netherland?, sample year: unknown, country: France/Netherlands[114], *in whole cheese

- Co-contamination: not reported
- Further contamination of animal products reported in the present book/article[114]: Cheese (blue-veined cheese) PA and MA, literature[114]; Cheese (goat cheese), PA, literature[114]; Cheese (hard cheese), PA, PAT, MA, and STG, literature[114]; Milk powder, AFB_1 and AFM_1, literature[114]

Penicillic Acid

incidence: 2/18*, conc. range: 45–210 µg/kg, ∅ conc.: 128 µg/kg, sample origin: local retail dealers and manufacturers, France/Netherland?, sample year: unknown, country: France/Netherlands[114], *in whole cheese

- Co-contamination: not reported
- Further contamination of animal products reported in the present book/article[114]: Cheese (blue-veined cheese), PA and MA, literature[114]; Cheese (goat cheese), PAT, literature[114]; Cheese (hard cheese), PA, PAT, MA, and STG, literature[114]; Milk powder, AFB_1 and AFM_1, literature[114]

see also **Cheese (hard cheese)**

Cheese (goat-sheep cheese) may contain the following mycotoxins:

Aspergillus **Toxins**

Aflatoxin M_1

incidence: 1/6*, conc.: 0.050 µg/kg, sample origin: sales centers in Apulia (region), Italy, sample year: unknown, country: Italy[43], *unripened/short-term ripened (<1 month) incidence: 3/5*, conc. range: 0.055–0.140 μg/kg, ∅ conc.: 0.085 μg/kg, sample origin: sales centers in Apulia (region), Italy, sample year: unknown, country: Italy[43], *medium-term ripened (1–3 months)

incidence: 1/5*, conc.: 0.075 μg/kg, sample origin: sales centers in Apulia (region), Italy, sample year: unknown, country: Italy[43], *long-term ripened (3–12 months)

- Co-contamination: not reported
- Further contamination of animal products reported in the present book/article[43]: Cheese (cow cheese), Cheese (goat cheese), Cheese (sheep cheese), AFM_1, literature[43]

Cheese (Gorgonzola cheese) may contain the following mycotoxins:

Aspergillus and *Penicillium* Toxins

Ochratoxin A

incidence: 23/54, conc. range: 0.2–3.0 μg/kg, sample origin: market, Italy, sample year: unknown, country: Italy[121]

- Co-contamination: not reported
- Further contamination of animal products reported in the present book/article[121]: Cheese (Roquefort cheese), OTA, literature[121]

Penicillium Toxins

Mycophenolic Acid

incidence: 3/12, conc. range: 10–100 μg/kg (3 sa.), sample origin: commercial, sample year: unknown, country: France[118]

- Co-contamination: not reported
- Further contamination of animal products reported in the present book/article[118]: Cheese, Cheese (Bleu des Causses cheese), Cheese (blue-veined cheese), Cheese (Roquefort cheese), MA, literature[118]

For detailed information see the article.

Roquefortine C

incidence: 2/2, conc. range: 140–220 μg/kg, ∅ conc.: 170 μg/kg, sample origin: Italy, imported and purchased in Canada, sample year: unknown, country: Canada[122]

- Co-contamination: not reported
- Further contamination of animal products reported in the present book/article[122]: Cheese (blue-veined cheese), IFC A, IFC B, and ROQ C, literature[122]; Cheese (Stilton cheese), IFC A and ROQ C, literature[122]

For detailed information see the article.

incidence: 16/16*, conc. range: 70–480 μg/kg, ∅ conc.: 188 μg/kg, sample origin: supplied by the Consortium for the Protection of Gorgonzola cheese, Italy, sample year: unknown, country: Italy[124], *Gorgonzola "dolce"

incidence: 4/4*, conc. range: 150–1440 μg/kg, ∅ conc.: 925 μg/kg, sample origin: supplied by the Consortium for the Protection of Gorgonzola cheese, Italy, sample year: unknown, country: Italy[124], *Gorgonzola "naturale"

- Co-contamination: not reported
- Further contamination of animal products reported in the present book/article[124]: Cheese (Danablu cheese), Cheese (Edelpilzkäse), Cheese (Roquefort cheese), Cheese (Stilton cheese), ROQ C, literature[124]

incidence: 1/1, conc.: 950 μg/kg, sample origin: unknown, sample year: unknown, country: Netherlands[126]

- Co-contamination: not reported
- Further contamination of animal products reported in the present book/article[126]: Cheese (Danablu cheese), Cheese (Roquefort cheese), ROQ C, literature[126]

see also **Cheese (blue-veined cheese)**

Cheese (Gouda cheese) may contain the following mycotoxins:

Aspergillus Toxins

Aflatoxin M_1

incidence: 1/7, conc.: 0.2 μg/kg, sample origin: Holland, imported and purchased in grocery stores in Washington, DC (capital) area, USA, sample year: before 1985, country: USA[30]

- Co-contamination: not reported
- Further contamination of animal products reported in the present book/article[30]: Cheese (Brie cheese), Cheese (Camembert cheese), Cheese (Cheshire cheese), Cheese (Edam cheese), Cheese (fondue cheese), Cheese (Gruyere cheese), AFM_1, literature[30]

incidence: 9/9, conc. range: 0.039–0.087 μg/kg, ∅ conc.: 0.063 μg/kg, sample origin: Netherlands, imported and purchased in Japan, sample year: 1981–1983, country: Japan[120]

- Co-contamination: not reported
- Further contamination of animal products reported in the present book/article[120]: Cheese, Cheese (blue-veined cheese), Cheese (Brie cheese), Cheese (butter cheese), Cheese (Camembert cheese), Cheese (Cheddar cheese), Cheese (cream cheese), Cheese (Edam cheese), Cheese (Havarti cheese), Cheese (Maribo cheese), Cheese (Mozzarella cheese), Cheese (Samsoe cheese), AFM_1, literature[120]

incidence: 1/1, conc.: 0.051 μg/kg, sample origin: unknown, sample year: unknown, country: Japan[128]

- Co-contamination: not reported
- Further contamination of animal products reported in the present book/article[128]: Cheese (Camembert cheese), Cheese (cream cheese), Cheese (Münster cheese), AFM_1, literature[128]

see also **Cheese**

Cheese (Grana Padano cheese) may contain the following mycotoxins:

Aspergillus Toxins

Aflatoxin M_1

incidence: 219/223, conc. range: 0.005–0.100 μg/kg (203 sa.), 0.101–0.250 μg/kg (15 sa.), 0.37 μg/kg (1 sa.), sample origin: dairies in 11 provinces of the Po valley, Italy, sample year: 1991–1994, country: Italy[140]

- Co-contamination: not reported
- Further contamination of animal products reported in the present book/article[140]: not reported

incidence: 25/25*, conc. range: 0.111–0.413 μg/kg, ∅ conc.: 0.246 μg/kg, sample origin: province of the Po valley, Italy, sample year: November–December 2003, country: Italy[484], *made from naturally contaminated cow milk

- Co-contamination: not reported
- Further contamination of animal products reported in the present book/article[484]: Cheese curd, Cheese whey, Milk (cow milk), AFM_1, literature[484]

see also **Cheese**

Cheese (Graviera cheese) may contain the following mycotoxins:

Aspergillus Toxins

Aflatoxin M_1

incidence: 5/5, conc. range: 0.51–0.250 μg/kg (3 sa.), 0.251–0.400 μg/kg (2 sa.), sample origin: local stores in Ardahan, Igdir, and Kars (cities), Turkey, sample year: unknown, country: Turkey[132]

- Co-contamination: not reported
- Further contamination of animal products reported in the present book/article[132]: Cheese (Civil cheese), Cheese (Kashar cheese), Cheese (white cheese), AFM_1, literature[132]

Cheese (Gruyere cheese) may contain the following mycotoxins:

Aspergillus Toxins

Aflatoxin M_1

incidence: 1/7, conc.: 0.2 μg/kg, sample origin: Switzerland, imported and purchased in grocery stores in Washington, DC (capital) area, USA, sample year: before 1985, country: USA[30]

- Co-contamination: not reported
- Further contamination of animal products reported in the present book/article[30]: Cheese (Brie cheese), Cheese (Camembert cheese), Cheese (Cheshire cheese), Cheese (Edam cheese), Cheese (fondue cheese), Cheese (Gouda cheese), AFM_1, literature[30]

Cheese (Halloumi cheese) see **Cheese (white cheese)**

Cheese (hard cheese) may contain the following mycotoxins:

Aspergillus Toxins

Aflatoxin M_1

incidence: 57/77, conc. range: 0.1–0.50 μg/kg (42 sa.), 0.51–1.5 μg/kg (15 sa., maximum: 1.30 μg/kg), Ø conc.: 0.43 μg/kg, sample origin: unknown, sample year: September 1972–December 1974, country: Germany[28]

- Co-contamination: not reported
- Further contamination of animal products reported in the present book/article[28]: Cheese (Camembert cheese), Cheese (fresh cheese), Cheese (processed cheese), Milk (cow milk), Milk powder, Yogurt, AFM_1, literature[28]

incidence: 4*/41**, conc. range: 0.0795–0.389 μg/kg, Ø conc.: 0.257 μg/kg, sample origin: cheese factories, Italy, sample year: 2005, country: Italy[31], *contaminated sa. produced with milk from intensive farms, **hard goat cheese

- Co-contamination: not reported
- Further contamination of animal products reported in the present book/article[31]: Milk (goat milk), AFM_1, literature[31]

incidence: 19/50, conc. range: 0.0516–0.182 μg/kg, Ø conc.: 0.13224 μg/kg, sample origin: supermarkets in Alexandria (city), Egypt, sample year: February 2008–March 2009, country: Egypt[396]

- Co-contamination: not reported
- Further contamination of animal products reported in the present book/article[396]: Cheese (processed cheese), Cheese (soft cheese), Milk (cow milk), AFM_1, literature[396]

incidence: 10/10*, conc. range: 0.08–2.23 μg/kg, sample origin: supermarkets, Serbia, sample year: May–June 2013, country: Serbia[516], *commercial hard cheese sa.

- Co-contamination: not reported
- Further contamination of animal products reported in the present book/article[516]: Cheese (white cheese), AFM_1, literature[516]

Sterigmatocystin

incidence: 3/66*, conc. range: 7.5–17.5 μg/kg, Ø conc.: 10.8 μg/kg, sample origin: retail shops, Czechoslovakia, sample year: unknown, country: Czechoslovakia[112], *different kinds of hard cheeses

- Co-contamination: not reported
- Further contamination of animal products reported in the present book/article[112]: not reported

For detailed information see the article.

incidence: 9/39*, conc. range: 5–600 μg/kg**, Ø conc.: 80.7 μg/kg**, sample origin: warehouses, Netherlands, sample year: 1976–1977?, country: Netherlands[113], *visibly moldy Edam and Gouda cheeses, **outer surface layer (1 cm deep)

- Co-contamination: not reported
- Further contamination of animal products reported in the present book/article[113]: not reported

incidence: 3/48, conc. range: 45–330 μg/kg*, Ø conc.: 167 μg/kg*, sample origin: local retail dealers and manufacturers, France/ Netherland?, sample year: unknown, country: France/ Netherlands[114], *in peripheral zone of the cheese (2 cm deep)

incidence: 3/48, conc. range: 45–330 μg/kg*, Ø conc.: 167 μg/kg*, sample origin: local retail dealers and manufacturers, France/ Netherland?, sample year: unknown, country: France/ Netherlands[114], *in central zone of the cheese

- Co-contamination: not reported
- Further contamination of animal products reported in the present book/article[114]: Cheese (blue cheese) PA and MA, literature[114]; Cheese (goat cheese), PA and PAT, literature[114]; Cheese (hard cheese), PA, PAT, and MA, literature[114]; Milk powder, AFB_1 and AFM_1, literature[114]

Aspergillus and *Penicillium* Toxins

Ochratoxin A

incidence: 6/40*, conc. range: 1.62–54.07 μg/kg, Ø conc.: 14.9 μg/kg, sample origin: retail shops, northern Italy, sample year: summer 2011, country: Italy[474], *commercial grated cheese sa.

- Co-contamination: not reported
- Further contamination of animal products reported in the present book/article[474]: not reported

Patulin

incidence: 1/48, conc.: 90 μg/kg*, sample origin: local retail dealers and manufacturers, France/Netherland?, sample year: unknown, country: France/Netherlands[114], *in peripheral zone of the cheese (2 cm deep)

incidence: 1/48, conc.: 90 μg/kg*, sample origin: local retail dealers and manufacturers, France/Netherland?, sample year: unknown, country: France/Netherlands[114], *in central zone of the cheese

incidence: 4/39*, conc. range: 45–335 μg/kg**, Ø conc.: 155 μg/kg**, sample origin: local retail dealers and manufacturers, France/ Netherland?, sample year: unknown, country: France/ Netherlands[114], *mold-ripened hard cheese, **in whole cheese

- Co-contamination: not reported
- Further contamination of animal products reported in the present book/article[114]: Cheese (blue-veined cheese) PA and

MA, literature[114]; Cheese (goat cheese), PA and PAT, literature[114]; Cheese (hard cheese), PA, MA, and STG, literature[114]; Milk powder, AFB_1 and AFM_1, literature[114]

Penicillic Acid

incidence: 4/48, conc. range: 45–950 µg/kg*, Ø conc.: 380 µg/kg*, sample origin: local retail dealers and manufacturers, France/Netherland?, sample year: unknown, country: France/Netherlands[114], *in peripheral zone of the cheese (2 cm deep)

incidence: 2/48, conc. range: tr-45 µg/kg*, sample origin: local retail dealers and manufacturers, France/Netherland?, sample year: unknown, country: France/Netherlands[114], *in central zone of the cheese

incidence: 4/48, conc. range: 45–385 µg/kg*, Ø conc.: 178 µg/kg*, sample origin: local retail dealers and manufacturers, France/Netherland?, sample year: unknown, country: France/Netherlands[114], *in whole cheese

incidence: 5/39*, conc. range: tr-710 µg/kg**, sample origin: local retail dealers and manufacturers, France/Netherland?, sample year: unknown, country: France/Netherlands[114], *mold-ripened hard cheese, **in whole cheese

- Co-contamination: not reported
- Further contamination of animal products reported in the present book/article[114]: Cheese (blue-veined cheese) PA and MA, literature[114]; Cheese (goat cheese), PA and PAT, literature[114]; Cheese (hard cheese), PAT, MA, and STG, literature[114]; Milk powder, AFB_1 and AFM_1, literature[114]

Penicillium Toxins

Mycophenolic Acid

incidence: 4/48, conc. range: 60–280 µg/kg*, Ø conc.: 155 µg/kg*, sample origin: local retail dealers and manufacturers, France/Netherland?, sample year: unknown, country: France/Netherlands[114], *in peripheral zone of the cheese (2 cm deep)

incidence: 4/48, conc. range: 60–280 µg/kg*, Ø conc.: 155 µg/kg*, sample origin: local retail dealers and manufacturers, France/Netherland?, sample year: unknown, country: France/Netherlands[114], *in whole cheese

incidence: 7/39*, conc. range: 50–2900 µg/kg**, Ø conc.: 854 µg/kg**, sample origin: local retail dealers and manufacturers, France/Netherland?, sample year: unknown, country: France/Netherlands[114], *mold-ripened hard cheese, **in whole cheese

- Co-contamination: not reported
- Further contamination of animal products reported in the present book/article[114]: Cheese (blue-veined cheese) PA and MA, literature[114]; Cheese (goat cheese), PA and PAT, literature[114]; Cheese (hard cheese), PA, PAT, and STG, literature[114]; Milk powder, AFB_1 and AFM_1, literature[114]

see also **Cheese**

Cheese (Hard Roumy cheese) may contain the following mycotoxins:

Fusarium Toxins

Zearalenone

incidence: 5/20, conc. range: 4.8–13.1 µg/kg, Ø conc.: 10.4 µg/kg, sample origin: shops and supermarkets in Alexandria (province), Egypt, sample year: unknown, country: Egypt[141]

- Co-contamination: not reported
- Further contamination of animal products reported in the present book/article[141]: Beef meat, Beefburger, Cheese (Karish cheese), Meat, Milk, Milk powder, Sausage, ZEA, literature[141]

see also **Cheese (Ras cheese)**

Cheese (Harzer cheese) see **Cheese**

Cheese (Havarti cheese) may contain the following mycotoxins:

Aspergillus Toxins

Aflatoxin M_1

incidence: 3/3, conc. range: 0.125–0.388 µg/kg, Ø conc.: 0.290 µg/kg, sample origin: Denmark, imported and purchased in Japan, sample year: 1981–1983, country: Japan[120]

- Co-contamination: not reported
- Further contamination of animal products reported in the present book/article[120]: Cheese, Cheese (blue-veined cheese), Cheese (Brie cheese), Cheese (butter cheese), Cheese (Camembert cheese), Cheese (Cheddar cheese), Cheese (cream cheese), Cheese (Edam cheese), Cheese (Gouda cheese), Cheese (Maribo cheese), Cheese (Mozzarella cheese), Cheese (Samsoe cheese), AFM_1, literature[120]

Cheese (Karish cheese) may contain the following mycotoxins:

Fusarium Toxins

Zearalenone

incidence: 6/25*, conc. range: 2.2–11.2 µg/kg, Ø conc.: 8.9 µg/kg, sample origin: shops and supermarkets in Alexandria (province), Egypt, sample year: unknown, country: Egypt[141]

- Co-contamination: not reported
- Further contamination of animal products reported in the present book/article[141]: Beef meat, Beefburger, Cheese (Hard Roumy cheese), Meat, Milk, Milk powder, Sausage, ZEA, literature[141]

Cheese (Kashar cheese) may contain the following mycotoxins:

Aspergillus Toxins

Aflatoxin M_1

incidence: 47/53, conc. range: <0.001 µg/kg (8 sa.), 0.011–0.050 µg/kg (2 sa.), 0.051–0.100 µg/kg (7 sa.), 0.101–0.250 µg/kg (23 sa.), >0.250 µg/kg (7 sa.), sample origin: markets in Ankara (capital), Turkey, sample year: September 2002–2003, country: Turkey[11]

- Co-contamination: not reported
- Further contamination of animal products reported in the present book/article[11]: Butter, Cheese (cream cheese), Cheese (white cheese), AFM_1, literature[11]

incidence: 10/20, conc. range: 0.025–0.050 µg/kg (1 sa.), 0.051–0.100 µg/kg (4 sa.), 0.101–0.150 µg/kg (3 sa.), 0.151–0.250 µg/kg (1 sa.), 0.388 µg/kg (1 sa.), Ø conc.: 0.119 µg/kg, sample origin: supermarkets in Adana (city), Turkey, sample year: unknown, country: Turkey[12]

- Co-contamination: not reported
- Further contamination of animal products reported in the present book/article[12]: Butter, Milk (UHT milk), Cheese (white cheese), AFM_1, literature[12]

incidence: 85/100, conc. range: 0.051–0.150 µg/kg (42 sa.), 0.151–0.250 µg/kg (9 sa.), 0.251–0.450 µg/kg (13 sa.), 0.451–0.650 µg/kg (8 sa.), 0.651–0.800 µg/kg (9 sa.), >0.800 µg/kg (4 sa.), sample origin: markets in various districts of Ankara (capital), Turkey, sample year: 2001–2002, country: Turkey[35]

- Co-contamination: not reported
- Further contamination of animal products reported in the present book/article[35]: Cheese (processed cheese), Cheese (Tulum cheese), Cheese (white cheese), AFM_1, literature[35]

incidence: 4/30*, conc. range: 0.0511–0.0745 µg/kg, ∅ conc.: 0.0624 µg/kg, sample origin: local markets in Kars (district), Turkey, sample year: winter months 2003, country: Turkey[130], *Kars Kashar Cheese

- Co-contamination: not reported
- Further contamination of animal products reported in the present book/article[130]: Cheese (Cecil cheese), AFM_1, literature[130]

incidence: 18/25*, conc. range: 0.51–0.250 µg/kg (12 sa.), 0.251–0.400 µg/kg (4 sa.), >0.400 µg/kg (2 sa.), sample origin: local stores in Ardahan, Igdir, and Kars (cities), Turkey, sample year: unknown, country: Turkey[132], *fresh Kashar

incidence: 14/25*, conc. range: 0.51–0.250 µg/kg (9 sa.), 0.251–0.400 µg/kg (4 sa.), >0.400 µg/kg (1 sa.), sample origin: local stores in Ardahan, Igdir, and Kars (cities), Turkey, sample year: unknown, country: Turkey[132], *old Kashar

- Co-contamination: not reported
- Further contamination of animal products reported in the present book/article[132]: Cheese (Civil cheese), Cheese (Graviera cheese), Cheese (white cheese), AFM_1, literature[132]

incidence: 6/14*, conc. range: 0.007–0.068 µg/kg, ∅ conc.: 0.0228 µg/kg, sample origin: retail markets in Erzurum (city), Turkey, sample year: unknown, country: Turkey[133], *Kaşar cheese

- Co-contamination: not reported
- Further contamination of animal products reported in the present book/article[133]: Cheese (Civil cheese), Cheese (Lor cheese), Cheese (Tulum cheese), Cheese (white cheese), AFM_1, literature[133]

incidence: 12/200, conc. range: 0.120–0.800 µg/kg, ∅ conc.: 0.272 µg/kg, sample origin: different Turkish provinces, Turkey, sample year: January 2001–February 2002, country: Turkey[135]

- Co-contamination: not reported
- Further contamination of animal products reported in the present book/article[135]: Cheese (cream cheese), Cheese (white cheese), AFM_1, literature[135]

incidence: 6/8, conc. range: 0.018–0.1243 µg/kg, ∅ conc.: 0.0584 µg/kg, sample origin: different markets, Turkey, sample year: unknown, country: Turkey[142]

- Co-contamination: not reported
- Further contamination of animal products reported in the present book/article[142]: Cheese (Tulum cheese), Cheese (white cheese), AFM_1, literature[142]

incidence: 109/132, conc. range: 0.050–0.100 µg/kg (26 sa.), 0.101–0.250 µg/kg (47 sa.), 0.251–0.500 µg/kg (26 sa.), >0.500 µg/kg (10 sa., maximum: 0.690 µg/kg), sample origin: supermarkets in Edirne, Izmir, Konya, Tekirdag (cities), and Istanbul (capital), Turkey, sample year: September 2007–January 2008, country: Turkey[143]

- Co-contamination: not reported
- Further contamination of animal products reported in the present book/article[143]: Milk (UHT milk), AFM_1, literature[143]

incidence: 8/20, conc. range: 0.012–0.3695 µg/kg, sample origin: Kayseri (city), Turkey, sample year: January–March 2010, country: Turkey/Kyrgyzstan[144]

- Co-contamination: not reported
- Further contamination of animal products reported in the present book/article[144]: Cheese (Tulum cheese), Cheese (white cheese), Dairy desserts, Milk, Yogurt, AFM_1, literature[144]

incidence: 144/147, conc. range: 0.015–3.774 µg/kg, ∅ conc.: 0.0273 µg/kg, sample origin: cities in the East Black Sea, Middle Black Sea, and West Black Sea region, Turkey, sample year: October–November 2008, February–March and May–June 2009, country: Turkey[348]

- Co-contamination: not reported
- Further contamination of animal products reported in the present book/article[348]: not reported

For detailed information see the article.

incidence: 12/30, conc. range: 0.05–0.10 µg/kg (4 sa.), 0.11–0.20 µg/kg (1 sa.), 0.21–0.30 µg/kg (2 sa.), 0.31–0.40 µg/kg (2 sa.), 0.41–0.50 µg/kg (1 sa.), >0.50 µg/kg (2 sa., maximum: 1.15 µg/kg), ∅ conc.: 0.25 µg/kg, sample origin: markets and bazaars in Istanbul (capital), Turkey, sample year: unknown, country: Turkey[446]

- Co-contamination: not reported
- Further contamination of animal products reported in the present book/article[446]: Cheese (Tulum cheese), Cheese (white cheese), AFM_1, literature[446]

see also **Cheese**

Cheese (Lancashire cheese) may contain the following mycotoxins:

Aspergillus Toxins

Aflatoxin M_1

incidence: 11/11, conc. range: 0.02–0.05 µg/kg (1 sa.), 0.06–0.1 µg/kg (5 sa.), 0.11–0.2 µg/kg (4 sa.), 0.21 µg/kg (1 sa.), sample origin: retail outlets, England and Wales, sample year: unknown, country: UK[131]

- Co-contamination: not reported
- Further contamination of animal products reported in the present book/article[131]: Cheese (Cheddar cheese), Cheese (Cheshire cheese), Cheese (Double Gloucester cheese), Cheese (Leicester cheese), Cheese (Wensleydale cheese), Milk (cow milk), Milk (infant formula), Yogurt, AFM_1, literature[131]

Cheese (Le Roitelet cheese) see **Cheese**

Cheese (Leicester cheese) may contain the following mycotoxins:

Aspergillus Toxins

Aflatoxin M_1

incidence: 13/13, conc. range: 0.02–0.05 µg/kg (6 sa.), 0.06–0.1 µg/kg (7 sa., maximum: 0.09 µg/kg), sample origin: retail outlets, England and Wales, sample year: unknown, country: UK[131]

- Co-contamination: not reported
- Further contamination of animal products reported in the present book/article[131]: Cheese (Cheddar cheese), Cheese (Cheshire cheese), Cheese (Double Gloucester cheese), Cheese (Lancashire cheese), Cheese (Wensleydale cheese), Milk (cow milk), Milk (infant formula), Yogurt, AFM_1, literature[131]

Cheese (Lighvan cheese) may contain the following mycotoxins:

Aspergillus **Toxins**

Aflatoxin M$_1$

incidence: 49/75, conc. range: 0.030–0.313 µg/kg, sample origin: dairy ranches, supermarkets, and retail outlets in Esfahan, Shiraz, and Tabriz (cities) as well as Tehran (capital), Iran, sample year: 2008, country: Iran[440]

- Co-contamination: not reported
- Further contamination of animal products reported in the present book/article[440]: Ayran, Kashk, Milk (cow milk), Milk (goat milk), Milk (sheep milk), Yogurt, AFM_1, literature[440]

incidence: 10/37, conc. range: 0.0705–0.203 µg/kg, sample origin: supermarkets and retail outlets in Rafsanjan (city), Iran, sample year: winter and spring 2012, country: Iran[488]

- Co-contamination: not reported
- Further contamination of animal products reported in the present book/article[488]: Cheese (white cheese), AFM_1, literature[488]

For detailed information see the article.

Cheese (Lor cheese) may contain the following mycotoxins:

Aspergillus **Toxins**

Aflatoxin M$_1$

incidence: 2/6, conc. range: 0.013–0.019 µg/kg, ∅ conc.: 0.01595 µg/kg, sample origin: retail markets in Erzurum (city), Turkey, sample year: unknown, country: Turkey[133]

- Co-contamination: not reported
- Further contamination of animal products reported in the present book/article[133]: Cheese (Civil cheese), Cheese (Kashar cheese), Cheese (Tulum cheese), Cheese (White cheese), AFM_1, literature[133]

Cheese (Maasdam cheese) see **Cheese**

Cheese (Manchego cheese) see **Cheese curd** and **Milk (sheep milk)**

Cheese (Maribo cheese) may contain the following mycotoxins:

Aspergillus **Toxins**

Aflatoxin M$_1$

incidence: 3/3, conc. range: 0.087–0.412 µg/kg, ∅ conc.: 0.264 µg/kg, sample origin: Denmark, imported and purchased in Japan, sample year: 1981–1983, country: Japan[120]

- Co-contamination: not reported
- Further contamination of animal products reported in the present book/article[120]: Cheese, Cheese (blue-veined cheese), Cheese (Brie cheese), Cheese (butter cheese), Cheese (Camembert cheese), Cheese (Cheddar cheese), Cheese (cream cheese), Cheese (Edam cheese), Cheese (Gouda cheese), Cheese (Havarti cheese), Cheese (Mozzarella cheese), Cheese (Samsoe cheese), AFM_1, literature[120]

see also **Cheese**

Cheese (Mihalic cheese) see **Cheese**

Cheese (Minas cheese) may contain the following mycotoxins:

Aspergillus **Toxins**

Aflatoxin M$_1$

incidence: 6/6*, conc. range: >0.003–0.010 µg/kg (3 sa.), 0.011–0.050 µg/kg (1 sa.), 0.051–0.100 µg/kg (2 sa.), sample origin: Ribeirão Preto (municipality), São Paulo (state), Brazil, sample year: unknown, country: Brazil/USA[44], *Minas Frescal light cheese

incidence: 7/7*, conc. range: >0.003–0.010 µg/kg (1 sa.), 0.011–0.050 µg/kg (3 sa.), 0.051–0.100 µg/kg (1 sa.), 0.101–0.250 µg/kg (1 sa.), 0.251–0.500 µg/kg (1 sa.), sample origin: Ribeirão Preto (municipality), São Paulo (state), Brazil, sample year: unknown, country: Brazil/USA[44], *Minas Frescal cheese

incidence: 6/6*, conc. range: >0.003–0.010 µg/kg (4 sa.), 0.011–0.050 µg/kg (2 sa.), sample origin: Ribeirão Preto (municipality), São Paulo (state), Brazil, sample year: unknown, country: Brazil/USA[44], *Minas Padrão cheese

- Co-contamination: not reported
- Further contamination of animal products reported in the present book/article[44]: Cheese (Prato cheese), Dairy beverages, Yogurt, AFM_1, literature[44]

incidence: 4/7*, conc. range: 0.03–0.18 µg/kg, ∅ conc.: 0.08 µg/kg, sample origin: Mercado Central in Belo Horizonte (city), Minas Gerais (state), Brazil, sample year: 1996–1998, country: Brazil[145], *fresh Minas cheese

incidence: 11/18*, conc. range: 0.02–1.7 µg/kg, ∅ conc.: 0.36 µg/kg, sample origin: Mercado Central in Belo Horizonte (city), Minas Gerais (state), Brazil, sample year: 1996–1998, country: Brazil[145], *Canastra

incidence: 41/50*, conc. range: 0.02–6.92 µg/kg, ∅ conc.: 0.62 µg/kg, sample origin: Mercado Central in Belo Horizonte (city), Minas Gerais (state), Brazil, sample year: 1996–1998, country: Brazil[145], *standard Minas cheese

- Co-contamination: not reported
- Further contamination of animal products reported in the present book/article[145]: not reported

incidence: 11/20*, conc. range: 0.011–0.050 µg/kg (8 sa.), 0.051–0.100 µg/kg (1 sa.), 0.101–0.250 µg/kg (1 sa.), 0.251–0.310 µg/kg (1 sa.), sample origin: grocery stores in Ribeirão Preto (municipality), São Paulo (state), Brazil, sample year: 2010, country: Brazil/USA[453], *Minas Frescal light cheese

incidence: 25/30*, conc. range: 0.011–0.050 µg/kg (12 sa.), 0.051–0.100 µg/kg (5 sa.), 0.101–0.250 µg/kg (5 sa.), 0.251–0.310 µg/kg (3 sa.), sample origin: grocery stores in Ribeirão Preto (municipality), São Paulo (state), Brazil, sample year: 2010, country: Brazil/USA[453], *Minas Frescal cheese

incidence: 3/8*, conc. range: 0.011–0.050 µg/kg (2 sa.), 0.051–0.100 µg/kg (1 sa.), sample origin: grocery stores in Ribeirão Preto (municipality), São Paulo (state), Brazil, sample year: 2010, country: Brazil/USA[453], *Minas Padrão cheese

- Co-contamination: not reported
- Further contamination of animal products reported in the present book/article[453]: Dairy beverages, Yogurt, AFM_1, literature[453]

Cheese (mold-ripened) may contain the following mycotoxins:

Penicillium Toxins

Mycophenolic Acid

incidence: 2/2, conc. range: 14–120 µg/kg, ∅ conc.: 67 µg/kg, sample origin: Netherlands, sample year: 2013, country: Netherlands[275]

- Co-contamination: 2 sa. co-contaminated with MA and ROQ C
- Further contamination of animal products reported in the present book/article[275]: Cheese (mold-ripened), ROQ C, literature[275]

Roquefortine C

incidence: 2/2, conc. range: 172–552 µg/kg, ∅ conc.: 362 µg/kg, sample origin: Netherlands, sample year: 2013, country: Netherlands[275]

- Co-contamination: 2 sa. co-contaminated with MA and ROQ C
- Further contamination of animal products reported in the present book/article[275]: Cheese (mold-ripened), MA, literature[275]

Cheese (Moravian Block cheese) see **Cheese**

Cheese (Mozzarella cheese) may contain the following mycotoxins:

Aspergillus Toxins

Aflatoxin M_1

incidence: 4/4, conc. range: 0.181–0.433 µg/kg, ∅ conc.: 0.334 µg/kg, sample origin: Denmark, imported and purchased in Japan, sample year: 1981–1983, country: Japan[120]

incidence: 5/5, conc. range: 0.028–0.252 µg/kg, ∅ conc.: 0.091 µg/kg, sample origin: West Germany, imported and purchased in Japan, sample year: 1981–1983, country: Japan[120]

- Co-contamination: not reported
- Further contamination of animal products reported in the present book/article[120]: Cheese, Cheese (blue-veined cheese), Cheese (Brie cheese), Cheese (butter cheese), Cheese (Camembert cheese), Cheese (Cheddar cheese), Cheese (cream cheese), Cheese (Edam cheese), Cheese (Gouda cheese), Cheese (Havarti cheese), Cheese (Maribo cheese), Cheese (Samsoe cheese), AFM_1, literature[120]

see also **Cheese**

Cheese (Münster cheese) may contain the following mycotoxins:

Aspergillus Toxins

Aflatoxin M_1

incidence: 1/1, conc.: 0.448 µg/kg, sample origin: unknown, sample year: unknown, country: Japan[128]

- Co-contamination: not reported
- Further contamination of animal products reported in the present book/article[128]: Cheese (Camembert cheese), Cheese (cream cheese), Cheese (Gouda cheese), AFM_1, literature[128]

Cheese (Naboulsi cheese) see **Cheese (white cheese)**

Cheese (Oaxaca cheese) may contain the following mycotoxins:

Aspergillus Toxins

Aflatoxin M_1

incidence: 16/30, conc. range: 0.01–43.99 µg/kg, ∅ conc.: 5.9 µg/kg, sample origin: cheese makers in Veracruz (state), southeastern Mexico, sample year: March–April 2016, country: Mexico/France[560]

- Co-contamination: 2 sa. co-contaminated with AFM_1 and AFM_2; 14 sa. contaminated solely with AFM_1
- Further contamination of animal products reported in the present book/article[560]: Cheese (Oaxaca cheese), AFM_2, literature[560]

Aflatoxin M_2

incidence: 3/30, conc. range: 0.68–3.62 µg/kg, ∅ conc.: 2.36 µg/kg, sample origin: cheese makers in Veracruz (state), southeastern Mexico, sample year: March–April 2016, country: Mexico/France[560]

- Co-contamination: 2 sa. co-contaminated with AFM_1 and AFM_2; 1 sa. contaminated solely with AFM_2
- Further contamination of animal products reported in the present book/article[560]: Cheese (Oaxaca cheese), AFM_1, literature[560]

Cheese (Parmesan cheese) may contain the following mycotoxins:

Aspergillus Toxins

Aflatoxin M_1

incidence: 18/200, conc. range: 0.035–0.190 µg/kg, sample origin: Modena (district), Italy, sample year: 1991, country: Italy[146]

- Co-contamination: not reported
- Further contamination of animal products reported in the present book/article[146]: not reported

incidence: 40/88, conc. range: >0.010–0.019 µg/kg (14 sa.), 0.020–0.250 µg/kg (24 sa.), >0.250 µg/kg (2 sa., maximum: 0.660 µg/kg), ∅ conc.: 0.0548 µg/kg, sample origin: groceries and supermarkets in Minas Gerais (state), Brazil, sample year: March–December 2004, country: Brazil[273]

- Co-contamination: not reported
- Further contamination of animal products reported in the present book/article[273]: not reported

incidence: 10/10*, conc. range: 0.172–1.207 µg/kg, ∅ conc.: 0.651 µg/kg, sample origin: cheese factories and retail outlets, northern Italy, sample year: unknown, country: Italy[497], *ripening time: 8–24 months

- Co-contamination: not reported
- Further contamination of animal products reported in the present book/article[497]: Cheese (semihard cheese), Cheese (soft cheese), AFM_1, literature[497]

incidence: 18/30*, conc. range: <0.25 µg/kg (10 sa.), 0.25–0.5 µg/kg (5 sa.), 0.5–2.5 µg/kg (3 sa., maximum: 0.69 µg/kg), sample origin: Niterói, Rio de Janeiro, and Seropédica (cities) in metropolitan region of Rio de Janeiro (city), Brazil, sample year: January–March 2011, country: Brazil[529], *grated Parmesan cheese sa.

- Co-contamination: not reported
- Further contamination of animal products reported in the present book/article[529]: not reported

see also **Cheese** and **Cheese (semihard cheese)**

Cheese (Parmigiano Reggiano cheese) see **Milk (cow milk)**

Cheese (Prato cheese) may contain the following mycotoxins:

Aspergillus Toxins

Aflatoxin M_1

incidence: 6/6, conc. range: >0.003–0.010 µg/kg (3 sa.), 0.011–0.050 µg/kg (2 sa.), 0.101–0.250 µg/kg (1 sa.), sample origin: Ribeirão Preto (municipality), São Paulo (state), Brazil, sample year: unknown, country: Brazil/USA[44]

- Co-contamination: not reported
- Further contamination of animal products reported in the present book/article[44]: Cheese (Minas cheese), Dairy beverages, Yogurt, AFM_1, literature[44]

Cheese (processed cheese) may contain the following mycotoxins:

Aspergillus Toxins

Aflatoxin M_1

incidence: 53/134, conc. range: 0.1–0.50 µg/kg (51 sa.), 0.51–1.0 µg/kg (2 sa., maximum: 0.55 µg/kg), ∅ conc.: 0.26 µg/kg, sample origin: unknown, sample year: September 1972–December 1974, country: Germany[28]

- Co-contamination: not reported
- Further contamination of animal products reported in the present book/article[28]: Cheese (Camembert cheese), Cheese (fresh cheese), Cheese (hard cheese), Milk (cow milk), Milk powder, Yogurt, AFM_1, literature[28]

incidence: 79/100, conc. range: 0.051–0.150 µg/kg (30 sa.), 0.151–0.250 µg/kg (24 sa.), 0.251–0.450 µg/kg (16 sa.), 0.451–0.650 µg/kg (9 sa.), sample origin: markets in various districts of Ankara (capital), Turkey, sample year: 2001–2002, country: Turkey[35]

- Co-contamination: not reported
- Further contamination of animal products reported in the present book/article[35]: Cheese (Kashar cheese), Cheese (Tulum cheese), Cheese (white cheese), AFM_1, literature[35]

incidence: 11/50, conc. range: 0.0518–0.054 µg/kg, ∅ conc.: 0.05252 µg/kg, sample origin: supermarkets in Alexandria (city), Egypt, sample year: February 2008–March 2009, country: Egypt[396]

- Co-contamination: not reported
- Further contamination of animal products reported in the present book/article[396]: Cheese (hard cheese), Cheese (soft cheese), Milk (cow milk), AFM_1, literature[396]

see also **Cheese (blue-veined cheese)**

Cheese (Ras cheese) may contain the following mycotoxins:

Aspergillus Toxins

Aflatoxin B_1

incidence: 1/10, conc.: 10 µg/kg, sample origin: Egyptians markets, Egypt, sample year: 1999–2000, country: Egypt[138]

- Co-contamination: not reported
- Further contamination of human breast milk and animal products reported in the present book/article[138]: Cheese (Domiati cheese), Milk (cow milk), Milk powder, AFM_1, literature[138]; Cheese (Ras cheese), AFG_1 and AFM_1, literature[138]; Milk (human breast milk), AFM_1 and OTA, literature[138]

Aflatoxin G_1

incidence: 1/10, conc.: 4 µg/kg, sample origin: Egyptians markets, Egypt, sample year: 1999–2000, country: Egypt[138]

- Co-contamination: not reported
- Further contamination of human breast milk and animal products reported in the present book/article[138]: Cheese (Domiati cheese), Milk (cow milk), Milk powder, AFM_1, literature[138]; Cheese (Ras cheese), AFB_1 and AFM_1, literature[138]; Milk (human breast milk), AFM_1 and OTA, literature[138]

Aflatoxin M_1

incidence: 2/10, conc. range: 3–6 µg/kg, ∅ conc.: 4.6 µg/kg, sample origin: Egyptians markets, Egypt, sample year: 1999–2000, country: Egypt[138]

- Co-contamination: not reported
- Further contamination of human breast milk and animal products reported in the present book/article[138]: Cheese (Domiati cheese), Milk (cow milk), Milk powder, AFM_1, literature[138]; Cheese (Ras cheese), AFB_1 and AFG_1, literature[138]; Milk (human breast milk), AFM_1 and OTA, literature[138]

Sterigmatocystin

incidence: 20/50*, conc. range: 16.0–37.7** µg/kg, ∅ conc.: 25.7 µg/kg, sample origin: local markets of Cairo (governorate), Egypt, sample year: unknown, country: Egypt[147], *outer surface layer (1 cm thick) and inner layer, **only 1 sa. contained 62.8 µg/kg

incidence: 7/25*, conc. range: 18.8–23.7 µg/kg, ∅ conc.: 21.3 µg/kg, sample origin: local markets of Giza (governorate), Egypt, sample year: unknown, country: Egypt[147], *outer surface layer (1 cm thick) and inner layer

incidence: 8/25*, conc. range: 10.0–37.0 µg/kg, ∅ conc.: 23.5 µg/kg, sample origin: local markets of Kalubia (governorate), Egypt, sample year: unknown, country: Egypt[147], *outer surface layer (1 cm thick) and inner layer

- Co-contamination: not reported
- Further contamination of animal products reported in the present book/article[147]: not reported

see also **Cheese (Hard Roumy cheese)**

Cheese (Ricotta cheese) see **Cheese (sheep cheese)**

Cheese (Roquefort cheese) may contain the following mycotoxins:

Aspergillus and ***Penicillium*** Toxins

Ochratoxin A

incidence: 7/14, conc. range: 0.1–1.4 µg/kg, sample origin: market, Italy, sample year: unknown, country: Italy[121]

- Co-contamination: not reported
- Further contamination of animal products reported in the present book/article[121]: Cheese (Gorgonzola cheese), OTA, literature[121]

Penicillium Toxins

Mycophenolic Acid

incidence: 21/25, conc. range: 10–100 µg/kg (1 sa.), 100–1000 µg/kg (2 sa.), 1000–5000 µg/kg (10 sa.), 5000–10,000 µg/kg (5 sa.), ≤14,300 µg/kg (3 sa.), sample origin: commercial, sample year: unknown, country: France[118]

incidence: 2/2*, conc. range: 1000–5000 µg/kg (2 sa.), sample origin: commercial, sample year: unknown, country: France[118], *melted Roquefort cheese

- Co-contamination: not reported
- Further contamination of animal products reported in the present book/article[118]: Cheese, Cheese (Bleu des Causses cheese), Cheese (blue-veined cheese), Cheese (Gorgonzola cheese), MA, literature[118]

For detailed information see the article

Roquefortine C

incidence: 3/3, conc. range: 50–690 µg/kg, ∅ conc.: 420 µg/kg, sample origin: purchased in France and Greece, sample year: unknown, country: Italy[124]

- Co-contamination: not reported
- Further contamination of animal products reported in the present book/article[124]: Cheese (Danablu cheese), Cheese (Edelpilzkäse), Cheese (Gorgonzola cheese), Cheese (Stilton cheese), ROQ C, literature[124]

incidence: 1/1, conc.: 705 µg/kg, sample origin: unknown, sample year: unknown, country: Netherlands[126]

- Co-contamination: not reported
- Further contamination of animal products reported in the present book/article[126]: Cheese (Danablu cheese), Cheese (Gorgonzola cheese), ROQ C, literature[126]

see also **Cheese** and **Cheese (blue-veined cheese)**

Cheese (Samsoe cheese) may contain the following mycotoxins:

Aspergillus Toxins

Aflatoxin M$_1$

incidence: 5/5, conc. range: 0.070–0.504 µg/kg, ∅ conc.: 0.214 µg/kg, sample origin: Denmark, imported and purchased in Japan, sample year: 1981–1983, country: Japan[120]

- Co-contamination: not reported
- Further contamination of animal products reported in the present book/article[120]: Cheese, Cheese (blue-veined cheese), Cheese (Brie cheese), Cheese (butter cheese), Cheese (Camembert cheese), Cheese (Cheddar cheese), Cheese (cream cheese), Cheese (Edam cheese), Cheese (Gouda cheese), Cheese (Havarti cheese), Cheese (Maribo cheese), Cheese (Mozzarella cheese), AFM$_1$, literature[120]

Cheese (semihard cheese) may contain the following mycotoxins:

Aspergillus Toxins

Aflatoxin M$_1$

incidence: 8/8*, conc. range: 0.115–0.794 µg/kg, ∅ conc.: 0.325 µg/kg, sample origin: cheese factories and retail outlets, northern Italy, sample year: unknown, country: Italy[497], *Parmesan and Caciotta cheese sa. (1–3 months ripening time)

- Co-contamination: not reported
- Further contamination of animal products reported in the present book/article[497]: Cheese (Parmesan cheese), Cheese (soft cheese), AFM$_1$, literature[497]

Aspergillus and ***Penicillium*** Toxins

Ochratoxin A

incidence: 6/32*, conc. range: 18.4–146.0 µg/kg**, sample origin: cellars of cheese producers, Italy, sample year: unknown, country: Italy[495], *handmade semihard cheese sa., **in the interior

incidence: 6/32*, conc. range: 1.0–262.2 µg/kg**, sample origin: cellars of cheese producers, Italy, sample year: unknown, country: Italy[495], *handmade semihard cheese sa., **in the rind

- Co-contamination: For detailed information see the article.
- Further contamination of animal products reported in the present book/article[495]: Cheese (semihard cheese), PAT, literature[495]

Patulin

incidence: 1/32*, conc.: 26.6 µg/kg**, sample origin: cellars of cheese producers, Italy, sample year: unknown, country: Italy[495], *handmade semihard cheese sa., **in the interior

incidence: 8/32*, conc. range: 15.4–460.8 µg/kg**, sample origin: cellars of cheese producers, Italy, sample year: unknown, country: Italy[495], *handmade semihard cheese sa., **in the rind

- Co-contamination: For detailed information see the article.
- Further contamination of animal products reported in the present book/article[495]: Cheese (semihard cheese), OTA, literature[495]

see also **Cheese**

Cheese (sheep cheese) may contain the following mycotoxins:

Aspergillus Toxins

Aflatoxin M$_1$

incidence: 4/30*, conc. range: 0.021–0.101 µg/kg, ∅ conc.: 0.0483 µg/kg, sample origin: sheep and dairy farms or market in Western Sicily (island, autonomous region), Italy, sample year: November 2001–June 2002, country: Italy[37], *sheep cheese and ricotta

- Co-contamination: not reported
- Further contamination of animal products reported in the present book/article[37]: Milk (sheep milk), AFM$_1$, literature[37]

For detailed information see the article.

incidence: 5/31*, conc. range: 0.060–0.170 µg/kg, ∅ conc.: 0.105 µg/kg, sample origin: sales centers in Apulia (region), Italy, sample year: unknown, country: Italy[43], *unripened/short-term ripened (<1 month)

incidence: 3/32*, conc. range: 0.055–0.190 µg/kg, ∅ conc.: 0.1217 µg/kg, sample origin: sales centers in Apulia (region), Italy, sample year: unknown, country: Italy[43], *medium-term ripened (1–3 months)

incidence: 4/31*, conc. range: 0.050–0.215 µg/kg, ∅ conc.: 0.1113 µg/kg, sample origin: sales centers in Apulia (region), Italy, sample year: unknown, country: Italy[43], *long-term ripened (3–12 months)

- Co-contamination: not reported
- Further contamination of animal products reported in the present book/article[43]: Cheese (cow cheese), Cheese (goat cheese), Cheese (goat-sheep cheese), AFM$_1$, literature[43]

Cheese (Shelal cheese) see Cheese (white cheese)

Cheese (skin bag cheese) see Cheese

Cheese (soft cheese) may contain the following mycotoxins:

Aspergillus **Toxins**

Aflatoxin M_1

incidence: 8/23*, conc. range: 0.015–0.050 µg/kg (4 sa.), 0.051–0.100 µg/kg (3 sa.), 0.160 µg/kg (1 sa.), sample origin: dairy farms in Lombardy (administrative region), Italy, sample year: March–October 1996, country: Italy[36], *soft goat cheese and soft goat cheese with aromatic herbs

- Co-contamination: not reported
- Further contamination of animal products reported in the present book/article[36]: Cheese (goat cheese), Milk (goat milk), AFM_1, literature[36]

incidence: 20/50, conc. range: 0.052–0.0876 µg/kg, ∅ conc.: 0.07063 µg/kg, sample origin: supermarkets in Alexandria (city), Egypt, sample year: February 2008–March 2009, country: Egypt[396]

- Co-contamination: not reported
- Further contamination of animal products reported in the present book/article[396]: Cheese (hard cheese), Cheese (processed cheese), Milk (cow milk), AFM_1, literature[396]

incidence: 26/70*, conc. range: 0.020–0.050 µg/kg (13 sa.), 0.050–0.500 µg/kg (13 sa., maximum: 0.276 µg/kg), ∅ conc.: 0.085 µg/kg, sample origin: main convenience stores and supermarkets, Costa Rica, sample year: 2012–2014, country: Costa Rica[411], *fresh soft cheese sa.

- Co-contamination: not reported
- Further contamination of animal products reported in the present book/article[411]: Cream (sour), AFM_2, literature[411]; Milk (cow milk), AFM_1, literature[411]

incidence: 6/6*, conc. range: 0.131–0.977 µg/kg, ∅ conc.: 0.400 µg/kg, sample origin: cheese factories and retail outlets, northern Italy, sample year: unknown, country: Italy[497], *soft fresh cheese included Mozzarella and Crescenza

- Co-contamination: not reported
- Further contamination of animal products reported in the present book/article[497]: Cheese (Parmesan cheese), Cheese (semihard cheese), AFM_1, literature[497]

see also **Cheese**

Cheese (Stilton cheese) may contain the following mycotoxins:

Penicillium **Toxins**

Isofumigaclavine A

incidence: 2/2, conc. range: tr?-90 µg/kg, sample origin: England, imported and purchased in Canada, sample year: unknown, country: Canada[122]

- Co-contamination: For detailed information see the article.
- Further contamination of animal products reported in the present book/article[122]: Cheese (blue-veined cheese), IFC A, IFC B, and ROQ C, literature[122]; Cheese (Gorgonzola cheese) ROQ C, literature[122]; Cheese (Stilton cheese), ROQ C, literature[122]

For detailed information see the article.

Roquefortine C

incidence: 2/2, conc. range: 620–3400 µg/kg, ∅ conc.: 1580 µg/kg, sample origin: England, imported and purchased in Canada, sample year: unknown, country: Canada[122]

- Co-contamination: For detailed information see the article.
- Further contamination of animal products reported in the present book/article[122]: Cheese (blue-veined cheese), IFC A, IFC B, and ROQ C, literature[122]; Cheese (Gorgonzola cheese) ROQ C, literature[122]; Cheese (Stilton cheese), IFC A, literature[122]

For detailed information see the article.

incidence: 2/2, conc. range: 210–650 µg/kg, ∅ conc.: 430 µg/kg, sample origin: purchased in Great Britain and Ireland, sample year: unknown, country: Italy[124]

- Co-contamination: not reported
- Further contamination of animal products reported in the present book/article[124]: Cheese (Danablu cheese), Cheese (Edelpilzkäse), Cheese (Gorgonzola cheese), Cheese (Roquefort cheese), ROQ C, literature[124]

see also **Cheese (blue-veined cheese)**

Cheese (Surk cheese) may contain the following mycotoxins:

Aspergillus **Toxins**

Aflatoxin M_1

incidence: 72/120*, conc. range: 0.010–0.050 µg/kg (1 sa.), 0.051–0.150 µg/kg (38 sa.), 0.151–0.250 µg/kg (18 sa.), 0.251–0.800 µg/kg (13 sa.), 0.801–1.043 µg/kg (3 sa.), ∅ conc.: 0.2213 µg/kg, sample origin: retail markets of Antioch (city), Turkey, sample year: June–November 2006, country: Turkey[148], *moldy Surk cheese sa.

- Co-contamination: not reported
- Further contamination of animal products reported in the present book/article[148]: not reported

Cheese (Tulum cheese) may contain the following mycotoxins:

Aspergillus **Toxins**

Aflatoxin M_1

incidence: 81/100, conc. range: 0.051–0.150 µg/kg (41 sa.), 0.151–0.250 µg/kg (16 sa.), 0.251–0.450 µg/kg (18 sa.), 0.451–0.650 µg/kg (5 sa.), >0.800 µg/kg (1 sa.), sample origin: markets in various districts of Ankara (capital), Turkey, sample year: 2001–2002, country: Turkey[35]

- Co-contamination: not reported
- Further contamination of animal products reported in the present book/article[35]: Cheese (Kashar cheese), Cheese (processed cheese), Cheese (white cheese), AFM_1, literature[35]

incidence: 7/11, conc. range: 0.011–0.202 µg/kg, ∅ conc.: 0.07405 µg/kg, sample origin: retail markets in Erzurum (city), Turkey, sample year: unknown, country: Turkey[133]

- Co-contamination: not reported
- Further contamination of animal products reported in the present book/article[133]: Cheese (Civil cheese), Cheese (Kashar cheese), Cheese (Lor cheese), Cheese (white cheese), AFM_1, literature[133]

incidence: 6/8, conc. range: 0.012–0.3314 µg/kg, ∅ conc.: 0.1496 µg/kg, sample origin: different markets, Turkey, sample year: unknown, country: Turkey[142]

- Co-contamination: not reported
- Further contamination of animal products reported in the present book/article[142]: Cheese (Kashar cheese), Cheese (white cheese), AFM_1, literature[142]

incidence: 16/20, conc. range: 0.013–0.378 µg/kg, sample origin: Kayseri (city), Turkey, sample year: January–March 2010, country: Turkey/Kyrgyzstan[144]

- Co-contamination: not reported
- Further contamination of animal products reported in the present book/article[144]: Cheese (Kashar cheese), Cheese (white cheese), Dairy desserts, Milk, Yogurt, AFM_1, literature[144]

incidence: 11/20, conc. range: 0.05–0.10 µg/kg (2 sa.), 0.11–0.20 µg/kg (1 sa.), 0.21–0.30 µg/kg (3 sa.), 0.31–0.40 µg/kg (2 sa.), 0.41–0.50 µg/kg (1 sa.), >0.50 µg/kg (2 sa., maximum: 1.36 µg/kg), ∅ conc.: 0.38 µg/kg, sample origin: markets and bazaars in Istanbul (capital), Turkey, sample year: unknown, country: Turkey[446]

- Co-contamination: not reported
- Further contamination of animal products reported in the present book/article[446]: Cheese (Kashar cheese), Cheese (white cheese), AFM_1, literature[446]

Cheese (Urfa cheese) may contain the following mycotoxins:

Aspergillus Toxins

Aflatoxin M_1

incidence: 36/127*, conc. range: 0.050–0.250 µg/kg (23 sa.), 0.251–0.500 µg/kg (6 sa.), 0.501–0.750 µg/kg (5 sa.), >0.750 µg/kg (2 sa., maximum: 0.77097 µg/kg), ∅ conc.: 0.2537 µg/kg, sample origin: supermarkets and retail outlets in Sanliurfa (province), Turkey, sample year: January–March 2008, country: Turkey[461]

- Co-contamination: not reported
- Further contamination of animal products reported in the present book/article[461]: not reported

see also **Cheese**

Cheese (Van otlu cheese) may contain the following mycotoxins:

Aspergillus Toxins

Aflatoxin M_1

incidence: 52/60, conc. range: 0.151–0.250 µg/kg (4 sa.), 0.251–0.450 µg/kg (3 sa.), 0.451–0.650 µg/kg (7 sa.), 0.651–0.800 µg/kg (4 sa.), >0.800 µg/kg (34 sa., maximum: 7.26 µg/kg), sample origin: retail outlets (partly with insufficient storage conditions) in Hakari and Van (cities), Turkey, sample year: September 2002–May 2003, country: Turkey[149]

- Co-contamination: not reported
- Further contamination of animal products reported in the present book/article[149]: Cheese (white cheese), AFM_1, literature[149]

see also **Cheese**

Cheese (Wensleydale cheese) may contain the following mycotoxins:

Aspergillus Toxins

Aflatoxin M_1

incidence: 11/11, conc. range: 0.02–0.05 µg/kg (2 sa.), 0.06–0.1 µg/kg (5 sa.), 0.11–0.2 µg/kg (3 sa.), 0.22 µg/kg (1 sa.), sample origin: retail outlets, England and Wales, sample year: unknown, country: UK[131]

- Co-contamination: not reported
- Further contamination of animal products reported in the present book/article[131]: Cheese (Cheddar cheese), Cheese (Cheshire cheese), Cheese (Double Gloucester cheese), Cheese (Lancashire cheese), Cheese (Leicester cheese), Milk (cow milk), Milk (infant formula), Yogurt, AFM_1, literature[131]

Cheese (white cheese) may contain the following mycotoxins:

Aspergillus Toxins

Aflatoxin M_1

incidence: 121/186, conc. range: 0.011–0.050 µg/kg (2 sa.), 0.051–0.100 µg/kg (12 sa.), 0.101–0.250 µg/kg (72 sa.), >0.250 µg/kg (35 sa., maximum: 4.89 µg/kg), sample origin: Istanbul (city), Turkey, sample year: 2001, country: Turkey[10]

- Co-contamination: not reported
- Further contamination of animal products reported in the present book/article[10]: Butter, AFM_1, literature[10]

incidence: 86/94, conc. range: <0.001 µg/kg (3 sa.), 0.001–0.010 µg/kg (5 sa.), 0.011–0.050 µg/kg (9 sa.), 0.051–0.100 µg/kg (21 sa.), 0.101–0.250 µg/kg (36 sa.), >0.250 µg/kg (12 sa.), sample origin: markets in Ankara (capital), Turkey, sample year: September 2002–2003, country: Turkey[11]

- Co-contamination: not reported
- Further contamination of animal products reported in the present book/article[11]: Butter, Cheese (cream cheese), Cheese (Kashar cheese), AFM_1, literature[11]

incidence: 16/20, conc. range: 0.051–0.100 µg/kg (4 sa.), 0.101–0.150 µg/kg (7 sa.), 0.151–0.250 µg/kg (4 sa.), 0.263 µg/kg (1 sa.), ∅ conc.: 0.142 µg/kg, sample origin: supermarkets in Adana (city), Turkey, sample year: unknown, country: Turkey[12]

- Co-contamination: not reported
- Further contamination of animal products reported in the present book/article[12]: Butter, Cheese (Kashar cheese), Milk (UHT milk), AFM_1, literature[12]

incidence: 82/100, conc. range: 0.051–0.150 µg/kg (44 sa.), 0.151–0.250 µg/kg (11 sa.), 0.251–0.450 µg/kg (18 sa.), 0.451–0.650 µg/kg (7 sa.), 0.651–0.800 µg/kg (2 sa.), sample origin: markets in various districts of Ankara (capital), Turkey, sample year: 2001–2002, country: Turkey[35]

- Co-contamination: not reported
- Further contamination of animal products reported in the present book/article[35]: Cheese (Kashar cheese), Cheese (processed cheese), Cheese (Tulum cheese), AFM_1, literature[35]

incidence: 17/25, conc. range: 0.51–0.250 µg/kg (10 sa.), 0.251–0.400 µg/kg (7 sa.), sample origin: local stores in Ardahan, Igdir, and Kars (cities), Turkey, sample year: unknown, country: Turkey[132]

- Co-contamination: not reported
- Further contamination of animal products reported in the present book/article[132]: Cheese (Civil cheese), Cheese (Graviera cheese), Cheese (Kashar cheese), AFM_1, literature[132]

incidence: 9/23, conc. range: 0.011–0.106 µg/kg, ∅ conc.: 0.02808 µg/kg, sample origin: retail markets in Erzurum (city), Turkey, sample year: unknown, country: Turkey[133]

- Co-contamination: not reported
- Further contamination of animal products reported in the present book/article[133]: Cheese (Civil cheese), Cheese (Kashar cheese), Cheese (Lor cheese), Cheese (Tulum cheese), AFM_1, literature[133]

incidence: 10/200, conc. range: 0.100–0.600 µg/kg, ∅ conc.: 0.253 µg/kg, sample origin: different Turkish provinces, Turkey, sample year: January 2001–February 2002, country: Turkey[135]

- Co-contamination: not reported
- Further contamination of animal products reported in the present book/article[135]: Cheese (cream cheese), Cheese (Kashar cheese), AFM_1, literature[135]

incidence: 93/116, conc. range: 0.050–0.150 µg/kg (32 sa.), 0.151–0.250 µg/kg (28 sa.), 0.251–0.450 µg/kg (21 sa.), 0.451–0.650 µg/kg (8 sa.), >0.651 µg/kg (4 sa., maximum: 0.7445 µg/kg), ∅ conc.: 0.2477 µg/kg, sample origin: supermarkets and retail outlets in Esfahan and Yazd (provinces), central part of Iran, sample year: October 2007–May 2008, country: Iran[136]

- Co-contamination: not reported
- Further contamination of animal products reported in the present book/article[136]: Cheese (cream cheese), AFM_1, literature[136]

incidence: 5/8, conc. range: 0.0221–0.2082 µg/kg, ∅ conc.: 0.1042 µg/kg, sample origin: different markets, Turkey, sample year: unknown, country: Turkey[142]

- Co-contamination: not reported
- Further contamination of animal products reported in the present book/article[142]: Cheese (Kashar cheese), Cheese (Tulum cheese), AFM_1, literature[142]

incidence: 14/20, conc. range: 0.0156–0.1546 µg/kg, sample origin: Kayseri (city), Turkey, sample year: January–March 2010, country: Turkey/Kyrgyzstan[144]

- Co-contamination: not reported
- Further contamination of animal products reported in the present book/article[144]: Cheese (Kashar cheese), Cheese (Tulum cheese), Dairy desserts, Milk, Yogurt, AFM_1, literature[144]

incidence: 31/50*, conc. range: 0.100–0.150 µg/kg (5 sa.), 0.151–0.250 µg/kg (6 sa.), 0.251–0.450 µg/kg (5 sa.), 0.451–0.650 µg/kg (2 sa.), >0.800 µg/kg (13 sa., maximum: 5.20 µg/kg), sample origin: retail outlets (partly with insufficient storage conditions) in Hakari and Van (cities), Turkey, sample year: September 2002–May 2003, country: Turkey[149], *white pickled cheese

- Co-contamination: not reported
- Further contamination of animal products reported in the present book/article[149]: Cheese (Van otlu cheese), AFM_1, literature[149]

incidence: 3/4*, conc. range: 0.19–0.48 µg/kg, ∅ conc.: 0.34 µg/kg, sample origin: local dairy factories in Sabratha and Zuwarah (provinces), north-west of Libya, sample year: July–August 2002, country: UK[150], *fresh white soft cheese

incidence: 5/7*, conc. range: 0.11–0.52 µg/kg, ∅ conc.: 0.29 µg/kg, sample origin: local dairy factories in Az zawiyha (province), north-west of Libya, sample year: July–August 2002, country: UK[150], *fresh white soft cheese

incidence: 3/5*, conc. range: 0.15–0.43 µg/kg, ∅ conc.: 0.25 µg/kg, sample origin: local dairy factories in Az iziyah (province), north-west of Libya, sample year: July–August 2002, country: UK[150], *fresh white soft cheese

incidence: 4/4*, conc. range: 0.16–0.35 µg/kg, ∅ conc.: 0.21 µg/kg, sample origin: local dairy factories in Tripoli (capital), Libya, sample year: July–August 2002, country: UK[150], *fresh white soft cheese

- Co-contamination: not reported
- Further contamination of animal products reported in the present book/article[150]: Milk (cow milk), AFM_1, literature[150]

incidence: 159/193*, conc. range: 0.050–0.250 µg/kg (108 sa.), 0.251–0.500 µg/kg (27 sa.), 0.501–0.750 µg/kg (15 sa.), >0.750 µg/kg (9 sa., maximum: 0.860 µg/kg), ∅ conc.: 0.2846 µg/kg, sample origin: Erzurum (province), Turkey, sample year: April–June 2006, country: Turkey[151], *white brined cheese

- Co-contamination: not reported
- Further contamination of animal products reported in the present book/article[151]: not reported

incidence: 6/10, conc. range: 0.0058–0.0212 µg/kg, ∅ conc.: 0.01586 µg/kg, sample origin: Sistan and Baluchestan (provinces), southeast of Iran, sample year: summer and winter 2015, country: Iran[417]

- Co-contamination: not reported
- Further contamination of animal products reported in the present book/article[417]: Butter, Milk, Yogurt, AFM_1, literature[417]

For detailed information see the article.

incidence: 40/45, conc. range: 0.051–0.100 µg/kg (11 sa.), 0.101–0.250 µg/kg (19 sa.), 0.251–0.500 µg/kg (7 sa.), >0.500 µg/kg (3 sa., maximum: 0.600 µg/kg), sample origin: various districts in Burdur (city), Turkey, sample year: 2008, country: Turkey[418]

- Co-contamination: not reported
- Further contamination of animal products reported in the present book/article[418]: Butter, Ice cream, Milk (cow milk), Milk powder, Yogurt, AFM_1, literature[418]

incidence: 38/53*, conc. range: <0.050 µg/kg (12 sa.), 0.050–0.100 µg/kg (8 sa.), 0.100–0.250 µg/kg (5 sa.), >0.250 µg/kg (13 sa., maximum: 0.315 µg/kg), ∅ conc.: 0.130 µg/kg, sample origin: small local farms, Lebanon, sample year: unknown, country: Lebanon[433], *included baladi, double cream, Feta, halloumi, low salt-akkawi, naboulsi, shelal, and other cheeses

incidence: 21/38, conc. range: <0.050 µg/kg (11 sa.), 0.050–0.100 (10 sa., maximum: 0.0771 µg/kg), ∅ conc.: 0.048 µg/kg, sample origin: Lebanese dairy industries, Lebanon, sample year: unknown, country: Lebanon[433]

incidence: 16/20*, conc. range: <0.050 µg/kg (16 sa., maximum: 0.00395 µg/kg), ∅ conc.: 0.00253 µg/kg, sample origin: imported from different European countries and purchased at supermarkets in Beirut (capital), Lebanon, sample year: unknown, country: Lebanon[433], *white and yellow cheese

- Co-contamination: not reported
- Further contamination of animal products reported in the present book/article[433]: not reported

For detailed information see the article.

incidence: 59/72, conc. range: 0.030–1.200 µg/kg, sample origin: retail stores and supermarkets in Esfahan, Shiraz, and Yazd (cities), as well as Tehran (capital), Iran, sample year: winter and summer 2009, country: Iran[435]

- Co-contamination: not reported
- Further contamination of animal products reported in the present book/article[435]: Butter, Ice cream, Milk, Yogurt, AFM_1, literature[435]

For detailed information see the article.

incidence: 18/30, conc. range: 0.05–0.10 µg/kg (3 sa.), 0.11–0.20 µg/kg (4 sa.), 0.21–0.30 µg/kg (3 sa.), 0.31–0.40 µg/kg (2 sa.), 0.41–0.50 µg/kg (2 sa.), >0.50 µg/kg (4 sa., maximum: 2.52 µg/kg), ∅ conc.: 0.42 µg/kg, sample origin: markets and bazaars in Istanbul (capital), Turkey, sample year: unknown, country: Turkey[446]

- Co-contamination: not reported
- Further contamination of animal products reported in the present book/article[446]: Cheese (Kashar cheese), Cheese (Tulum cheese), AFM_1, literature[446]

incidence: 93/119, conc. range: <0.050 µg/kg (38 sa.), >0.050–0.150 µg/kg (41 sa.), >0.151 µg/kg (14 sa., maximum: 0.5954 µg/kg), ∅ conc.: 0.1891 µg/kg, sample origin: main districts of Punjab (province), Pakistan, sample year: November 2010–April 2011, country: Pakistan[451]

- Co-contamination: not reported
- Further contamination of animal products reported in the present book/article[451]: Butter, Cheese (cream cheese), Milk, Yogurt, AFM_1, literature[451]

incidence: 29/45, conc. range: 0.0933–0.309 µg/kg, sample origin: supermarkets and retail outlets in Rafsanjan (city), Iran, sample year: winter and spring 2012, country: Iran[488]

- Co-contamination: not reported
- Further contamination of animal products reported in the present book/article[488]: Cheese (Lighvan cheese), AFM_1, literature[488]

For detailed information see the article.

incidence: 10/23*, conc. range: 0.13–0.55 µg/kg, sample origin: supermarkets, Serbia, sample year: May–June 2013, country: Serbia[516], *commercial white cheese

incidence: 9/21*, conc. range: 0.13–0.22 µg/kg, sample origin: households in the Bačka and the Srem counties, northern Serbia, sample year: May–June 2013, country: Serbia[516], *not commercially available white cheese sa.

- Co-contamination: not reported
- Further contamination of animal products reported in the present book/article[516]: Cheese (hard cheese), AFM_1, literature[516]

incidence: 50/50*, conc. range: 0.04041–0.13089 µg/kg, ∅ conc.: 0.1032 µg/kg, sample origin: city markets in Şanlıurfa (city) and locally produced dairy products, Turkey., sample year: January–February 2012, country: Turkey[523], *white pickled cheese

- Co-contamination: not reported
- Further contamination of animal products reported in the present book/article[523]: Milk (cow milk), Milk (UHT milk), Yogurt, AFM_1, literature[523]

incidence: 20/25, conc. range: 0.0554–0.3740 µg/kg, sample origin: dairy factories in Tehran (capital), Iran, sample year: winter 2008, country: Iran[524]

incidence: 10/25, conc. range: 0.0409–0.2151 µg/kg, sample origin: dairy factories in Tehran (capital), Iran, sample year: summer 2009, country: Iran[524]

- Co-contamination: not reported
- Further contamination of animal products reported in the present book/article[524]: not reported

incidence: 93/197*, conc. range: 0.02–0.640 µg/kg, ∅ conc.: 0.047 µg/kg, sample origin: control points such as import, primary storage, and market, Cyprus, sample year: 2004–2013, country: Cyprus[542], *Halloumi cheese

- Co-contamination: not reported
- Further contamination of animal products reported in the present book/article[542]: Cheese, Cheese (Anari cheese), Ice cream, Milk, Milk powder, Yogurt, AFM_1, literature[542]

see also **Cheese**

Cheese curd may contain the following mycotoxins:

Aspergillus Toxins

Aflatoxin M_1

incidence: 2/14, conc. range: 0.051–0.127 µg/kg, ∅ conc.: 0.089 µg/kg, sample origin: small food-processing plants in central part of Slovenia, sample year: November 2004–January 2005, country: Slovenia[40]

- Co-contamination: not reported
- Further contamination of animal products reported in the present book/article[40]: Cheese, AFM_1, literature[40]

incidence: 1/1*, conc.: 1.12 µg/kg, sample origin: local dairies, sample year: unknown, country: France/UK[129], *fate of AFM_1 during the manufacture of Camembert cheese

- Co-contamination: not reported
- Further contamination of animal products reported in the present book/article[129]: Cheese (Camembert cheese), Cheese whey, Milk (cow milk), AFM_1, literature[129]

incidence: 74?/82*, conc. range: 0.005–0.010 µg/kg (4 sa.), >0.010–0.020 µg/kg (11 sa.), >0.020–0.050 µg/kg (58 sa.), 0.5223 µg/kg (1 sa.), sample origin: farms in the Castilla-La Mancha region, southeast Spain, sample year: autumn/winter 2007 and 2008, country: Spain[153], *curd from Manchego cheese (ewe milk) production

- Co-contamination: not reported
- Further contamination of animal products reported in the present book/article[153]: Milk (sheep milk), AFM_1, literature[153]

incidence: 25/25*, conc. range: 0.093–0.309 µg/kg, ∅ conc.: 0.180 µg/kg, sample origin: province of the Po valley, Italy, sample year: November–December 2003, country: Italy[484], *made from naturally contaminated cow milk

- Co-contamination: not reported
- Further contamination of animal products reported in the present book/article[484]: Cheese (Grana Padano cheese), Cheese whey, Milk (cow milk), AFM_1, literature[484]

see also **Cheese**

Cheese whey may contain the following mycotoxins:

Aspergillus Toxins

Aflatoxin M_1

incidence: 1/1*, conc.: 0.12 µg/kg, sample origin: local dairies, sample year: unknown, country: France/UK[129], *fate of AFM_1 during the manufacture of Camembert cheese

- Co-contamination: not reported
- Further contamination of animal products reported in the present book/article[129]: Cheese (Camembert cheese), Cheese curd, Milk (cow milk), AFM_1, literature[129]

incidence: 25/25*, conc. range: 0.017–0.061 µg/kg, ∅ conc.: 0.034 µg/kg, sample origin: province of the Po valley, Italy, sample year: November–December 2003, country: Italy[484], *made from naturally contaminated cow milk

- Co-contamination: not reported
- Further contamination of animal products reported in the present book/article[484]: Cheese (Grana Padano cheese), Cheese curd, Milk (cow milk), AFM_1, literature[484]

Chicken gizzard may contain the following mycotoxins:

Aspergillus Toxins

Aflatoxin B_1

incidence: 2/50, conc. range: 0.81–1.34 µg/kg, Ø conc.: 1.07 µg/kg, sample origin: industrial locations of chicken production and slaughtering facilities, Maputo (capital), Mozambique, sample year: May–June 2016, country: Mozambique[517]

incidence: 9/30, conc. range: 0.68–2.12 µg/kg, Ø conc.: 1.04 µg/kg, sample origin: smallholder families or informal producers, Maputo (capital), Mozambique, sample year: May–June 2016, country: Mozambique[517]

- Co-contamination: not reported
- Further contamination of animal products in the present book/article[517]: Chicken liver, AFB_1, literature[517]

Aspergillus and *Penicillium* Toxins

Ochratoxin A

incidence: 15/90, conc. range: 0.25–9.94 µg/kg, Ø conc.: 2.18 µg/kg, sample origin: slaughterhouses in northern and central Serbia, sample year: 2009–2010, country: Serbia[366]

- Co-contamination: not reported
- Further contamination of animal products in the present book/article[366]: Chicken kidney, Chicken liver, Pig kidney, Pig liver, Pig serum, OTA, literature[366]

Fusarium Toxins

Zearalenone

incidence: 20/20, conc. range: 39.9–84.9 µg/kg, Ø conc.: 71.6 µg/kg, sample origin: Zhejiang (province), east China, sample year: 2017–2018, country: China[326]

- Co-contamination: not reported
- Further contamination of animal products reported in the present book/article[326]: Eggs, AFB_1, AFB_2, AFG_2, DON, 3-AcDON, 15-AcDON, and ZEA, literature[326]; Chicken heart, Chicken liver, ZEA, literature[326]; see also Chicken gizzard, Chicken heart, Chicken liver, Eggs, chapter: Further Mycotoxins and Microbial Metabolites

Chicken heart may contain the following mycotoxins:

Fusarium Toxins

Zearalenone

incidence: 20/20, conc. range: 49.3–87.5 µg/kg, Ø conc.: 64.6 µg/kg, sample origin: Zhejiang (province), east China, sample year: 2017–2018, country: China[326]

- Co-contamination: not reported
- Further contamination of animal products reported in the present book/article[326]: Eggs, AFB_1, AFB_2, AFG_2, DON, 3-AcDON, 15-AcDON, and ZEA, literature[326]; Chicken gizzard, Chicken liver, ZEA, literature[326]; see also Chicken gizzard, Chicken heart, Chicken liver, Eggs, chapter: Further Mycotoxins and Microbial Metabolites

Chicken kidney may contain the following mycotoxins:

Aspergillus Toxins

Aflatoxin B_1

incidence: 89/199*, conc. range: 0.02–0.049 µg/kg (40 sa.), 0.05–0.09 µg/kg (24 sa.), 0.1–0.49 µg/kg (23 sa.), 0.5–0.99 µg/kg (2 sa.), sample origin: commercial broiler farms in Faisalabad (city) region, Pakistan, sample year: November 2006–April 2007, country: Pakistan[467], *healthy broiler birds

incidence: 46/65*, conc. range: 0.02–0.049 µg/kg (11 sa.), 0.05–0.09 µg/kg (6 sa.), 0.1–0.49 µg/kg (12 sa.), 0.5–0.99 µg/kg (3 sa.), 1.0–1.99 µg/kg (10 sa.), 2.0–2.99 µg/kg (2 sa.), 3.0–3.99 µg/kg (2 sa.), sample origin: commercial broiler farms in Faisalabad (city) region, Pakistan, sample year: November 2006–April 2007, country: Pakistan[467], *sick broiler birds

- Co-contamination: not reported
- Further contamination of animal products in the present book/article[467]: Chicken kidney, AF/S, literature[467]; Chicken liver, Chicken muscle, AFB_1 and AF/S, literature[467]

Aflatoxin/s

incidence: 89/199*, conc. range: 0.02–0.049 µg/kg (25 sa.), 0.05–0.09 µg/kg (10 sa.), 0.1–0.49 µg/kg (44 sa.), 0.5–0.99 µg/kg (7 sa.), 1.0–1.99 µg/kg (3 sa.), sample origin: commercial broiler farms in Faisalabad (city) region, Pakistan, sample year: November 2006–April 2007, country: Pakistan[467], *healthy broiler birds

incidence: 55/65*, conc. range: 0.02–0.049 µg/kg (6 sa.), 0.05–0.09 µg/kg (10 sa.), 0.1–0.49 µg/kg (16 sa.), 0.5–0.99 µg/kg (4 sa.), 1.0–1.99 µg/kg (7 sa.), 2.0–2.99 µg/kg (3 sa.), 3.0–3.99 µg/kg (4 sa.), 4.0–4.99 µg/kg (4 sa.), 5.0–5.99 µg/kg (1 sa.), sample origin: commercial broiler farms in Faisalabad (city) region, Pakistan, sample year: November 2006–April 2007, country: Pakistan[467], *sick broiler birds

- Co-contamination: not reported
- Further contamination of animal products in the present book/article[467]: Chicken kidney, AFB_1, literature[467]; Chicken liver, Chicken muscle, AFB_1 and AF/S, literature[467]

Aspergillus and *Penicillium* Toxins

Ochratoxin A

incidence: 17/90, conc. range: 0.1–7.02 µg/kg, Ø conc.: 1.19 µg/kg, sample origin: slaughterhouses in northern and central Serbia, sample year: 2009–2010, country: Serbia[366]

- Co-contamination: not reported
- Further contamination of animal products in the present book/article[366]: Chicken gizzard, Chicken liver, Pig kidney, Pig liver, Pig serum, OTA, literature[366]

Chicken liver may contain the following mycotoxins:

Aspergillus Toxins

Aflatoxin B_1

incidence: 122/225, conc. range: 0.003–35.45 µg/kg, sample origin: retail or fresh markets of five regions of Bangkok (capital), Thailand, sample year: March–May, November 1996–January 1997, country: Thailand[189]

- Co-contamination: not reported
- Further contamination of animal products in the present book/article[189]: Chicken muscle, AFB_1, literature[189]

incidence: 36/50, conc. range: 0.30–16.36 μg/kg, ∅ conc.: 1.87 μg/kg, sample origin: shopping center and supermarket in Tabriz (city), Iran, sample year: unknown, country: Iran[397]
- Co-contamination: not reported
- Further contamination of animal products reported in the present book/article[397]: Eggs, AFB_1, literature[397]

incidence: 151/199*, conc. range: 0.02–0.049 μg/kg (50 sa.), 0.05–0.09 μg/kg (42 sa.), 0.1–0.49 μg/kg (49 sa.), 0.5–0.99 μg/kg (5 sa.), 1.0–1.99 μg/kg (4 sa.), 2.0–2.99 μg/kg (1 sa.), sample origin: commercial broiler farms in Faisalabad (city) region, Pakistan, sample year: November 2006–April 2007, country: Pakistan[467], *healthy broiler birds

incidence: 58/65*, conc. range: 0.02–0.049 μg/kg (12 sa.), 0.05–0.09 μg/kg (12 sa.), 0.1–0.49 μg/kg (9 sa.), 0.5–0.99 μg/kg (4 sa.), 1.0–1.99 μg/kg (6 sa.), 2.0–2.99 μg/kg (6 sa.), 3.0–3.99 μg/kg (2 sa.), 4.0–4.99 μg/kg (2 sa.), 5.0–5.99 μg/kg (2 sa.), 6.0–6.99 μg/kg (1 sa.), 7.0–7.99 μg/kg (2 sa.), sample origin: commercial broiler farms in Faisalabad (city) region, Pakistan, sample year: November 2006–April 2007, country: Pakistan[467], *sick broiler birds
- Co-contamination: not reported
- Further contamination of animal products in the present book/article[467]: Chicken kidney, Chicken muscle, AFB_1 and AF/S, literature[467]; Chicken liver, AF/S, literature[467]

incidence: 19/70, conc. range: 0.61–2.48 μg/kg, ∅ conc.: 1.35 μg/kg, sample origin: industrial locations of chicken production and slaughtering facilities, Maputo (capital), Mozambique, sample year: May–June 2016, country: Mozambique[517]

incidence: 20/30, conc. range: 0.57–3.80 μg/kg, ∅ conc.: 1.73 μg/kg, sample origin: smallholder families or informal producers, Maputo (capital), Mozambique, sample year: May–June 2016, country: Mozambique[517]
- Co-contamination: not reported
- Further contamination of animal products in the present book/article[517]: Chicken gizzard, AFB_1, literature[517]

incidence: 1/3, conc.: 0.36 μg/kg, sample origin: local supermarket, China, sample year: unknown, country: China[538]
- Co-contamination: not reported
- Further contamination of animal products reported in the present book/article[538]: Chicken liver, CIT and OTA, literature[538]; Chicken muscle, AFB_1 and CIT, literature[538]; Pig liver, AFB_1, AFB_2, AFM_1, CIT, and PAT, literature[538]; Pig muscle, AFB_1, OTA, and STG, literature[538]

Aflatoxin/s

incidence: 153/199*, conc. range: 0.02–0.049 μg/kg (37 sa.), 0.05–0.09 μg/kg (18 sa.), 0.1–0.49 μg/kg (59 sa.), 0.5–0.99 μg/kg (23 sa.), 1.0–1.99 μg/kg (9 sa.), 2.0–2.99 μg/kg (6 sa.), 3.0–3.99 μg/kg (1 sa.), s sample origin: commercial broiler farms in Faisalabad (city) region, Pakistan, sample year: November 2006–April 2007, country: Pakistan[467], *healthy broiler birds

incidence: 62/65*, conc. range: 0.02–0.049 μg/kg (2 sa.), 0.05–0.09 μg/kg (9 sa.), 0.1–0.49 μg/kg (22 sa.), 0.5–0.99 μg/kg (3 sa.), 1.0–1.99 μg/kg (4 sa.), 2.0–2.99 μg/kg (3 sa.), 3.0–3.99 μg/kg (3 sa.), 4.0–4.99 μg/kg (2 sa.), 5.0–5.99 μg/kg (4 sa.), 6.0–6.99 μg/kg (1 sa.), ≥7.0 μg/kg (2 sa., maximum: 13.83 μg/kg), sample origin: commercial broiler farms in Faisalabad (city) region, Pakistan, sample year: November 2006–April 2007, country: Pakistan[467], *sick broiler birds
- Co-contamination: not reported
- Further contamination of animal products in the present book/article[467]: Chicken kidney, Chicken muscle, AFB_1 and AF/S, literature[467]; Chicken liver, AFB_1, literature[467]

Aspergillus and *Penicillium* Toxins

Citrinin

incidence: 1/3, conc.: 0.89 μg/kg, sample origin: local supermarket, China, sample year: unknown, country: China[538]
- Co-contamination: not reported
- Further contamination of animal products reported in the present book/article[538]: Chicken liver, AFB_1 and OTA, literature[538]; Chicken muscle, AFB_1 and CIT, literature[538]; Pig liver, AFB_1, AFB_2, AFM_1, CIT, and PAT, literature[538]; Pig muscle, AFB_1, OTA, and STG, literature[538]

Ochratoxin A

incidence: 23/90, conc. range: 0.14–3.90 μg/kg, ∅ conc.: 1.62 μg/kg, sample origin: slaughterhouses in northern and central Serbia, sample year: 2009–2010, country: Serbia[366]
- Co-contamination: not reported
- Further contamination of animal products in the present book/article[366]: Chicken gizzard, Chicken kidney, Pig kidney, Pig liver, Pig serum, OTA, literature[366]

incidence: 1/3, conc.: 1.05 μg/kg, sample origin: local supermarket, China, sample year: unknown, country: China[538]
- Co-contamination: not reported
- Further contamination of animal products reported in the present book/article[538]: Chicken liver, AFB_1 and CIT, literature[538]; Chicken muscle, AFB_1 and CIT, literature[538]; Pig liver, AFB_1, AFB_2, AFM_1, CIT, and PAT, literature[538]; Pig muscle, AFB_1, OTA, and STG, literature[538]

Fusarium Toxins

Zearalenone

incidence: 30/30, conc. range: 40.0–74.0 μg/kg, ∅ conc.: 62.9 μg/kg, sample origin: Zhejiang (province), east China, sample year: 2017–2018, country: China[326]
- Co-contamination: not reported
- Further contamination of animal products reported in the present book/article[326]: Eggs, AFB_1, AFB_2, AFG_2, DON, 3-AcDON, 15-AcDON, and ZEA, literature[326]; Chicken gizzard, Chicken heart, ZEA, literature[326]; see also Chicken gizzard, Chicken heart, Chicken liver, Eggs, chapter: Further Mycotoxins and Microbial Metabolites

see also **Chicken meat**

Chicken meat may contain the following mycotoxins:

Aspergillus Toxins

Aflatoxin/s

incidence: 40/115*, conc. range: ≤8.01 μg/kg**, sample origin: markets, slaughterhouses, and shops from central areas in Punjab (region), Pakistan, sample year: December 2011–March 2012, country: Pakistan/Malaysia[460], *included wings, chest, legs, and liver of chicken broilers, chicken layers and domestic chicken, **AFB_1, AFB_2, AFG_1, and AFG_2
- Co-contamination: not reported
- Further contamination of animal products in the present book/article[460]: Chicken meat, OTA and ZEA, literature[460]; Eggs, AF/S, OTA, and ZEA, literature[460]

Aspergillus and *Penicillium* Toxins

Ochratoxin A

incidence: 1/12*, conc.: 0.38 μg/kg, sample origin: 12 regions, Czech Republic, sample year: 2011–2013, country: Czech Republic[101], *fresh meat

- Co-contamination: not reported
- Further contamination of animal products reported in the present book/article[101]: Pig kidney, Pork meat, OTA, literature[101]

incidence: 36/65, conc. range: LOD-0.99 μg/kg (36 sa., maximum: 0.18 μg/kg), sample origin: retail shops, Denmark, sample year: 1995, country: Denmark[102]

- Co-contamination: not reported
- Further contamination of animal products reported in the present book/article[102]: Duck liver, Duck meat, Goose liver, Goose meat, Pork meat, Turkey liver, Turkey meat, OTA, literature[102]

incidence: 47/115*, conc. range: ≤4.70 μg/kg, sample origin: markets, slaughterhouses, and shops from central areas in Punjab (region), Pakistan, sample year: December 2011–March 2012, country: Pakistan/Malaysia[460], *included wings, chest, legs, and liver of chicken broilers, chicken layers and domestic chicken

- Co-contamination: not reported
- Further contamination of animal products in the present book/article[460]: Chicken meat, AF/S and ZEA, literature[460]; Eggs, AF/S, OTA, and ZEA, literature[460]

Fusarium Toxins

Zearalenone

incidence: 60/115*, conc. range: ≤5.10 μg/kg, sample origin: markets, slaughterhouses, and shops from central areas in Punjab (region), Pakistan, sample year: December 2011–March 2012, country: Pakistan/Malaysia[460], *included wings, chest, legs, and liver of chicken broilers, chicken layers and domestic chicken

- Co-contamination: not reported
- Further contamination of animal products in the present book/article[460]: Chicken meat, AF/S and OTA, literature[460]; Eggs, AF/S, OTA, and ZEA, literature[460]

Chicken muscle may contain the following mycotoxins:

Aspergillus Toxins

Aflatoxin B_1

incidence: 97/225, conc. range: 0.024–24.34 μg/kg, sample origin: retail or fresh markets of five regions of Bangkok (capital), Thailand, sample year: March–May, November 1996–January 1997, country: Thailand[189]

- Co-contamination: not reported
- Further contamination of animal products in the present book/article[189]: Chicken liver, AFB_1, literature[189]

incidence: 43/199*, conc. range: 0.02–0.049 μg/kg (29 sa.), 0.05–0.09 μg/kg (11 sa.), 0.1–0.49 μg/kg (3 sa.), sample origin: commercial broiler farms in Faisalabad (city) region, Pakistan, sample year: November 2006–April 2007, country: Pakistan[467], *healthy broiler birds

incidence: 36/65*, conc. range: 0.02–0.049 μg/kg (8 sa.), 0.05–0.09 μg/kg (4 sa.), 0.1–0.49 μg/kg (15 sa.), 0.5–0.99 μg/kg (8 sa.), 1.0–1.99 μg/kg (1 sa.), sample origin: commercial broiler farms in Faisalabad (city) region, Pakistan, sample year: November 2006–April 2007, country: Pakistan[467], *sick broiler birds

- Co-contamination: not reported
- Further contamination of animal products in the present book/article[467]: Chicken kidney, Chicken liver, AFB_1 and AF/S, literature[467]; Chicken muscle, AF/S, literature[467]

incidence: 1/2, conc.: 0.53 μg/kg, sample origin: local supermarket, China, sample year: unknown, country: China[538]

- Co-contamination: not reported
- Further contamination of animal products reported in the present book/article[538]: Chicken liver, AFB_1, CIT, and OTA, literature[538]; Chicken muscle, CIT, literature[538]; Pig liver, AFB_1, AFB_2, AFM_1, CIT, and PAT, literature[538]; Pig muscle, AFB_1, OTA, and STG, literature[538]

Aflatoxin/s

incidence: 46/199*, conc. range: 0.02–0.049 μg/kg (18 sa.), 0.05–0.09 μg/kg (9 sa.), 0.1–0.49 μg/kg (17 sa.), 0.5–0.99 μg/kg (2 sa.), sample origin: commercial broiler farms in Faisalabad (city) region, Pakistan, sample year: November 2006–April 2007, country: Pakistan[467], *healthy broiler birds

incidence: 39/65*, conc. range: 0.02–0.049 μg/kg (7 sa.), 0.05–0.09 μg/kg (3 sa.), 0.1–0.49 μg/kg (13 sa.), 0.5–0.99 μg/kg (7 sa.), 1.0–1.99 μg/kg (6 sa.), 2.0–2.99 μg/kg (3 sa.), sample origin: commercial broiler farms in Faisalabad (city) region, Pakistan, sample year: November 2006–April 2007, country: Pakistan[467], *sick broiler birds

- Co-contamination: not reported
- Further contamination of animal products in the present book/article[467]: Chicken kidney, Chicken liver, AFB_1 and AF/S, literature[467]; Chicken muscle, AFB_1, literature[467]

Aspergillus and *Penicillium* Toxins

Citrinin

incidence: 1/2, conc.: 1.21 μg/kg, sample origin: local supermarket, China, sample year: unknown, country: China[538]

- Co-contamination: not reported
- Further contamination of animal products reported in the present book/article[538]: Chicken liver, AFB_1, CIT, and OTA, literature[538]; Chicken muscle, AFB_1, literature[538]; Pig liver, AFB_1, AFB_2, AFM_1, CIT, and PAT, literature[538]; Pig muscle, AFB_1, OTA, and STG, literature[538]

Ochratoxin A

incidence: 2/50, conc. range: 12.45–21 μg/kg, ∅ conc.: 17 μg/kg, sample origin: markets, slaughterhouses, and supermarkets, Egypt, sample year: unknown, country: Egypt[98]

- Co-contamination: not reported
- Further contamination of animal products in the present book/article[98]: Cow liver, Pig kidney, OTA, literature[98]

Fusarium Toxins

Type A Trichothecenes

T-2 Toxin

incidence: 6/36, conc. range: 0.0704–0.0904 μg/kg, ∅ conc.: 0.0804 μg/kg, sample origin: local markets and supermarkets, Chongqing (city), China, sample year: unknown, country: China[103]

- Co-contamination: not reported
- Further contamination of animal products in the present book/article[103]: Pig fat, DON and T-2, literature[103]; Pig muscle, T-2, literature[103]

Chicken nuggets may contain the following mycotoxins:

Aspergillus **and** ***Penicillium*** **Toxins**

Ochratoxin A

incidence: 1/1, conc.: 0.09 µg/kg, sample origin: supermarkets and fast-food chains in Quebec (city), Canada, sample year: September–October 2008, country: Canada[87]

incidence: 1/1, conc.: 0.04 µg/kg, sample origin: supermarkets and fast-food chains in Calgary (city), Canada, sample year: September–October 2009, country: Canada[87]

- Co-contamination: not reported
- Further contamination of animal products reported in the present book/article[87]: Beefburger, Butter (peanut butter), Chickenburger, Hot-dog, Ice cream, Meat, Meat products, Pork meat, Sausage, OTA, literature[87]

Chickenburger may contain the following mycotoxins:

Aspergillus **and** ***Penicillium*** **Toxins**

Ochratoxin A

incidence: 1/1, conc.: 0.32 µg/kg, sample origin: supermarkets and fast-food chains in Quebec (city), Canada, sample year: September–October 2008, country: Canada[87]

incidence: 1/1, conc.: 0.23 µg/kg, sample origin: supermarkets and fast-food chains in Calgary (city), Canada, sample year: September–October 2009, country: Canada[87]

- Co-contamination: not reported
- Further contamination of animal products reported in the present book/article[87]: Beefburger, Butter (peanut butter), Chicken nuggets, Hot-dog, Ice cream, Meat, Meat products, Pork meat, Sausage, OTA, literature[87]

Cod see Fish

Coppa may contain the following mycotoxins:

Aspergillus **and** ***Penicillium*** **Toxins**

Ochratoxin A

incidence: 5/18, conc. range: ≤0.24 µg/kg, Ø conc.: 0.12 µg/kg, sample origin: retail outlets in the Emilia region, northern Italy, sample year: 2001–2002, country: Italy[152]

- Co-contamination: not reported
- Further contamination of animal products reported in the present book/article[152]: Ham, Pig muscle, Sausage (salami), Sausage (Würstel), OTA, literature[152]

Cow liver may contain the following mycotoxins:

Aspergillus **and** ***Penicillium*** **Toxins**

Ochratoxin A

incidence: 1/50, conc.: 14 µg/kg, sample origin: markets, slaughterhouses, and supermarkets, Egypt, sample year: unknown, country: Egypt[98]

- Co-contamination: not reported
- Further contamination of animal products in the present book/article[98]: Chicken muscle, Pig kidney, OTA, literature[98]

Cream may contain the following mycotoxins:

Aspergillus **Toxins**

Aflatoxin M_1

incidence: 3/7, conc. range: ≤0.003–0.005 µg/kg (3 sa., maximum: 0.005 µg/kg), sample origin: dairies in different areas on Sicily (island, autonomous region), Italy, sample year: January–June 2012, country: Italy[508]

- Co-contamination: not reported
- Further contamination of animal products reported in the present book/article[508]: Cheese, Milk (cow milk), Milk (goat milk), Milk (sheep milk), AFM_1, literature[508]

incidence: 1/4, conc.: 0.10 µg/kg*, sample origin: local stores, USA/Taiwan, sample year: unknown, country: USA/Taiwan[534], *imported table cream sa.

- Co-contamination: not reported
- Further contamination of animal products reported in the present book/article[534]: Milk, Milk (infant formula), AFM_1, literature[534]; Milk powder, AFB_1 and AFB_2, literature[534]

Aflatoxin/s

incidence: 5/5*, conc. range: 0.065–4.905 µg/kg**, Ø conc.: 1.550 µg/kg**, sample origin: groceries and supermarkets, Turkey, sample year: 2003, country: Turkey[95], *Cream Chantilly, **AFS total

- Co-contamination: not reported
- Further contamination of animal products reported in the present book/article[95]: Milk powder, AF/S, literature[95]

For detailed composition see the article.

Cream (sour) may contain the following mycotoxins:

Aspergillus **Toxins**

Aflatoxin M_2

incidence: 3/70* conc. range: ~LOD, sample origin: main convenience stores and supermarkets, Costa Rica, sample year: 2012–2014, country: Costa Rica[411], *sour cream sa. contained 9–18 g/100 g fat

- Co-contamination: not reported
- Further contamination of animal products reported in the present book/article[411]: Cheese (soft cheese), Milk (cow milk), AFM_1, literature[411]

Cremonese sausage see Sausage

Curd see Cheese curd

Custard powder may contain the following mycotoxins:

Aspergillus **Toxins**

Aflatoxin/s

incidence: 2/6, conc. range: 0.17–1.20 µg/kg, Ø conc.: 0.685 µg/kg, sample origin: sales outlets, Qatar, sample year: October 2002, country: Qatar[76], *AFB_1, AFB_2, AFG_1, and AFG_2

- Co-contamination: not reported
- Further contamination of animal products reported in the present book/article[76]: Butter (peanut butter), AF/S, literature[76]; Custard powder, DON, literature[76]

Fusarium Toxins

Type B Trichothecenes

Deoxynivalenol

incidence: 1/6, conc.: 86.43 µg/kg, sample origin: sales outlets, Qatar, sample year: October 2002, country: Qatar[76]

- Co-contamination: not reported
- Further contamination of animal products reported in the present book/article[76]: Butter (peanut butter), Custard powder, AF/S, literarture[76]

Dairy beverages may contain the following mycotoxins:

Aspergillus Toxins

Aflatoxin M_1

incidence: 6/6, conc. range: 0.011–0.050 µg/kg, sample origin: Ribeirão Preto (municipality), São Paulo (state), Brazil, sample year: unknown, country: Brazil/USA[44]

- Co-contamination: not reported
- Further contamination of animal products reported in the present book/article[44]: Cheese (Minas cheese), Cheese (Prato cheese), Yogurt, AFM_1, literature[44]

incidence: 1/8*, conc.: 0.057 µg/l, sample origin: retail shops in Marang and Kuala Terengganu (town and city), Malaysia, sample year: July 2013?, country: Malaysia[163], *milk-based 3-in-1 beverages

incidence: 4/7*, conc. range: 0.0037–0.0619 µg/l, ∅ conc.: 0.0196 µg/l, sample origin: retail shops in Marang and Kuala Terengganu (town and city), Malaysia, sample year: July 2013?, country: Malaysia[163], *cultured milk drink

- Co-contamination: not reported
- Further contamination of animal products reported in the present book/article[163]: Cheese, Milk, Milk (cow milk), Milk powder, Yogurt, AFM_1, literature[163]

incidence: 7/9*, conc. range: 0.011–0.050 µg/kg (7 sa.), sample origin: grocery stores in Ribeirão Preto (municipality), São Paulo (state), Brazil, sample year: 2010, country: Brazil/USA[453], *whole dairy drink

incidence: 3/3*, conc. range: 0.011–0.050 µg/kg (3 sa.), sample origin: grocery stores in Ribeirão Preto (municipality), São Paulo (state), Brazil, sample year: 2010, country: Brazil/USA[453], *semi-skimmed dairy drink

- Co-contamination: not reported
- Further contamination of animal products reported in the present book/article[453]: Cheese (Minas cheese), Yogurt, AFM_1, literature[453]

Dairy desserts may contain the following mycotoxins:

Aspergillus Toxins

Aflatoxin M_1

incidence: 6/10*, conc. range: 0.0025–0.0783 µg/kg, sample origin: Kayseri (city), Turkey, sample year: January–March 2010, country: Turkey/Kyrgyzstan[144], *Keskul (milk pudding containing coconut)

incidence: 5/10*, conc. range: 0.0015–0.043 µg/kg, sample origin: Kayseri (city), Turkey, sample year: January–March 2010, country: Turkey/Kyrgyzstan[144], *Muhallebi (milk pudding containing starch)

incidence: 6/10*, conc. range: 0.0024–0.030 µg/kg, sample origin: Kayseri (city), Turkey, sample year: January–March 2010, country: Turkey/Kyrgyzstan[144], *Kazandibi (milk pudding slightly burned on the bottom)

incidence: 9/10*, conc. range: 0.0018–0.080 µg/kg, sample origin: Kayseri (city), Turkey, sample year: January–March 2010, country: Turkey/Kyrgyzstan[144], *Sutlac (milk pudding containing rice)

- Co-contamination: not reported
- Further contamination of animal products reported in the present book/article[144]: Cheese (Kashar cheese), Cheese (Tulum cheese), Cheese (white cheese), Milk, Yogurt, AFM_1, literature[144]

incidence: 15/21*, conc. range: ≤0.2061 µg/kg, sample origin: patisseries in Burdur (city), Turkey, sample year: unknown, country: Turkey[154], *Muhallebi and Sutlac are traditional Turkish milk pudding desserts

incidence: 10/10*, conc. range: 0.0124–0.3529 µg/kg, ∅ conc.: 0.0835 µg/kg, sample origin: patisseries in Burdur (city), sample year: unknown, country: Turkey[154], *Gullac is a traditional Turkish milk pudding dessert

- Co-contamination: not reported
- Further contamination of animal products reported in the present book/article[154]: Ice cream, AFM_1, literature[154]

Dairy drinks see **Dairy beverages**

Dairy products may contain the following mycotoxins:

Aspergillus Toxins

Aflatoxin B_1

incidence: 1/22, conc.: 6.4 µg/kg, sample origin: local marketing institutions, GDR, sample year: unknown, country: GDR[22]

- Co-contamination: not reported
- Further contamination of animal products reported in the present book/article[22]: Butter (peanut butter), AFB_1, AFB_2, AFG_1, and AFG_2, literature[22]

incidence: 2/23*, conc. range: 10.0–20.0 µg/kg, ∅ conc.: 15.0 µg/kg, sample origin: local market in Karnal (city), India, sample year: September 1974–February 1975, country: India[27], *indigenous dairy products: Barfi, Khoa, and Paneer

- Co-contamination: not reported
- Further contamination of animal products reported in the present book/article[27]: Cheese, AFB_1, AFB_2, and AFG_1, literature[27]; Dairy products, AFG_1, literature[27]; Milk, Milk (buffalo milk), Milk (cow milk), Milk powder, AFM_1, literature[27]

Aflatoxin G_1

incidence: 1/23*, conc.: pr, sample origin: local market in Karnal (city), India, sample year: September 1974–February 1975, country: India[27], *indigenous dairy products: Barfi, Khoa, and Paneer

- Co-contamination: not reported
- Further contamination of animal products reported in the present book/article[27]: Cheese, AFB_1, AFB_2, and AFG_1, literature[2]; Dairy products, AFB_1, literature[27]; Milk, Milk (buffalo milk), Milk (cow milk), Milk powder, AFM_1, literature[27]

Aflatoxin M_1

incidence: 5*/8**, conc. range: ≤0.340 µg/kg, sample origin: retailers and traders in Dagoretti (low-income area), Kenya, sample year: November 2013–October 2014, country: Kenya/Sweden[228], *>0.050 µg/kg, **lala = locally fermented cow milk incidence: 8*/27**, conc. range: ≤0.160 µg/kg, sample origin: supermarket in Westlands (middle- to high-income area), Nairobi (capital), Kenya, sample year: November 2013–October 2014, country: Kenya/Sweden[228], *>0.050 µg/kg, **lala = locally fermented cow milk

- Co-contamination: not reported
- Further contamination of animal products reported in the present book/article[228]: Milk (cow), Milk (UHT milk), Yogurt, AFM_1, literature[228]

incidence: 3/10*, conc. range: 0.6–15 µg/kg**, sample origin: retail market in Anantapur (city), Andhra Pradesh (state), southern India, sample year: 2001–2002, country: India/UK[256], *milk confectionery, **authors pointed out the high levels of AFM_1 in milk (g of dry milk/ml of solution)

- Co-contamination: not reported
- Further contamination of animal products reported in the present book/article[256]: Milk, Milk (buffalo milk), Milk (cow milk), Milk powder, AFM_1, literature[256]

incidence: 66/104*, conc. range: >0–0.16 µg/l (18 sa.), 0.16–0.32 µg/l (6 sa.), 0.32–0.5 µg/l (42 sa.), sample origin: supermarkets in Heilongjiang (province), northeast China, sample year: March–May 2008, country: China/Russia/Korea[283], *including madzoon (fermented yogurt like milk product), pure milk, and milk beverage

- Co-contamination: not reported
- Further contamination of animal products reported in the present book/article[283]: Cheese, Milk (cow milk), Milk powder, AFM_1, literature[283]

incidence: 6/12*, conc. range: 0.013–0.067 µg/l, ∅ conc.: 0.035 µg/l, sample origin: directly from producer or from supermarkets and drugstores, southern Brazil, sample year: unknown, country: Brazil[407], *powdered milk-based products for young children from 1 to 5 years

- Co-contamination: not reported
- Further contamination of animal products reported in the present book/article[407]: Milk (infant formula), AFM_1, literature[407]

incidence: 3/30*, conc. range: 0.002–0.005 µg/kg ∅ conc.: 0.0045 µg/kg, sample origin: local market, China/USA, sample year: unknown, country: China/USA[457], *included milk, milk beverages, and yogurt products

- Co-contamination: 1 sa. co-contaminated with AFM_1, OTA, and OT-α; 2 sa. contaminated solely with AFM_1
- Further contamination of animal products reported in the present book/article[457]: Dairy products, FB_1 and OTA, literature[457]; see also Dairy products, chapter: Further Mycotoxins and Microbial Metabolites

For detailed information see the article.

Aspergillus and *Penicillium* Toxins

Ochratoxin A

incidence: 19/195, conc. range: LOD/LOQ-0.9 µg/kg (19 sa., maximum: 0.8600 µg/kg), sample origin: Germany, sample year: 1995–1998, country: EU[215]

incidence: 15/85, conc. range: LOD/LOQ-0.9 µg/kg (15 sa., maximum: 0.119 µg/kg), sample origin: Norway, sample year: 1998, country: EU[215]

- Co-contamination: not reported
- Further contamination of animal products reported in the present book/article[215]: Meat, Pig kidney, Pig liver, Pork meat, Sausage, OTA, literature[215]

incidence: 1/30*, conc.: 0.010 µg/kg**, sample origin: local market, China/USA, sample year: unknown, country: China/USA[457], included milk* **, milk beverages*, and yogurt products*

- Co-contamination: 1 sa. co-contaminated with AFM_1, OTA, and OT-α
- Further contamination of animal products reported in the present book/article[457]: Dairy products, AFM_1 and OTA, literature[457]; see also Dairy products, chapter: Further Mycotoxins and Microbial Metabolites

For detailed information see the article.

Fusarium Toxins

Fumonisin/s

Fumonisin B_1

incidence: 1/30*, conc.: 0.11 µg/kg**, sample origin: local market, China/USA, sample year: unknown, country: China/USA[457], *included milk, milk beverages**, and yogurt products

- Co-contamination: not reported
- Further contamination of animal products reported in the present book/article[457]: Dairy products, AFM_1 and OTA, literature[457]; see also Dairy products, chapter: Further Mycotoxins and Microbial Metabolites

For detailed information see the article.
see also **Milk**

Desserts see **Dairy desserts**

Dicentrarchus labrax see **Fish**

Doogh see **Ayran**

Dried milk see **Milk powder**

Duck liver may contain the following mycotoxins:

Aspergillus and *Penicillium* Toxins

Ochratoxin A

incidence: 4/7, conc. range: LOD-0.99 µg/kg (4 sa., maximum: 0.16 µg/kg), sample origin: retail shops, Denmark, sample year: 1995, country: Denmark[102]

- Co-contamination: not reported
- Further contamination of animal products reported in the present book/article[102]: Chicken meat, Duck meat, Goose liver, Goose meat, Pork meat, Turkey liver, Turkey meat, OTA, literature[102]

Duck meat may contain the following mycotoxins:

Aspergillus and *Penicillium* Toxins

Ochratoxin A

incidence: 11/19, conc. range: LOD-0.09 µg/kg (11 sa., maximum: 0.09 µg/kg), sample origin: retail shops, Denmark, sample year: 1995, country: Denmark[102]

- Co-contamination: not reported
- Further contamination of animal products reported in the present book/article[102]: Chicken meat, Duck liver, Goose liver, Goose meat, Pork meat, Turkey liver, Turkey meat, OTA, literature[102]

Eggs may contain the following mycotoxins:

Alternaria **Toxins**

Alternariol Methyl Ether

incidence: 2/5*, conc. range: 0.72–1.31 µg/kg, Ø conc.: 1.02 µg/kg, sample origin: vegetable markets, grain shops, farmer's markets, or rural household in 5 provinces (Sixth China Total Diet Study), China, sample year: unknown, country: China[536], *eggs and egg products

- Co-contamination: not reported
- Further contamination of animal products reported in the present book/article[536]: Aquatic food, ALT, AME, BEA, ENB, ENB_1, and TA, literature[536]; Eggs, BEA, ENA, ENA_1, ENB, ENB_1, and TA, literature[536]; Meat, ENB, literature[536]; Milk, BEA, ENB, and TA, literature[536]; see also Aquatic food, Eggs, chapter: Further Mycotoxins and Microbial Metabolites

Tenuazonic Acid

incidence: 4/5*, conc. range: 0.26–27.73 µg/kg, sample origin: vegetable markets, grain shops, farmer's markets, or rural household in 5 provinces (Sixth China Total Diet Study), China, sample year: unknown, country: China[536], *eggs and egg products

- Co-contamination: not reported
- Further contamination of animal products reported in the present book/article[536]: Aquatic food, ALT, AME, BEA, ENB, ENB_1, and TA, literature[536]; Eggs, AME, BEA, ENA, ENA_1, ENB, and ENB_1, literature[536]; Meat, ENB, literature[536]; Milk, BEA, ENB, and TA, literature[536]; see also Aquatic food, Eggs, chapter: Further Mycotoxins and Microbial Metabolites

Aspergillus **Toxins**

Aflatoxin B_1

incidence: 1/7, conc.: <1 µg/kg, sample origin: local supermarkets in Almeria (city), Spain, sample year: unknown, country: Spain[155]

- Co-contamination: 1 sa. co-contaminated with AFB_1 and AFB_2
- Further contamination of animal products reported in the present book/article[155]: Eggs, AFB_2, AFG_1, AFG_2, and BEA, literature[155]

incidence: 1/72, conc.: 168 µg/kg, sample origin: local markets and supermarkets in Jiangsu (province), China, sample year: 2015–2016, country: China[326]

incidence: 2/40, conc. range: 2.60–6.55 µg/kg, Ø conc.: 4.58 µg/kg, sample origin: local markets and supermarkets in Zhejiang (province), China, sample year: 2015–2016, country: China[326]

incidence: 1/40, conc.: 1.46 µg/kg, sample origin: local markets and supermarkets in Shanghai (municipality), China, sample year: 2015–2016, country: China[326]

- Co-contamination: not reported
- Further contamination of animal products reported in the present book/article[326]: Eggs, AFB_2, AFG_2, DON, 3-AcDON, 15-AcDON, and ZEA, literature[326]; Chicken gizzard, Chicken heart, Chicken liver, ZEA, literature[326]; see also Chicken gizzard, Chicken heart, Chicken liver, Eggs, chapter: Further Mycotoxins and Microbial Metabolites

incidence: 29/50*, conc. range: 0.31–1.40 µg/kg, Ø conc.: 0.74 µg/kg, sample origin: shopping center and supermarket in Tabriz (city), Iran, sample year: unknown, country: Iran[397], *farm eggs with brand

incidence: 37/60*, conc. range: 0.30–1.37 µg/kg, Ø conc.: 0.52 µg/kg, sample origin: shopping center and supermarket in Tabriz (city), Iran, sample year: unknown, country: Iran[397], *farm eggs without brand

incidence: 21/40*, conc. range: 0.34–2.35 µg/kg, Ø conc.: 0.94 µg/kg, sample origin: Tabriz (city), Iran, sample year: unknown, country: Iran[397], *domestic eggs

- Co-contamination: not reported
- Further contamination of animal products reported in the present book/article[397]: Chicken liver, AFB_1, literature[397]

incidence: 1/12, conc.: 0.3 µg/kg, sample origin: local supermarkets, China, sample year: unknown, country: China[535]

- Co-contamination: 1 sa. co-contaminated with AFB_1, ZEA, and ß-ZEL
- Further contamination of animal products reported in the present book/article[535]: Eggs, AFG_2, DON, 15-AcDON, and ZEA, literature[535]; see also Eggs, chapter: Further Mycotoxins and Microbial Metabolites

Aflatoxin B_2

incidence: 2/7, conc. range: <1 µg/kg, sample origin: local supermarkets in Almeria (city), Spain, sample year: unknown, country: Spain[155]

- Co-contamination: 1 sa. co-contaminated with AFB_1 and AFB_2; 1 sa. co-contaminated with AFB_2 and AFG_1
- Further contamination of animal products reported in the present book/article[155]: Eggs, AFB_1, AFG_1, AFG_2, and BEA, literature[155]

incidence: 0/72, conc.: no contamination, sample origin: local markets and supermarkets in Jiangsu (province), China, sample year: 2015–2016, country: China[326]

incidence: 1/40, conc.: 1.89 µg/kg, sample origin: local markets and supermarkets in Zhejiang (province), China, sample year: 2015–2016, country: China[326]

incidence: 1/40, conc.: 1.25 µg/kg, sample origin: local markets and supermarkets in Shanghai (municipality), China, sample year: 2015–2016, country: China[326]

- Co-contamination: not reported
- Further contamination of animal products reported in the present book/article[326]: Eggs, AFB_1, AFG_2, DON, 3-AcDON, 15-AcDON, and ZEA, literature[326]; Chicken gizzard, Chicken heart, Chicken liver, ZEA, literature[326]; see also Chicken gizzard, Chicken heart, Chicken liver, Eggs, chapter: Further Mycotoxins and Microbial Metabolites

Aflatoxin G_1

incidence: 3/7, conc. range: <1 µg/kg, sample origin: local supermarkets in Almeria (city), Spain, sample year: unknown, country: Spain[155]

- Co-contamination: 1 sa. co-contaminated with AFB_2 and AFG_1; 1 sa. co-contaminated with AFG_1 and BEA; 1 sa. contaminated solely with AFG_1
- Further contamination of animal products reported in the present book/article[155]: Eggs, AFB_1, AFB_2, AFG_2, and BEA, literature[155]

Aflatoxin G_2

incidence: 1/7, conc.: <2 µg/kg, sample origin: local supermarkets in Almeria (city), Spain, sample year: unknown, country: Spain[155]

- Co-contamination: not reported
- Further contamination of animal products reported in the present book/article[155]: Eggs, AFB_1, AFB_2, AFG_1, and BEA, literature[155]

incidence: 1/72, conc.: 1.47 µg/kg, sample origin: local markets and supermarkets in Jiangsu (province), China, sample year: 2015–2016, country: China[326]

incidence: 0/40, conc.: no contamination, sample origin: local markets and supermarkets in Zhejiang (province), China, sample year: 2015–2016, country: China[326]

incidence: 0/40, conc.: no contamination, sample origin: local markets and supermarkets in Shanghai (municipality), China, sample year: 2015–2016, country: China[326]

- Co-contamination: not reported
- Further contamination of animal products reported in the present book/article[326]: Eggs, AFB_1, AFB_2, DON, 3-AcDON, 15-AcDON, and ZEA, literature[326]; Chicken gizzard, Chicken heart, Chicken liver, ZEA, literature[326]; see also Chicken gizzard, Chicken heart, Chicken liver, Eggs, chapter: Further Mycotoxins and Microbial Metabolites

incidence: 1/12, conc.: <LOQ, sample origin: local supermarkets, China, sample year: unknown, country: China[535]

- Co-contamination: 1 sa. co-contaminated with AFG_2 and ß-ZEL
- Further contamination of animal products reported in the present book/article[535]: Eggs, AFB_1, DON, 15-AcDON, and ZEA, literature[535]; see also Eggs, chapter: Further Mycotoxins and Microbial Metabolites

Aflatoxin/s

incidence: 28/62, conc. range: 0.002–7.2 µg/kg*, sample origin: poultry farms, Cameroon, sample year: rainy season 1991–1995, country: Cameroon[156], *AFB_1, AFB_2, AFB_{2a}, AFG_1, and AFM_1

- Co-contamination: not reported
- Further contamination of human breast milk and animal products reported in the present book/article[156]: Milk (cow milk), Milk (human breast milk), AFM_1, literature[156]

For detailed information see the article.

incidence: 5/40*, conc. range: 0.20–5.80 µg/kg**, sample origin: local markets, Jordan, sample year: January–May 2007, country: Jordan[157], *Hubbard eggs, **AFB_1, AFB_2, AFG_1, AFG_2, and AFM_1

- Co-contamination: not reported
- Further contamination of animal products reported in the present book/article[157]: Beef meat, Meat, Milk, AF/S, literature[157]

For detailed information see the article.

incidence: 22/80*, conc. range: ≤4.46 µg/kg, sample origin: markets, slaughterhouses, and shops from central areas in Punjab (region), Pakistan, sample year: December 2011–March 2012, country: Pakistan/Malaysia[460], *included farm and domestic eggs

- Co-contamination: not reported
- Further contamination of animal products in the present book/article[460]: Chicken meat, AF/S, OTA, and ZEA, literature[460]; Eggs, OTA and ZEA, literature[460]

Sterigmatocystin

incidence: 10/45*, conc. range: 0.5–3608 µg/kg, sample origin: local markets in Zhejiang (province), China, sample year: unknown, country: China[73], *fresh eggs

- Co-contamination: not reported
- Further contamination of animal products in the present book/article[73]: Milk, AFM_1, literature[73]

Aspergillus and *Penicillium* Toxins

Ochratoxin A

incidence: 28/80*, conc. range: ≤2.98 µg/kg, sample origin: markets, slaughterhouses, and shops from central areas in Punjab (region), Pakistan, sample year: December 2011–March 2012, country: Pakistan/Malaysia[460], *included farm and domestic eggs

- Co-contamination: not reported
- Further contamination of animal products in the present book/article[460]: Chicken meat, AF/S, OTA, and ZEA, literature[460]; Eggs, AF/S and ZEA, literature[460]

Fusarium Toxins

Depsipeptides

Beauvericin

incidence: 1/7, conc.: <2 µg/kg, sample origin: local supermarkets in Almeria (city), Spain, sample year: unknown, country: Spain[155]

- Co-contamination: 1 sa. co-contaminated with AFG_1 and BEA
- Further contamination of animal products reported in the present book/article[155]: Eggs, AFB_1, AFB_2, AFG_1, and AFG_2, literature[155]

incidence: ?/3*, conc. range: nd-<1 µg/kg, sample origin: residue control program of Finland, sample year: 2004, country: Finland[158], *organic eggs

incidence: ?/11*, conc. range: nd-<1 µg/kg, sample origin: residue control program of Finland, sample year: 2004, country: Finland[158], *barn eggs

incidence: ?/48*, conc. range: nd-<1 µg/kg, sample origin: residue control program of Finland, sample year: 2004, country: Finland[158], *cage eggs

incidence: ?/32*, conc. range: nd-<1 µg/kg, sample origin: residue control program of Finland, sample year: 2005, country: Finland[158], *cage eggs

incidence: ?/138*, conc. range: nd-1.3 µg/kg, sample origin: local grocery markets, Finland, sample year: 2005, country: Finland[158], *organic egg yolk

incidence: ?/112*, conc. range: nd-<1 µg/kg, sample origin: local grocery markets, Finland, sample year: 2005, country: Finland[158], *barn egg yolk

incidence: ?/117*, conc. range: nd-<1 µg/kg, sample origin: local grocery markets, Finland, sample year: 2005, country: Finland[158], *cage egg yolk

- Co-contamination: not reported
- Further contamination of animal products reported in the present book/article[158]: Eggs, ENA, ENA_1, ENB, and ENB_1, literature[158]

incidence: 3/5*, conc. range: 1.28–6.70 µg/kg, sample origin: vegetable markets, grain shops, farmer's markets, or rural household in 5 provinces (Sixth China Total Diet Study), China, sample year: unknown, country: China[536], *eggs and egg products

- Co-contamination: not reported
- Further contamination of animal products reported in the present book/article[536]: Aquatic food, ALT, AME, BEA, ENB, ENB_1, and TA, literature[536]; Eggs, AME, ENA, ENA_1, ENB, ENB_1, and TA, literature[536]; Meat, ENB, literature[536]; Milk, BEA, ENB, and TA, literature[536]; see also Aquatic food, Eggs, chapter: Further Mycotoxins and Microbial Metabolites

Enniatin A

incidence: ?/138*, conc. range: nd-0.07 µg/kg, sample origin: local grocery markets, Finland, sample year: 2005, country: Finland[158], *organic egg yolk

incidence: ?/112*, conc. range: nd-1.3 µg/kg, sample origin: local grocery markets, Finland, sample year: 2005, country: Finland[158], *barn egg yolk

incidence: ?/117*, conc. range: nd-<0.03 µg/kg, sample origin: local grocery markets, Finland, sample year: 2005, country: Finland[158], *cage egg yolk

- Co-contamination: not reported
- Further contamination of animal products reported in the present book/article[158]: Eggs, BEA, ENA_1, ENB, and ENB_1, literature[158]

incidence: 3/5*, conc. range: 0.20–1.55 µg/kg, sample origin: vegetable markets, grain shops, farmer's markets, or rural household in 5 provinces (Sixth China Total Diet Study), China, sample year: unknown, country: China[536], *eggs and egg products

- Co-contamination: not reported
- Further contamination of animal products reported in the present book/article[536]: Aquatic food, ALT, AME, BEA, ENB, ENB_1, and TA, literature[536]; Eggs, AME, BEA, ENA_1, ENB, ENB_1, and TA, literature[536]; Meat, ENB, literature[536]; Milk, BEA, ENB, and TA, literature[536]; see also Aquatic food, Eggs, chapter: Further Mycotoxins and Microbial Metabolites

Enniatin A_1

incidence: ?/138*, conc. range: nd-<0.42 µg/kg, sample origin: local grocery markets, Finland, sample year: 2005, country: Finland[158], *organic egg yolk

incidence: ?/112*, conc. range: nd-7.5 µg/kg, sample origin: local grocery markets, Finland, sample year: 2005, country: Finland[158], *barn egg yolk

incidence: ?/117*, conc. range: nd-<0.42 µg/kg, sample origin: local grocery markets, Finland, sample year: 2005, country: Finland[158], *cage egg yolk

- Co-contamination: not reported
- Further contamination of animal products reported in the present book/article[158]: Eggs, BEA, ENA, ENB, and ENB_1, literature[158]

incidence: 2/5*, conc. range: 0.30–0.51 µg/kg, ∅ conc.: 0.41 µg/kg, sample origin: vegetable markets, grain shops, farmer's markets, or rural household in 5 provinces (Sixth China Total Diet Study), China, sample year: unknown, country: China[536], *eggs and egg products

- Co-contamination: not reported
- Further contamination of animal products reported in the present book/article[536]: Aquatic food, ALT, AME, BEA, ENB, ENB_1, and TA, literature[536]; Eggs, AME, BEA, ENA, ENB, ENB_1, and TA, literature[536]; Meat, ENB, literature[536]; Milk, BEA, ENB, and TA, literature[536]; see also Aquatic food, Eggs, chapter: Further Mycotoxins and Microbial Metabolites

Enniatin B

incidence: ?/3*, conc. range: nd-0.7 µg/kg, sample origin: residue control program of Finland, sample year: 2004, country: Finland[158], *organic eggs

incidence: ?/3*, conc. range: nd-<0.4 µg/kg, sample origin: residue control program of Finland, sample year: 2005, country: Finland[158], *organic eggs

incidence: ?/11*, conc. range: nd-0.7 µg/kg, sample origin: residue control program of Finland, sample year: 2004, country: Finland[158], *barn eggs

incidence: ?/15*, conc. range: nd-<0.4 µg/kg, sample origin: residue control program of Finland, sample year: 2005, country: Finland[158], *barn eggs

incidence: ?/48*, conc. range: nd-1 µg/kg, sample origin: residue control program of Finland, sample year: 2004, country: Finland[158], *cage eggs

incidence: ?/32*, conc. range: nd-0.5 µg/kg, sample origin: residue control program of Finland, sample year: 2005, country: Finland[158], *cage eggs

incidence: ?/138*, conc. range: nd-1.5 µg/kg, sample origin: local grocery markets, Finland, sample year: 2005, country: Finland[158], *organic egg yolk

incidence: ?/112*, conc. range: nd-3.8 µg/kg, sample origin: local grocery markets, Finland, sample year: 2005, country: Finland[158], *barn egg yolk

incidence: ?/117*, conc. range: nd-1.8 µg/kg, sample origin: local grocery markets, Finland, sample year: 2005, country: Finland[158], *cage egg yolk

- Co-contamination: not reported
- Further contamination of animal products reported in the present book/article[158]: Eggs, BEA, ENA, ENA_1, and ENB_1, literature[158]

incidence: 4/5*, conc. range: 0.15–0.61 µg/kg, sample origin: vegetable markets, grain shops, farmer's markets, or rural household in 5 provinces (Sixth China Total Diet Study), China, sample year: unknown, country: China[536], *eggs and egg products

- Co-contamination: not reported
- Further contamination of animal products reported in the present book/article[536]: Aquatic food, ALT, AME, BEA, ENB, ENB_1, and TA, literature[536]; Eggs, AME, BEA, ENA, ENA_1, ENB_1, and TA, literature[536]; Meat, ENB, literature[536]; Milk, BEA, ENB, and TA, literature[536]; see also Aquatic food, Eggs, chapter: Further Mycotoxins and Microbial Metabolites

Enniatin B_1

incidence: ?/3*, conc. range: nd-<1.12 µg/kg, sample origin: residue control program of Finland, sample year: 2004, country: Finland[158], *organic eggs

incidence: ?/48*, conc. range: nd-<1.12 µg/kg, sample origin: residue control program of Finland, sample year: 2004, country: Finland[158], *cage eggs

incidence: ?/138*, conc. range: nd-<1.12 µg/kg, sample origin: local grocery markets, Finland, sample year: 2005, country: Finland[158], *organic egg yolk

incidence: ?/112*, conc. range: nd-7.0 µg/kg, sample origin: local grocery markets, Finland, sample year: 2005, country: Finland[158], *barn egg yolk

incidence: ?/117*, conc. range: nd-<1.12 µg/kg, sample origin: local grocery markets, Finland, sample year: 2005, country: Finland[158], *cage egg yolk

- Co-contamination: not reported
- Further contamination of animal products reported in the present book/article[158]: Eggs, BEA, ENA, ENA_1, and ENB, literature[158]

incidence: 4/5*, conc. range: 0.18–0.50 µg/kg, sample origin: vegetable markets, grain shops, farmer's markets, or rural household in 5 provinces (Sixth China Total Diet Study), China, sample year: unknown, country: China[536], *eggs and egg products

- Co-contamination: not reported
- Further contamination of animal products reported in the present book/article[536]: Aquatic food, ALT, AME, BEA, ENB, ENB_1, and TA, literature[536]; Eggs, AME, BEA, ENA, ENA_1, ENB, and TA, literature[536]; Meat, ENB, literature[536]; Milk, BEA, ENB, and TA, literature[536]; see also Aquatic food, Eggs, chapter: Further Mycotoxins and Microbial Metabolites

Type B Trichothecenes

Deoxynivalenol

incidence: 22/72, conc. range: 2.01–1600 µg/kg, ∅ conc.: 96.2 µg/kg, sample origin: local markets and supermarkets in Jiangsu (province), China, sample year: 2015–2016, country: China[326]

incidence: ~19/40, conc. range: 4.68–135 µg/kg, ∅ conc.: 44 µg/kg, sample origin: local markets and supermarkets in Zhejiang (province), China, sample year: 2015–2016, country: China[326]

incidence: 20/40, conc. range: <2–88.9 µg/kg, ∅ conc.: 43 µg/kg, sample origin: local markets and supermarkets in Shanghai (municipality), China, sample year: 2015–2016, country: China[326]

- Co-contamination: not reported
- Further contamination of animal products reported in the present book/article[326]: Eggs, AFB_1, AFB_2, AFG_2, 3-AcDON, 15-AcDON, and ZEA, literature[326]; Chicken gizzard, Chicken heart, Chicken liver, ZEA, literature[326]; see also Chicken gizzard, Chicken heart, Chicken liver, Eggs, chapter: Further Mycotoxins and Microbial Metabolites

incidence: 6/12, conc. range: <LOQ-11.3 µg/kg, sample origin: local supermarkets, China, sample year: unknown, country: China[535]

- Co-contamination: 4 sa. co-contaminated with DON and ZEA; 1 sa. co-contaminated with DON, 15-AcDON, and ZEA; 1 sa. contaminated solely with DON
- Further contamination of animal products reported in the present book/article[535]: Eggs, AFB_1, AFG_2, 15-AcDON, and ZEA, literature[535]; see also Eggs, chapter: Further Mycotoxins and Microbial Metabolites

3-Acetyldeoxynivalenol

incidence: 3/72, conc. range: 16.8–89.1 µg/kg, ∅ conc.: 42.9 µg/kg, sample origin: local markets and supermarkets in Jiangsu (province), China, sample year: 2015–2016, country: China[326]

incidence: 0/40, conc.: no contamination, sample origin: local markets and supermarkets in Zhejiang (province), China, sample year: 2015–2016, country: China[326]

incidence: 1/40, conc.: 5.94 µg/kg, sample origin: local markets and supermarkets in Shanghai (municipality), China, sample year: 2015–2016, country: China[326]

- Co-contamination: not reported
- Further contamination of animal products reported in the present book/article[326]: Eggs, AFB_1, AFB_2, AFG_2, DON, 15-AcDON, and ZEA, literature[326]; Chicken gizzard, Chicken heart, Chicken liver, ZEA, literature[326]; see also Chicken gizzard, Chicken heart, Chicken liver, Eggs, chapter: Further Mycotoxins and Microbial Metabolites

15-Acetyldeoxynivalenol

incidence: 25/72, conc. range: <5–664 µg/kg, ∅ conc.: 43.6 µg/kg, sample origin: local markets and supermarkets in Jiangsu (province), China, sample year: 2015–2016, country: China[326]

incidence: ~18/40, conc. range: <5–152 µg/kg, ∅ conc.: 29.5 µg/kg, sample origin: local markets and supermarkets in Zhejiang (province), China, sample year: 2015–2016, country: China[326]

incidence: 19/40, conc. range: 12.7–155 µg/kg, ∅ conc.: 62.4 µg/kg, sample origin: local markets and supermarkets in Shanghai (municipality), China, sample year: 2015–2016, country: China[326]

- Co-contamination: not reported
- Further contamination of animal products reported in the present book/article[326]: Eggs, AFB_1, AFB_2, AFG_2, DON, 3-AcDON, and ZEA, literature[326]; Chicken gizzard, Chicken heart, Chicken liver, ZEA, literature[326]; see also Chicken gizzard, Chicken heart, Chicken liver, Eggs, chapter: Further Mycotoxins and Microbial Metabolites

incidence: 1/12, conc.: 9.6 µg/kg, sample origin: local supermarkets, China, sample year: unknown, country: China[535]

- Co-contamination: not reported
- Further contamination of animal products reported in the present book/article[535]: Eggs, AFB_1, AFG_2, DON, and ZEA, literature[535]; see also Eggs, chapter: Further Mycotoxins and Microbial Metabolites

Other ***Fusarium*** Toxins

Zearalenone

incidence: 32/72, conc. range: 0.30–418 µg/kg, ∅ conc.: 29.1 µg/kg, sample origin: local markets and supermarkets in Jiangsu (province), China, sample year: 2015–2016, country: China[326]

incidence: ~20/40, conc. range: 0.25–986 µg/kg, ∅ conc.: 29.7 µg/kg, sample origin: local markets and supermarkets in Zhejiang (province), China, sample year: 2015–2016, country: China[326]

incidence: 18/40, conc. range: 1.54–390 µg/kg, ∅ conc.: 33.3 µg/kg, sample origin: local markets and supermarkets in Shanghai (municipality), China, sample year: 2015–2016, country: China[326]

- Co-contamination: not reported
- Further contamination of animal products reported in the present book/article[326]: Eggs, AFB_1, AFB_2, AFG_2, DON, 3-AcDON, and 15-AcDON, literature[326]; Chicken gizzard, Chicken heart, Chicken liver, ZEA, literature[326]; see also Chicken gizzard, Chicken heart, Chicken liver, Eggs, chapter: Further Mycotoxins and Microbial Metabolites

incidence: 26/80*, conc. range: ≤3.60 µg/kg, sample origin: markets, slaughterhouses, and shops from central areas in Punjab (region), Pakistan, sample year: December 2011–March 2012, country: Pakistan/Malaysia[460], *included farm and domestic eggs

- Co-contamination: not reported
- Further contamination of animal products in the present book/article[460]: Chicken meat, AF/S, OTA, and ZEA, literature[460]; Eggs, AF/S and OTA, literature[460]

incidence: 7/12, conc. range: 0.9–21.6 µg/kg, ∅ conc.: 7.4 µg/kg, sample origin: local supermarkets, China, sample year: unknown, country: China[535]

- Co-contamination: 4 sa. co-contaminated with DON and ZEA; 1 sa. co-contaminated with DON, 15-AcDON, and ZEA; 1 sa. co-contaminated with AFB_1, ZEA, and ß-ZEL; 1 sa. contaminated solely with ZEA
- Further contamination of animal products reported in the present book/article[535]: Eggs, AFB_1, AFG_2, DON, and 15-AcDON, literature[535]; see also Eggs, chapter: Further Mycotoxins and Microbial Metabolites

***Ethomolosa* sp.** see Fish

Fish may contain the following mycotoxins:

Aspergillus Toxins

Aflatoxin B$_1$

incidence: 5/20*, conc. range: tr-moderately high, sample origin: local markets and homes in Njala (town), Sierra Leone, sample year: unknown, country: Sierra Leone[159], *moldy smoke-dried fish ("Bonga" = *Ethomolosa* sp.)

- Co-contamination: 2 sa. co-contaminated with AFB_1, AFG_1, and OTA; 1 sa. co-contaminated with AFB_1, AFG_1, AFG_2, and OTA; 1 sa. co-contaminated with AFB_1, AFG_1, and AFG_2; 1 sa. co-contaminated with AFB_1 and AFG_1
- Further contamination of animal products reported in the present book/article[159]: Fish, AFG_1, AFG_2, and OTA, literature[159]

Aflatoxin B$_2$

incidence: 1/7*, conc.: 1.2 µg/kg, sample origin: supermarkets in Shanghai (city), China, sample year: unknown, country: China/Belgium[522], *muscle of fresh fish

- Co-contamination: not reported
- Further contamination of animal products reported in the present book/article[522]: Fish, Seafood, OTA and ZEA, literature[522]

Aflatoxin G$_1$

incidence: 6/20*, conc. range: tr-moderately high, sample origin: local markets and homes in Njala (town), Sierra Leone, sample year: unknown, country: Sierra Leone[159], *moldy smoke-dried fish ("Bonga" = *Ethomolosa* sp.)

- Co-contamination: 2 sa. co-contaminated with AFB_1, AFG_1, and OTA; 1 sa. co-contaminated with AFB_1, AFG_1, AFG_2, and OTA; 1 sa. co-contaminated with AFB_1, AFG_1, and AFG_2; 1 sa. co-contaminated with AFB_1 and AFG_1; 1 sa. contaminated solely with AFG_1
- Further contamination of animal products reported in the present book/article[159]: Fish, AFB_1, AFG_2, and OTA, literature[159]

Aflatoxin G$_2$

incidence: 2/20*, conc. range: tr-medium, sample origin: local markets and homes in Njala (town), Sierra Leone, sample year: unknown, country: Sierra Leone[159], *moldy smoke-dried fish ("Bonga" = *Ethomolosa* sp.)

- Co-contamination: 1 sa. co-contaminated with AFB_1, AFG_1, AFG_2, and OTA; 1 sa. co-contaminated with AFB_1, AFG_1, and AFG_2
- Further contamination of animal products reported in the present book/article[159]: Fish, AFB_1, AFG_1, and OTA, literature[159]

Aflatoxin/s

incidence: 10/25*, conc. range: >9.9–20.4 µg/kg, sample origin: markets in nine districts, Zambia, sample year: 2016–2017, country: Zambia/Nigeria/USA[338], **Petrocephalus*, **AFB_1, AFB_2, AFG_1, and AFG_2

incidence: 6/35*, conc. range: >9.9–17.2 µg/kg, sample origin: markets in nine districts, Zambia, sample year: 2016–2017, country: Zambia/Nigeria/USA[338], **Limnothrissa*, **AFB_1, AFB_2, AFG_1, and AFG_2

- Co-contamination: not reported
- Further contamination of animal products reported in the present book/article[338]: Insects, AF/S, literature[338]

For detailed information see the article.

Aspergillus* and *Penicillium Toxins

Ochratoxin A

incidence: 3/20*, conc. range: tr-moderately high, sample origin: local markets and homes in Njala (town), Sierra Leone, sample year: unknown, country: Sierra Leone[159], *moldy smoke-dried fish ("Bonga" = *Ethomolosa* sp.)

- Co-contamination: 2 sa. co-contaminated with AFB_1, AFG_1, and OTA; 1 sa. co-contaminated with AFB_1, AFG_1, AFG_2, and OTA
- Further contamination of animal products reported in the present book/article[159]: Fish, AFB_1, AFG_1, and AFG_2, literature[159]

incidence: 2/7*, conc. range: 0.5–1.4 µg/kg, ∅ conc.: 0.95 µg/kg, sample origin: supermarkets in Shanghai (city), China, sample year: unknown, country: China/Belgium[522], *muscle of fresh fish

- Co-contamination: not reported
- Further contamination of animal products reported in the present book/article[522]: Fish, AFB_2 and ZEA, literature[522]; Seafood, OTA and ZEA, literature[522]

Fusarium Toxins

Depsipeptides

Enniatin A$_1$

incidence: 5/10*, conc. range: 1.7–6.9 µg/kg, ∅ conc.: 4.3 µg/kg, sample origin: produced in hatcheries and purchased from supermarkets in Valencia (city), Spain, sample year: unknown, country: Spain[527], *muscle of sea bass (*Dicentrarchus labrax*)

incidence: 3/10*, conc. range: 2.1–7.5 µg/kg, ∅ conc.: 4.0 µg/kg, sample origin: produced in hatcheries and purchased from supermarkets in Valencia (city), Spain, sample year: unknown, country: Spain[527], *muscle of sea bream (*Sparus aurata*)

- Co-contamination: not reported
- Further contamination of animal products reported in the present book/article[527]: Fish, ENB and ENB_1, literature[527]

incidence: 1/1*, conc.: 1.51 µg/kg, sample origin: imported from China and purchased from supermarkets in Valencia (city), Spain, sample year: unknown, country: Spain[528], **Tilapia*

incidence: 0/1*, conc.: no contamination, sample origin: imported from Vietnam and purchased from supermarkets in Valencia (city), Spain, sample year: unknown, country: Spain[528], **Panga*

incidence: 0/1*, conc.: no contamination, sample origin: fished in the northwest Atlantic and purchased from supermarkets in Valencia (city), Spain, sample year: unknown, country: Spain[528], *mackerel incidence: 0/1*, conc.: no contamination, sample origin: fished in the southeast Atlantic and purchased from supermarkets in Valencia (city), Spain, sample year: unknown, country: Spain[528], *hake

incidence: 0/1*, conc.: no contamination, sample origin: fished in the northwest Atlantic and purchased from supermarkets in Valencia (city), Spain, sample year: unknown, country: Spain[528], *cod

- Co-contamination: not reported
- Further contamination of animal products reported in the present book/article[528]: Fish, ENB and ENB_1, literature[528]

Enniatin B

incidence: 1/10*, conc.: 7.0 µg/kg, sample origin: markets in Valencia (city), Spain, sample year: unknown, country: Spain[489], *Gula substitute (fish protein)

- Co-contamination: 1 sa. co-contaminated with FUS-X and ENB
- Further contamination of animal products reported in the present book/article[489]: Fish, FUS-X, literature[489]

For detailed information see the article.

incidence: 9/10*, conc. range: 1.3–44.6 µg/kg, Ø conc.: 12.8 µg/kg, sample origin: produced in hatcheries and purchased from supermarkets in Valencia (city), Spain, sample year: unknown, country: Spain[527], *muscle of sea bass (*Dicentrarchus labrax*)

incidence: 4/10*, conc. range: 1.3–21.6 µg/kg, Ø conc.: 14.9 µg/kg, sample origin: produced in hatcheries and purchased from supermarkets in Valencia (city), Spain, sample year: unknown, country: Spain[527], *muscle of sea bream (*Sparus aurata*)

- Co-contamination: not reported
- Further contamination of animal products reported in the present book/article[527]: Fish, ENA_1 and ENB_1, literature[527]

incidence: 1/1*, conc.: 5.35 µg/kg, sample origin: imported from China and purchased from supermarkets in Valencia (city), Spain, sample year: unknown, country: Spain[528], **Tilapia*

incidence: 1/1*, conc.: 1.26 µg/kg, sample origin: imported from Vietnam and purchased from supermarkets in Valencia (city), Spain, sample year: unknown, country: Spain[528], **Panga*

incidence: 0/1*, conc.: no contamination, sample origin: fished in the northwest Atlantic and purchased from supermarkets in Valencia (city), Spain, sample year: unknown, country: Spain[528], *mackerel

incidence: 0/1*, conc.: no contamination, sample origin: fished in the southeast Atlantic and purchased from supermarkets in Valencia (city), Spain, sample year: unknown, country: Spain[528], *hake

incidence: 0/1*, conc.: no contamination, sample origin: fished in the northwest Atlantic and purchased from supermarkets in Valencia (city), Spain, sample year: unknown, country: Spain[528], *cod

- Co-contamination: not reported
- Further contamination of animal products reported in the present book/article[528]: Fish, ENA_1 and ENB_1, literature[528]

Enniatin B_1

incidence: 7/10*, conc. range: 1.4–31.5 µg/kg, Ø conc.: 10.2 µg/kg, sample origin: produced in hatcheries and purchased from supermarkets in Valencia (city), Spain, sample year: unknown, country: Spain[527], *muscle of sea bass (*Dicentrarchus labrax*)

incidence: 3/10*, conc. range: 7.1–19.0 µg/kg, Ø conc.: 12.7 µg/kg, sample origin: produced in hatcheries and purchased from supermarkets in Valencia (city), Spain, sample year: unknown, country: Spain[527], *muscle of sea bream (*Sparus aurata*)

- Co-contamination: not reported
- Further contamination of animal products reported in the present book/article[527]: Fish, ENA_1 and ENB, literature[527]

incidence: 1/1*, conc.: 2.20 µg/kg, sample origin: imported from China and purchased from supermarkets in Valencia (city), Spain, sample year: unknown, country: Spain[528], **Tilapia*

incidence: 0/1*, conc.: no contamination, sample origin: imported from Vietnam and purchased from supermarkets in Valencia (city), Spain, sample year: unknown, country: Spain[528], **Panga*

incidence: 0/1*, conc.: no contamination, sample origin: fished in the northwest Atlantic and purchased from supermarkets in Valencia (city), Spain, sample year: unknown, country: Spain[528], *mackerel

incidence: 0/1*, conc.: no contamination, sample origin: fished in the southeast Atlantic and purchased from supermarkets in Valencia (city), Spain, sample year: unknown, country: Spain[528], *hake

incidence: 0/1*, conc.: no contamination, sample origin: fished in the northwest Atlantic and purchased from supermarkets in Valencia (city), Spain, sample year: unknown, country: Spain[528], *cod

- Co-contamination: not reported
- Further contamination of animal products reported in the present book/article[528]: Fish, ENA_1 and ENB, literature[528]

Type B Trichothecenes

Fusarenon-X (4-Acetylnivalenol)

incidence: 2/10*, conc. range: 4.0 µg/kg, Ø conc.: 4.0 µg/kg, sample origin: markets in Valencia (city), Spain, sample year: unknown, country: Spain[489], *Gula substitute (fish protein)

- Co-contamination: 1 sa. co-contaminated with FUS-X and ENB; 1 sa. contaminated solely with FUS-X
- Further contamination of animal products reported in the present book/article[489]: Fish, ENB, literature[489]

For detailed information see the article.

Other ***Fusarium*** Toxins

Zearalenone

incidence: 2/7*, conc. range: 11.2–14.8 µg/kg, Ø conc.: 13.0 µg/kg, sample origin: supermarkets in Shanghai (city), China, sample year: unknown, country: China/Belgium[522], *muscle of fresh fish

- Co-contamination: not reported
- Further contamination of animal products reported in the present book/article[522]: Fish, AFB_2 and OTA, literature[522]; Seafood, OTA and ZEA, literature[522]

see also **Meat**

Fish oil may contain the following mycotoxins:

Aspergillus Toxins

Aflatoxin B_1

incidence: 1/1, conc.: 0.25 µg/kg, sample origin: local markets in Baoding, Shijiazhuang, and Tangshan (areas), Hebei (province), China, sample year: July–August 2011, country: China[109]

- Co-contamination: not reported
- Further contamination of animal products reported in the present book/article[109]: not reported

Fish products may contain the following mycotoxins:

Aspergillus Toxins

Aflatoxin/s
incidence: 73/107, Ø conc.: 3.8 µg/kg*, sample origin: greater Manila (capital) area, Philippines, sample year: unknown, country: Philippines[70], *AF of pos. sa.?

- Co-contamination: not reported
- Further contamination of animal products reported in the present book/article[70]: Butter (peanut butter), Meat products, AF/S, literature[70]

Fish protein see **Fish**

Fish/shrimp may contain the following mycotoxins:

Aspergillus Toxins

Aflatoxin/s
incidence: 7/139*, conc. range: ≤772 µg/kg**, Ø conc.: 166 µg/kg**, sample origin: market retailers, street vendors, wholesalers, and distributors, Thailand, sample year: 1967–1969, country: USA/Hong Kong/Thailand[162], *dried fish/shrimp, **AFB_1, AFB_2, AFG_1, and AFG_2

- Co-contamination: not reported
- Further contamination of animal products reported in the present book/article[162]: not reported

Food may contain the following mycotoxins:

Aspergillus Toxins

Aflatoxin M_1
incidence: 17/17*, conc. range: 0.077–0.844 µg/kg, Ø conc.: 0.350 µg/kg, sample origin: Lucknow (city), India, sample year: unknown, country: India[91], *infant milk food

- Co-contamination: not reported
- Further contamination of animal products reported in the present book/article[91]: Milk (cow milk), Milk (infant formula), AFM_1, literature[91]

Goose liver may contain the following mycotoxins:

Aspergillus and ***Penicillium*** Toxins

Ochratoxin A
incidence: 4/12, conc. range: LOD-0.99 µg/kg (4 sa., maximum: 0.06 µg/kg), sample origin: retail shops, Denmark, sample year: 1995, country: Denmark[102]

- Co-contamination: not reported
- Further contamination of animal products reported in the present book/article[102]: Chicken meat, Duck liver, Duck meat, Goose meat, Pork meat, Turkey liver, Turkey meat, OTA, literature[102]

Goose meat may contain the following mycotoxins:

Aspergillus and ***Penicillium*** Toxins

Ochratoxin A
incidence: 5/12, conc. range: LOD-0.99 µg/kg (5 sa., maximum: 0.10 µg/kg), sample origin: retail shops, Denmark, sample year: 1995, country: Denmark[102]

- Co-contamination: not reported
- Further contamination of animal products reported in the present book/article[102]: Chicken meat, Duck liver, Duck meat, Goose liver, Pork meat, Turkey liver, Turkey meat, OTA, literature[102]

Gula see **Fish**

Gullac see **Dairy desserts**

Hake see **Fish**

Ham may contain the following mycotoxins:

Aspergillus Toxins

Aflatoxin B_1
incidence: 2/25*, conc. range: 0.95–1.06 µg/kg, Ø conc.: 1.01 µg/kg, sample origin: markets, fairs, or directly from the producing household, Primorje, Zagorje, Medimurje, Slavonia, and Baranja (regions), Croatia, sample year: 2011–2014, country: Croatia/Italy[500], *Istrian prosciutto sa.

incidence: 2/27*, conc. range: 0.92–0.96 µg/kg, Ø conc.: 0.94 µg/kg, sample origin: markets, fairs, or directly from the producing household, Primorje, Zagorje, Medimurje, Slavonia, and Baranja (regions), Croatia, sample year: 2011–2014, country: Croatia/Italy[500], *Dalmatian prosciutto sa.

incidence: 2/44*, conc. range: 0.89–0.97 µg/kg, Ø conc.: 0.93 µg/kg, sample origin: markets, fairs, or directly from the producing household, Primorje, Zagorje, Medimurje, Slavonia, and Baranja (regions), Croatia, sample year: 2011–2014, country: Croatia/Italy[500], *Slavonian ham sa.

- Co-contamination: not reported
- Further contamination of animal products reported in the present book/article[500]: Bacon, Ham, OTA, literature[500]; Sausage, Sausage (liver sausage), AFB_1 and OTA, literature[500]

Aspergillus and ***Penicillium*** Toxins

Ochratoxin A
incidence: 12/30*, conc. range: ≤28.42 µg/kg, Ø conc.: 4.06 µg/kg, sample origin: retail outlets in the Emilia region, northern Italy, sample year: 2001–2002, country: Italy[152], *dry-cured ham sa.

incidence: 1/12*, conc.: 0.05 µg/kg, sample origin: retail outlets in the Emilia region, northern Italy, sample year: 2001–2002, country: Italy[152], *cooked ham sa.

- Co-contamination: not reported
- Further contamination of animal products reported in the present book/article[152]: Coppa, Pig muscle, Sausage (salami), Sausage (Würstel), OTA, literature[152]

incidence: 17/21*, conc. range: 0.04–0.70 µg/kg (3 sa.), 0.70–1.00 µg/kg (6 sa.), >1.00–2.00 µg/kg (7 sa.), 2.20 µg/kg (1 sa.), sample origin: manufactures, Italy, sample year: unknown, country: Italy[165], *ham sa. aged 6 months (lean muscle part)

incidence: 18/21*, conc. range: 0.04–0.70 µg/kg (6 sa.), 0.70–1.00 µg/kg (5 sa.), >1.00–2.00 µg/kg (7 sa.), sample origin: manufactures, Italy, sample year: unknown, country: Italy[165], *ham sa. aged 12 months (lean muscle part)

- Co-contamination: not reported
- Further contamination of animal products reported in the present book/article[165]: not reported

incidence: 2/5*, conc. range: 0.28–1.52 μg/kg, ∅ conc.: 0.90 μg/kg, sample origin: market, Italy, sample year: unknown, country: Italy[166], *dry-cured ham (inner part)

incidence: 3/5*, conc. range: 0.11–7.28 μg/kg, ∅ conc.: 2.67 μg/kg, sample origin: market, Italy, sample year: unknown, country: Italy[166], *dry-cured ham (outer part, 10-mm thickness)

incidence: 0/5*, conc.: no contamination, sample origin: market, Italy, sample year: unknown, country: Italy[166], *smoked ham (inner part)

incidence: 2/5*, conc. range: 5.20–6.20 μg/kg, ∅ conc.: 5.70 μg/kg, sample origin: market, Italy, sample year: unknown, country: Italy[166], *smoked ham (outer part, 10-mm thickness)

- Co-contamination: not reported
- Further contamination of animal products reported in the present book/article[166]: not reported

incidence: 40/40*, conc. range: 0.61–4.11 μg/kg, ∅ conc.: 1.86 μg/kg, sample origin: slaughterhouses and manufacturing plants, northern Italy, sample year: 2007–2010, country: Italy[167], *dry-cured ham

- Co-contamination: not reported
- Further contamination of animal products reported in the present book/article[167]: Pig muscle, OTA, literature[167]

incidence: 10/20*, conc. range: 2.0–160.9 μg/kg, ∅ conc.: 27.1 μg/kg, sample origin: company, Spain, sample year: unknown, country: Spain[168], *dry-cured Iberian ham after 6 months of ripening (superficial portion, up to 0.5 cm from the surface)

incidence: 3/20*, conc. range: 2.0–28.4 μg/kg, ∅ conc.: 16.3 μg/kg, sample origin: company, Spain, sample year: unknown, country: Spain[168], *dry-cured Iberian ham after 6 months of ripening (deep portion, 0.5–1.0 cm beneath the surface)

- Co-contamination: not reported
- Further contamination of animal products reported in the present book/article[168]: not reported

incidence: 5/25*, conc. range: 0.98–9.42 μg/kg, sample origin: markets, fairs, or directly from the producing household, Primorje, Zagorje, Medimurje, Slavonia, and Baranja (regions), Croatia, sample year: 2011–2014, country: Croatia/Italy[500], *Istrian prosciutto sa.

incidence: 4/27*, conc. range: 1.56–3.16 μg/kg, sample origin: markets, fairs, or directly from the producing household, Primorje, Zagorje, Medimurje, Slavonia, and Baranja (regions), Croatia, sample year: 2011–2014, country: Croatia/Italy[500], *Dalmatian prosciutto sa.

incidence: 6/44*, conc. range: 0.97–1.29 μg/kg, sample origin: markets, fairs, or directly from the producing household, Primorje, Zagorje, Medimurje, Slavonia, and Baranja (regions), Croatia, sample year: 2011–2014, country: Croatia/Italy[500], *Slavonian ham sa.

incidence: 3/9*, conc. range: 2.23–9.95 μg/kg, sample origin: markets, fairs, or directly from the producing household, Primorje, Zagorje, Medimurje, Slavonia, and Baranja (regions), Croatia, sample year: 2011–2014, country: Croatia/Italy[500], *other prosciutto sa.

- Co-contamination: not reported
- Further contamination of animal products reported in the present book/article[500]: Bacon, OTA, literature[500]; Ham, AFB_1, literature[500]; Sausage, Sausage (liver sausage), AFB_1 and OTA, literature[500]

Hot-dog may contain the following mycotoxins:

Aspergillus Toxins

Aflatoxin B_1

incidence: 1/25, conc.: 5 μg/kg, sample origin: local companies in Cairo (capital), Egypt, sample year: unknown, country: Egypt[2]

- Co-contamination: 1 sa. co-contaminated with AFB1 and AFB_2
- Further contamination of animal products in the present book/article[2]: Beefburger, AFB_1, literature[2]; Hot-dog, AFB_2, literature[2]; Kubeba, Meat, Sausage, AFB_1 and AFB_2, literature[2]

Aflatoxin B_2

incidence: 1/25, conc.: 2 μg/kg, sample origin: local companies in Cairo (capital), Egypt, sample year: unknown, country: Egypt[2]

- Co-contamination: 1 sa. co-contaminated with AFB1 and AFB_2
- Further contamination of animal products in the present book/article[2]: Beefburger, Hot-dog, AFB_1, literature[2]; Kubeba, Meat, Sausage, AFB_1 and AFB_2, literature[2]

Aspergillus and *Penicillium* Toxins

Ochratoxin A

incidence: 1/1, conc.: 0.38 μg/kg, sample origin: supermarkets and fast-food chains in Quebec (city), Canada, sample year: September–October 2008, country: Canada[87]

incidence: 1/1, conc.: 0.41 μg/kg, sample origin: supermarkets and fast-food chains in Calgary (city), Canada, sample year: September–October 2009, country: Canada[87]

- Co-contamination: not reported
- Further contamination of animal products reported in the present book/article[87]: Beefburger, Butter (peanut butter), Chicken nuggets, Chickenburger, Ice cream, Meat, Meat products, Pork meat, Sausage, literature[87]

Ice cream may contain the following mycotoxins:

Aspergillus Toxins

Aflatoxin M_1

incidence: 6/16, conc. range: ≤0.02618 μg/kg, sample origin: patisseries in Burdur (city), sample year: unknown, country: Turkey[154]

- Co-contamination: not reported
- Further contamination of animal products reported in the present book/article[154]: Dairy desserts, AFM_1, literature[154]

incidence: 34/45, conc. range: 0.005–0.010 μg/kg (2 sa.), 0.011–0.025 μg/kg (13 sa.), 0.026–0.050 μg/kg (14 sa.), 0.051–0.100 μg/kg (4 sa.), 0.136 μg/kg (1 sa.), sample origin: various districts in Burdur (city), Turkey, sample year: 2008, country: Turkey[418]

- Co-contamination: not reported
- Further contamination of animal products reported in the present book/article[418]: Butter, Cheese (white cheese), Milk (cow milk), Milk powder, Yogurt, AFM_1, literature[418]

incidence: 10/45, conc. range: 0.0012–0.103 μg/kg, sample origin: supermarkets in Babol (city), northern Iran, sample year: autumn 2010, country: Iran[419]

- Co-contamination: not reported
- Further contamination of animal products reported in the present book/article[419]: not reported

For detailed information see the article.

incidence: 34/60, conc. range: 0.0149–0.1474 µg/l, sample origin: retail markets in Isfahan (province), Iran, sample year: February 2011–2012, country: Iran[424]

- Co-contamination: not reported
- Further contamination of animal products reported in the present book/article[424]: Cheese, Yogurt, AFM_1, literature[424]

For detailed information see the article.

incidence: 25/36, conc. range: 0.015–0.132 µg/kg, sample origin: retail stores and supermarkets in Esfahan, Shiraz, and Yazd (cities), as well as Tehran (capital), Iran, sample year: winter and summer 2009, country: Iran[435]

- Co-contamination: not reported
- Further contamination of animal products reported in the present book/article[435]: Butter, Cheese (white cheese), Milk, Yogurt, AFM_1, literature[435]

For detailed information see the article.

incidence: 10/94, conc. range: 0.002–0.033 µg/kg, ∅ conc.: 0.02 µg/kg, sample origin: control points such as import, primary storage, and market, Cyprus, sample year: 2004–2013, country: Cyprus[542]

- Co-contamination: not reported
- Further contamination of animal products reported in the present book/article[542]: Cheese, Cheese (Anari cheese), Cheese (white cheese), Milk, Milk powder, Yogurt, AFM_1, literature[542]

Aspergillus and ***Penicillium*** Toxins

Ochratoxin A

incidence: 1/1, conc.: 0.02 µg/l, sample origin: supermarkets and fast-food chains in Quebec (city), Canada, sample year: September–October 2008, country: Canada[87]

incidence: 1/1, conc.: 0.03 µg/l, sample origin: supermarkets and fast-food chains in Calgary (city), Canada, sample year: September–October 2009, country: Canada[87]

- Co-contamination: not reported
- Further contamination of animal products reported in the present book/article[87]: Beefburger, Butter (peanut butter), Chicken nuggets, Chickenburger, Hot-dog, Meat, Meat products, Pork meat, Sausage, OTA, literature[87]

Infant formula

see **Baby food**, **Food** and **Milk (infant formula)**

Infant cereal

see **Infant food**

Infant food

may contain the following mycotoxins:

Aspergillus Toxins

Aflatoxin B_1

incidence: 57/91*, conc. range: <LOD? (15 sa.), ≤3.11 µg/kg** (42 sa.), sample origin: households or purchased in pharmacies and specialized organic feeding shops in Pamplona (city), Navarra (chartered community), Spain, sample year: unknown, country: Spain[96], *infant cereals partly with milk and partly gluten free, **outlier

- Co-contamination: not reported
- Further contamination of animal products reported in the present book/article[96]: Infant food, AFB_2, AFG_1, and AFG_2, literature[96]

For detailed information see the article.

Aflatoxin B_2

incidence: 36/91*, conc. range: ≤0.41 µg/kg** (36 sa.), sample origin: households or purchased in pharmacies and specialized organic feeding shops in Pamplona (city), Navarra (chartered community), Spain, sample year: unknown, country: Spain[96], *infant cereals partly with milk and partly gluten free, **outlier

- Co-contamination: not reported
- Further contamination of animal products reported in the present book/article[96]: Infant food, AFB_1, AFG_1, and AFG_2, literature[96]

For detailed information see the article.

Aflatoxin G_1

incidence: 60/91*, conc. range: <LOD? (29 sa.), ≤0.42 µg/kg** (31 sa.), sample origin: households or purchased in pharmacies and specialized organic feeding shops in Pamplona (city), Navarra (chartered community), Spain, sample year: unknown, country: Spain[96], *infant cereals partly with milk and partly gluten free, **outlier

- Co-contamination: not reported
- Further contamination of animal products reported in the present book/article[96]: Infant food, AFB_1, AFB_2, and AFG_2, literature[96]

For detailed information see the article.

Aflatoxin G_2

incidence: 39/91*, conc. range: <LOD? (29 sa.), ≤0.07 µg/kg** (10 sa.), sample origin: households or purchased in pharmacies and specialized organic feeding shops in Pamplona (city), Navarra (chartered community), Spain, sample year: unknown, country: Spain[96], *infant cereals partly with milk and partly gluten free, **outlier

- Co-contamination: not reported
- Further contamination of animal products reported in the present book/article[96]: Infant food, AFB_1, AFB_2, and AFG_1, literature[96]

For detailed information see the article.

Insects

may contain the following mycotoxins:

Alternaria Toxins

Alternariol

incidence: 5/6*, conc. range: <15 µg/kg, sample origin: local pet shop and the research group RADIUS, Belgium, sample year: unknown, country: Belgium/UK[549], *lesser mealworm = *Alphitobius diaperinus* (edible insect)

incidence: 0/3*, conc.: no contamination, sample origin: local pet shop and the research group RADIUS, Belgium, sample year: unknown, country: Belgium/UK[549], *black soldier = *Hermetia illucens* (edible insect)

incidence: 0/6*, conc.: no contamination, sample origin: local pet shop and the research group RADIUS, Belgium, sample year: unknown, country: Belgium/UK[549], *house cricket = *Acheta domesticus* (edible insect)

incidence: 0/1*, conc.: no contamination, sample origin: local pet shop and the research group RADIUS, Belgium, sample year: unknown, country: Belgium/UK[549], *grasshopper = *Caelifera* sp. (edible insect)

- Co-contamination: not reported

- Further contamination of animal products reported in the present book/article[549]: Insects, AME, HT-2, NIV, ROQ C, and ZEA, literature[549]

Alternariol Methyl Ether

incidence: 0/6*, conc.: no contamination, sample origin: local pet shop and the research group RADIUS, Belgium, sample year: unknown, country: Belgium/UK[549], *lesser mealworm = *Alphitobius diaperinus* (edible insect)

incidence: 0/3*, conc.: no contamination, sample origin: local pet shop and the research group RADIUS, Belgium, sample year: unknown, country: Belgium/UK[549], *black soldier = *Hermetia illucens* (edible insect)

incidence: 3/6*, conc. range: <100 μg/kg, sample origin: local pet shop and the research group RADIUS, Belgium, sample year: unknown, country: Belgium/UK[549], *house cricket = *Acheta domesticus* (edible insect)

incidence: 0/1*, conc.: no contamination, sample origin: local pet shop and the research group RADIUS, Belgium, sample year: unknown, country: Belgium/UK[549], *grasshopper = *Caelifera* sp. (edible insect)

- Co-contamination: not reported
- Further contamination of animal products reported in the present book/article[549]: Insects, AOH, HT-2, NIV, ROQ C, and ZEA, literature[549]

Aspergillus Toxins

Aflatoxin/s

incidence: 20/49*, conc. range: >9.9–24.4 μg/kg**, sample origin: markets in nine districts, Zambia, sample year: 2016–2017, country: Zambia/Nigeria/USA[338], *speckled emperor = *Gynanis maja* (edible insect), **AFB_1, AFB_2, AFG_1, and AFG_2

incidence: 24/44*, conc. range: >9.9–25.1 μg/kg**, sample origin: markets in nine districts, Zambia, sample year: 2016–2017, country: Zambia/Nigeria/USA[338], *African emperor moth = *Gonimbrasia zambesina* and mopane worm = *Gonimbrasia belina* (edible insects) **AFB_1, AFB_2, AFG_1, and AFG_2

incidence: 4/4*, conc. range: 16–36.8 μg/kg**, ∅ conc.: 24 μg/kg**, sample origin: markets in nine districts, Zambia, sample year: 2016–2017, country: Zambia/Nigeria/USA[338], *termite = *Macrotermes falciger* (edible insect), **AFB_1, AFB_2, AFG_1, and AFG_2

- Co-contamination: not reported
- Further contamination of animal products reported in the present book/article[338]: Fish, AF/S, literature[338]

For detailed information see the article.

Fusarium Toxins

Type A Trichothecenes

HT-2 Toxin

incidence: 3/6*, conc. range: >15 μg/kg, sample origin: local pet shop and the research group RADIUS, Belgium, sample year: unknown, country: Belgium/UK[549], *lesser mealworm = *Alphitobius diaperinus* (edible insect)

incidence: 0/3*, conc.: no contamination, sample origin: local pet shop and the research group RADIUS, Belgium, sample year: unknown, country: Belgium/UK[549], *black soldier = *Hermetia illucens* (edible insect)

incidence: 0/6*, conc.: no contamination, sample origin: local pet shop and the research group RADIUS, Belgium, sample year: unknown, country: Belgium/UK[549], *house cricket = *Acheta domesticus* (edible insect)

incidence: 0/1*, conc.: no contamination, sample origin: local pet shop and the research group RADIUS, Belgium, sample year: unknown, country: Belgium/UK[549], *grasshopper = *Caelifera* sp. (edible insect)

- Co-contamination: not reported
- Further contamination of animal products reported in the present book/article[549]: Insects, AME, AOH, NIV, ROQ C, and ZEA, literature[549]

Type B Trichothecenes

Nivalenol

incidence: 0/6*, conc.: no contamination, sample origin: local pet shop and the research group RADIUS, Belgium, sample year: unknown, country: Belgium/UK[549], *lesser mealworm = *Alphitobius diaperinus* (edible insect)

incidence: 0/3*, conc.: no contamination, sample origin: local pet shop and the research group RADIUS, Belgium, sample year: unknown, country: Belgium/UK[549], *black soldier = *Hermetia illucens* (edible insect)

incidence: 0/6*, conc.: no contamination, sample origin: local pet shop and the research group RADIUS, Belgium, sample year: unknown, country: Belgium/UK[549], *house cricket = *Acheta domesticus* (edible insect)

incidence: 1/1*, conc.: 3–6 μg/kg, sample origin: local pet shop and the research group RADIUS, Belgium, sample year: unknown, country: Belgium/UK[549], *grasshopper = *Caelifera* sp. (edible insect)

- Co-contamination: not reported
- Further contamination of animal products reported in the present book/article[549]: Insects, AME, AOH, HT-2, ROQ C, and ZEA, literature[549]

Other ***Fusarium*** Toxins

Zearalenone

incidence: 0/6*, conc.: no contamination, sample origin: local pet shop and the research group RADIUS, Belgium, sample year: unknown, country: Belgium/UK[549], *lesser mealworm = *Alphitobius diaperinus* (edible insect)

incidence: 3/3*, conc. range: 60 μg/kg, sample origin: local pet shop and the research group RADIUS, Belgium, sample year: unknown, country: Belgium/UK[549], *black soldier = *Hermetia illucens* (edible insect)

incidence: 0/6*, conc.: no contamination, sample origin: local pet shop and the research group RADIUS, Belgium, sample year: unknown, country: Belgium/UK[549], *house cricket = *Acheta domesticus* (edible insect)

incidence: 0/1*, conc.: no contamination, sample origin: local pet shop and the research group RADIUS, Belgium, sample year: unknown, country: Belgium/UK[549], *grasshopper = *Caelifera* sp. (edible insect)

- Co-contamination: not reported
- Further contamination of animal products reported in the present book/article[549]: Insects, AME, AOH, HT-2, NIV, and ROQ C, literature[549]

Penicillium Toxins

Roquefortine C

incidence: 1/6*, conc.: <30 µg/kg, sample origin: local pet shop and the research group RADIUS, Belgium, sample year: unknown, country: Belgium/UK[549], *lesser mealworm = *Alphitobius diaperinus* (edible insect)

incidence: 0/3*, conc.: no contamination, sample origin: local pet shop and the research group RADIUS, Belgium, sample year: unknown, country: Belgium/UK[549], *black soldier = *Hermetia illucens* (edible insect)

incidence: 0/6*, conc.: no contamination, sample origin: local pet shop and the research group RADIUS, Belgium, sample year: unknown, country: Belgium/UK[549], *house cricket = *Acheta domesticus* (edible insect)

incidence: 0/1*, conc.: no contamination, sample origin: local pet shop and the research group RADIUS, Belgium, sample year: unknown, country: Belgium/UK[549], *grasshopper = *Caelifera* sp. (edible insect)

- Co-contamination: not reported
- Further contamination of animal products reported in the present book/article[549]: Insects, AME, AOH, HT-2, NIV, and ZEA, literature[549]

Joints may contain the following mycotoxins:

Aspergillus and ***Penicillium*** Toxins

Ochratoxin A

incidence: 14/40*, conc. range: ≤0.18 µg/kg, sample origin: CMA, Germany, sample year: unknown, country: Germany[1], *raw cured joints

- Co-contamination: not reported
- Further contamination of animal products in the present book/article[1]: Beef meat, Meat products, Pig kidney, Pig liver, Pork meat, Sausage (beef sausage), Sausage (blood sausage), Sausage (Bologna sausage), Sausage (liver sausage), Sausage (poultry sausage), Sausage (raw sausage), OTA, literature[1]

Kashk may contain the following mycotoxins:

Aspergillus Toxins

Aflatoxin B_1

incidence: 1/41, conc.: 0.64 µg/kg, sample origin: supermarkets and traditional bazaars in Khorramabad (city), Iran, sample year: February–April 2008, country: Iran[97]

- Co-contamination: not reported
- Further contamination of animal products in the present book/article[97]: not reported

Aflatoxin M_1

incidence: 9/14, conc. range: 0.0629–0.09244 µg/kg, ∅ conc.: 0.0775 µg/kg, sample origin: rural regions in Hamadan, Ilam, Kermanshah, and Kurdistan (provinces), Iran, sample year: winter and summer seasons 2014, country: Iran[328]

- Co-contamination: not reported
- Further contamination of animal products reported in the present book/article[328]: Ayran, Cheese, Milk (cow milk), Milk (goat milk), Milk (sheep milk), Tarkhineh, Yogurt, AFM_1, literature[328]

incidence: 34/64*, conc. range: 0.028–0.285 µg/kg, sample origin: dairy ranches, supermarkets, and retail outlets in Esfahan, Shiraz, and Tabriz (cities) as well as Tehran (capital), Iran, sample year: 2008, country: Iran[440], *industrial Kashk

incidence: 19/61*, conc. range: 0.046–0.291 µg/kg, sample origin: dairy ranches, supermarkets, and retail outlets in Esfahan, Shiraz, and Tabriz (cities) as well as Tehran (capital), Iran, sample year: 2008, country: Iran[440], *traditional Kashk

- Co-contamination: not reported
- Further contamination of animal products reported in the present book/article[440]: Ayran, Cheese (Lighvan cheese), Milk (cow milk), Milk (goat milk), Milk (sheep milk), Yogurt, AFM_1, literature[440]

Kashk is a boiled yogurt which is afterwards sun-dried for dehydration.

see also **Yogurt**

Kazandibi see **Dairy desserts**

Keskul see **Dairy desserts**

Khoa see **Dairy products**

Kubeba may contain the following mycotoxins:

Aspergillus Toxins

Aflatoxin B_1

incidence: 1/25, conc.: 150 µg/kg, sample origin: local companies in Cairo (capital), Egypt, sample year: unknown, country: Egypt[2]

- Co-contamination: 1 sa. co-contaminated with AFB1 and AFB_2
- Further contamination of animal products in the present book/article[2]: Beefburger, AFB_1, literature[2]; Hot-dog, Meat, Sausage, AFB_1 and AFB_2, literature[2]; Kubeba, AFB_2, literature[2]

Aflatoxin B_2

incidence: 1/25, conc.: 25 µg/kg, sample origin: local companies in Cairo (capital), Egypt, sample year: unknown, country: Egypt[2]

- Co-contamination: 1 sa. co-contaminated with AFB1 and AFB_2
- Further contamination of animal products in the present book/article[2]: Beefburger, Kubeba, AFB_1, literature[2]; Hot-dog, Meat, Sausage, AFB_1 and AFB_2, literature[2]

Kubeba is a mixture of black pepper, coriander, garlic, minced meat, rice, and various vegetables used in Egyptian households.

Lala see **Dairy products**

Limnothrissa sp. see **Fish**

Luncheon meat see **Beef meat** and **Meat**

Mackerel see **Fish**

Madzoon see **Dairy products**

Meat may contain the following mycotoxins:

Aspergillus Toxins

Aflatoxin B_1

incidence: 1/25*, conc.: 4 µg/kg, sample origin: local companies in Cairo (capital), Egypt, sample year: unknown, country: Egypt[2], *luncheon meat

- Co-contamination: 1 sa. co-contaminated with AFB_1 and AFB_2
- Further contamination of animal products in the present book/article[2]: Beefburger, AFB_1, literature[2]; Hot-dog, Kubeba, Sausage, AFB_1 and AFB_2, literature[2]; Meat, AFB_2, literature[2]

incidence: 7/50*, conc. range: 0.5–11.1 µg/kg, Ø conc.: 3 µg/kg, sample origin: Assiut city markets, Egypt, sample year: unknown, country: Egypt[191], *luncheon meat

- Co-contamination: 1 sa. co-contaminated with AFB_1 and AFG_1; 6 sa. contaminated solely with AFB_1
- Further contamination of animal products in the present book/article[191]: Meat, AFG_1, literature[191]

Aflatoxin B_2

incidence: 1/25*, conc.: 2 µg/kg, sample origin: local companies in Cairo (capital), Egypt, sample year: unknown, country: Egypt[2], *luncheon meat

- Co-contamination: 1 sa. co-contaminated with AFB_1 and AFB_2
- Further contamination of animal products in the present book/article[2]: Beefburger, Meat, AFB_1, literature[2]; Hot-dog, Kubeba, Sausage, AFB_1 and AFB_2, literature[2]

Aflatoxin G_1

incidence: 1/50*, conc.: 3.2 µg/kg, sample origin: Assiut city markets, Egypt, sample year: unknown, country: Egypt[191], *luncheon meat

- Co-contamination: 1 sa. co-contaminated with AFB_1 and AFG_1
- Further contamination of animal products in the present book/article[191]: Meat, AFB_1, literature[191]

Aflatoxin/s

incidence: 6/60*, conc. range: 0.15–5.10 µg/kg**, sample origin: local markets, Jordan, sample year: January–May 2007, country: Jordan[157], *locally produced cow, goat, and sheep meat, **AFB_1, AFB_2, AFG_1, AFG_2, AFM_1, and AFM_2

- Co-contamination: not reported
- Further contamination of animal products reported in the present book/article[157]: Beef meat, Eggs, Milk, AF/S, literature[157]

For detailed information see the article.

Aspergillus and *Penicillium* Toxins

Ochratoxin A

incidence: 1/1*, conc.: 0.01 µg/kg, sample origin: supermarkets and fast-food chains in Quebec (city), Canada, sample year: September–October 2008, country: Canada[87], *luncheon meat, canned

incidence: 1/1*, conc.: 0.01 µg/kg, sample origin: supermarkets and fast-food chains in Calgary (city), Canada, sample year: September–October 2009, country: Canada[87], *luncheon meat, canned

incidence: 0/1*, conc.: no contamination, sample origin: supermarkets and fast-food chains in Quebec (city), Canada, sample year: September–October 2008, country: Canada[87], *organ meats

incidence: 1/1*, conc.: 0.01 µg/kg, sample origin: supermarkets and fast-food chains in Calgary (city), Canada, sample year: September–October 2009, country: Canada[87], *organ meats

- Co-contamination: not reported
- Further contamination of animal products reported in the present book/article[87]: Beefburger, Butter (peanut butter), Chicken nuggets, Chickenburger, Hot-dog, Ice cream, Meat products, Pork meat, Sausage, OTA, literature[87]

incidence: 6/6*, conc. range: 0.1–2.2 µg/kg, sample origin: Tunisia, sample year: unknown, country: Tunisia/France[105], *meat and fish

- Co-contamination: not reported
- Further contamination of animal products reported in the present book/article[105]: not reported

incidence: 9/116*, conc. range: LOD/LOQ-0.9 µg/kg (9 sa., maximum: 0.1360 µg/kg), sample origin: Germany, sample year: 1995–1998, country: EU[215], *meat from mammals other than marine

- Co-contamination: not reported
- Further contamination of animal products reported in the present book/article[215]: Dairy products, Pig kidney, Pig liver, Pork meat, Sausage, OTA, literature[215]

Fusarium Toxins

Depsipeptides

Enniatin B

incidence: 2/5*, conc. range: 0.13–0.24 µg/kg, Ø conc.: 0.19 µg/kg, sample origin: vegetable markets, grain shops, farmer's markets, or rural household in 5 provinces (Sixth China Total Diet Study), China, sample year: unknown, country: China[536], *meat and meat products

- Co-contamination: not reported
- Further contamination of animal products reported in the present book/article[536]: Aquatic food, ALT, AME, BEA, ENB, ENB_1, and TA, literature[536]; Eggs, AME, BEA, ENA, ENA_1, ENB, ENB_1, and TA, literature[536]; Milk, BEA, ENB, and TA, literature[536]; see also Aquatic food, Eggs, chapter: Further Mycotoxins and Microbial Metabolites

Other *Fusarium* Toxins

Zearalenone

incidence: 6/25*, conc. range: 3.3–13.2 µg/kg, Ø conc.: 6.3 µg/kg, sample origin: shops and supermarkets in Alexandria (province), Egypt, sample year: unknown, country: Egypt[141]; *frozen meat

incidence: 5/20*, conc. range: 1.9–9.9 µg/kg, Ø conc.: 7.2 µg/kg, sample origin: shops and supermarkets in Alexandria (province), Egypt, sample year: unknown, country: Egypt[141]; *minced meat

incidence: 4/20*, conc. range: 1.3–7.5 µg/kg, Ø conc.: 6.4 µg/kg, sample origin: shops and supermarkets in Alexandria (province), Egypt, sample year: unknown, country: Egypt[141]; *luncheon meat

- Co-contamination: not reported
- Further contamination of animal products reported in the present book/article[141]: Beef meat, Beefburger, Cheese (Hard Roumy cheese), Cheese (Karish cheese), Milk, Milk powder, Sausage, ZEA, literature[141]

Meat products may contain the following mycotoxins:

Aspergillus Toxins

Aflatoxin B_1

incidence: 6/50*, conc. range: <LOQ-3.0 μg/kg, sample origin: individual producer or Croatian meat industrial facilities, Croatia, sample year: unknown, country: Croatia[481], *included different dry-meat products

- Co-contamination: 2 sa. co-contaminated with AFB_1, CIT, and OTA; 4 sa. contaminated solely with AFB_1
- Further contamination of animal products in the present book/article[481]: Meat products, CIT and OTA, literature[481]; Sausage, AFB_1, CIT, and OTA, literature[481]

For detailed information see the article.

Aflatoxin/s

incidence: 20/98, Ø conc.: 1.0 μg/kg*, sample origin: greater Manila (capital) area, Philippines, sample year: unknown, country: Philippines[70], *AF of pos. sa.?

- Co-contamination: not reported
- Further contamination of animal products reported in the present book/article[70]: Butter (peanut butter), Fish products, AF/S, literature[70]

Aspergillus and *Penicillium* Toxins

Citrinin

incidence: 3/50*, conc. range: <LOQ-1.3 μg/kg, sample origin: individual producer or Croatian meat industrial facilities, Croatia, sample year: unknown, country: Croatia[481], *included different dry-meat products (for detailed information see the article)

- Co-contamination: For detailed information see the article.
- Further contamination of animal products in the present book/article[481]: Meat products, AFB_1 and OTA, literature[481]; Sausage, AFB_1, CIT, and OTA, literature[481]

Ochratoxin A

incidence: 3/21*, conc. range: ≤0.04 μg/kg, sample origin: CMA, Germany, sample year: unknown, country: Germany[1], *cooked meat products

- Co-contamination: not reported
- Further contamination of animal products in the present book/article[1]: Beef meat, Joints, Pig kidney, Pig liver, Pork meat, Sausage (beef sausage), Sausage (blood sausage), Sausage (Bologna sausage), Sausage (liver sausage), Sausage (poultry sausage), Sausage (raw sausage), OTA, literature[1]

incidence: 0/1*, conc.: no contamination, sample origin: supermarkets and fast-food chains in Quebec (city), Canada, sample year: September–October 2008, country: Canada[87], *dinners = cereal + vegetable + meat

incidence: 1/1*, conc.: 0.01 μg/kg, sample origin: supermarkets and fast-food chains in Calgary (city), Canada, sample year: September–October 2009, country: Canada[87], *dinners = cereal + vegetable + meat

- Co-contamination: not reported
- Further contamination of animal products reported in the present book/article[87]: Beefburger, Butter (peanut butter), Chicken nuggets, Chickenburger, Hot-dog, Ice cream, Meat, Pork meat, Sausage, OTA, literature[87]

incidence: 23/50*, conc. range: <LOQ-≤7.83 μg/kg, sample origin: individual producer or Croatian meat industrial facilities, Croatia, sample year: unknown, country: Croatia[481], *included different dry-meat products

- Co-contamination: 2 sa. co-contaminated with AFB_1, CIT, and OTA; 3 sa. co-contaminated with CIT and OTA; 18 sa. contaminated solely with OTA
- Further contamination of animal products in the present book/article[481]: Meat products, AFB_1 and CIT, literature[481]; Sausage, AFB_1, CIT, and OTA, literature[481]

For detailed information see the article.

see also **Meat**

Milk may contain the following mycotoxins:

Alternaria Toxins

Tenuazonic Acid

incidence: 3/5*, conc. range: 1.93–3.00 μg/kg, sample origin: vegetable markets, grain shops, farmer's markets, or rural household in 5 provinces (Sixth China Total Diet Study), China, sample year: unknown, country: China[536], *milk and dairy products

- Co-contamination: not reported
- Further contamination of animal products reported in the present book/article[536]: Aquatic food, ALT, AME, BEA, ENB, ENB_1, and TA, literature[536]; Eggs, AME, BEA, ENA, ENA_1, ENB, ENB_1, and TA, literature[536]; Meat and Meat products, ENB, literature[536]; Milk, BEA and ENB, literature[536]; see also Aquatic food, Eggs, chapter: Further Mycotoxins and Microbial Metabolites

Aspergillus Toxins

Aflatoxin M_1

incidence: 1/14*, conc.: tr, sample origin: PAU dairy farm, India, sample year: September 1974–February 1975, country: India[27], *bulk milk

incidence: 0/21*, conc.: no contamination, sample origin: local market, India, sample year: September 1974–February 1975, country: India[27], *bulk milk

- Co-contamination: not reported
- Further contamination of animal products reported in the present book/article[27]: Cheese, AFB_1, AFB_2, and AFG_1, literature[27]; Dairy products, AFB_1 and AFG_1, literature[27]; Milk (buffalo milk), Milk (cow milk), Milk powder, AFM_1, literature[27]

incidence: 1/10*, conc.: 0.0108 μg/l, sample origin: supermarkets and street milkmen in Bursa Province, Turkey, sample year: unknown, country: Turkey[33], *raw and pasteurized milk sa.

- Co-contamination: not reported
- Further contamination of animal products reported in the present book/article[33]: Cheese, AFM_1, literature[33]

incidence: 42/57, conc. range: 0.025–0.95 μg/kg, Ø conc.: 0.08 μg/kg, sample origin: small markets in several urban Shanghai (city) areas, China, sample year: April 1988–1989, country: China/USA[46]

- Co-contamination: not reported
- Further contamination of animal products reported in the present book/article[46]: Butter (peanut butter), Pig liver, AFB_1, literature[46]; Milk powder, AFM_1, literature[46]

incidence: 5/19*, conc. range: 0.01–0.02 μg/l, Ø conc.: 0.015 μg/l, sample origin: market, Cyprus, sample year: 1993/1995/1996, country: Cyprus[50], *full pasteurized milk

incidence: 1/4*, conc.: 0.01 μg/l, sample origin: market, Cyprus, sample year: 1993/1995/1996, country: Cyprus[50], *light pasteurized milk

incidence: 3/8*, conc. range: 0.01–0.04 μg/l, Ø conc.: 0.02 μg/l, sample origin: market, Cyprus, sample year: 1993/1995/1996, country: Cyprus[50], *skimmed pasteurized milk

- Co-contamination: not reported
- Further contamination of animal products reported in the present book/article[50]: Butter (peanut butter), AFB_1, AFB_2, AFG_1, and AFG_2, literature[50]; Milk (cow milk), AFM_1, literature[50]

incidence: 11/30?, conc. range: 0.002–0.03 μg/kg, sample origin: local markets in Zhejiang (province), China, sample year: unknown, country: China[73]

- Co-contamination: not reported
- Further contamination of animal products in the present book/article[73]: Eggs, STG, literature[73]

incidence: 43/50*, conc. range: 0.001–0.030 μg/l, sample origin: Kayseri (city), Turkey, sample year: January–March 2010, country: Turkey/Kyrgyzstan[144], *raw milk

- Co-contamination: not reported
- Further contamination of animal products reported in the present book/article[144]: Cheese (Kashar cheese), Cheese (Tulum cheese), Cheese (white cheese), Dairy desserts, Yogurt, AFM_1, literature[144]

incidence: 3/6*, conc. range: 0.0036–0.1005 μg/l, ∅ conc.: 0.0399 μg/l, sample origin: retail shops in Marang and Kuala Terengganu (town and city), Malaysia, sample year: July 2013?, country: Malaysia[163], *milk, condensed and sweetened

- Co-contamination: not reported
- Further contamination of animal products reported in the present book/article[163]: Cheese, Dairy beverages, Milk (cow milk), Milk powder, Yogurt, AFM_1, literature[163]

incidence: 2/24*, conc. range: 0.02–0.03 μg/l, ∅ conc.: 0.025 μg/l, sample origin: supermarket in Madrid (capital), Spain, sample year: January–June 1993, country: Spain[194], *whole pasteurized milk

- Co-contamination: not reported
- Further contamination of animal products reported in the present book/article[194]: Milk (UHT milk), AFM_1, literature[194]

incidence: 70/82*, conc. range: 0.005–0.010 μg/l (42 sa.), 0.011–0.020 μg/l (18 sa.), 0.021–0.050 μg/l (10 sa.), sample origin: supermarkets, Greece, sample year: December 1999–May 2000, country: Greece[195], *pasteurized milk

incidence: 43/54*, conc. range: 0.005–0.010 μg/l (18 sa.), 0.011–0.020 μg/l (15 sa.), 0.021–0.050 μg/l (10 sa.), sample origin: supermarkets, Greece, sample year: December 2000–May 2001, country: Greece[195], *pasteurized milk

incidence: 14/15*, conc. range: 0.005–0.010 μg/l (2 sa.), 0.011–0.020 μg/l (4 sa.), 0.021–0.050 μg/l (6 sa.), >0.050 μg/l (2 sa.), sample origin: supermarkets, Greece, sample year: December 1999–May 2000, country: Greece[195], *concentrated milk

- Co-contamination: not reported
- Further contamination of animal products reported in the present book/article[195]: Milk (cow milk), Milk (goat milk), Milk (sheep milk), Milk (UHT milk), AFM_1, literature[195]

incidence: 7/95*, conc. range: 0.020–0.040 μg/l, sample origin: Madrid (capital), Spain, sample year: 1983, country: Spain[200], *included raw, pasteurized, sterilized, concentrated milk

- Co-contamination: not reported
- Further contamination of animal products reported in the present book/article[200]: not reported

incidence: 7/22, conc. range: 0.2–0.5 μg/kg (6 sa.), >0.5 μg/kg (1 sa.)*, sample origin: unknown, sample year: 1993–1995, country: Uruguay[206], *imported milk powder lot

- Co-contamination: not reported
- Further contamination of animal products reported in the present book/article[206]: not reported

incidence: 16/17*, conc. range: 0.003–0.010 μg/l (6 sa.), 0.011–0.020 μg/l (6 sa.), 0.021–0.030 μg/l (2 sa.), 0.031–0.040 μg/l (2 sa., maximum: 0.034 μg/l), sample origin: supermarkets in Florence (city), Italy, sample year: unknown, country: Italy[217], *pasteurized milk

- Co-contamination: not reported
- Further contamination of animal products reported in the present book/article[217]: Milk (UHT milk), Yogurt, AFM_1, literature[217]

incidence: 78/78*, ∅ conc.: 0.2305 μg/l, sample origin: supermarkets in Babol (city), Iran, sample year: winter 2006, country: Iran[218], *pasteurized milk

incidence: 33/33*, ∅ conc.: 0.22166 μg/l, sample origin: supermarkets in Babol (city), Iran, sample year: winter 2006, country: Iran[218], *sterilized milk

- Co-contamination: not reported
- Further contamination of animal products reported in the present book/article[218]: not reported

incidence: 3/71*, conc. range: 0.71–3.37 μg/l, ∅ conc.: 2.26 μg/l, sample origin: production units and retail outlets in Jabalpur (city), India, sample year: unknown, country: India[220], *fresh milk

- Co-contamination: not reported
- Further contamination of animal products reported in the present book/article[220]: not reported

incidence: 5/110*, conc. range: 0.038–0.079 μg/l, ∅ conc.: 0.0574 μg/l, sample origin: Minas Gerais (state), Brazil, sample year: 1996–1997, country: Brazil[221], *including vitaminized, skimmed, and full sterilized (UHT) milk

- Co-contamination: not reported
- Further contamination of animal products reported in the present book/article[221]: not reported

incidence: 5/5, conc. range: 0.16–1.07 μg/l, ∅ conc.: 0.534 μg/l, sample origin: unknown, sample year: unknown, country: USA[222]

- Co-contamination: not reported
- Further contamination of animal products reported in the present book/article[222]: not reported

incidence: 11/22, conc. range: 0.010–0.250 μg/l, sample origin: state-owned farms, Poland, sample year: unknown, country: Poland[233]

- Co-contamination: not reported
- Further contamination of animal products reported in the present book/article[233]: not reported

incidence: 136*/159, conc. range: 0.001–0.010 μg/l (125 sa.), >0.010–0.050 μg/I (9 sa.), >0.050 μg/l (2 sa., maximum: 0.10851 μg/l), ∅ conc.: 0.01019 μg/l, sample origin: 4 large Italian cities, Italy, sample year: 1995, country: Italy[234], *112 UHT and 24 pasteurized milk sa.

- Co-contamination: not reported
- Further contamination of animal products reported in the present book/article[234]: Milk powder, Yogurt, AFM_1, literature[234]

incidence: 46/46*, conc. range: 0.010–0.210 μg/l, sample origin: individual suppliers and bulk milk tankers in different parts of the Netherlands, sample year: unknown, country: Netherlands[246], *raw milk after skimming

- Co-contamination: not reported
- Further contamination of animal products reported in the present book/article[246]: not reported

incidence: 16/44, conc. range: 0.6–15 µg/kg (11 sa.)*, 16–30 µg/kg (5 sa.)*, sample origin: Hyderabad (city) in Andhra Pradesh (state), southern India, sample year: 2001–2002, country: India/UK[256], *authors pointed out the high levels of AFM_1 in milk (packaged)

- Co-contamination: not reported
- Further contamination of animal products reported in the present book/article[256]: Dairy products, Milk (buffalo milk), Milk (cow milk), Milk powder, AFM_1, literature[256]

incidence: 4/25*, conc. range: 0.25–0.84** µg/l, Ø conc.: 0.62 µg/l, sample origin: Cairo (governorate), Egypt, sample year: August 1996, country: Egypt[261], *raw milk, **from El-Sayeda area

incidence: 8/50*, conc. range: 0.35–3.72** µg/l, Ø conc.: 1.23 µg/l, sample origin: El-Giza (governorate), Egypt, sample year: August 1996, country: Egypt[261], *raw milk, **from Bolak area

- Co-contamination: not reported
- Further contamination of animal products reported in the present book/article[261]: not reported

incidence: 4/12*, conc. range: 0.010–0.049 µg/kg (3 sa.), conc.: ≥0.050 µg/kg (1 sa.), sample origin: supermarkets in Dohar (capital), Qatar, sample year: unknown, country: Qatar/Italy[277], *formula milk based on cow (9 sa.) and goat milk (3 sa.)

- Co-contamination: For detailed information see the article.
- Further contamination of animal products reported in the present book/article[277]: Milk (infant formula), AFM_1, OTA, T-2, and ZEA, literature[277]

incidence: 10/10*, conc. range: 0.14939–0.26482 µg/kg, Ø conc.: 0.19591 µg/kg, sample origin: Jordanian market, Jordan, sample year: 2014–2015, country: Jordan[278], *evaporated milk

- Co-contamination: not reported
- Further contamination of animal products reported in the present book/article[278]: Milk (camel milk), Milk (cow milk), Milk (goat milk), Milk (infant formula), Milk (sheep milk), Milk powder, AFM_1, literature[278]

incidence: 83/116*, conc. range: 0.0058–0.5285 µg/l, Ø conc.: 0.0738 µg/l, sample origin: supermarkets and retail shops in Esfahan and Yazd (provinces), Iran, sample year: December 2008–June 2009, country: Iran[288], *pasteurized milk

- Co-contamination: not reported
- Further contamination of animal products reported in the present book/article[288]: Milk (UHT milk), AFM_1, literature[288]

incidence: 7/143*, conc. range: 0.10–0.40 µg/l, sample origin: stores in Lisbon (capital), Portugal, sample year: January 1984–August 1985, country: Portugal[292], *pasteurized milk

- Co-contamination: not reported
- Further contamination of animal products reported in the present book/article[292]: not reported

incidence: 7/10*, conc. range: 0.010–0.020 µg/kg, sample origin: supermarkets in São Paulo (city), Brazil, sample year: September–November 2006, country: Brazil[305], *pasteurized milk

- Co-contamination: not reported
- Further contamination of animal products reported in the present book/article[305]: Milk (UHT milk), Milk powder, AFM_1, literature[305]

incidence: 70/84*, conc. range: 0.005–0.010 µg/l (13 sa.), 0.011–0.025 µg/l (27 sa.), 0.026–0.050 µg/l (12 sa.), 0.050–0.080 µg/l (12 sa.), >0.080 µg/l (6 sa., maximum: 0.11093 µg/l), Ø conc.: 0.0324 µg/l, sample origin: supermarkets in Prishtina (city), Kosovo, sample year: July–December 2013, country: Serbia[401], *pasteurized milk

- Co-contamination: not reported
- Further contamination of animal products reported in the present book/article[401]: Milk (UHT milk), AFM_1, literature[401]

incidence: 3/22*, conc. range: tr-0.011 µg/l, sample origin: small farms and dairies, Constantine (city), Algeria, sample year: February–October 2011, country: Algeria/France[406], *raw milk

incidence: 4/11*, conc. range: tr, sample origin: imported, Constantine (city), Algeria, sample year: February–October 2011, country: Algeria/France[406], *reconstituted milk

- Co-contamination: not reported
- Further contamination of animal products reported in the present book/article[406]: Milk powder, AFM_1, literature[406]

incidence: 40/49*, conc. range: 0.0033–0.0961 µg/l, Ø conc.: 0.0233 µg/l, sample origin: Sistan and Baluchestan (provinces), south-east of Iran, sample year: summer and winter 2015, country: Iran[417], *pasteurized milk

- Co-contamination: not reported
- Further contamination of animal products reported in the present book/article[417]: Butter, Cheese (white cheese), Yogurt, AFM_1, literature[417]

For detailed information see the article.

incidence: 23/23*, conc. range: 0.018–3.072 µg/l, Ø conc.: 0.27 µg/l, sample origin: retail stores, South Africa, sample year: winter, country: South Africa[427], *included different milk varieties

incidence: 20/22*, conc. range: 0.001–0.112 µg/l, Ø conc.: 0.025 µg/l, sample origin: retail stores, South Africa, sample year: summer, country: South Africa[427], *included different milk varieties

- Co-contamination: not reported
- Further contamination of animal products reported in the present book/article[427]: not reported

incidence: 26/64, conc. range: 0.005–0.050 µg/l (15 sa.), >0.050 µg/l (11 sa.), sample origin: supermarkets, Lebanon, sample year: unknown, country: Lebanon[432]

- Co-contamination: not reported
- Further contamination of animal products reported in the present book/article[432]: Yogurt, AFM_1, literature[432]

incidence: 66/91*, conc. range: 0.013–0.250 µg/l, sample origin: retail stores and supermarkets in Esfahan, Shiraz, and Yazd (cities), as well as Tehran (capital), Iran, sample year: winter and summer 2009, country: Iran[435], *pasteurized milk

- Co-contamination: not reported
- Further contamination of animal products reported in the present book/article[435]: Butter, Cheese (white cheese), Ice cream, Yogurt, AFM_1, literature[435]

For detailed information see the article.

incidence: 13/15, conc. range: 0.0121–0.3012 µg/l, Ø conc.: 0.1189 µg/l, sample origin: local markets and supermarkets, China, sample year: unknown, country: China[448]

- Co-contamination: not reported
- Further contamination of animal products reported in the present book/article[448]: Milk (infant formula), Milk powder, AFM_1, literature[448]

incidence: 76/107, conc. range: <0.050 µg/l (32 sa.), >0.050–0.150 µg/l (24 sa.), >0.151 µg/l (20 sa., maximum: 0.8454 µg/l), ∅ conc.: 0.2122 µg/l, sample origin: main districts of Punjab (province), Pakistan, sample year: November 2010–April 2011, country: Pakistan[451]

- Co-contamination: not reported
- Further contamination of animal products reported in the present book/article[451]: Butter, Cheese (cream cheese), Cheese (white cheese), Yogurt, AFM_1, literature[451]

incidence: 54/54*, conc. range: 0.178–0.820 µg/l, sample origin: markets in Goa (state), India, sample year: unknown, country: India[464], *included raw pasteurized and UHT milk (packaged)

- Co-contamination: not reported
- Further contamination of animal products reported in the present book/article[464]: Milk (infant formula), AFM_1, literature[464]

incidence: 59/90*, conc. range: <0.010 µg/l (8 sa.), 0.011–0.025 µg/l (13 sa.), 0.026–0.050 µg/l (10 sa.), ≥0.050 µg/l (28 sa., maximum: 0.131 µg/kg), sample origin: dairy farm, Gilan (province), northern Iran, sample year: autumn and winter 2011, country: Iran[471]

- Co-contamination: not reported
- Further contamination of animal products reported in the present book/article[471]: not reported

incidence: 20/86*, conc. range: 0.0034–0.0327 µg/l, sample origin: storage place in Kayseri (city, province), Turkey, sample year: unknown, country: Turkey[480], *raw milk

- Co-contamination: not reported
- Further contamination of animal products reported in the present book/article[480]: not reported

incidence: 137/175*, conc. range: 0.002–1.6 µg/l, ∅ conc.: 0.176 µg/l, sample origin: districts in Punjab (province), Pakistan, sample year: unknown, country: Pakistan/USA[511], *local shop milk

incidence: 25/40*, conc. range: 0.003–1.9 µg/l, ∅ conc.: 0.47 µg/l, sample origin: districts in Punjab (province), Pakistan, sample year: unknown, country: Pakistan/USA[511], *household milk

incidence: 15/17*, conc. range: 0.002–0.794 µg/l, ∅ conc.: 0.11 µg/l, sample origin: districts in Punjab (province), Pakistan, sample year: unknown, country: Pakistan/USA[511], *dairy farm milk

- Co-contamination: not reported
- Further contamination of animal products reported in the present book/article[511]: not reported

incidence: 1/27, conc.: 0.41 µg/kg*, sample origin: local stores, USA/Taiwan, sample year: unknown, country: USA/Taiwan[534], *imported condensed milk sa.

- Co-contamination: not reported
- Further contamination of animal products reported in the present book/article[534]: Cream, Milk (infant formula), AFM_1, literature[534]; Milk powder, AFB_1 and AFB_2, literature[534]

incidence: 367/896, conc. range: 0.002–0.68 µg/kg, ∅ conc.: 0.023 µg/kg, sample origin: control points such as import, primary storage, and market, Cyprus, sample year: 2004–2013, country: Cyprus[542]

- Co-contamination: not reported
- Further contamination of animal products reported in the present book/article[542]: Cheese, Cheese (Anari cheese), Cheese (white cheese), Ice cream, Milk powder, Yogurt, AFM_1, literature[542]

incidence: 23/826*, conc. range: 0.0052–0.0266 µg/l, sample origin: different regions of Kosovo, sample year: April–October 2009 and February–November 2010, country: Serbia/Italy[548], *raw milk sa.

- Co-contamination: not reported
- Further contamination of animal products reported in the present book/article[548]: Milk (UHT milk), AFM_1, literature[548]

For detailed information see the article.

Aflatoxin/s

incidence: 143/192, conc. range: >0.125–0.50 µg/kg* (80 sa.), –1.0 µg/kg* (30 sa.), –1.5 µg/kg* (18 sa.), –2.0 µg/kg* (8 sa.), –3.0 µg/kg* (5 sa.), –4.5 µg/kg* (1 sa.), –6.0 µg/kg* (1 sa.), sample origin: different Ecuadorian regions, Ecuador, sample year: April 1992–October 1994, country: Switzerland[99], *AFB_1, AFB_2, AFG_1, and AFG_2

- Co-contamination: not reported
- Further contamination of animal products reported in the present book/article[99]: not reported

incidence: 5/60*, conc. range: 0.16–5.23 µg/kg**, sample origin: local markets, Jordan, sample year: January–May 2007, country: Jordan[157], *cow, goat, and sheep milk, **AFB_1, AFB_2, AFG_1, AFG_2, and AFM_1

- Co-contamination: not reported
- Further contamination of animal products reported in the present book/article[157]: Beef meat, Eggs, Meat, AF/S, literature[157]

For detailed information see the article.

Aspergillus and ***Penicillium*** Toxins

Ochratoxin A

incidence: 0/20* **, conc.: no contamination, sample origin: retail markets, Italy, sample year: unknown, country: Italy[224], *20 bovine milk sa., **conventional

incidence: 3/63* **, conc. range: 0.07–0.11 µg/l, sample origin: organic farms, Italy, sample year: unknown, country: Italy[224], *39 bovine milk sa., 15 goat milk sa., and 9 sheep milk sa., **organic

- Co-contamination: not reported
- Further contamination of animal products reported in the present book/article[224]: not reported

Fusarium Toxins

Depsipeptides

Beauvericin

incidence: 1/5*, conc.: 0.16 µg/kg, sample origin: vegetable markets, grain shops, farmer's markets, or rural household in 5 provinces (Sixth China Total Diet Study), China, sample year: unknown, country: China[536], *milk and dairy products

- Co-contamination: not reported
- Further contamination of animal products reported in the present book/article[536]: Aquatic food, ALT, AME, BEA, ENB, ENB_1, and TA, literature[536]; Eggs, AME, BEA, ENA, ENA_1, ENB, ENB_1, and TA, literature[536]; Meat and Meat products, ENB, literature[536]; Milk, ENB and TA, literature[536]; see also Aquatic food, Eggs, chapter: Further Mycotoxins and Microbial Metabolites

Enniatin B

incidence: 1/5*, conc.: 0.26 µg/kg, sample origin: vegetable markets, grain shops, farmer's markets, or rural household in 5 provinces (Sixth China Total Diet Study), China, sample year: unknown, country: China[536], *milk and dairy products

- Co-contamination: not reported
- Further contamination of animal products reported in the present book/article[536]: Aquatic food, ALT, AME, BEA, ENB,

ENB_1, and TA, literature[536]; Eggs, AME, BEA, ENA, ENA_1, ENB, ENB_1, and TA, literature[536]; Meat and Meat products, ENB, literature[536]; Milk, BEA and TA, literature[536]; see also Aquatic food, Eggs, chapter: Further Mycotoxins and Microbial Metabolites

Other ***Fusarium*** Toxins

Zearalenone

incidence: 4/20*, conc. range: 2.9–10.1 μg/l, ∅ conc.: 6.9 μg/l, sample origin: shops and supermarkets in Alexandria (province), Egypt, sample year: unknown, country: Egypt[141], *raw milk

incidence: 3/20*, conc. range: 1.2–7.2 μg/kg, ∅ conc.: 5.0 μg/kg, sample origin: shops and supermarkets in Alexandria (province), Egypt, sample year: unknown, country: Egypt[141], *pasteurized milk

incidence: 5/20*, conc. range: 1.6–9.3 μg/kg, ∅ conc.: 4.4 μg/kg, sample origin: shops and supermarkets in Alexandria (province), Egypt, sample year: unknown, country: Egypt[141], *condensed milk

- Co-contamination: not reported
- Further contamination of animal products reported in the present book/article[141]: Beef meat, Beefburger, Cheese (Hard Roumy cheese), Cheese (Karish cheese), Meat, Milk powder, Sausage, ZEA, literature[141]

Milk (bovine milk) see Milk (cow milk)

Milk (buffalo milk) may contain the following mycotoxins:

Aspergillus Toxins

Aflatoxin M_1

incidence: 1/25, conc.: 10.0 μg/l, sample origin: dairy herd from PAU, India, sample year: September 1974–February 1975, country: India[27]

- Co-contamination: not reported
- Further contamination of animal products reported in the present book/article[27]: Cheese, AFB_1, AFB_2, and AFG_1, literature[27]; Dairy products, AFB_1 and AFG_1, literature[27]; Milk, Milk (cow milk), Milk powder, AFM_1, literature[27]

incidence: 32/38*, ∅ conc.: 0.076 μg/l, sample origin: villages around Anand (town) and herds of GAU of Anand campus, India, sample year: unknown, country: India[198], *raw individual buffalo milk

incidence: 27/28*, ∅ conc.: 0.0.74 μg/l, sample origin: villages around Anand (town) and herds of GAU of Anand campus, India, sample year: unknown, country: India[198], *raw bulk buffalo milk

- Co-contamination: not reported
- Further contamination of animal products reported in the present book/article[198]: Milk (cow milk), AFM_1, literature[198]

incidence: 4/43*, conc. range: 0.004–0.039 μg/kg, sample origin: Campania and Calabria (regions), southern Italy, sample year: 2002, country: Italy[225], *raw buffalo milk

incidence: 20/47*, conc. range: 0.004–0.043 μg/kg, sample origin: Campania and Calabria (regions), southern Italy, sample year: 2003, country: Italy[225], *raw buffalo milk

incidence: 18/59*, conc. range: 0.005–0.023 μg/kg, sample origin: Campania and Calabria (regions), southern Italy, sample year: 2004, country: Italy[225], *raw buffalo milk

incidence: 18/58*, conc. range: 0.004–0.676 μg/kg, sample origin: Campania and Calabria (regions), southern Italy, sample year: 2005, country: Italy[225], *raw buffalo milk

- Co-contamination: not reported
- Further contamination of animal products reported in the present book/article[225]: Milk (cow milk), Milk (sheep milk), AFM_1, literature[225]

incidence: 32/50, conc. range: 0.010–0.050 μg/l (8 sa.), 0.051–0.100 μg/l (10 sa.), 0.101–0.150 μg/l (6 sa.), 0.151–0.200 μg/l (4 sa.), 0.201–0.250 μg/l (3 sa.), 0.270 μg/l (1 sa.), sample origin: different areas in the Ismailia Governorate, Egypt, sample year: summers 2003–2004, country: Egypt/Germany/USA[226]

- Co-contamination: not reported
- Further contamination of animal products reported in the present book/article[226]: Milk (camel milk), Milk (cow milk), Milk (goat milk), AFM_1, literature[226]

incidence: 29/75*, conc. range: 0.005–0.020 μg/l (15 sa.), 0.021–0.050 μg/l (8 sa.), >0.050 μg/l (6 sa.), ∅ conc.: 0.0319 μg/l, sample origin: different parts of Khuzestan (province), Iran, sample year: November 2007–December 2008, country: Iran[227], *water buffalo milk

- Co-contamination: not reported
- Further contamination of animal products reported in the present book/article[227]: Milk (camel milk), Milk (cow milk), Milk (goat milk), Milk (sheep milk), AFM_1, literature[227]

incidence: 30/90*, conc. range: 0.0393–0.3426 μg/l, ∅ conc.: 0.137 μg/l, sample origin: dairy farms in Cattle Town which supplied fresh milk shops, Landhi (town), Karachi (city), Pakistan, sample year: December 2002–May 2003, country: Pakistan[229], *fresh buffalo milk

- Co-contamination: not reported
- Further contamination of animal products reported in the present book/article[229]: Milk (UHT milk), AFM_1, literature[229]

incidence: 189/352*, conc. range: 0.6–15 μg/kg (133 sa.)**, 16–30 μg/kg (40 sa.)**, 31–50 μg/kg (16 sa., maximum: 48 μg/kg)**, sample origin: peri-urban (Hyderabad, city) and rural (Anantapur, city) areas in Andhra Pradesh (state), southern India, sample year: 2001–2002, country: India/UK[256], *raw buffalo milk, **authors pointed out the high levels of AFM_1 in milk

- Co-contamination: not reported
- Further contamination of animal products reported in the present book/article[256]: Dairy products, Milk, Milk (cow milk), Milk powder, AFM_1, literature[256]

incidence: 22/48, conc. range: >LOD-0.050 μg/kg (16 sa.), >0.05 μg/kg (6 sa., maximum: 0.137 μg/kg), sample origin: milking sites and farmhouses from major cities in Punjab, Pakistan, sample year: November 2009–April 2010, country: Pakistan/Spain[450]

incidence: 24/46, conc. range: >LOD-0.050 μg/kg (14 sa.), >0.05 μg/kg (10 sa., maximum: 0.350 μg/kg), sample origin: milking sites and farmhouses from major cities in NWFP, Pakistan, sample year: November 2009–April 2010, country: Pakistan/Spain[450]

- Co-contamination: not reported
- Further contamination of animal products reported in the present book/article[450]: Milk (cow milk), AFM_1, literature[450]

incidence: 34/126*, conc. range: <0.008–0.032 μg/l, ∅ conc.: 0.003 μg/l, sample origin: Afyonkarahisar (province), Turkey, sample year: August 2012–January 2013, country: Turkey[463], *raw buffalo milk

- Co-contamination: not reported
- Further contamination of animal products reported in the present book/article[463]: not reported

incidence: 8/10*, conc. range: 0.0191–0.2095 µg/l, ∅ conc.: 0.0932 µg/l, sample origin: Shush (city) in Khuzestan (province), Turkey, sample year: winter 2012, country: Turkey[465], *raw buffalo milk

incidence: 20/30*, conc. range: 0.0195–0.4227 µg/l, ∅ conc.: 0.1091 µg/l, sample origin: Shush (city) in Khuzestan (province), Turkey, sample year: spring 2012, country: Turkey[465], *raw buffalo milk

incidence: 18/20*, conc. range: 0.0127–0.3673 µg/l, ∅ conc.: 0.1384 µg/l, sample origin: Shush (city) in Khuzestan (province), Turkey, sample year: summer 2012, country: Turkey[465], *raw buffalo milk

- Co-contamination: not reported
- Further contamination of animal products reported in the present book/article[465]: Milk (cow milk), AFM_1, literature[465]

see also **Milk (cow milk)**

Milk (buttermilk) may contain the following mycotoxins:

Aspergillus Toxins

Aflatoxin M_1

incidence: 50/50, conc. range: 0.04797–2.02711 µg/kg, ∅ conc.: 1.07974 µg/kg, sample origin: Jordan, sample year: November–February 2012, country: Jordan[313]

- Co-contamination: not reported
- Further contamination of human breast milk and animal products reported in the present book/article[313]: Milk (cow milk), Milk (human breast milk), AFM_1, literature[313]

Milk (camel milk) may contain the following mycotoxins:

Aspergillus Toxins

Aflatoxin M_1

incidence: 9/25, conc. range: 0.010–0.050 µg/l (4 sa.), 0.051–0.100 µg/l (2 sa.), 0.101–0.150 µg/l (1 sa.), 0.151–0.200 µg/l (1 sa.), 0.210 µg/l (1 sa.), sample origin: different areas in the Ismailia Governorate, Egypt, sample year: summers 2003–2004, country: Egypt/Germany/USA[226]

- Co-contamination: not reported
- Further contamination of animal products reported in the present book/article[226]: Milk (buffalo milk), Milk (cow milk), Milk (goat milk), AFM_1, literature[226]

incidence: 5/40, conc. range: 0.005–0.020 µg/l (3 sa.), 0.021–0.050 µg/l (2 sa.), ∅ conc.: 0.0190 µg/l, sample origin: different parts of Khuzestan (province), Iran, sample year: November 2007–December 2008, country: Iran[227]

- Co-contamination: not reported
- Further contamination of animal products reported in the present book/article[227]: Milk (buffalo milk), Milk (cow milk), Milk (goat milk), Milk (sheep milk), AFM_1, literature[227]

incidence: 6/20, conc. range: 0.25–0.8 µg/l, ∅ conc.: 0.46 µg/l, sample origin: farms in the vicinity of Abu Dhabi (capital), UAE, sample year: unknown, country: UAE/UK[230]

- Co-contamination: not reported
- Further contamination of human breast milk and animal products reported in the present book/article[230]: Milk (human breast milk), AFM_1, literature[230]

incidence: 10/10*, conc. range: 0.02357–0.09652 µg/kg, ∅ conc.: 0.03715 µg/kg, sample origin: Jordanian market, Jordan, sample year: 2014–2015, country: Jordan[278], *fresh camel milk

- Co-contamination: not reported
- Further contamination of animal products reported in the present book/article[278]: Milk, Milk (cow milk), Milk (goat milk), Milk (infant formula), Milk (sheep milk), Milk powder, AFM_1, literature[278]

incidence: 5/124, conc. range: 0.010–0.050 µg/kg (5 sa., maximum: 0.019 µg/kg), sample origin: Yazd (province), Iran, sample year: 2014, country: Iran[437]

- Co-contamination: not reported
- Further contamination of animal products reported in the present book/article[437]: Milk (cow milk), Milk (goat milk), Milk (sheep milk), AFM_1, literature[437]

For detailed information see the article.

incidence: 92/117*, conc. range: 0.0025–0.050 µg/l (8 sa.), 0.050–0.100 µg/l (27 sa.), 0.101–0.200 µg/l (23 sa.), >0.200 µg/l (34 sa., maximum: 0.3986 µg/l), ∅ conc.: 0.16472 µg/l, sample origin: El-Asha (governorate), eastern Saudi Arabia, sample year: February–April 2013, country: Egypt/Saudi Arabia[532], *raw camel milk

- Co-contamination: not reported
- Further contamination of animal products reported in the present book/article[532]: not reported

Milk (cow milk) may contain the following mycotoxins:

Aspergillus Toxins

Aflatoxin B_1

incidence: 4/290*, conc. range: ≥0.05–0.42 µg/l, sample origin: supermarkets in Mexico City (capital), Mexico, sample year: April 1996–1998, country: Mexico[231], *pasteurized and ultra-pasteurized cow milk

- Co-contamination: not reported
- Further contamination of animal products reported in the present book/article[231]: see also Milk (cow milk), chapter: Further Mycotoxins and Microbial Metabolites

Aflatoxin M_1

incidence: 3/21, conc. range: 3.3–13.3 µg/l, sample origin: dairy herd from PAU, India, sample year: September 1974–February 1975, country: India[27]

- Co-contamination: not reported
- Further contamination of animal products reported in the present book/article[27]: Cheese, AFB_1, AFB_2, and AFG_1, literature[27]; Dairy products, AFB_1 and AFG_1, literature[27]; Milk, Milk (buffalo milk), Milk powder, AFM_1, literature[27]

incidence: 93/134*, conc. range: 0.05–0.33 µg/l, sample origin: unknown, sample year: November–May in 1972–1974, country: Germany[28], *"wintermilk" (with supplementary feeding)

incidence: 25/126*, conc. range: 0.05–0.28 µg/l, sample origin: unknown, sample year: June–October in 1972–1974, country: Germany[28], *"summermilk" (without supplementary feeding)

- Co-contamination: not reported
- Further contamination of animal products reported in the present book/article[28]: Cheese (Camembert cheese), Cheese (fresh cheese), Cheese (hard cheese), Cheese (processed cheese), Milk powder, Yogurt, AFM_1, literature[28]

incidence: 31/225, conc. range: 0.001–0.010 µg/l, sample origin: West Germany, imported from Italy, sample year: 1984, country: Italy[32]

incidence: 11/77, conc. range: 0.001–0.010 μg/l (10 sa.), 0.011–0.050 μg/l (1 sa.), sample origin: France, imported from Italy, sample year: 1984, country: Italy[32]

incidence: 70/276, conc. range: 0.001–0.010 μg/l (39 sa.), 0.011–0.050 μg/l (24 sa.), 0.051–0.100 μg/l (5 sa.), 0.101–0.200 μg/l (2 sa.), sample origin: Lombardy plain, Italy, sample year: spring–summer 1985, country: Italy[32]

- Co-contamination: not reported
- Further contamination of animal products reported in the present book/article[32]: Cheese, AFM_1, literature[32]

incidence: 176/177*, conc. range: ≤0.0687 μg/kg, sample origin: local and imported sa. purchased in supermarkets in Kuwait City (capital), Kuwait, sample year: January 2005–March 2007, country: Kuwait[41], *111 local (0.0049–0.0687 μg/kg) and 66 imported (≤0.0603 μg/kg) fresh cow milk sa.

- Co-contamination: not reported
- Further contamination of human breast milk and animal products reported in the present book/article[41]: Cheese, Milk (human breast milk), Milk (UHT milk), Milk powder, AFM_1, literature[41]

For detailed information see the article.

incidence: 3/71*, conc. range: 0.03–0.04 μg/l, ∅ conc.: 0.035 μg/l, sample origin: farms, Cyprus, sample year: 1993/1995/1996, country: Cyprus[50], *raw cow milk

- Co-contamination: not reported
- Further contamination of animal products reported in the present book/article[50]: Butter (peanut butter), AFB_1, AFB_2, AFG_1, and AFG_2, literature[50]; Milk, AFM_1, literature[50]

incidence: 4/12, conc. range: 0.028–0.164 μg/kg, ∅ conc.: 0.086 μg/kg, sample origin: Lucknow (city), India, sample year: unknown, country: India[91]

- Co-contamination: not reported
- Further contamination of animal products reported in the present book/article[91]: Food, Milk (infant formula), AFM_1, literature[91]

incidence: 128/128*, conc. range: 0.031–0.113 μg/kg, ∅ conc.: 0.0722 μg/kg, sample origin: Tehran (capital), Iran, sample year: April–September 2005, country: Iran[92], *pasteurized liquid cow milk

- Co-contamination: not reported
- Further contamination of cereals and cereal products reported in the present book/article[92]: Milk (infant formula), AFM_1, literature[92]

incidence: 1/1*, conc.: 0.26 μg/l, sample origin: local dairies, sample year: unknown, country: France/UK[129], *fate of AFM_1 during the manufacture of Camembert cheese

- Co-contamination: not reported
- Further contamination of animal products reported in the present book/article[129]: Cheese (Camembert cheese), Cheese curd, Cheese whey, AFM_1, literature[129]

incidence: 18/38*, conc. range: 0.01–0.04 μg/kg (18 sa., maximum: 0.03 μg/kg), sample origin: retail outlets, England and Wales, sample year: summer 1994, country: UK[131], *skimmed and semi-skimmed cow milk

incidence: 19/36*, conc. range: 0.01–0.04 μg/kg (19 sa., maximum: 0.03 μg/kg), sample origin: retail outlets, England and Wales, sample year: winter 1994–1995, country: UK[131], *skimmed and semi-skimmed cow milk

incidence: 21/48*, conc. range: 0.01–0.04 μg/kg (20 sa.), 0.22 μg/kg (1 sa.), sample origin: retail outlets, England and Wales, sample year: summer 1994, country: UK[131], *full fat cow milk

incidence: 31/40*, conc. range: 0.01–0.04 μg/kg (31 sa., maximum: 0.02 μg/kg), sample origin: retail outlets, England and Wales, sample year: winter 1994–1995, country: UK[131], *full fat cow milk

- Co-contamination: not reported
- Further contamination of animal products reported in the present book/article[131]: Cheese (Cheddar cheese), Cheese (Cheshire cheese), Cheese (Double Gloucester cheese), Cheese (Lancashire cheese), Cheese (Leicester cheese), Cheese (Wensleydale cheese), Milk (infant formula), Yogurt, AFM_1, literature[131]

incidence: 3/15, conc. range: 5–8 μg/l, ∅ conc.: 6.3 μg/l, sample origin: Egyptians markets, Egypt, sample year: 1999–2000, country: Egypt[138]

- Co-contamination: not reported
- Further contamination of human breast milk and animal products reported in the present book/article[138]: Cheese (Domiati cheese), Milk powder, AFM_1, literature[138]; Cheese (Ras cheese), AFB_1, AFG_1, and AFM_1, literature[138]; Milk (human breast milk), AFM_1 and OTA, literature[138]

incidence: 7/9, conc. range: 0.03–0.73 μg/l, ∅ conc.: 0.31 μg/l, sample origin: local dairy factories in Sabratha and Zuwarah (provinces), north-west of Libya, sample year: July–August 2002, country: UK[150]

incidence: 12/17, conc. range: 0.08–0.28 μg/l, ∅ conc.: 0.12 μg/l, sample origin: local dairy factories in Az zawiyha (province), north-west of Libya, sample year: July–August 2002, country: UK[150]

incidence: 8/12, conc. range: 0.08–3.13 μg/l, ∅ conc.: 0.49 μg/l, sample origin: local dairy factories in Az iziyah (province), north-west of Libya, sample year: July–August 2002, country: UK[150]

incidence: 8/11, conc. range: 0.08–2.68 μg/l, ∅ conc.: 0.72 μg/l, sample origin: local dairy factories in Tripoli (capital), Libya, sample year: July–August 2002, country: UK[150]

- Co-contamination: not reported
- Further contamination of animal products reported in the present book/article[150]: Cheese (white cheese), AFM_1, literature[150]

incidence: 10/63, conc. range: 0.006–<0.05 μg/l (4 sa.), >0.05 μg/l (6 sa., maximum: 0.527 μg/l), sample origin: market or from a herd, Cameroon, sample year: 1991–1995, country: Cameroon[156]

- Co-contamination: not reported
- Further contamination of human breast milk and animal products reported in the present book/article[156]: Eggs, AF/S, literature[156]; Milk (human breast milk), AFM_1, literature[156]

incidence: 4/12, conc. range: 0.0073–0.0118 μg/kg, ∅ conc.: 0.0093 μg/kg, sample origin: retail shops in Marang and Kuala Terengganu (town and city), Malaysia, sample year: July 2013?, country: Malaysia[163]

- Co-contamination: not reported
- Further contamination of animal products reported in the present book/article[163]: Cheese, Dairy beverages, Milk, Milk powder, Yogurt, AFM_1, literature[163]

incidence: 25/26*, conc. range: 0.005–0.029 μg/l (6 sa.), 0.030–0.049 μg/l (2 sa.), 0.050–0.099 μg/l (12 sa.), 0.100–0.200 μg/l

(5 sa., maximum: 1.154 µg/l?), sample origin: supermarkets in Shanghai (city) and Beijing (capital), China, sample year: July-September? 2010, country: China[193], *pasteurized cow milk

- Co-contamination: not reported
- Further contamination of animal products reported in the present book/article[193]: Milk (UHT milk), AFM_1, literature[193]

incidence: 22/30*, conc. range: 0.005–0.010 µg/l (7 sa.), 0.011–0.020 µg/l (10 sa.), 0.021–0.050 µg/l (4 sa.), >0.050 µg/l (1 sa.), sample origin: different milk producers, Greece, sample year: December 1999–May 2000, country: Greece[195], *raw cow milk

incidence: 18/28*, conc. range: 0.005–0.010 µg/l (3 sa.), 0.011–0.020 µg/l (10 sa.), 0.021–0.050 µg/l (4 sa.), >0.050 µg/l (1 sa.), sample origin: supermarkets, Greece, sample year: December 2000–May 2001, country: Greece[195], *raw cow milk

incidence: 18/23*, conc. range: 0.005–0.010 µg/l (12 sa.), 0.011–0.020 µg/l (4 sa.), 0.021–0.050 µg/l (2 sa.), sample origin: milk factories, Greece, sample year: December 2000–May 2001, country: Greece[195], *bulk tank cow milk

- Co-contamination: not reported
- Further contamination of animal products reported in the present book/article[195]: Milk, Milk (goat milk), Milk (sheep milk), Milk (UHT milk), AFM_1, literature[195]

incidence: 66/67*, conc. range: >0–0.05 µg/l (9 sa.), >0.05–0.125 µg/l (16 sa.), >0.125–0.25 µg/l (19 sa.), >0.25–0.5 µg/l (5 sa.), >0.5 µg/l (17 sa.), sample origin: 50-liter milk cans, Thailand, sample year: June 1995–January 1996, country: Thailand[196], *raw cow milk

incidence: 63/63*, conc. range: >0.05–0.125 µg/l (13 sa.), >0.125–0.25 µg/l (16 sa.), >0.25–0.5 µg/l (14 sa.), >0.5 µg/l (20 sa.), sample origin: retail markets, Thailand, sample year: May 1995–January 1996, country: Thailand[196], *pasteurized cow milk

incidence: 60/60*, conc. range: >0.05–0.125 µg/l (12 sa.), >0.125–0.25 µg/l (19 sa.), >0.25–0.5 µg/l (26 sa.), >0.5 µg/l (3 sa.), sample origin: retail markets, Thailand, sample year: May 1995–January 1996, country: Thailand[196], *sterilized cow milk

- Co-contamination: not reported
- Further contamination of animal products reported in the present book/article[196]: Milk (UHT milk), Milk powder, AFM_1, literature[196]

incidence: 2/15*, conc. range: 0.0394–0.1012 µg/l, Ø conc.: 0.0703 µg/l, sample origin: supermarkets and drugstores in São Paulo (city), Brazil, sample year: April 1998–May 1999, country: Brazil[197], *pasteurized cow milk

- Co-contamination: not reported
- Further contamination of animal products reported in the present book/article[197]: Milk (UHT milk), AFM_1, literature[197]

incidence: 90/91*, Ø conc.: 0.092 µg/l, sample origin: retail outlets, Anand (town), India, sample year: unknown, country: India[198], *market cow milk (standardized)

incidence: 30/34*, Ø conc.: 0.143 µg/l, sample origin: villages around Anand (town) and herds of GAU of Anand campus, India, sample year: unknown, country: India[198], *raw individual cow milk

incidence: 31/32*, Ø conc.: 0.110 µg/l, sample origin: villages around Anand (town) and herds of GAU of Anand campus, India, sample year: unknown, country: India[198], *raw bulk cow milk

- Co-contamination: not reported
- Further contamination of animal products reported in the present book/article[198]: Milk (buffalo milk), AFM_1, literature[198]

incidence: 4/52*, conc. range: 0.073–0.370 µg/l, Ø conc.: 0.155 µg/l, sample origin: groceries and supermarkets in Campinas (city), Brazil, sample year: 1989–1990, country: Brazil[199], *pasteurized cow milk

- Co-contamination: not reported
- Further contamination of animal products reported in the present book/article[199]: not reported

incidence: 43/43*, conc. range: 0.0031–0.05 µg/l (30 sa.), 0.05–0.50 µg/l (13 sa., maximum: 0.1057 µg/l), Ø conc.: 0.0359 µg/l, sample origin: bulk tanks at dairy farms in Minas Gerais (state), Brazil, sample year: August 2009–February 2010 (dry period), country: Brazil[203], *raw cow milk

incidence: 43/43*, conc. range: 0.0017–0.05 µg/l (38 sa.), 0.05–0.50 µg/l (5 sa., maximum: 0.0709 µg/l), Ø conc.: 0.0171 µg/l, sample origin: bulk tanks at dairy farms in Minas Gerais (state), Brazil, sample year: August 2009–February 2010 (transition period), country: Brazil[203], *raw cow milk

incidence: 43/43*, conc. range: 0.0002–0.05 µg/l (43 sa., maximum: 0.0249 µg/l), Ø conc.: 0.0055 µg/l, sample origin: bulk tanks at dairy farms in Minas Gerais (state), Brazil, sample year: August 2009–February 2010 (rainy period), country: Brazil[203], *raw cow milk

- Co-contamination: not reported
- Further contamination of animal products reported in the present book/article[203]: not reported

incidence: 3/5*, conc. range: 0.01–0.03 µg/l, Ø conc.: 0.02 µg/l, sample origin: groceries and supermarkets in Kuwait (capital), Kuwait, sample year: July–September 1998, country: Kuwait[205], *skimmed fresh cow milk

incidence: 5/7*, conc. range: 0.02–0.16 µg/l, Ø conc.: 0.068 µg/l, sample origin: groceries and supermarkets in Kuwait (capital), Kuwait, sample year: July–September 1998, country: Kuwait[205], *full cream fresh cow milk

incidence: 5/9*, conc. range: 0.20–0.21 µg/l, Ø conc.: 0.206 µg/l, sample origin: private local producers, outskirts of Kuwait (capital), Kuwait, sample year: July–September 1998, country: Kuwait[205], *fresh cow milk

- Co-contamination: not reported
- Further contamination of animal products reported in the present book/article[205]: not reported

incidence: 11/118*, conc. range: 0.01–0.04 µg/kg (9 sa.), 0.05–0.10 µg/kg (1 sa.), 0.18 µg/kg (1sa.), sample origin: farms throughout England and Wales, UK, sample year: 1988, country: UK[207], *farm cow milk

incidence: 26/127*, conc. range: 0.01–0.04 µg/kg (20 sa.), 0.05–0.10 µg/kg (5 sa.), 0.16 µg/kg (1 sa.), sample origin: farms throughout England and Wales, UK, sample year: 1989, country: UK[207], *farm cow milk

incidence: 13/79*, conc. range: 0.01–0.04 µg/kg (10 sa.), 0.05–0.10 µg/kg (3 sa., maximum: 0.09 µg/kg), sample origin: farms throughout England and Wales, UK, sample year: later than 1989, country: UK[207], *farm cow milk

- Co-contamination: not reported
- Further contamination of animal products reported in the present book/article[207]: not reported

incidence: 0/40*, conc.: no contamination, sample year: unknown, country: UK[208], *conventional retail cow milk

incidence: 3/40*, conc. range: 0.01–0.021 µg/l, sample year: unknown, country: UK[208], *conventional farm-gate cow milk incidence: 0/10*, conc.: no contamination, sample year: unknown, country: UK[208], *organic retail cow milk

incidence: 0/10*, conc.: no contamination, sample year: unknown, country: UK[208], *organic farm-gate cow milk

- Co-contamination: not reported
- Further contamination of animal products reported in the present book/article[208]: not reported

incidence: 48/54*, conc. range: 0.001–0.117 µg/l, sample origin: local markets of Rabat (capital), Morocco, sample year: February–April 2006, country: Morocco/Spain[209], *pasteurized cow milk

- Co-contamination: not reported
- Further contamination of animal products reported in the present book/article[209]: not reported

incidence: 624/624*, conc. range: <0.045 µg/l (390 sa.), 0.045–0.050 µg/l (123 sa.), 0.050–0.080 µg/l (94 sa.), >0.080 µg/l (17 sa.), sample origin: supermarkets in Shiraz (city), Iran, sample year: April–September 2003, country: Iran[210], *pasteurized cow milk

- Co-contamination: not reported
- Further contamination of animal products reported in the present book/article[210]: not reported

incidence: 60/94*, conc. range: <0.025 µg/l (30 sa.), >0.025 µg/l (23 sa.), >0.05 µg/l (7 sa., maximum: 0.07 µg/l), sample origin: dairy farms in the Villa Maria basin, Cordoba (province), Argentina, sample year: March–September 2007, country: Argentina[211], *raw whole cow milk from bulk tanks

- Co-contamination: not reported
- Further contamination of animal products reported in the present book/article[211]: not reported

incidence: 9/264*, conc. range: 0.005–<0.008 µg/l (6 sa.), 0.008–0.026 µg/l (3 sa.), sample origin: dairy farms in Aquitaine, Brittany, Poitou-Charentes, and Pays de Loire (regions), western France, sample year: February–March and September–October 2003, country: France[212], *raw bulk cow milk

- Co-contamination: not reported
- Further contamination of animal products reported in the present book/article[212]: Milk (cow milk), OTA, literature[212]

incidence: 65/113*, conc. range: 0.005–0.025 µg/l, ∅ conc.: 0.00853 µg/l, sample origin: dairy farms in Yogyakarta (province), Indonesia, sample year: 2006, country: Indonesia/Austria[213], *fresh cow milk

- Co-contamination: not reported
- Further contamination of animal products reported in the present book/article[213]: not reported

incidence: 25/31*, conc. range: 0.005–0.020 µg/l (19 sa.), 0.021–0.040 µg/l (4 sa.), 0.041–0.050 µg/l (2 sa.), sample origin: Portuguese dairy factories, Portugal, sample year: 1999, country: Portugal[219], *raw cow milk

incidence: 47/104*, conc. range: 0.005–0.020 µg/l (36 sa.), 0.021–0.040 µg/l (11 sa.), sample origin: Portuguese dairy factories, Portugal, sample year: 2000, country: Portugal[219], *raw cow milk

incidence: 74/107*, conc. range: 0.005–0.020 µg/l (62 sa.), 0.021–0.040 µg/l (8 sa.), 0.041–0.050 µg/l (1 sa.), 0.051–0.080 µg/l (3 sa.), sample origin: Portuguese dairy factories, Portugal, sample year: 2001, country: Portugal[219], *raw cow milk

incidence: 100/118*, conc. range: 0.005–0.020 µg/l (45 sa.), 0.021–0.040 µg/l (26 sa.), 0.041–0.050 µg/l (6 sa.), 0.051–0.080 µg/l (23 sa.), sample origin: Portuguese dairy factories, Portugal, sample year: 2002, country: Portugal[219], *raw cow milk

incidence: 78/104*, conc. range: 0.005–0.020 µg/l (58 sa.), 0.021–0.040 µg/l (10 sa.), 0.041–0.050 µg/l (2 sa.), 0.051–0.080 µg/l (8 sa.), sample origin: Portuguese dairy factories, Portugal, sample year: 2003, country: Portugal[219], *raw cow milk

incidence: 70/134*, conc. range: 0.005–0.020 µg/l (34 sa.), 0.021–0.040 µg/l (19 sa.), 0.041–0.050 µg/l (2 sa.), 0.051–0.080 µg/l (15 sa.), sample origin: Portuguese dairy factories, Portugal, sample year: 2004, country: Portugal[219], *raw cow milk

- Co-contamination: not reported
- Further contamination of animal products reported in the present book/article[219]: not reported

incidence: 3/20*, conc. range: 0.007–0.014 µg/kg, sample origin: Campania and Calabria (regions), southern Italy, sample year: 2002, country: Italy[225], *raw cow milk

incidence: 22/42*, conc. range: 0.006–0.244 µg/kg, sample origin: Campania and Calabria (regions), southern Italy, sample year: 2003, country: Italy[225], *raw cow milk

incidence: 33/59*, conc. range: 0.006–0.770 µg/kg, sample origin: Campania and Calabria (regions), southern Italy, sample year: 2004, country: Italy[225], *raw cow milk

incidence: 68/114*, conc. range: 0.004–1.262 µg/kg, sample origin: Campania and Calabria (regions), southern Italy, sample year: 2005, country: Italy[225], *raw cow milk

incidence: 5/9*, conc. range: 0.004–0.064 µg/kg, sample origin: Campania and Calabria (regions), southern Italy, sample year: 2002, country: Italy[225], *heat-treated cow milk (pasteurized or UHT milk)

incidence: 6/11*, conc. range: 0.004–0.020 µg/kg, sample origin: Campania and Calabria (regions), southern Italy, sample year: 2003, country: Italy[225], *heat-treated cow milk (pasteurized or UHT milk)

incidence: 15/23*, conc. range: 0.006–0.029 µg/kg, sample origin: Campania and Calabria (regions), southern Italy, sample year: 2004, country: Italy[225], *heat-treated cow milk (pasteurized or UHT milk)

incidence: 28/50*, conc. range: 0.005–0.088 µg/kg, sample origin: Campania and Calabria (regions), southern Italy, sample year: 2005, country: Italy[225], *heat-treated cow milk (pasteurized or UHT milk)

- Co-contamination: not reported
- Further contamination of animal products reported in the present book/article[225]: Milk (buffalo milk), Milk (sheep milk), AFM_1, literature[225]

incidence: 26/50, conc. range: 0.010–0.050 µg/l (9 sa.), 0.051–0.100 µg/l (9 sa.), 0.101–0.150 µg/l (4 sa.), 0.151–0.200 µg/l (2 sa.), 0.201–0.250 µg/l (2 sa., maximum: 0.220 µg/l), sample origin: different areas in the Ismailia Governorate, Egypt, sample year: summers 2003–2004, country: Egypt/Germany/USA[226]

- Co-contamination: not reported
- Further contamination of animal products reported in the present book/article[226]: Milk (buffalo milk), Milk (camel milk), Milk (goat milk), AFM_1, literature[226]

incidence: 59/75, conc. range: 0.005–0.020 µg/l (18 sa.), 0.021–0.050 µg/l (14 sa.), >0.050 µg/l (27 sa.), ∅ conc.: 0.0601 µg/l, sample origin: different parts of Khuzestan (province), Iran, sample year: November 2007–December 2008, country: Iran[227]

- Co-contamination: not reported

- Further contamination of animal products reported in the present book/article[227]: Milk (buffalo milk), Milk (camel milk), Milk (goat milk), Milk (sheep milk), AFM_1, literature[227]

incidence: 5*/13**, conc. range: ≤0.088 µg/kg, sample origin: retailers and traders in Dagoretti (low-income area), Kenya, sample year: November 2013–October 2014, country: Kenya/Sweden[228], *>0.050 µg/kg, **boiled cow milk

incidence: 0*/0**, conc.: not analyzed, sample origin: supermarket in Westlands (middle- to high-income area), Nairobi (capital), Kenya, sample year: November 2013–October 2014, country: Kenya/Sweden[228], *>0.050 µg/kg, **boiled cow milk

incidence: 46*/62**, conc. range: ≤1.100 µg/kg, sample origin: retailers and traders in Dagoretti (low-income area), Kenya, sample year: November 2013–October 2014, country: Kenya/Sweden[228], *>0.050 µg/kg, **raw cow milk

incidence: 0*/0**, conc.: not analyzed, sample origin: supermarket in Westlands (middle- to high-income area), Nairobi (capital), Kenya, sample year: November 2013–October 2014, country: Kenya/Sweden[228], *>0.050 µg/kg, **raw cow milk

incidence: 13*/18**, conc. range: ≤0.740 µg/kg, sample origin: retailers and traders in Dagoretti (low-income area), Kenya, sample year: November 2013–October 2014, country: Kenya/Sweden[228], *>0.050 µg/kg, **pasteurized cow milk

incidence: 26*/53**, conc. range: ≤0.210 µg/kg, sample origin: supermarket in Westlands (middle- to high-income area), Nairobi (capital), Kenya, sample year: November 2013–October 2014, country: Kenya/Sweden[228], *>0.050 µg/kg, **pasteurized cow milk

- Co-contamination: not reported
- Further contamination of animal products reported in the present book/article[228]: Dairy products, Milk (UHT milk), Yogurt, AFM_1, literature[228]

incidence: 27/89*, conc. range: 0.025–0.1 µg/l (21 sa.), 0.1–0.5 µg/l (6 sa.), sample origin: dairy farms, Czechoslovakia, sample year: 1988–1989, country: Czechoslovakia[232], *raw cow milk

incidence: 16/314*, conc. range: 0.025–0.1 µg/l (16 sa.), sample origin: dairy plant, Czechoslovakia, sample year: 1988–1989, country: Czechoslovakia[232], *pasteurized cow milk

incidence: 28/168, conc. range: 0.025–0.1 µg/l (26 sa.), >0.1 µg/l (2 sa.), sample origin: dairy plant for production of infant and children milk-based foods, Czechoslovakia, sample year: autumn 1987, country: Czechoslovakia[232]

incidence: 18/208, conc. range: 0.025–0.1 µg/l (18 sa.), sample origin: dairy plant for production of infant and children milk-based foods, Czechoslovakia, sample year: spring 1988, country: Czechoslovakia[232]

- Co-contamination: not reported
- Further contamination of animal products reported in the present book/article[232]: not reported

incidence: 24/409*, conc. range: 0.02–0.05 µg/kg (10 sa.), 0.05–0.1 µg/kg, (6 sa.), >0.1 µg/kg (8 sa.), sample origin: farms in England, North Ireland, Scotland, sample year: 1981–1983, country: UK[235], *probably cow milk

- Co-contamination: not reported
- Further contamination of animal products reported in the present book/article[235]: Milk powder, AFM_1, literature[235]

incidence: 8/31*, conc. range: 0.005–0.024 µg/kg (7 sa.), 0.091 µg/kg (1 sa.), sample origin: dairy farms in southern Italy, sample year: unknown, country: Italy[236], *raw farm cow milk

incidence: 59/66*, conc. range: 0.004–0.024 µg/kg (43 sa.), 0.025–0.049 µg/kg (11 sa.), 0.050–0.099 µg/kg (4 sa.), 0.150 µg/kg (1 sa.), sample origin: dairy farms in southern Italy, sample year: unknown, country: Italy[236], *heat-treated cow milk

- Co-contamination: not reported
- Further contamination of animal products reported in the present book/article[236]: Milk powder, AFM_1, literature[236]

incidence: 205/5489, conc. range: 0.05–0.50 µg/l (200 sa.), >0.5 µg/l (5 sa.), sample origin: monitoring programs, France, sample year: 1978–1991, country: France[237]

incidence: 98/757*, conc. range: 0.05–0.50 µg/l (84 sa.), >0.5 µg/l (14 sa.), sample origin: dairy plant, France, sample year: 1979–1992, country: France[237], *bulk raw cow milk

- Co-contamination: not reported
- Further contamination of animal products reported in the present book/article[237]: not reported

incidence: 4/99*, conc. range: 0.10–0.13 µg/kg, sample origin: farms in northern Greece, sample year: November 1986–March 1987, country: Greece[238], *raw cow milk

incidence: 0/51*, conc.: no contamination, sample origin: dairy plants and retail shops, Greece, sample year: November 1986–March 1987, country: Greece[238], *pasteurized cow milk

- Co-contamination: not reported
- Further contamination of animal products reported in the present book/article[238]: not reported

incidence: 12/36, conc. range: 0.04–0.25 µg/l, sample origin: single farms, Germany, sample year: February–April, country: Germany[239]

incidence: 7/7, conc. range: 0.05–0.13 µg/l, sample origin: tanker bulk milk, Germany, sample year: February–March, country: Germany[239]

incidence: 9/12, conc. range: 0.04–0.08 µg/l, sample origin: tanker bulk milk, Germany, sample year: May, country: Germany[239]

incidence: 0/6, conc.: no contamination, sample origin: tanker bulk milk, Germany, sample year: June, country: Germany[239]

- Co-contamination: not reported
- Further contamination of animal products reported in the present book/article[239]: not reported

incidence: 79/419, conc. range: 0.02–0.54 µg/l, Ø conc.: 0.12 µg/l, sample origin: dairy plant "Weihenstephan," Germany, sample year: March–April 1976, country: Germany[240]

- Co-contamination: not reported
- Further contamination of animal products reported in the present book/article[240]: not reported

incidence: 39/70*, conc. range: 0.015–0.052 µg/l, Ø conc.: 0.031 µg/l, sample origin: supermarkets in Seoul (capital), Korea, sample year: 1997, country: Korea[241], *pasteurized cow milk

- Co-contamination: not reported
- Further contamination of animal products reported in the present book/article[241]: Milk (infant formula), Milk powder, Yogurt, AFM_1, literature[241]

incidence: 9/13, conc. range: 0.01–0.80 µg/l, Ø conc.: 0.2 µg/l, sample origin: college station and Texas Department of Health, Texas (state), USA, sample year: unknown, country: USA[242]

- Co-contamination: not reported
- Further contamination of animal products reported in the present book/article[242]: not reported

incidence: 9/67, conc. range: 0.05–0.10 µg/l, sample origin: unknown, sample year: spring 1980, country: Czechoslovakia[243]

incidence: 0/50, conc.: no contamination, sample origin: unknown, sample year: spring 1982, country: Czechoslovakia[243]

- Co-contamination: not reported
- Further contamination of animal products reported in the present book/article[243]: not reported

incidence: 84/105*, conc. range: 0.015–0.090 µg/l, ∅ conc.: 0.03 µg/l, sample origin: different regions, Netherlands, sample year: 1981, country: Netherlands[244], *raw and heated cow milk

- Co-contamination: not reported
- Further contamination of animal products reported in the present book/article[244]: not reported

incidence: 32/35*, conc. range: 0.0025–0.1770 µg/l, ∅ conc.: 0.01280 µg/l, sample origin: Athens (capital) market, Greece, sample year: winter 1995, country: Greece[245], *pasteurized cow milk

incidence: 13/18*, conc. range: 0.0012–0.0140 µg/l, ∅ conc.: 0.00179 µg/l, sample origin: Athens (capital) market, Greece, sample year: spring 1996, country: Greece[245], *pasteurized cow milk

incidence: 14/14*, conc. range: 0.0005–0.1700 µg/l, ∅ conc.: 0.0130 µg/l, sample origin: Athens (capital) market, Greece, sample year: summer 1996, country: Greece[245], *pasteurized cow milk

incidence: 13/14*, conc. range: 0.0025 µg/l?, ∅ conc.: 0.0025 µg/l?, sample origin: Athens (capital) market, Greece, sample year: autumn 1996, country: Greece[245], *pasteurized cow milk

- Co-contamination: not reported
- Further contamination of animal products reported in the present book/article[245]: not reported

incidence: 3/20*, conc. range: >0.010–0.067 µg/l, ∅ conc.: 0.0497 µg/l, sample origin: dairy farms in Andria, Gravina, Minervino, and Ruvo (city/towns and communes), Puglia (region), Italy, sample year: January–February 1999, country: Italy[247], *bulk cow milk in refrigerated tanks

incidence: 10/30*, conc. range: >0.010–0.138 µg/l, ∅ conc.: 0.0620 µg/l, sample origin: dairy farms in Altamura, Casamassima, and Gioia del Colle (city/towns and communes), Puglia (region), Italy, sample year: January–February 1999, country: Italy[247], *bulk cow milk in refrigerated tanks

incidence: 3/50*, conc. range: >0.010–0.091 µg/l, ∅ conc.: 0.0487 µg/l, sample origin: dairy farms in Castellana, Laterza, and Martina Franca (towns and communes/municipality), Puglia (region), Italy, sample year: January–February 1999, country: Italy[247], *bulk cow milk in refrigerated tanks

- Co-contamination: not reported
- Further contamination of animal products reported in the present book/article[247]: not reported

incidence: 33/48*, conc. range: 0.005–0.020 µg/kg (14 sa.), >0.020 < 0.050 µg/kg (11 sa.), ≥0.050 µg/kg (8 sa.), sample origin: milk tank on the farms or dairies in mid-eastern Sweden, sample year: January 1986, country: Sweden[248], *milk sa. resulted from fed grain treated with formic acid, 700 g/l

incidence: 26/46*, conc. range: 0.005–0.020 µg/kg (11 sa.), >0.020 < 0.050 µg/kg (6 sa.), ≥0.050 µg/kg (9 sa.), sample origin: milk tank on the farms or dairies in mid-eastern Sweden, sample year: January 1986, country: Sweden[248], *milk sa. resulted from fed grain treated with formic acid, 850 g/l

incidence: 4/4*, conc. range: 0.005–0.020 µg/kg (2 sa.), >0.02 < 0.050 µg/kg (1 sa.), ≥0.050 µg/kg (1 sa.), sample origin: milk tank on the farms or dairies in mid-eastern Sweden, sample year: January 1986, country: Sweden[248], *milk sa. resulted from fed grain treated with formic acid mixed with other acids

incidence: 29/48*, conc. range: 0.005–0.020 µg/kg (27 sa.), >0.020 < 0.050 µg/kg (2 sa.), sample origin: milk tank on the farms or dairies in mid-eastern Sweden, sample year: January 1986, country: Sweden[248], *milk sa. resulted from fed grain treated with propionic acid

incidence: 61/121*, conc. range: 0.005–0.020 µg/kg (59 sa.), >0.020 < 0.050 µg/kg (1 sa.), ≥0.050 µg/kg (1 sa.), sample origin: milk tank on the farms or dairies in mid-eastern Sweden, sample year: January 1986, country: Sweden[248], *milk sa. resulted from fed grain which was dried or unspecified

- Co-contamination: not reported
- Further contamination of animal products reported in the present book/article[248]: not reported

For detailed information see the article.

incidence: 5/21, conc. range: <0.16 µg/kg (3 sa.), ~0.16 µg/kg (2 sa.), sample origin: commercial outlets in Lydenburg and Middleburg (towns), Pietersburg and Witbank (cities), and Pretoria (capital), South Africa, sample year: unknown, country: South Africa[249]

- Co-contamination: not reported
- Further contamination of animal products reported in the present book/article[249]: not reported

incidence: 89/504, conc. range: 0.10–3.50 µg/l, ∅ conc.: 1.159 µg/l, sample origin: villages in Thrissur (district), India, sample year: September 1992–May 1993, country: India[250]

- Co-contamination: not reported
- Further contamination of animal products reported in the present book/article[250]: not reported

For detailed information see the article.

incidence: 1/100*, conc.: 0.2 µg/l, sample origin: São Paulo (state), Brazil, sample year: July 1979–September 1981, country: Brazil[251], *commercial cow milk (pasteurized)

incidence: 7/50*, conc. range: 0.1–1.68 µg/l, ∅ conc.: 0.59 µg/l, sample origin: farms in the Médio Vale do Paraiba (area), Brazil, sample year: July 1979–September 1981, country: Brazil[251], *farm cow milk

- Co-contamination: not reported
- Further contamination of animal products reported in the present book/article[251]: not reported

incidence: 50/85*, conc. range: ≤0.15 µg/l, sample origin: dairy farms in Assiut (province), Egypt, sample year: October 1999–February 2000, country: Egypt[252], *raw cow milk

- Co-contamination: not reported
- Further contamination of animal products reported in the present book/article[252]: not reported

For detailed information see the article.

incidence: 1/60, conc.: 0.050–0.100 µg/l, sample origin: different parts of Sweden, sample year: spring 1983, country: Sweden[253]

incidence: 13/13, conc. range: 0.005–0.036 µg/l, ∅ conc.: 0.017 µg/l, sample origin: different parts of Sweden, sample year: spring 1983, country: Sweden[253]

- Co-contamination: not reported
- Further contamination of animal products reported in the present book/article[253]: not reported

incidence: 5/77, conc. range: tr-0.38 μg/l*, sample origin: from farms and tank trucks, Czechoslovakia, sample year: unknown, country: Czechoslovakia[254], *only farm cow milk contaminated

- Co-contamination: not reported
- Further contamination of animal products reported in the present book/article[254]: not reported

incidence: 33/77, conc. range: tr-0.2 μg/l (22 sa.), 0.3–0.4 μg/l (7 sa.), 0.5–0.7 μg/l (3 sa.), >0.7 μg/l (1 sa.), sample origin: bottled in Alabama (state), USA, sample year: October–November 1977, country: USA[255]

incidence: 60/75, conc. range: tr-0.2 μg/l (38 sa.), 0.3–0.4 μg/l (16 sa.), 0.5–0.7 μg/l (2 sa.), >0.7 μg/l (4 sa.), sample origin: bottled in Georgia (state), USA, sample year: October–November 1977, country: USA[255]

incidence: 45/75, conc. range: tr-0.2 μg/l (28 sa.), 0.3–0.4 μg/l (11 sa.), 0.5–0.7 μg/l (6 sa.), sample origin: bottled in South Carolina (state), USA, sample year: October–November 1977, country: USA[255]

incidence: 53/75, conc. range: tr-0.2 μg/l (38 sa.), 0.3–0.4 μg/l (12 sa.), 0.5–0.7 μg/l (3 sa.), sample origin: bottled in North Carolina (state), USA, sample year: October–November 1977, country: USA[255]

- Co-contamination: not reported
- Further contamination of animal products reported in the present book/article[255]: not reported

incidence: 17/50, conc. range: 0.6–15 μg/kg*, sample origin: public sector dairy farm in Andhra Pradesh (state), southern India, sample year: 2001–2002, country: India/UK[256], *authors pointed out the high levels of AFM_1 in cow milk

- Co-contamination: not reported
- Further contamination of animal products reported in the present book/article[256]: Dairy products, Milk, Milk (buffalo milk), Milk powder, AFM_1, literature[256]

incidence: 64/64, conc. range: 0.0056–0.280 μg/l, Ø conc.: 0.0804 μg/l, sample origin: markets and farms, Korea, sample year: unknown, country: Korea[257]

- Co-contamination: not reported
- Further contamination of animal products reported in the present book/article[257]: not reported

incidence: 79/90*, conc. range: 0.0302–0.0636 μg/l, sample origin: milk producers, Turkey, sample year: unknown, country: Turkey[258], *raw cow milk

- Co-contamination: not reported
- Further contamination of animal products reported in the present book/article[258]: not reported

incidence: 25/31*, conc. range: 0.005–0.010 μg/l (17 sa.), 0.011–0.020 μg/l (2 sa.), 0.021–0.050 μg/l (6 sa.), sample origin: farms, Portugal, sample year: June–September 1999, country: Portugal[259], *raw cow milk

- Co-contamination: not reported
- Further contamination of animal products reported in the present book/article[259]: Milk (UHT milk), AFM_1, literature[259]

incidence: 10/42*, conc. range: 0.29505–1.9749 μg/l, Ø conc.: 0.68485 μg/l, sample origin: farms in 12 municipal districts of Paraná (state), Brazil, sample year: July 2001–November 2002, country: Brazil[260], *raw cow milk

- Co-contamination: not reported
- Further contamination of animal products reported in the present book/article[260]: not reported

incidence: 6/30*, conc. range: 0.0036–0.0106 μg/kg, sample origin: farms in Cracow (district), southern Poland, sample year: 1993–1994, country: Poland/Germany[262], *milk sa. collected during the indoor period of the cows feeding cycle

incidence: 37/157*, conc. range: <0.001 μg/kg (25 sa.), 0.001–0.05 μg/kg (12 sa., maximum: 0.025 μg/kg), sample origin: individual suppliers in Cracow (district), southern Poland, sample year: 1993–1994, country: Poland/Germany[262], *milk sa. collected during the indoor period of the cows feeding cycle

- Co-contamination: not reported
- Further contamination of animal products reported in the present book/article[262]: not reported

incidence: 28/36*, conc. range: 0.010–0.050 μg/l (18 sa.), 0.051–0.100 μg/l (5 sa.), 0.101–0.200 μg/l (3 sa.), 0.201–0.300 μg/l (2 sa., maximum : 0.251 μg/l), Ø conc.: 0.06 μg/l, sample origin: bakeries and supermarkets in São Paulo (city), Brazil, sample year: July–December 2004, country: Brazil[263], *pasteurized cow milk of grade A–C

- Co-contamination: not reported
- Further contamination of animal products reported in the present book/article[263]: Milk (cow milk), CPA, literature[263]; Milk (UHT milk), AFM_1, literature[263]

For detailed information see the article.

incidence: 13/13*, conc. range: 0.056–0.082 μg/l, sample origin: local market and vendor in Punjab (state), India, sample year: September–November 2010, country: India[264], *unpasteurized and pasteurized cow milk

- Co-contamination: not reported
- Further contamination of animal products reported in the present book/article[264]: not reported

For detailed information see the article.

incidence: 36/300*, conc. range: 0.12–1.00 μg/l, sample origin: different localities in Chennai (city), India, sample year: unknown, country: India[265], *milk from cows and buffaloes

- Co-contamination: not reported
- Further contamination of animal products reported in the present book/article[265]: not reported

For detailed information see the article.

incidence: 27/53*, conc. range: 0.005–0.146 μg/kg, sample origin: farms near Rome (capital), central Italy, sample year: February–June 1984, country: Italy[266], *whole raw milk of cows and buffaloes

incidence: 12/18*, conc. range: 0.005–0.030 μg/kg, Ø conc.: 0.014 μg/kg, sample origin: different packaging places, central Italy, sample year: January–September 1984, country: Italy[266], *commercial pasteurized or sterilized cow milk

incidence: 4/5, conc. range: 0.008–0.012 μg/kg, Ø conc.: 0.009 μg/kg, sample origin: dairy plants, central Italy, sample year: July 1984, country: Italy[266]

- Co-contamination: not reported
- Further contamination of animal products reported in the present book/article[266]: not reported

For detailed information see the article.

incidence: 25/26*, conc. range: 0.0134–0.1321 µg/l, sample origin: supermarkets in Bogotá (capital), Colombia, sample year: January–March 2004, country: Colombia[267], *pasteurized and homogenized cow milk

incidence: 19/25*, conc. range: 0.0107–0.0531 µg/l, sample origin: supermarkets in Bogotá (capital), Colombia, sample year: April–June 2004, country: Colombia[267], *pasteurized and homogenized cow milk

incidence: 15/29*, conc. range: 0.0137–0.1194 µg/l, sample origin: supermarkets in Bogotá (capital), Colombia, sample year: July–September 2004, country: Colombia[267], *pasteurized and homogenized cow milk

incidence: 24/40*, conc. range: 0.0155–0.2130 µg/l, sample origin: supermarkets in Bogotá (capital), Colombia, sample year: October–December 2004, country: Colombia[267], *pasteurized and homogenized cow milk

incidence: 16/18*, conc. range: 0.0112–0.1058 µg/l, sample origin: supermarkets in Bogotá, Colombia, sample year: January–March 2005, country: Colombia[267], *pasteurized and homogenized cow milk

incidence: 19/21*, conc. range: 0.0120–0.1149 µg/l, sample origin: supermarkets in Bogotá (capital), Colombia, sample year: April–June 2005, country: Colombia[267], *pasteurized and homogenized cow milk

incidence: 31/40*, conc. range: 0.0106–0.0865 µg/l, sample origin: supermarkets in Bogotá (capital), Colombia, sample year: July–September 2005, country: Colombia[267], *pasteurized and homogenized cow milk

incidence: 30/42*, conc. range: 0.0114–0.2889 µg/l, sample origin: supermarkets in Bogotá (capital), Colombia, sample year: October–December 2005, country: Colombia[267], *pasteurized and homogenized cow milk

- Co-contamination: not reported
- Further contamination of animal products reported in the present book/article[267]: not reported

incidence: 6/56*, conc. range: 0.012–0.030 µg/l, ∅ conc.: 0.016 µg/l, sample origin: Argentina, sample year: March–September 1999, country: Argentina[268], *farm cow milk

incidence: 8/16*, conc. range: 0.010–0.017 µg/l, ∅ conc.: 0.013 µg/l, sample origin: Argentina, sample year: March–September 1999, country: Argentina[268], *commercial pasteurized fluid cow milk

- Co-contamination: not reported
- Further contamination of animal products reported in the present book/article[268]: Milk powder, AFM_1, literature[268]

incidence: 207/208*, conc. range: 0.001–0.005 µg/l (29 sa.), 0.005 < 0.010 µg/l (100 sa.), 0.010 < 0.015 µg/l (60 sa.), 0.015 < 0.020 µg/l (15 sa.), 0.020 < 0.025 µg/l (2 sa.), 0.029 µg/l (1 sa.), ∅ conc.: 0.009 µg/l, sample origin: retail outlets, Japan, sample year: December 2001–February 2002, country: Japan[269], *included also UHT milk sa.

- Co-contamination: not reported
- Further contamination of animal products reported in the present book/article[269]: Milk (cow milk), AFM_2, literature[269]

incidence: 58/79*, conc. range: 0.015–0.050 µg/l (46 sa.), 0.050–0.500 µg/l (12 sa.), sample origin: supermarkets in Ribeirão Preto (municipality), São Paulo (state), Brazil, sample year: 1999–2000, country: Brazil[270], *pasteurized cow milk

- Co-contamination: not reported
- Further contamination of animal products reported in the present book/article[270]: Milk (UHT milk), AFM_1, literature[270]

incidence: 69/87, conc. range: 0.006–0.567 µg/l, sample origin: Turkey, sample year: unknown, country: Turkey[271]

- Co-contamination: not reported
- Further contamination of animal products reported in the present book/article[271]: not reported

incidence: 5/92*, conc. range: 0.014–0.0249 µg/l**, sample origin: dairy farms in Leon (province), Spain, sample year: autumn 2000–spring 2001, country: Spain[272], *raw milk, **determined by ELISA

incidence: 3/9*, conc. range: 0.0137–0.091 µg/l**, sample origin: dairy farms in Leon (province), Spain, sample year: autumn 2000–spring 2001, country: Spain[272], *raw milk, **determined by HPLC

- Co-contamination: not reported
- Further contamination of animal products reported in the present book/article[272]: not reported

incidence: 50/50*, conc. range: 0.00971–0.12979 µg/kg, ∅ conc.: 0.06891 µg/kg, sample origin: Jordanian market, Jordan, sample year: 2014–2015, country: Jordan[278], *fresh cow milk

incidence: 30/30*, conc. range: 0.01460–0.21678 µg/kg, ∅ conc.: 0.05945 µg/kg, sample origin: Jordanian market, Jordan, sample year: 2014–2015, country: Jordan[278], *pasteurized cow milk

- Co-contamination: not reported
- Further contamination of animal products reported in the present book/article[278]: Milk, Milk (camel milk), Milk (goat milk), Milk (infant formula), Milk (sheep milk), Milk powder, AFM_1, literature[278]

incidence: 145/290*, conc. range: ≥0.05 µg/l (117 sa.), ≥0.5 µg/l (28 sa., maximum: 8.35 µg/l**), sample origin: Mexico City (capital), Mexico, sample year: April 1996–July 1997, country: Mexico[281], *pasteurized and ultra-pasteurized cow milk, **sa. from Tizayuca (municipality), Hidalgo (state)

- Co-contamination: not reported
- Further contamination of animal products reported in the present book/article[281]: not reported

For detailed information see the article.

incidence: 75/85*, conc. range: <0.010 µg/l (11 sa.), 0.011–0.030 µg/l (3 sa.), 0.031–0.050 µg/l (13 sa.), 0.051–0.070 µg/l (29 sa.), 0.071–0.090 µg/l (12 sa.), >91 µg/l (7 sa., maximum: 0.1276 µg/l), sample origin: markets in various districts of Ankara (capital), Turkey, sample year: unknown, country: Turkey[282], *pasteurized cow milk

- Co-contamination: not reported
- Further contamination of animal products reported in the present book/article[282]: not reported

incidence: 12/12*, conc. range: 0.16–0.32 µg/l (4 sa.), 0.32–0.5 µg/l (8 sa.), sample origin: supermarkets in Heilongjiang (province), northeast China, sample year: March–May 2008, country: China/Russia/Korea[283], *raw cow milk

- Co-contamination: not reported
- Further contamination of animal products reported in the present book/article[283]: Cheese, Dairy products, Milk powder, AFM_1, literature[283]

incidence: 40/40, conc. range: 0.001–0.273 µg/l, sample origin: farms in North Carolina? (state), USA, sample year: unknown, country: USA[284]

- Co-contamination: not reported
- Further contamination of animal products reported in the present book/article[284]: not reported

incidence: 42/44*, conc. range: 0.22–6.90 µg/l, sample origin: dairy farms and vendors in Omdurman (city), Khartoum (capital), and Khartoum North, Sudan, sample year: 2009, country: Sudan[285], *bulk cow milk

- Co-contamination: not reported
- Further contamination of animal products reported in the present book/article[285]: not reported

incidence: 3/5*, conc. range: 0.11–0.25 µg/l, ∅ conc.: 0.16 µg/l, sample origin: dairy farms in al Bohoth, al Kiraiba, Atrah, Brakarat, and Hantoub in Wad Medani (city), Sudan, sample year: March 2009, country: Sudan[286], *bulk cow milk

- Co-contamination: not reported
- Further contamination of animal products reported in the present book/article[286]: Milk (cow milk), OTA, literature[286]

incidence: 48/200*, conc. range: 0.003–0.200 µg/l, ∅ conc.: 0.06309 µg/l, sample origin: farms in Konya (city), Turkey, sample year: unknown, country: Turkey[289], *raw cow milk

- Co-contamination: not reported
- Further contamination of animal products reported in the present book/article[289]: not reported

incidence: 50/50*, conc. range: <0.050 µg/l (19 sa.), 0.050–0.080 µg/l (22 sa.), >0.080 µg/l (9 sa., maximum: 0.259 µg/l), sample origin: supermarkets in Tabriz (city), Iran, sample year: July–December 2008, country: Iran[290], *pasteurized cow milk

- Co-contamination: not reported
- Further contamination of animal products reported in the present book/article[290]: not reported

incidence: 40/44*, conc. range: 0.002–0.01 µg/l (10 sa.), 0.01–0.05 µg/l (29 sa.), 0.083 µg/l (1 sa.), sample origin: supermarkets, convenience stores, and drug stores in 23 counties, Taiwan, sample year: June–August 2002, country: Taiwan[291], *fresh cow milk

- Co-contamination: not reported
- Further contamination of animal products reported in the present book/article[291]: Yogurt, AFM_1, literature[291]

incidence: 70/74*, conc. range: 0.020–0.690 µg/l, ∅ conc.: 0.143 µg/l, sample origin: north, south, and east of Syria, sample year: April 2005–2006, country: Syria[293], *raw cow milk

incidence: 10/10*, conc. range: 0.008–0.765 µg/l, ∅ conc.: 0.492 µg/l, sample origin: north, south, and east of Syria, sample year: April 2005–2006, country: Syria[293], *pasteurized cow milk

- Co-contamination: not reported
- Further contamination of animal products reported in the present book/article[293]: Milk (goat milk), Milk (sheep milk), Milk powder, AFM_1, literature[293]

incidence: 101*/158**, conc. range: 0.015–>0.410 µg/l, sample origin: traditional and industrial dairy farms in Hamedan (district), western Iran, sample year: 2000, country: Iran[294], *traditional husbandry, **raw cow milk

incidence: 18*/28**, conc. range: 0.016–>0.410 µg/l, sample origin: traditional and industrial dairy farms in Hamedan (district), western Iran, sample year: 2000, country: Iran[294], *industrial husbandry, **raw cow milk

- Co-contamination: not reported
- Further contamination of animal products reported in the present book/article[294]: not reported

For detailed information see the article.

incidence: 8/78*, conc. range: ≤0.066 µg/l, ∅ conc.: 0.021 µg/l, sample origin: farms in Emilia Romagna and Lombardy (administrative regions), Italy, sample year: October 2001–2002, country: Italy[295], *conventional farm tank cow milk

incidence: 38/78*, conc. range: ≤0.093 µg/l, ∅ conc.: 0.035 µg/l, sample origin: farms in Emilia Romagna and Lombardy (administrative regions), Italy, sample year: October 2001–2002, country: Italy[295], *organic farm tank cow milk

- Co-contamination: not reported
- Further contamination of animal products reported in the present book/article[295]: not reported

incidence: 168/168*, conc. range: 0.01–0.70 µg/l, ∅ conc.: 0.371 µg/l, sample origin: dairies, Pakistan, sample year: 2005, country: Pakistan[296], *raw cow milk

- Co-contamination: not reported
- Further contamination of animal products reported in the present book/article[296]: not reported

For detailed information see the article.

incidence: 3/50*, conc. range: 0.01–0.25 µg/kg, sample origin: local supermarkets in Beijing (capital), China, sample year: unknown, country: China[297], *liquid cow milk including whole, low fat, and skimmed milk as well as milk powder

- Co-contamination: not reported
- Further contamination of animal products reported in the present book/article[297]: not reported

incidence: 26/30*, conc. range: 0.02–0.28 µg/l, sample origin: dairy plants of Sarab (city), Iran, sample year: winter 2001, country: Iran[298], *raw cow milk

incidence: 20/27*, conc. range: 0.02–0.19 µg/l, sample origin: dairy plants of Sarab (city), Iran, sample year: spring 2001, country: Iran[298], *raw cow milk

incidence: 17/27*, conc. range: 0.015–0.14 µg/l, sample origin: dairy plants of Sarab (city), Iran, sample year: summer 2001, country: Iran[298], *raw cow milk

incidence: 22/27*, conc. range: 0.02–0.28 µg/l, sample origin: dairy plants of Sarab (city), Iran, sample year: autumn 2001, country: Iran[298], *raw cow milk

- Co-contamination: not reported
- Further contamination of animal products reported in the present book/article[298]: not reported

For detailed information see the article.

incidence: 20/60*, conc. range: 0.0054–0.03050 µg/l (15 sa.), 0.0612–0.30020 µg/l (5 sa.), ∅ conc.: 0.04834 µg/l, sample origin: Istanbul (province), Turkey, sample year: April–June 2007, country: Turkey[299], *raw cow milk

- Co-contamination: not reported
- Further contamination of human breast milk and animal products reported in the present book/article[299]: Milk (human breast milk), AFM_1, literature[299]

incidence: 144/144*, conc. range: <0.01 µg/l (36 sa.), 0.01–0.05 µg/l (107 sa.), 0.055 µg/l (1 sa.), sample origin: supermarkets and convenience stores in Taipei (capital), Taiwan, sample year: 2005, country: Taiwan[300], *included low and whole-fat cow milk

- Co-contamination: not reported
- Further contamination of animal products reported in the present book/article[300]: not reported

incidence: 317/332*, conc. range: 0.001–0.010 μg/kg (168 sa.), 0.011–0.050 μg/kg (121 sa.), 0.051–0.100 μg/kg (25 sa.), >0.100 μg/kg (3 sa., maximum: 0.406 μg/kg), sample origin: dairy farms in Emilia region, Italy, sample year: 1993–1999 (except 1997), country: Italy[301], *milk sa. destined for Parmigiano Reggiano cheese production

- Co-contamination: not reported
- Further contamination of animal products reported in the present book/article[301]: not reported

For detailed information see the article.

incidence: 80/80*, conc. range: 0.014–≤0.05 μg/l (42 sa.), 0.051–0.075 μg/l (27 sa.), 0.075–0.100 μg/l (10 sa.), 0.102 μg/l (1 sa.), sample origin: milk tanks of dairy farms in the central region of Thailand, sample year: April–May (summer) 2006, country: Thailand[302], *raw cow milk

incidence: 80/80*, conc. range: 0.022–≤0.05 μg/l (27 sa.), 0.051–0.075 μg/l (13 sa.), 0.075–0.100 μg/l (30 sa.), ≥0.101 μg/l (10 sa., maximum: 0.128 μg/l), sample origin: milk tanks of dairy farms in the central region of Thailand, sample year: September–October (rainy season) 2006, country: Thailand[302], *raw cow milk

incidence: 80/80*, conc. range: 0.028–≤0.05 μg/l (16 sa.), 0.051–0.075 μg/l (12 sa.), 0.075–0.100 μg/l (31 sa.), ≥0.101 μg/l (21 sa., maximum: 0.197 μg/l), sample origin: milk tanks of dairy farms in the central region of Thailand, sample year: January–February (winter) 2007, country: Thailand[302], *raw cow milk

- Co-contamination: not reported
- Further contamination of animal products reported in the present book/article[302]: not reported

incidence: 50/50*, conc. range: 0.012–≤0.05 μg/l (20 sa.), 0.051–0.075 μg/l (23 sa.), 0.075–0.100 μg/l (7 sa., maximum: 0.084 μg/l), sample origin: schools in the central region of Thailand, sample year: May–June (summer) 2006, country: Thailand[303], *pasteurized cow milk

incidence: 50/50*, conc. range: 0.019–≤0.05 μg/l (13 sa.), 0.051–0.075 μg/l (8 sa.), 0.075–0.100 μg/l (27 sa.), ≥0.101 μg/l (2 sa., maximum: 0.108 μg/l), sample origin: schools in the central region of Thailand, sample year: August–September (rainy season) 2006, country: Thailand[303], *pasteurized cow milk

incidence: 50/50*, conc. range: 0.025–≤0.05 μg/l (8 sa.), 0.051–0.075 μg/l (6 sa.), 0.075–0.100 μg/l (30 sa.), ≥0.101 μg/l (6 sa., maximum: 0.114 μg/l), sample origin: schools in the central region of Thailand, sample year: December 2006–January 2007 (winter), country: Thailand[303], *pasteurized cow milk

- Co-contamination: not reported
- Further contamination of animal products reported in the present book/article[303]: not reported

incidence: 196/196, conc. range: 0.019–0.126 μg/l, ∅ conc.: 0.07491 μg/l, sample origin: dairies in Khorasan (province), Iran, sample year: March–June 2008, country: Iran[304], *pasteurized cow milk

- Co-contamination: not reported
- Further contamination of animal products reported in the present book/article[304]: not reported

For detailed information see the article.

incidence: 13/22*, conc. range: >LOD (5 sa.), 0.02–0.05 μg/l (6 sa.), >0.05–<0.5 μg/l (2 sa.), sample origin: farms across region of Marília (city), Brazil, sample year: 2002–2003, country: Brazil[306], *raw cow milk

incidence: 32/43*, conc. range: >LOD (19 sa.), 0.02–0.05 μg/l (11 sa.), >0.05–<0.5 μg/l (2 sa.), sample origin: supermarkets in Marília and São Paulo (cities), Brazil, sample year: 2002–2003, country: Brazil[306], *pasteurized cow milk

- Co-contamination: not reported
- Further contamination of animal products reported in the present book/article[306]: Milk (UHT milk), AFM_1, literature[306]

incidence: 72/72*, conc. range: 0.0043–0.0918 μg/l, ∅ conc.: 0.02421 μg/l, sample origin: farms in various areas, Iran, sample year: 2005–2006, country: Iran[307], *raw cow milk

incidence: 72/72*, conc. range: 0.0051–0.0285 μg/l, ∅ conc.: 0.00873 μg/l, sample origin: supermarkets in and around Urmia (city), Iran, sample year: 2005–2006, country: Iran[307], *pasteurized cow milk

- Co-contamination: not reported
- Further contamination of animal products reported in the present book/article[307]: not reported

incidence: ?/98*, conc. range: ≤0.050 μg/l (61 sa.), 0.05–0.10 μg/l (29 sa.), 0.10–0.39 μg/l (8 sa.), sample origin: dairy plants in Hamedan, Gorgan, Rasht, Shiraz, and Tehran (regions), Iran, sample year: April 2003–May 2004, country: Iran[308], *raw cow milk

- Co-contamination: not reported
- Further contamination of animal products reported in the present book/article[308]: not reported

incidence: 23/23*, conc. range: 0.0087–0.0819 μg/kg, ∅ conc.: 0.0529 μg/kg, sample origin: local stores of Ardabil (city), Iran, sample year: spring 2006, country: Iran[309], *included raw, pasteurized, and sterilized cow milk sa.

incidence: 22/22*, conc. range: 0.0029–0.0559 μg/kg, ∅ conc.: 0.0174 μg/kg, sample origin: local stores of Ardabil (city), Iran, sample year: summer 2006, country: Iran[309], *included raw, pasteurized, and sterilized cow milk sa.

incidence: 22/22*, conc. range: 0.0078–0.0289 μg/kg, ∅ conc.: 0.0223 μg/kg, sample origin: local stores of Ardabil (city), Iran, sample year: autumn 2006, country: Iran[309], *included raw, pasteurized, and sterilized cow milk sa.

incidence: 23/23*, conc. range: 0.0058–0.0850 μg/kg, ∅ conc.: 0.0563 μg/kg, sample origin: local stores of Ardabil (city), Iran, sample year: winter 2006, country: Iran[309], *included raw, pasteurized, and sterilized cow milk sa.

- Co-contamination: not reported
- Further contamination of animal products reported in the present book/article[309]: not reported

For detailed information see the article.

incidence: 2/3*, conc. range: 0.010–0.020 μg/l (1 sa.), 0.021–0.050 μg/l (1 sa.), sample origin: supermarkets in Ankara (capital), Turkey, sample year: unknown, country: Turkey[310], *daily-pasteurized cow milk

- Co-contamination: not reported
- Further contamination of animal products reported in the present book/article[310]: Milk (UHT milk), AFM_1, literature[310]

incidence: 40/60, conc. range: 0.011–0.115 μg/l, ∅ conc.: 0.0487 μg/l, sample origin: Chaharmahal va Bakhtiari (province), central and southern Iran, sample year: December 2008–2009, country: Iran/USA[311]

- Co-contamination: not reported

- Further contamination of animal products reported in the present book/article[311]: Milk (goat milk), Milk (sheep milk), AFM_1, literature[311]

incidence: 7/12*, conc. range: 0.010–0.100 µg/l, ∅ conc.: 0.0529 µg/l, sample origin: traditional dairies of Fez (city), northern center of Morocco, sample year: autumn 2009, country: Morocco/France[312], *raw cow milk

incidence: 4/12*, conc. range: 0.010–0.090 µg/l, ∅ conc.: 0.0375 µg/l, sample origin: traditional dairies of Fez (city), northern center of Morocco, sample year: winter 2009–2010, country: Morocco/France[312], *raw cow milk

incidence: 0/12*, conc.: no contamination, sample origin: traditional dairies of Fez (city), northern center of Morocco, sample year: spring 2010, country: Morocco/France[312], *raw cow milk

incidence: 2/12*, conc. range: 0.010 µg/l, ∅ conc.: 0.010 µg/l, sample origin: traditional dairies of Fez (city), northern center of Morocco, sample year: summer 2010, country: Morocco/France[312], *raw cow milk

- Co-contamination: not reported
- Further contamination of animal products reported in the present book/article[312]: not reported

incidence: 50/50*, conc. range: 0.00705–0.12979 µg/kg, ∅ conc.: 0.05617 µg/kg, sample origin: Jordan, sample year: November–February 2012, country: Jordan[313], *raw cow milk

- Co-contamination: not reported
- Further contamination of human breast milk and animal products reported in the present book/article[313]: Milk (buttermilk), Milk (human breast milk), AFM_1, literature[313]

incidence: 37/54*, conc. range: 0.019–0.2034 µg/l, ∅ conc.: 0.0697 µg/l, sample origin: rural regions in Hamadan, Ilam, Kermanshah, and Kurdistan (provinces), Iran, sample year: winter and summer seasons 2014, country: Iran[328], *raw cow milk

- Co-contamination: not reported
- Further contamination of animal products reported in the present book/article[328]: Ayran, Cheese, Kashk, Milk (goat milk), Milk (sheep milk), Tarkhineh, Yogurt, AFM_1, literature[328]

incidence: 13/13, conc. range: 0.43–2.52 µg/kg, ∅ conc.: 1.23 µg/kg, sample origin: dairy farms in Khartoum (state), central Sudan, sample year: unknown, country: Sudan[393], *locally produced raw bulk cow milk

incidence: 22/22, conc. range: 0.09–2.17 µg/kg, ∅ conc.: 0.71 µg/kg, sample origin: dairy farms in Khartoum (state), central Sudan, sample year: unknown, country: Sudan[393], *commercially produced raw bulk cow milk

- Co-contamination: not reported
- Further contamination of animal products reported in the present book/article[393]: Milk powder, AFM_1, literature[393]

incidence: 19/50*, conc. range: 0.023–0.073 µg/l, ∅ conc.: 0.04974 µg/l, sample origin: supermarkets in Alexandria (city), Egypt, sample year: February 2008–March 2009, country: Egypt[396], *raw cow milk

- Co-contamination: not reported
- Further contamination of animal products reported in the present book/article[396]: Cheese (hard cheese), Cheese (processed cheese), Cheese (soft cheese), AFM_1, literature[396]

incidence: 200/214*, conc. range: ≤7.28 µg/l, ∅ conc.: 2.19 µg/l, sample origin: local shops, households, and dairy farms in Punjab (province), Pakistan, sample year: October 2012–September 2013, country: Australia/Pakistan[398], *bulk tank cow and buffalo milk from farmers

incidence: 95/98*, conc. range: ≤6.79 µg/l, ∅ conc.: 2.25 µg/l, sample origin: local shops, households, and dairy farms in Punjab (province), Pakistan, sample year: October 2012–September 2013, country: Australia/Pakistan[398], *bulk tank cow and buffalo milk from small collectors

incidence: 35/35*, conc. range: ≤4.93 µg/l, ∅ conc.: 2.36 µg/l, sample origin: local shops, households, and dairy farms in Punjab (province), Pakistan, sample year: October 2012–September 2013, country: Australia/Pakistan[398], *bulk tank cow and buffalo milk from large collectors

incidence: 138/138*, conc. range: ≤5.98 µg/l, ∅ conc.: 2.58 µg/l, sample origin: local shops, households, and dairy farms in Punjab (province), Pakistan, sample year: October 2012–September 2013, country: Australia/Pakistan[398], *bulk tank cow and buffalo milk from retailers

- Co-contamination: not reported
- Further contamination of animal products reported in the present book/article[398]: not reported

incidence: 484/520, conc. range: 0.002–0.261 µg/l, sample origin: dairy farms of southern Punjab (province), Pakistan, sample year: September 2013–August 2014, country: Pakistan/USA/UK[402]

- Co-contamination: not reported
- Further contamination of animal products reported in the present book/article[402]: not reported

For detailed information see the article.

incidence: 28/38*, conc. range: <0.005 µg/l** (2 sa.), 0.005–0.025 µg/l** (6 sa.), 0.025–0.050 µg/l** (3 sa.), >0.050 µg/l** (17 sa., maximum: 0.126 µg/l**), ∅ conc.: 0.0604 µg/l**, sample origin: Lebanese regions and markets, Lebanon, sample year: March–July 2010, country: Lebanon[404], raw cow* ** (35 sa.) and goat milk* (3 sa., not contaminated)

incidence: 17/25*, conc. range: <0.005 µg/l (2 sa.), 0.005–0.025 µg/l (7 sa.), 0.025–0.050 µg/l (4 sa.), >0.050 µg/l (4 sa., maximum: 0.0844 µg/l), ∅ conc.: 0.0306 µg/l**, sample origin: Lebanese regions and markets, Lebanon, sample year: March–July 2010, country: Lebanon[404], *local (14 sa.) and imported (11 sa.) pasteurized cow milk

- Co-contamination: not reported
- Further contamination of animal products reported in the present book/article[404]: Milk powder, AFM_1, literature[404]

For detailed information see the article.

incidence: 67/70*, conc. range: 0.020–0.050 µg/l (22 sa.), 0.050–0.500 µg/l (44 sa.), 0.629 µg/l (1 sa.), ∅ conc.: 0.135 µg/l, sample origin: main convenience stores and supermarkets, Costa Rica, sample year: 2012–2014, country: Costa Rica[411], *cow milk sa. contained 2 g/100 ml fat, were homogenized and pasteurized

- Co-contamination: not reported
- Further contamination of animal products reported in the present book/article[411]: Cheese (soft cheese), AFM_1, literature[411]; Cream (sour), AFM_2, literature[411]

incidence: 1538?/3635*, conc. range: 0.0066–0.0335 µg/kg (1062 sa.), >0.0335 µg/kg (372 sa.), >0.0500 µg/kg (105 sa., maximum: 0.4081 µg/kg), sample origin: Food and Veterinary Agency,

Macedonia, sample year: February 2013–January 2014, country: Macedonia[412], *raw cow milk

- Co-contamination: not reported
- Further contamination of animal products reported in the present book/article[412]: not reported

For detailed information see the article.

incidence: 41/45*, conc. range: 0.011–0.025 µg/l (10 sa.), 0.026–0.050 µg/l (15 sa.), 0.051–0.100 µg/l (16 sa., maximum: 0.083 µg/l), sample origin: various districts in Burdur (city), Turkey, sample year: 2008, country: Turkey[418], *raw cow milk

incidence: 30/45*, conc. range: 0.005–0.010 µg/l (16 sa.), 0.011–0.025 µg/l (5 sa.), 0.026–0.050 µg/l (7 sa.), 0.051–0.100 µg/l (2 sa., maximum: 0.057 µg/l), sample origin: various districts in Burdur (city), Turkey, sample year: 2008, country: Turkey[418], *pasteurized cow milk

- Co-contamination: not reported
- Further contamination of animal products reported in the present book/article[418]: Butter, Cheese (white cheese), Ice cream, Milk powder, Yogurt, AFM_1, literature[418]

incidence: 10/12*, conc. range: 0.003–3.81 µg/kg, ∅ conc.: 1.11 µg/kg, sample origin: Tizayuca (city and municipality), Hidalgo (state), Mexico, sample year: unknown, country: Mexico[421], *conventionally produced raw cow milk

incidence: 7/11*, conc. range: 0.023–7.66 µg/kg, ∅ conc.: 2.79 µg/kg, sample origin: Tuxpan (city and municipality), Veracruz (state), Mexico, sample year: unknown, country: Mexico[421], *organically produced raw cow milk

- Co-contamination: not reported
- Further contamination of animal products reported in the present book/article[421]: not reported

incidence: 31/37*, conc. range: 0.005–0.020 µg/l (15 sa.), 0.020–0.050 µg/l (14 sa.), >0.050 µg/l (2 sa., maximum: 0.165 µg/l), sample origin: farms across Esfahan (province), Iran, sample year: September–October 2009, country: Iran[422], *raw cow milk

incidence: 32/37*, conc. range: 0.005–0.020 µg/l (4 sa.), 0.020–0.050 µg/l (22 sa.), >0.050 µg/l (6 sa., maximum: 0.108 µg/l), sample origin: farms across Esfahan (province), Iran, sample year: September–October 2009, country: Iran[422], *pasteurized cow milk

- Co-contamination: not reported
- Further contamination of animal products reported in the present book/article[422]: not reported

incidence: 19/19*, conc. range: <0.025 µg/l (3 sa.), 0.026–0.050 µg/l (8 sa.), 0.051–0.075 µg/l (4 sa.), 0.076–0.100 µg/l (3 sa.), 0.104 µg/l (1 sa.), ∅ conc.: 0.0547 µg/l, sample origin: retail stores in Sari (city), Mazandaran (province), Iran, sample year: winter 2015, country: Iran[425], *pasteurized cow milk

incidence: 19/19*, conc. range: <0.025 µg/l (1 sa.), 0.026–0.050 µg/l (3 sa.), 0.051–0.075 µg/l (3 sa.), 0.076–0.100 µg/l (7 sa.), >0.100 µg/l (5 sa., maximum: 0.1066 µg/l), ∅ conc.: 0.0774 µg/l, sample origin: retail stores in Sari (city), Mazandaran (province), Iran, sample year: spring 2015, country: Iran[425], *pasteurized cow milk

incidence: 19/19*, conc. range: <0.025 µg/l (1 sa.), 0.026–0.050 µg/l (8 sa.), 0.051–0.075 µg/l (5 sa.), 0.076–0.100 µg/l (2 sa.), >0.100 µg/l (3 sa., maximum: 0.103 µg/l), ∅ conc.: 0.0613 µg/l, sample origin: retail stores in Sari (city), Mazandaran (province), Iran, sample year: summer 2015, country: Iran[425], *pasteurized cow milk

incidence: 19/19*, conc. range: 0.026–0.050 µg/l (6 sa.), 0.051–0.075 µg/l (3 sa.), 0.076–0.100 µg/l (7 sa.), >0.100 µg/l (3 sa., maximum: 0.1037 µg/l), ∅ conc.: 0.0709 µg/l, sample origin: retail stores in Sari (city), Mazandaran (province), Iran, sample year: autumn 2015, country: Iran[425], *pasteurized cow milk

- Co-contamination: not reported
- Further contamination of animal products reported in the present book/article[425]: not reported

incidence: 11/40*, conc. range: 0.0069–0.0697 µg/l, ∅ conc.: 0.0234 µg/l, sample origin: markets and supermarkets, Portugal, sample year: March–April 2011, country: Mexico[428], *pasteurized cow and UHT milk

- Co-contamination: not reported
- Further contamination of animal products reported in the present book/article[428]: not reported

incidence: 127/127*, conc. range: LOQ-0.050 µg/l (15 sa.), 0.050–0.100 µg/l (64 sa.), 0.100–0.200 µg/l (40 sa.), >0.200 µg/l (8 sa., maximum: 0.321 µg/l), ∅ conc.: 0.097 µg/l, sample origin: dairy farms in Qazvin (province), Iran, sample year: January–March 2014, country: Iran[436], *raw cow milk

incidence: 77/127*, conc. range: LOQ-0.050 µg/l (45 sa.), 0.050–0.100 µg/l (10 sa.), 0.100–0.200 µg/l (20 sa.), >0.200 µg/l (2 sa., maximum: 0.210 µg/l), sample origin: dairy farms in Qazvin (province), Iran, sample year: July–September 2014, country: Iran[436], *raw cow milk

- Co-contamination: not reported
- Further contamination of animal products reported in the present book/article[436]: not reported

incidence: 46/192*, conc. range: 0.010–0.050 µg/kg (45 sa.), 0.062 µg/kg (1 sa.), sample origin: Yazd (province), Iran, sample year: 2014, country: Iran[437], *cow milk sa. from industrial farms

incidence: 99/120*, conc. range: 0.010–0.050 µg/kg (52 sa.), 0.051–0.100 µg/kg (42 sa.), >0.100 µg/kg (5 sa., maximum: 0.198 µg/kg), sample origin: Yazd (province), Iran, sample year: 2014, country: Iran[437], *cow milk sa. from traditional farms

- Co-contamination: not reported
- Further contamination of animal products reported in the present book/article[437]: Milk (camel milk), Milk (goat milk), Milk (sheep milk), AFM_1, literature[437]

For detailed information see the article.

incidence: 74/88*, conc. range: 0.013–0.394 µg/l, sample origin: dairy ranches, supermarkets, and retail outlets in Esfahan, Shiraz, and Tabriz (cities) as well as Tehran (capital), Iran, sample year: 2008, country: Iran[440], *raw cow milk

- Co-contamination: not reported
- Further contamination of animal products reported in the present book/article[440]: Ayran, Cheese (Lighvan cheese), Kashk, Milk (goat milk), Milk (sheep milk), Yogurt, AFM_1, literature[440]

incidence: 50/50*, conc. range: <0.050 µg/l (19 sa.), 0.050–0.080 µg/l (22 sa.), >0.080 µg/l (9 sa., maximum: 0.259 µg/l), ∅ conc.: 0.05055 µg/l, sample origin: supermarkets in Tabriz (city), northwestern Iran, sample year: July–December 2008, country: Iran[441], *pasteurized cow milk

- Co-contamination: not reported
- Further contamination of animal products reported in the present book/article[441]: not reported

incidence: 20/63*, conc. range: LOQ-0.050 µg/kg (13 sa.), 0.050–0.250 µg/kg (6 sa.), 0.552 µg/kg (1 sa.), ∅ conc.: 0.082 µg/kg, sample origin: dairy plants in Adana (province), Turkey, sample year: autumn 2012, country: Turkey[444], *raw cow milk incidence: 19/47*, conc. range: LOQ-0.050 μg/kg (7 sa.), 0.050–0.250 μg/kg (7 sa.), 0.250–0.500 μg/kg (1 sa.), >0.500 μg/kg (4 sa., maximum: 1.101 μg/kg), Ø conc.: 0.275 μg/kg, sample origin: dairy plants in Adana (province), Turkey, sample year: winter 2012, country: Turkey[444], *raw cow milk

incidence: 11/33*, conc. range: LOQ-0.050 μg/kg (1 sa.), 0.050–0.250 μg/kg (10 sa., maximum: 0.150 μg/kg), Ø conc.: 0.099 μg/kg, sample origin: dairy plants in Adana (province), Turkey, sample year: spring 2012, country: Turkey[444], *raw cow milk

incidence: 3/33*, conc. range: LOQ-0.050 μg/kg (2 sa.), 0.102 μg/kg (1 sa.), Ø conc.: 0.055 μg/kg, sample origin: dairy plants in Adana (province), Turkey, sample year: summer 2012, country: Turkey[444], *raw cow milk

- Co-contamination: not reported
- Further contamination of animal products reported in the present book/article[444]: not reported

incidence: 68/80*, conc. range: 0.0050–0.0299 μg/l (55 sa.), 0.0300–0.0499 μg/l (4 sa.), 0.0500–0.0999 μg/l (9 sa.), sample origin: Beijing (province), China, sample year: September 2010, country: China[445], *raw cow milk sa. from large-scale farms, milk processing plants, and small farm cooperatives

incidence: 60/80*, conc. range: 0.0050–0.0299 μg/l (51 sa.), 0.0300–0.0499 μg/l (3 sa.), 0.0500–0.0999 μg/l (5 sa.), 0.100–0.1299 μg/l (1 sa.), sample origin: Hebei (province), China, sample year: September 2010, country: China[445], *raw cow milk sa. from large-scale farms, milk processing plants, and small farm cooperatives

incidence: 43/80*, conc. range: 0.0050–0.0299 μg/l (39 sa.), 0.0300–0.0499 μg/l (2 sa.), 0.0500–0.0999 μg/l (2 sa.), sample origin: Shanxi (province), China, sample year: September 2010, country: China[445], *raw cow milk sa. from large-scale farms, milk processing plants, and small farm cooperatives

incidence: 55/60*, conc. range: 0.0050–0.0299 μg/l (38 sa.), 0.0300–0.0499 μg/l (6 sa.), 0.0500–0.0999 μg/l (11 sa.), sample origin: Shanghai (province), China, sample year: September 2010, country: China[445], *raw cow milk sa. from large-scale farms and milk processing plants

incidence: 55/60*, conc. range: 0.0050–0.0299 μg/l (41 sa.), 0.0300–0.0499 μg/l (6 sa.), 0.0500–0.0999 μg/l (8 sa.), sample origin: Guangdong (province), China, sample year: September 2010, country: China[445], *raw cow milk sa. from large-scale farms, milk processing plants, and small farm cooperatives

- Co-contamination: not reported
- Further contamination of animal products reported in the present book/article[445]: not reported

For detailed information see the article.

incidence: 45/200*, conc. range: 0.0050–0.020 μg/l (38 sa.), 0.0201–0.050 μg/l (4 sa.), 0.051–0.060 μg/l (3 sa., maximum: 0.0596 μg/l), Ø conc.: 0.0153 μg/l, sample origin: 10 provinces in China, sample year: summer August 2010, country: China[449], *raw cow milk

- Co-contamination: not reported
- Further contamination of animal products reported in the present book/article[449]: not reported

For detailed information see the article.

incidence: 20/41, conc. range: >LOD-0.050 μg/kg (14 sa.), >0.05 μg/kg (6 sa., maximum: 0.062 μg/kg), sample origin: milking sites and farmhouses in major cities in Punjab, Pakistan, sample year: November 2009–April 2010, country: Pakistan/Spain[450]

incidence: 22/43, conc. range: >LOD-0.050 μg/kg (15 sa.), >0.05 μg/kg (7 sa., maximum: 0.084 μg/kg), sample origin: milking sites and farmhouses in major cities in NWFP, Pakistan, sample year: November 2009–April 2010, country: Pakistan/Spain[450]

- Co-contamination: not reported
- Further contamination of animal products reported in the present book/article[450]: Milk (buffalo milk), AFM_1, literature[450]

incidence: 26/30*, conc. range: 0.009–0.100 μg/l (18 sa.), 0.101–0.250 μg/l (4 sa.), 0.251–0.500 μg/l (4 sa., maximum: 0.437 μg/l), sample origin: supermarkets in Ribeirão Petro (municipality) in São Paulo (state), Brazil, sample year: 2010, country: Brazil/USA[452], *pasteurized cow milk

incidence: 6/6*, conc. range: 0.003–0.008 μg/l (1 sa.), 0.009–0.100 μg/l (5 sa.), sample origin: supermarkets in Ribeirão Petro (municipality) in São Paulo (state), Brazil, sample year: 2010, country: Brazil/USA[452], *whole cow milk with additives (vitamin D and iron)

incidence: 5/5*, conc. range: 0.003–0.008 μg/l (1 sa.), 0.009–0.100 μg/l (4 sa.), sample origin: supermarkets in Ribeirão Petro (municipality) in São Paulo (state), Brazil, sample year: 2010, country: Brazil/USA[452], *partially skim cow milk with additives (calcium and reduction of lactose)

incidence: 6/6*, conc. range: 0.003–0.008 μg/l (2 sa.), 0.009–0.100 μg/l (4 sa.), sample origin: supermarkets in Ribeirão Petro (municipality) in São Paulo (state), Brazil, sample year: 2010, country: Brazil/USA[452], *skim cow milk with additives (vitamin A and D, calcium, and fiber)

- Co-contamination: not reported
- Further contamination of animal products reported in the present book/article[452]: Milk (UHT milk), Milk powder, AFM_1, literature[452]

incidence: 24/30*, conc. range: ≤0.2374 μg/kg, Ø conc.: 0.0804 μg/kg, sample origin: dairy farms in Beijing (capital), China, sample year: April 2012, country: China[456], *raw cow milk

incidence: 4/12*, conc. range: ≤0.0460 μg/kg, Ø conc.: 0.0323 μg/kg, sample origin: supermarkets in Beijing (capital), China, sample year: April 2012, country: China[456], *fluid cow milk

- Co-contamination: For detailed information see the article
- Further contamination of animal products reported in the present book/article[456]: Milk (cow milk), OTA and ZEA, literature[456]; Milk powder, AFM_1, OTA, and ZEA, literature[456]; see also Milk (cow milk), Milk powder, chapter: Further Mycotoxins and Microbial Metabolites

incidence: 26/65*, conc. range: 0.009–0.069 μg/l, Ø conc.: 0.030 μg/l, sample origin: households from employees of the University of São Paulo at Pirassununga (municipality), Brazil, sample year: June 2011–March 2012, country: Brazila[458], *fluid cow milk

- Co-contamination: not reported
- Further contamination of animal products reported in the present book/article[458]: Cheese, Milk powder, AFM_1, literature[458]

incidence: 41/45*, conc. range: 0.051–0.075 μg/l (30 sa.), >0.075 μg/l (11 sa.), sample origin: local farms in Kars (city) vicinity, Turkey, sample year: May–July 2006, country: Turkey[462], *raw cow milk

- Co-contamination: not reported
- Further contamination of animal products reported in the present book/article[462]: Milk (UHT milk), AFM_1, literature[462]

incidence: 4/10*, conc. range: 0.0045–0.0141 µg/l, ∅ conc.: 0.0061 µg/l, sample origin: Shush (city) in Khuzestan (province), Turkey, sample year: winter 2012, country: Turkey[465], *raw cow milk

incidence: 20/30*, conc. range: 0.0036–0.1009 µg/l, ∅ conc.: 0.0326 µg/l, sample origin: Shush (city) in Khuzestan (province), Turkey, sample year: spring 2012, country: Turkey[465], *raw cow milk

incidence: 20/20*, conc. range: 0.0232–0.4195 µg/l, ∅ conc.: 0.1149 µg/l, sample origin: Shush (city) in Khuzestan (province), Turkey, sample year: summer 2012, country: Turkey[465], *raw cow milk

- Co-contamination: not reported
- Further contamination of animal products reported in the present book/article[465]: Milk buffalo, AFM_1, literature[465]

incidence: 22/22*, conc. range: 0.05–0.50 µg/kg (21 sa.), 0.70 µg/kg (1 sa.), ∅ conc.: 0.35 µg/kg, sample origin: produced in large dairies and purchased in supermarkets, Serbia, sample year: February–May 2013, country: Serbia[468], *pasteurized cow milk

incidence: 13/13*, conc. range: 0.05–0.50 µg/kg (8 sa.), >0.50 µg/kg (5 sa., maximum: 1.20 µg/kg), ∅ conc.: 0.42 µg/kg, sample origin: produced in small dairies and purchased in supermarkets, Vojvodina (autonomous province), northern Serbia, sample year: February–May 2013, country: Serbia[468], *pasteurized cow milk

incidence: 6/6*, conc. range: 0.005–0.05 µg/kg (5 sa.), 0.08 µg/kg (1 sa.), ∅ conc.: 0.036 µg/kg, sample origin: produced in large dairies and purchased in supermarkets, Serbia, sample year: February–May 2013, country: Serbia[468], *organic cow milk

incidence: 38/40*, conc. range: 0.005–0.05 µg/kg (8 sa.), 0.05–0.50 µg/kg (25 sa.), >0.50 µg/kg (5 sa., maximum: 0.90 µg/kg), ∅ conc.: 0.19 µg/kg, sample origin: produced on small milking farms, Vojvodina (autonomous province), northern Serbia, sample year: February–May 2013, country: Serbia[468], *raw cow milk

- Co-contamination: not reported
- Further contamination of human breast milk and animal products reported in the present book/article[468]: Milk (donkey milk), Milk (goat milk), Milk (human breast milk), Milk (UHT milk), AFM_1, literature[468]

For detailed information see the article.

incidence: 93/100*, conc. range: LOD-0.05 µg/l (72 sa.), 0.05 < ~0.1 µg/kg (21 sa.), sample origin: central and southern cattle ranches, South Korea, sample year: January 2008, country: South Korea[469], *raw cow milk

incidence: 66/100*, conc. range: LOD-0.05 µg/l (62 sa.), 0.05 < ~0.1 µg/kg (4 sa.), sample origin: central and southern cattle ranches, South Korea, sample year: January 2009, country: South Korea[469], *raw cow milk

- Co-contamination: not reported
- Further contamination of animal products reported in the present book/article[469]: not reported

For detailed information see the article.

incidence: 10/59*, conc. range: >LOD-0.05 µg/l, sample origin: cattle ranches in the central of South Korea, sample year: August 2007, country: South Korea[475], *raw cow milk

incidence: 8/35*, conc. range: >LOD-0.05 µg/l, sample origin: cattle ranches in the south east of South Korea, sample year: August 2007, country: South Korea[475], *raw cow milk

incidence: 1/6*, conc.: 0.05–0.5 µg/kg (1 sa.), sample origin: cattle ranches in the south west of South Korea, sample year: August 2007, country: South Korea[475], *raw cow milk

- Co-contamination: not reported
- Further contamination of animal products reported in the present book/article[475]: not reported

incidence: 73/144*, conc. range: 0.01–0.25 µg/l, ∅ conc.: 0.08 µg/l, sample origin: industrial dairy farms in Qazvin (province), Iran, sample year: summer 2012, country: Iran[476], *raw bulk cow milk

incidence: 90/144*, conc. range: 0.03–0.22 µg/l, ∅ conc.: 0.12 µg/l, sample origin: industrial dairy farms in Qazvin (province), Iran, sample year: winter 2012, country: Iran[476], *raw bulk cow milk

- Co-contamination: not reported
- Further contamination of animal products reported in the present book/article[476]: not reported

incidence: 3/21*, conc. range: 0.01075–0.0400 µg/l, ∅ conc.: 0.0185 µg/l, sample origin: dairy farms, sale points, bazaars, and markets in Faisalabad (city), Pakistan, sample year: unknown, country: Pakistan[482], *single fresh cow milk

- Co-contamination: not reported
- Further contamination of animal products reported in the present book/article[482]: Butter, Yogurt, AFM_1, literature[482]

For detailed information see the article.

incidence: 13/40*, conc. range: LOD-LOQ (6 sa.), >0.015 µg/kg (7 sa., maximum: 0.029 µg/kg), sample origin: cattle ranches in Sichuan and Chongqing City (provinces), southwestern China, sample year: July–September 2014, country: China[483], *raw cow milk

- Co-contamination: not reported
- Further contamination of animal products reported in the present book/article[483]: not reported

incidence: 25/25, conc. range: 0.030–0.098 µg/kg, ∅ conc.: 0.056 µg/kg, sample origin: Po valley (province), Italy, sample year: November–December 2003, country: Italy[484]

- Co-contamination: not reported
- Further contamination of animal products reported in the present book/article[484]: Cheese (Grana Padano cheese), Cheese curd, Cheese whey, AFM_1, literature[484]

incidence: 31/37*, conc. range: ≤2.007 µg/l, ∅ conc.: 0.386 µg/l, sample origin: household farmers in Kibaoni, VETA, Sabasaba, and Majengo (locations) in Singida (municipality) as well as Amani (location) in Manyoni (township), Tanzania, sample year: February 2014, country: Tanzania[487], *raw cow milk

- Co-contamination: not reported
- Further contamination of animal products reported in the present book/article[487]: not reported

incidence: 122/217*, conc. range: <LOQ (48 sa.), ≥0.010–0.050 µg/kg (57 sa.), 0.051–0.500 µg/kg (16 sa.), 0.725 µg/kg (1 sa.), sample origin: Araçtuba (milk-producing region) in São Paulo (state), Brazil, sample year: July 2009–September 2010, country: Brazil[490], *raw cow milk

incidence: 156/214*, conc. range: <LOQ (44 sa.), ≥0.010–0.050 µg/kg (69 sa.), 0.051–0.500 µg/kg (41 sa.), >0.500 µg/kg (2 sa., maximum: 0.708 µg/kg), sample origin: Bauru (milk-producing region) in São Paulo (state), Brazil, sample year: July 2009–September 2010, country: Brazil[490], *raw cow milk

incidence: 56/204*, conc. range: <LOQ (38 sa.), ≥0.010–0.050 µg/kg (14 sa.), 0.051–0.500 µg/kg (4 sa., maximum: 0.224 µg/kg), sample origin: Vale do Paraiba (milk-producing region) in São

Paulo (state), Brazil, sample year: July 2009–September 2010, country: Brazil[490], *raw cow milk
- Co-contamination: not reported
- Further contamination of animal products reported in the present book/article[490]: not reported

incidence: 100/100*, conc. range: 0.0013–0.068 µg/l, ∅ conc.: 0.0765 µg/l, sample origin: dairy farm in Urmia (city), northwest Iran, sample year: March–May, country: Iran[492], *raw cow milk
- Co-contamination: not reported
- Further contamination of animal products reported in the present book/article[492]: not reported

incidence: 45/45, conc. range: 0.0077–0.0683 µg/l, ∅ conc.: 0.0257 µg/l, sample origin: dairy farm in Trakya region, Turkey, sample year: March 2005, country: Turkey[493]

incidence: 8/45, conc. range: 0.0050–0.0200 µg/l, ∅ conc.: 0.0071 µg/l, sample origin: dairy farm in Trakya region, Turkey, sample year: June 2005, country: Turkey[493]

incidence: 7/45*, conc. range: ≤0.0058 µg/l, ∅ conc.: 0.0051 µg/l, sample origin: milk producers selling to dairy plants in Trakya region, Turkey, sample year: June 2005, country: Turkey[493], *randomly obtained sa.
- Co-contamination: not reported
- Further contamination of animal products reported in the present book/article[493]: not reported

incidence: 11/30*, conc. range: 0.010–0.645 µg/l, ∅ conc.: 0.144 µg/l, sample origin: dairy farms, São Paulo (state), Brazil, sample year: October 2005–February 2006, country: Brazil[494], *raw cow milk from grade B dairy farms

incidence: 10/20*, conc. range: 0.010–0.127 µg/l, ∅ conc.: 0.060 µg/l, sample origin: dairy farms, São Paulo (state), Brazil, sample year: October 2005–February 2006, country: Brazil[494], *raw cow milk from grade C dairy farms
- Co-contamination: not reported
- Further contamination of animal products reported in the present book/article[494]: Milk (cow milk), CPA, literature[494]

incidence: 29/87*, conc. range: 0.010–0.020 µg/l (5 sa.), 0.021–0.050 µg/l (16 sa.), ≥0.050 µg/l (8 sa., maximum: 0.0845 µg/l), ∅ conc.: 0.0402 µg/l, sample origin: 9 provinces, Sri Lanka, sample year: unknown, country: Sri Lanka[496], *farm gate raw cow milk
- Co-contamination: not reported
- Further contamination of animal products reported in the present book/article[496]: not reported

For detailed information see the article.

incidence: 56/59*, conc. range: 0.034–0.211 µg/l, sample origin: supermarkets and bulk tank milk from dairy plants in Esfahan and Shahr-e Kord (cities), Iran, sample year: September 2006–2007 (winter), country: Iran[502], *raw, pasteurized, and UHT cow milk

incidence: 53/59*, conc. range: 0.012–0.152 µg/l, sample origin: supermarkets and bulk tank milk from dairy plants in Esfahan and Shahr-e Kord (cities), Iran, sample year: September 2006–2007 (spring), country: Iran[502], *raw, pasteurized, and UHT cow milk

incidence: 49/59*, conc. range: 0.013–0.123 µg/l, sample origin: supermarkets and bulk tank milk from dairy plants in Esfahan and Shahr-e Kord (cities), Iran, sample year: September 2006–2007 (summer), country: Iran[502], *raw, pasteurized, and UHT cow milk

incidence: 55/59*, conc. range: 0.024–0.218 µg/l, sample origin: supermarkets and bulk tank milk from dairy plants in Esfahan and Shahr-e Kord (cities), Iran, sample year: September 2006–2007 (autumn), country: Iran[502], *raw, pasteurized, and UHT cow milk
- Co-contamination: not reported
- Further contamination of animal products reported in the present book/article[502]: not reported

For detailed information see the article.

incidence: 36/72, conc. range: 0.010–0.401 µg/l, ∅ conc.: 0.182 µg/l, sample origin: dairy farms in Kerman (province), Iran, sample year: April 2006–March 2007, country: Iran[504]
- Co-contamination: not reported
- Further contamination of animal products reported in the present book/article[504]: not reported

For detailed information see the article.

incidence: 15/15*, conc. range: <0.010 µg/l (2 sa.), 0.011–0.025 µg/l (13 sa., maximum: 0.017 µg/l), ∅ conc.: 0.0131 µg/l, sample origin: supermarkets in Fariman (city), Iran, sample year: July 2012, country: Iran[506], *pasteurized cow milk

incidence: 15/15*, conc. range: <0.010 µg/l (2 sa.), 0.011–0.025 µg/l (13 sa., maximum: 0.023 µg/l), ∅ conc.: 0.0156 µg/l, sample origin: supermarkets in Fariman (city), Iran, sample year: August 2012, country: Iran[506], *pasteurized cow milk

incidence: 15/15*, conc. range: 0.011–0.025 µg/l (1 sa.), 0.026–0.050 µg/l (1 sa.), >0.050 µg/l (13 sa., maximum: 0.064 µg/l), ∅ conc.: 0.054 µg/l, sample origin: supermarkets in Fariman (city), Iran, sample year: September 2012, country: Iran[506], *pasteurized cow milk
- Co-contamination: not reported
- Further contamination of animal products reported in the present book/article[506]: not reported

incidence: 40/40*, conc. range: 0.079–0.205 µg/l, ∅ conc.: 0.170 µg/l, sample origin: milk-collecting centers in Qom (province), Iran, sample year: January 2013, country: Iran[507], *raw cow milk
- Co-contamination: not reported
- Further contamination of animal products reported in the present book/article[507]: not reported

incidence: 6/12, conc. range: ≤0.003–0.005 µg/l (5 sa.), 0.010 µg/l (1 sa.), sample origin: different areas on Sicily (island, autonomous region), Italy, sample year: January–June 2012, country: Italy[508]
- Co-contamination: not reported
- Further contamination of animal products reported in the present book/article[508]: Cheese, Cream, Milk (goat milk), Milk (sheep milk), AFM_1, literature[508]

For detailed information see the article.

incidence: 60/60*, conc. range: 0.002–0.064 µg/l, ∅ conc.: 0.01616 µg/l, sample origin: retail stores in Mashhad (city), northeastern Iran, sample year: June 2011, country: Iran[509], *pasteurized cow milk
- Co-contamination: not reported
- Further contamination of animal products reported in the present book/article[509]: not reported

incidence: 19/90*, conc. range: 0.011–0.100 µg/l, sample origin: Corum (province), Turkey, sample year: October 2012–March 2015, country: Turkey[510], *raw cow milk
- Co-contamination: not reported
- Further contamination of animal products reported in the present book/article[510]: Ayran, Yogurt, AFM_1, literature[510]

incidence: 68/120*, conc. range: 0.004–0.3523 µg/l, sample origin: traditional and semi-industrial cattle farms in Babol (city),

northern Iran, sample year: winter 2006, country: Iran[514], *raw cow milk
- Co-contamination: not reported
- Further contamination of animal products reported in the present book/article[514]: not reported

For detailed information see the article.

incidence: 8/8*, conc. range: 0.01–1.44** µg/kg, ∅ conc.: 0.49 µg/kg, sample origin: private farm as well as settlements in the Bačka and the Srem county, Serbia, sample year: February–April 2013, country: Serbia/Czech Republic[515], *raw cow milk, **domestically produced milk sa.

incidence: 12/12*, conc. range: 0.01–0.30 µg/kg, ∅ conc.: 0.19 µg/kg, sample origin: produced in Bosnia-Herzegowina (country) and Serbia (country) purchased in Novi Sad (city), northern Serbia, sample year: February/April/May 2013, country: Serbia/Czech Republic[515], *pasteurized cow milk with 2.2–2.8% milk fat
- Co-contamination: not reported
- Further contamination of animal products reported in the present book/article[515]: Milk (UHT milk), AFM_1, literature[515]

For detailed information see the article.

incidence: 29/45*, conc. range: 0.02–0.05 µg/l (7 sa.), 0.05–0.1 µg/l (5 sa.), 0.1–0.5 µg/l (11 sa.), >0.5 µg/l (6 sa., maximum: 3.8 µg/l), sample origin: milk producers in Karnataka and Tamil Nadu (states), India, sample year: April–July 2011, country: India[518], *raw cow milk

incidence: 3/7*, conc. range: >0.5 µg/l (3 sa., maximum: 3.8 µg/l), sample origin: milk producers in Karnataka and Tamil Nadu (states), India, sample year: April–July 2011, country: India[518], *pasteurized cow milk
- Co-contamination: not reported
- Further contamination of animal products reported in the present book/article[518]: Milk (UHT milk), AFM_1, literature[518]

For detailed information see the article.

incidence: 36/38*, conc. range: 0.00082–0.12570 µg/kg, sample origin: city markets in Şanlıurfa (city) and locally produced dairy products, Turkey, sample year: January–February 2012, country: Turkey[523], *raw cow milk
- Co-contamination: not reported
- Further contamination of animal products reported in the present book/article[523]: Cheese (white cheese), Milk (UHT milk), Yogurt, AFM_1, literature[523]

incidence: 41/60*, conc. range: 0.003–0.104 µg/kg, sample origin: Novi Sad (city), Serbia, sample year: April/July/September/December 2014, country: Serbia[526], *pasteurized cow and UHT milk
- Co-contamination: not reported
- Further contamination of animal products reported in the present book/article[526]: not reported

For detailed information see the article.

incidence: 34/54*, conc. range: 0.05003–0.08524 µg/l, ∅ conc.: 0.05639 µg/l, sample origin: milk collecting centers, Ilam (province), Iran, sample year: February 2011–August 2012, country: Iran[530], *raw cow milk sa. from traditional farms

incidence: 19/48*, conc. range: 0.05019–0.06071 µg/l, ∅ conc.: 0.05315 µg/l, sample origin: raw milk tankers, Ilam (province), Iran, sample year: February 2011–August 2012, country: Iran[530], *raw cow milk sa. from industrial farms

incidence: 10/42*, conc. range: 0.05003–0.06266 µg/l, ∅ conc.: 0.05439 µg/l, sample origin: supermarkets and retail shops, Ilam (province), Iran, sample year: February 2011–August 2012, country: Iran[530], *pasteurized cow milk sa.
- Co-contamination: not reported
- Further contamination of animal products reported in the present book/article[530]: not reported

For detailed information see the article.

incidence: 1/4*, conc.: 0.054 µg/kg, sample origin: local cattle ranches, China, sample year: unknown, country: China[531], *raw cow milk

incidence: 2/5*, conc. range: 0.038–0.102 µg/kg, ∅ conc.: 0.070 µg/kg, sample origin: major manufactures, China, sample year: unknown, country: China[531], *4 pasteurized cow milks and 1 pasteurized goat milk (not contaminated)
- Co-contamination: not reported
- Further contamination of animal products reported in the present book/article[531]: Milk (UHT milk), Milk powder, AFM_1, literature[531]

incidence: 16/18*, conc. range: 0.010–0.420 µg/l, sample origin: dairy farms in Hangzhou, Jinhua, Nanjing, and Shanghai (cities), China, sample year: winter 2011–2012, country: China[533], *raw cow milk

incidence: 14/18*, conc. range: 0.011–0.098 µg/l, sample origin: dairy farms in Hangzhou, Jinhua, Nanjing, and Shanghai (cities), China, sample year: spring 2012, country: China[533], *raw cow milk

incidence: 8/18*, conc. range: 0.011–0.082 µg/l, sample origin: dairy farms in Hangzhou, Jinhua, Nanjing, and Shanghai (cities), China, sample year: summer 2012, country: China[533], *raw cow milk

incidence: 5/18*, conc. range: 0.016–0.076 µg/l, sample origin: dairy farms in Hangzhou, Jinhua, Nanjing, and Shanghai (cities), China, sample year: autumn 2012, country: China[533], *raw cow milk
- Co-contamination: not reported
- Further contamination of animal products reported in the present book/article[533]: not reported

incidence: 65/88*, conc. range: LOD<25 µg/l** (22 sa.), 25 < 50 µg/l** (25 sa.), >50 µg/l** (18 sa., maximum: 0.07806 µg/l**), sample origin: dairy farms in semiarid area, Assiut (governorate), Egypt, sample year: April–October 2013, country: South Africa/Egypt/Saudi Arabia[544], *milk sa. from rural and commercial farms, **tested by ELISA

incidence: 34/50*, conc. range: LOD<25 µg/l** (22 sa.), 25 < 50 µg/l** (6 sa.), >50 µg/l** (6 sa., maximum: 0.120 µg/l**), sample origin: dairy farms in semiarid area, Ngaka Modiri Molema (district), South Africa, sample year: April–October 2013, country: South Africa/Egypt/Saudi Arabia[544], *milk sa. from rural and commercial farms, **tested by ELISA
- Co-contamination: not reported
- Further contamination of animal products reported in the present book/article[544]: not reported

For HPLC results of South African milk sa. see the article.

incidence: 20/20*, conc. range: <LOQ (1 sa.), 0.02–0.19 µg/kg (19 sa.), sample origin: local markets and vendors in Assiut Governorate, Upper Egypt, sample year: summer 2014, country: Egypt/Turkey[545], *raw milk

- Co-contamination: not reported
- Further contamination of animal products reported in the present book/article[545]: not reported

incidence: 92/360*, conc. range: 0.05–1.5 µg/l, sample origin: Rio de Janeiro State, Brazil, sample year: January 2011–2014, country: Brazil/Argentina[546], *raw and pasteurized milk sa.

- Co-contamination: not reported
- Further contamination of animal products reported in the present book/article[546]: not reported

incidence: 120/131*, conc. range: 0.005–0.0499 µg/l (42 sa.), 0.050–0.0995 µg/l (23 sa.), 0.100–0.1999 µg/l (12 sa.), >0.200 µg/l (43 sa., maximum: 0.352 µg/l), sample origin: Henan, Hubei, and Hunan (provinces), central China, sample year: November 2016–February 2017, country: China[547], *pasteurized milk sa.

- Co-contamination: not reported
- Further contamination of animal products reported in the present book/article[547]: Milk (UHT milk), AFM_1, literature[547]

incidence: 110/110*, conc. range: ≤0.05 µg/l (9 sa.), >0.05–≤0.1 µg/l (49 sa.), >0.1–≤0.5 µg/l (33 sa.), >0.5–≤1 µg/l (10 sa.), >1–≤2 µg/l (4 sa.), >2 µg/l (5 sa., maximum: 4.98 µg/l), sample origin: dairy farmers and milk traders in Oromia (region), Greater Addis Ababa (capital), Ethiopia, sample year: September 2014–February 2015, country: Ethiopia/Kenya[550], *raw milk sa.

- Co-contamination: not reported
- Further contamination of animal products reported in the present book/article[550]: not reported

For detailed information see the article.

incidence: 32/45*, conc. range: 0.005–0.164 µg/l, sample origin: dairy farms, Costa Rica, sample year: 2014, country: Costa Rica[551], *fresh milk sa.

incidence: 34/73*, conc. range: 0.017–0.989 µg/l, sample origin: dairy farms, Costa Rica, sample year: 2015, country: Costa Rica[551], *fresh milk sa.

incidence: 8/27*, conc. range: 0.014–0.109 µg/l, sample origin: dairy farms, Costa Rica, sample year: 2016, country: Costa Rica[551], *fresh milk sa.

incidence: 24/38*, conc. range: 0.013–0.334 µg/l, sample origin: dairy farms, Costa Rica, sample year: 2017, country: Costa Rica[551], *fresh milk sa.

- Co-contamination: not reported
- Further contamination of animal products reported in the present book/article[551]: not reported

For detailed information see the article.

incidence: 28/52*, conc. range: 0.1–0.25 µg/l (4 sa.), 0.26–0.49 µg/l (3 sa.), ≥0.50 µg/l (19 sa.), >1.00 µg/l (2 sa., maximum: 3.385 µg/l), sample origin: small rural farms, Concórdia (city), western region of Santa Catarina (state), south of Brazil, sample year: November 2014–January 2015, country: Brazil[552], *fresh milk sa.

- Co-contamination: not reported
- Further contamination of animal products reported in the present book/article[552]: not reported

incidence: 4/102*, conc. range: 0.020 µg/l (1 sa.), 0.051–0.5 µg/l (3 sa., maximum: 0.142 µg/l), sample origin: retail stores and supermarkets, Penang (state), Malaysia, sample year: February 2011–2016, country: Malaysia/Jordan[553], *fresh milk sa.

- Co-contamination: not reported
- Further contamination of animal products reported in the present book/article[553]: not reported

incidence: 81/84*, conc. range: 0.01–0.76 µg/l, ∅ conc.: 0.38 µg/l, sample origin: retail stores and supermarkets, local area/market in Hyderabad (city) region, Sindh (province), Pakistan, sample year: unknown, country: Pakistan[554], *different kinds of milk sa.

- Co-contamination: not reported
- Further contamination of animal products reported in the present book/article[554]: not reported

incidence: 62/160, conc. range: 0.003–0.293 µg/l, ∅ conc.: 0.037 µg/l, sample origin: cooling tanks from dairy farms, Santa Fe (province), Argentina, sample year: September 2012–August 2013, country: Argentina[555]

- Co-contamination: not reported
- Further contamination of animal products reported in the present book/article[555]: not reported

incidence: 36/1668, conc. range: 0.018–0.208 µg/l, sample origin: northwest Italy, sample year: 2012–2014, country: Italy[556]

- Co-contamination: not reported
- Further contamination of animal products reported in the present book/article[556]: not reported

For detailed information see the article.

incidence: 24/36*, conc. range: 0.009–0.026 µg/l, sample origin: supermarkets, small retail shops, small groceries, and specialized suppliers, Bologna (city), Emilia Romagna (administrative region), north Italy, sample year: April–June 2013, country: Italy[558], *conventional

incidence: 11/22*, conc. range: 0.009–0.026 µg/l, sample origin: supermarkets, small retail shops, small groceries, and specialized suppliers, Bologna (city), Emilia Romagna (administrative region), north Italy, sample year: April–June 2013, country: Italy[558], *organic

- Co-contamination: not reported
- Further contamination of animal products reported in the present book/article[558]: not reported

For detailed information see the article.

incidence: 288/288*, conc. range: 0.005–0.010 µg/kg (196 sa.), 0.011–0.019 µg/kg (81 sa.), 0.020–0.0255 µg/kg (11 sa., maximum: 0.025 µg/kg), sample origin: different suppliers, Marche (region), central Italy, sample year: 2012, country: Italy[559], *raw milk sa.

- Co-contamination: not reported
- Further contamination of animal products reported in the present book/article[559]: not reported

For detailed information see the article.

incidence: 6/6*, conc. range: 0.003–0.05 µg/kg (4 sa.), 0.05–0.1 µg/kg (2 sa.), sample origin: supermarkets, South Korea, sample year: October 2013–March 2014, country: South Korea[561], *fat-free milk sa.

incidence: 18/22*, conc. range: 0.003–0.05 µg/kg (15 sa.), 0.05–0.1 µg/kg (3 sa.), sample origin: supermarkets, South Korea, sample year: October 2013–March 2014, country: South Korea[561], *low-fat milk sa.

incidence: 11/13*, conc. range: 0.001–0.003 µg/kg (6 sa.), 0.003–0.05 µg/kg (5 sa.), sample origin: supermarkets, South Korea, sample year: October 2013–March 2014, country: South Korea[561], *organic milk sa.

incidence: 53/67*, conc. range: 0.003–0.05 µg/kg (49 sa.), 0.05–0.1 µg/kg (4 sa.), sample origin: supermarkets, South Korea, sample year: October 2013–March 2014, country: South Korea[561], *conventional? milk sa.

- Co-contamination: not reported
- Further contamination of animal products reported in the present book/article[561]: Cheese, Yogurt, AFM_1, literature[561]

incidence: 79/82*, conc. range: 0.020–3.090 µg/l, Ø conc.: 0.3897 µg/l, sample origin: different areas in Karachi (city), Pakistan, sample year: April–August 2016 (summer), country: Pakistan[563], *liquid milk sa.

incidence: 64/74*, conc. range: 0.016–1.520 µg/l, Ø conc.: 0.2926 µg/l, sample origin: different areas in Karachi (city), Pakistan, sample year: September 2016–February 2017 (winter), country: Pakistan[563], *liquid milk sa.

- Co-contamination: not reported
- Further contamination of animal products reported in the present book/article[563]: not reported

incidence: 96/96*, conc. range: 0.0154–4.563 µg/kg, Ø conc.: 0.2903 µg/kg, sample origin: on-farm milk kiosks, dairy shops, milk ATMs (milk dispensing machines), and street vendors, Nairobi (capital), Kenya, sample year: unknown, country: Kenya/Sweden[565], *raw milk sa.

- Co-contamination: not reported
- Further contamination of animal products reported in the present book/article[565]: not reported

Aflatoxin M_2

incidence: 25/52*, conc. range: 0.0003–0.0034 µg/l, sample origin: retail outlets, Japan, sample year: December 2001–February 2002, country: Japan[269], *as well as UHT milk sa.

- Co-contamination: not reported
- Further contamination of animal products reported in the present book/article[269]: Milk (cow milk), AFM_1, literature[269]

Aflatoxins (M_1, M_2)

incidence: 23/26*, conc. range: 0.10–1.43 µg/l, Ø conc.: 0.593 µg/l, sample origin: Florida (state), USA, sample year: unknown, country: USA[314], *raw milk (probably cow milk)

- Co-contamination: not reported
- Further contamination of animal products reported in the present book/article[314]: not reported

Aspergillus and *Penicillium* Toxins

Cyclopiazonic Acid

incidence: 3/20*, conc. range: 4.5–8.3 µg/l, Ø conc.: 6.33 µg/l, sample origin: local supermarkets in different regions of Italy, sample year: unknown, country: Italy[223], *pasteurized cow milk

- Co-contamination: not reported
- Further contamination of animal products reported in the present book/article[223]: not reported

incidence: 2/36*, conc. range: 6.4–9.7 µg/l, Ø conc.: 8.05 µg/l, sample origin: bakeries and supermarkets in São Paulo (city), Brazil, sample year: July–December 2004, country: Brazil[263], *pasteurized cow milk of grade A–C

- Co-contamination: not reported
- Further contamination of animal products reported in the present book/article[263]: Milk (cow milk), Milk (UHT milk), AFM_1, literature[263]

incidence: 1/30*, conc.: 6.4 µg/l, sample origin: dairy farms, São Paulo (state), Brazil, sample year: October 2005–February 2006, country: Brazil[494], *raw milk from grade B dairy farms

incidence: 2/20*, conc. range: 8.8–9.1 µg/l, Ø conc.: 8.9 µg/l, sample origin: dairy farms, São Paulo (state), Brazil, sample year: October 2005–February 2006, country: Brazil[494], *raw milk from grade C dairy farms

- Co-contamination: not reported
- Further contamination of animal products reported in the present book/article[494]: Milk (cow milk), AFM_1, literature[494]

Ochratoxin A

incidence: 3/264*, conc. range: 0.005–0.0066 µg/l, sample origin: dairy farms in Aquitaine, Brittany, Poitou-Charentes, and Pays de Loire (regions), western France, sample year: February–March and September–October 2003, country: France[212], *raw bulk cow milk

- Co-contamination: not reported
- Further contamination of animal products reported in the present book/article[212]: Milk (cow milk), AFM_1, literature[212]

incidence: 1/5*, conc.: 2.73 µg/l, sample origin: dairy farms in al Bohoth, al Kiraiba, Atrah, Brakarat, and Hantoub in Wad Medani (city), Sudan, sample year: March 2009, country: Sudan[286], *bulk cow milk

- Co-contamination: not reported
- Further contamination of animal products reported in the present book/article[286]: Milk (cow milk), AFM_1, literature[286]

incidence: 2/12*, conc. range: 0.010–0.040 µg/l, sample origin: dairy in Lycksele (locality), north of Sweden, sample year: June–July 1991, country: Sweden[315], *tank truck cow milk

incidence: 1/6*, conc.: 0.010–0.040 µg/l, sample origin: dairy in Östersund (city), middle of Sweden, sample year: June 1991, country: Sweden[315], *tank truck cow milk

incidence: 1/6*, conc.: 0.010–0.040 µg/l, sample origin: dairy in Uppsala (city), middle south of Sweden, sample year: June 1991, country: Sweden[315], *tank truck cow milk

incidence: 0/6*, conc.: no contamination, sample origin: dairy in Visby (locality on the island of Gotland), Sweden, sample year: June 1991, country: Sweden[315], *tank truck cow milk

incidence: 1/6*, conc.: 0.010–0.040 µg/l, sample origin: dairy in Malmö (city), south of Sweden, sample year: June 1991, country: Sweden[315], *tank truck cow milk

- Co-contamination: not reported
- Further contamination of human breast milk and animal products reported in the present book/article[315]: Milk (human breast milk), OTA, literature[315]; see also Milk (human breast milk), chapter: Further Mycotoxins and Microbial Metabolites

incidence: 6/40*, conc. range: 0.011–0.058 µg/l, sample origin: groceries in different counties, Norway, sample year: January–July 1995–1996, country: Norway[316], *conventional

incidence: 5/47*, conc. range: 0.015–0.028 µg/l, dairy in Skarnes (administrative center), Hedmark (county), Norway, sample year: June 1997–February 1998, country: Norway[316], *organic

- Co-contamination: not reported
- Further contamination of animal products reported in the present book/article[316]: not reported

incidence: 29/30*, conc. range: ≤0.0841 µg/kg, Ø conc.: 0.0567 µg/kg, sample origin: dairy farms in Beijing (capital), China, sample year: April 2012, country: China[456], *raw cow milk

incidence: 11/12*, conc. range: ≤0.0579 µg/kg, Ø conc.: 0.0268 µg/kg, sample origin: supermarkets in Beijing (capital), China, sample year: April 2012, country: China[456], *fluid cow milk

- Co-contamination: For detailed information see the article

- Further contamination of animal products reported in the present book/article[456]: Milk (cow milk), AFM_1 and ZEA, literature[456]; Milk powder, AFM_1, OTA, and ZEA, literature[456]; see also Milk (cow milk), Milk powder, chapter: Further Mycotoxins and Microbial Metabolites

Fusarium Toxins

Fumonisin/s

Fumonisin B_1
incidence: 1/155*, conc. 1.29 µg/l, sample origin: Wisconsin (state), USA, sample year: March–May 1993, country: USA[317], *raw cow milk
- Co-contamination: not reported
- Further contamination of animal products reported in the present book/article[317]: not reported

incidence: 6/8*, conc. range: 0.26–0.43 µg/kg, ∅ conc.: 0.33 µg/kg, sample origin: retail shops, Italy, sample year: unknown, country: Italy[442], *included conventionally produced raw, fresh whole, and high quality bovine milk

incidence: 2/2*, conc. range: 0.28–0.38 µg/kg, ∅ conc.: 0.33 µg/kg, sample origin: retail shops, Italy, sample year: unknown, country: Italy[442], *included organically produced bovine milk
- Co-contamination: not reported
- Further contamination of animal products reported in the present book/article[442]: not reported

Other ***Fusarium*** Toxins

Zearalenone
incidence: 7/30*, conc. range: ≤0.0458 µg/kg, ∅ conc.: 0.0149 µg/kg, sample origin: dairy farms in Beijing (capital), China, sample year: April 2012, country: China[456], *raw cow milk

incidence: 2/12*, conc. range: ≤0.0283 µg/kg, ∅ conc.: 0.0205 µg/kg, sample origin: supermarkets in Beijing (capital), China, sample year: April 2012, country: China[456], *fluid cow milk
- Co-contamination: For detailed information see the article
- Further contamination of animal products reported in the present book/article[456]: Milk (cow milk), AFM_1 and OTA, literature[456]; Milk powder, AFM_1, OTA, and ZEA, literature[456]; see also Milk (cow milk), Milk powder, chapter: Further Mycotoxins and Microbial Metabolites

see also **Milk** and **Milk (buffalo milk)**

Milk (donkey milk) may contain the following mycotoxins:

Aspergillus Toxins

Aflatoxin M_1
incidence: 3/5, conc. range: 0.005–0.05 µg/kg (3 sa., maximum: 0.035 µg/kg), ∅ conc.: 0.02 µg/kg, sample origin: small milking farms, Serbia, sample year: February–May 2013, country: Serbia[468]
- Co-contamination: not reported
- Further contamination of human breast milk and animal products reported in the present book/article[468]: Milk (cow milk), Milk (goat milk), Milk (human breast milk), Milk (UHT milk), AFM_1, literature[468]

Milk (ewe milk) see **Milk (sheep milk)**

Milk (goat milk) may contain the following mycotoxins:

Aspergillus Toxins

Aflatoxin M_1
incidence: 7/100*, conc. range: ≥0.005–0.010 µg/l (4 sa.), >0.010–0.020 µg/l (3 sa., maximum: 0.0125 µg/l), ∅ conc.: 0.010 µg/l, sample origin: Sardinian flocks in Mediterranean area, Italy, sample year: 2003, country: Italy[31], *bulk tank goat milk

incidence: 29/108*, conc. range: ≥0.005–0.010 µg/l (11 sa.), >0.010–0.020 µg/l (9 sa.), >0.020–0.030 µg/l (7 sa.), >0.030–0.040 µg/l (2 sa., maximum: 0.0361 µg/l), ∅ conc.: 0.0151 µg/l, sample origin: Sardinian flocks in Mediterranean area, Italy, sample year: 2004, country: Italy[31], *bulk tank goat milk

incidence: 20/28*, conc. range: ≥0.005–0.010 µg/l (5 sa.), >0.010–0.020 µg/l (6 sa.), >0.020–0.030 µg/l (7 sa.), >0.030–0.040 µg/l (2 sa.), ∅ conc.: 0.0177 µg/l, sample origin: Sardinian flocks in Mediterranean area, Italy, sample year: 2004, country: Italy[31], *intensive animal husbandry system

incidence: 9/80*, conc. range: ≥0.005–0.010 µg/l (6 sa.), >0.010–0.020 µg/l (3 sa.), ∅ conc.: 0.009 µg/l, sample origin: Sardinian flocks in Mediterranean area, Italy, sample year: 2004, country: Italy[31], *extensive animal husbandry system

incidence: 2/36*, ∅ conc.: 0.0099 µg/l, sample origin: Sardinian flocks in Mediterranean area, Italy, sample year: 2003, country: Italy[31], *bulk tank goat milk from extensive farms during 2 lactations

incidence: 1/36*, conc.: 0.0081 µg/l, sample origin: Sardinian flocks in Mediterranean area, Italy, sample year: 2004, country: Italy[31], *bulk tank goat milk from extensive farms during 2 lactations
- Co-contamination: not reported
- Further contamination of animal products reported in the present book/article[31]: Cheese (hard cheese), AFM_1, literature[31]

incidence: 25/80, conc. range: 0.004–0.010 µg/l (14 sa.), 0.011–0.020 µg/l (10 sa.), 0.037 µg/l (1 sa.), sample origin: dairy farms in provinces of Lombardy (administrative region), Italy, sample year: March–October 1996, country: Italy[36]
- Co-contamination: not reported
- Further contamination of animal products reported in the present book/article[36]: Cheese (soft cheese), Cheese (goat cheese), AFM_1, literature[36]

incidence: 4/10*, conc. range: 0.011–0.020 µg/l (2 sa.), 0.021–0.050 µg/l (2 sa.), sample origin: different milk producers, Greece, sample year: December 1999–May 2000, country: Greece[195], *raw goat milk

incidence: 8/12*, conc. range: 0.005–0. 010 µg/l (7 sa.), 0.011–0.020 µg/l (1 sa.), sample origin: different milk producers, Greece, sample year: December 2000–May 2001, country: Greece[195], *raw goat milk
- Co-contamination: not reported
- Further contamination of animal products reported in the present book/article[195]: Milk, Milk (cow milk), Milk (sheep milk), Milk (UHT milk), AFM_1, literature[195]

incidence: 28?/50, conc. range: 0.010–0.050 µg/l (15 sa.), 0.051–0.100 µg/l (6 sa.), 0.101–0.150 µg/l (3 sa.), 0.151–0.200 µg/l (3 sa.), 0.230 µg/l (1 sa.), sample origin: different areas in the Ismailia Governorate, Egypt, sample year: summers 2003–2004, country: Egypt/Germany/USA[226]
- Co-contamination: not reported

- Further contamination of animal products reported in the present book/article[226]: Milk (buffalo milk), Milk (camel milk), Milk (cow milk), AFM_1, literature[226]

incidence: 19/60, conc. range: 0.005–0.020 µg/l (7 sa.), 0.021–0.050 µg/l (8 sa.), >0.050 µg/l (4 sa.), ∅ conc.: 0.0301 µg/l, sample origin: different parts of Khuzestan (province), Iran, sample year: November 2007–December 2008, country: Iran[227]
- Co-contamination: not reported
- Further contamination of animal products reported in the present book/article[227]: Milk (buffalo milk), Milk (camel milk), Milk (cow milk), Milk (sheep milk), AFM_1, literature[227]

incidence: 20/20*, conc. range: 0.02025–0.12589 µg/kg, ∅ conc.: 0.06025 µg/kg, sample origin: Jordanian market, Jordan, sample year: 2014–2015, country: Jordan[278], *fresh goat milk
- Co-contamination: not reported
- Further contamination of animal products reported in the present book/article[278]: Milk, Milk (camel milk), Milk (cow milk), Milk (infant formula), Milk (sheep milk), Milk powder, AFM_1, literature[278]

incidence: 7/11*, conc. range: 0.008–0.054 µg/l, ∅ conc.: 0.019 µg/l, sample origin: north, south, and east of Syria, sample year: April 2005–2006, country: Syria[293], *raw goat milk
- Co-contamination: not reported
- Further contamination of animal products reported in the present book/article[293]: Milk (cow milk), Milk (sheep milk), Milk powder, AFM_1, literature[293]

incidence: 17/48, conc. range: 0.015–0.061 µg/l, ∅ conc.: 0.0318 µg/l, sample origin: Chaharmahal va Bakhtiari (province), central and southern Iran, sample year: December 2008–2009, country: Iran/USA[311]
- Co-contamination: not reported
- Further contamination of animal products reported in the present book/article[311]: Milk (cow milk), Milk (sheep milk), AFM_1, literature[311]

incidence: 7/12*, conc. range: 0.010–0.050 µg/l (4 sa.), 0.051–0.100 µg/l (1 sa.), 0.101–0.200 µg/l (2 sa.), ∅ conc.: 0.072 µg/l, sample origin: bakeries and supermarkets in Campinas (city), São Paulo (state), Brazil, sample year: October–December 2004, March–May 2005, country: Brazil[319], *pasteurized goat milk
- Co-contamination: not reported
- Further contamination of animal products reported in the present book/article[319]: Milk (UHT milk), Milk powder, AFM_1, literature[319]

incidence: 18/25*, conc. range: 0.0251–0.1016 µg/l, ∅ conc.: 0.0497 µg/l, sample origin: rural regions in Hamadan, Ilam, Kermanshah, and Kurdistan (provinces), Iran, sample year: winter and summer seasons 2014, country: Iran[328], *raw goat milk
- Co-contamination: not reported
- Further contamination of animal products reported in the present book/article[328]: Ayran, Cheese, Kashk, Milk (cow milk), Milk (sheep milk), Tarkhineh, Yogurt, AFM_1, literature[328]

incidence: 33/164, conc. range: 0.010–0.050 µg/kg (18 sa.), 0.051–0.100 µg/kg (14 sa.), 0.126 µg/kg (1 sa.), sample origin: Yazd (province), Iran, sample year: 2014, country: Iran[437]
- Co-contamination: not reported
- Further contamination of animal products reported in the present book/article[437]: Milk (camel milk), Milk (cow milk), Milk (sheep milk), AFM_1, literature[437]

For detailed information see the article.

incidence: 28/65*, conc. range: 0.013–0.055 µg/l, sample origin: dairy ranches, supermarkets, and retail outlets in Esfahan, Shiraz, and Tabriz (cities) as well as Tehran (capital), Iran, sample year: 2008, country: Iran[440], *raw goat milk
- Co-contamination: not reported
- Further contamination of animal products reported in the present book/article[440]: Ayran, Cheese (Lighvan cheese), Kashk, Milk (cow milk), Milk (sheep milk), Yogurt, AFM_1, literature[440]

incidence: 8/10, conc. range: 0.005–0.05 µg/kg (4 sa.), 0.05–0.50 µg/kg (4 sa., maximum: 0.24 µg/kg), ∅ conc.: 0.08 µg/kg, sample origin: small milking farms, Serbia, sample year: February–May 2013, country: Serbia[468]
- Co-contamination: not reported
- Further contamination of human breast milk and animal products reported in the present book/article[468]: Milk (cow milk), Milk (donkey milk), Milk (human breast milk), Milk (UHT milk), AFM_1, literature[468]

incidence: 1/3, conc. range: ≤0.003–0.005 µg/l (1 sa.), sample origin: different areas on Sicily (island, autonomous region), Italy, sample year: January–June 2012, country: Italy[508]
- Co-contamination: not reported
- Further contamination of animal products reported in the present book/article[508]: Cheese, Cream, Milk (cow milk), Milk (sheep milk), AFM_1, literature[508]

For detailed information see the article.

incidence: 29/85*, conc. range: 0.0031–0.005 µg/kg (7 sa.), 0.0051–0.010 µg/kg (10 sa.), 0.0101–0.015 µg/kg (5 sa.), 0.0151–0.025 µg/kg (4 sa.), 0.0251–0.040 µg/kg (2 sa.), 0.0401–0.050 µg/kg (1 sa.), sample origin: Veneto (region in Belluno province), Trentino Alto Adige (autonomous region in Trento province), and Friuli Venezia Giulia (region), northeastern Italy, sample year: 2005–2006, country: Italy[512], *raw bulk goat milk
- Co-contamination: not reported
- Further contamination of animal products reported in the present book/article[512]: not reported

incidence: 9/88, conc. range: ≥0.008–0.020 µg/l (2 sa.)*, >0.020–0.050 µg/l (5 sa.)*, >0.050 µg/l (2 sa., maximum: 0.13816 µg/l)*, ∅ conc.: 0.04721 µg/l, sample origin: Sardinia (island), Italy, sample year: 2010–2013, country: Italy[557], *in year 2013
- Co-contamination: not reported
- Further contamination of animal products reported in the present book/article[557]: Milk (sheep milk), AFM_1, literature[557]

For detailed information see the article.

see also **Milk, Milk (cow milk)**, and **Milk (sheep milk)**

Milk (human breast milk) see page 111

Milk (infant formula) may contain the following mycotoxins:

Aspergillus **Toxins**

Aflatoxin B_1

incidence: 55/63*, conc. range: 0.10–6.04 µg/kg, sample origin: pharmacies and supermarkets in Ankara (capital), Turkey, sample year: 2003–2004, country: Turkey[89], *baby food and infant formula
- Co-contamination: not reported

- Further contamination of animal products reported in the present book/article[89]: Milk (infant formula), AFM_1 and OTA, literature[89]

For detailed information please see the article.

incidence: 27/30*, conc. range: <LOQ-15.15 µg/kg, ∅ conc.: 2.959 µg/kg, sample origin: Iranian market, Iran, sample year: 2014–2015, country: Iran[279], *sa. consisted of rice and milk powder or skimmed milk as well as other constituents

incidence: 4/16*, conc. range: <LOQ-1.8 µg/kg, ∅ conc.: 0.579 µg/kg, sample origin: Iranian market, Iran, sample year: 2014–2015, country: Iran[279], *sa. consisted of wheat and milk powder or skimmed milk as well as other constituents

incidence: 2/2*, conc. range: 1.03–2.50 µg/kg, ∅ conc.: 1.76 µg/kg, sample origin: Iranian market, Iran, sample year: 2014–2015, country: Iran[279], *sa. consisted of cereal grain mixtures and milk powder as well as other constituents

- Co-contamination: not reported
- Further contamination of animal products reported in the present book/article[279]: not reported

Aflatoxin M_1

incidence: 23/63*, conc. range: 0.06–0.32 µg/kg, sample origin: pharmacies and supermarkets in Ankara (capital), Turkey, sample year: 2003–2004, country: Turkey[89], *baby food and infant formula

- Co-contamination: not reported
- Further contamination of animal products reported in the present book/article[89]: Milk (infant formula), AFB_1 and OTA, literature[89]

For detailed information see the article.

incidence: 38/40*, conc. range: 0.065–1.012 µg/kg, ∅ conc.: 0.267 µg/kg, sample origin: Lucknow (city), India, sample year: unknown, country: India[91], *milk-based cereal weaning food

incidence: 17/18, conc. range: 0.143–0.770 µg/kg, ∅ conc.: 0.326 µg/kg, sample origin: Lucknow (city), India, sample year: unknown, country: India[91]

- Co-contamination: not reported
- Further contamination of animal products reported in the present book/article[91]: Food, Milk (cow milk), AFM_1, literature[91]

incidence: 72/80*, conc. range: 0.003–0.035 µg/kg, sample origin: Tehran (capital), Iran, sample year: April–September 2005, country: Iran[92], *milk-based cereal weaning food

incidence: 116/120*, conc. range: 0.001–0.014 µg/kg, sample origin: Tehran (capital), Iran, sample year: April–September 2005, country: Iran[92]

- Co-contamination: not reported
- Further contamination of animal products reported in the present book/article[92]: Milk (cow milk), AFM_1, literature[92]

incidence: 1/6*, conc.: 0.016 µg/kg or µg/l, sample origin: supermarkets and drug stores in Corum (city), Turkey, sample year: January–February 2011, country: Turkey[93], *infant formula (for infants aged up to 4–6 months)

- Co-contamination: not reported
- Further contamination of animal products reported in the present book/article[93]: Milk (infant formula), OTA, literature[93]

incidence: 2/36*, conc. range: 0.016–0.020 µg/kg or µg/l, ∅ conc.: 0.018 µg/kg or µg/l, sample origin: supermarkets and drug stores in Corum (city), Turkey, sample year: January–February 2011, country: Turkey[93], *follow-on formula (for infants aged from 6 months)

- Co-contamination: not reported
- Further contamination of animal products reported in the present book/article[93]: Milk (infant formula), OTA, literature[93]

incidence: 2/20*, conc. range: 0.017–0.022 µg/kg or µg/l, ∅ conc.: 0.020 µg/kg or µg/l, sample origin: supermarkets and drug stores in Corum (city), Turkey, sample year: January–February 2011, country: Turkey[93], *toddler formula (for young children aged 12–36 months)

- Co-contamination: 1 sa. co-contaminated with AFM_1 and OTA; 1 sa. contaminated solely with AFM_1
- Further contamination of animal products reported in the present book/article[93]: Milk (infant formula), OTA, literature[93]

incidence: 4/62, conc. range: 0.02–0.05 µg/kg* (4 sa., maximum: 0.05 µg/kg*), sample origin: retail outlets, England and Wales, sample year: unknown, country: UK[131], *on dry weight basis

- Co-contamination: not reported
- Further contamination of animal products reported in the present book/article[131]: Cheese (Cheddar cheese), Cheese (Cheshire cheese), Cheese (Double Gloucester cheese), Cheese (Lancashire cheese), Cheese (Leicester cheese), Cheese (Wensleydale cheese), Milk (cow milk), Yogurt, AFM_1, literature[131]

incidence: 31/50*, conc. range: 0.0055–0.0201 µg/kg, ∅ conc.: 0.0089 µg/kg, sample origin: markets in Ankara (capital), Turkey, sample year: unknown, country: Turkey[201], *follow-on milks for babies

incidence: 1/34, conc.: 0.0061 µg/kg, sample origin: markets in Ankara (capital), Turkey, sample year: unknown, country: Turkey[201]

- Co-contamination: not reported
- Further contamination of animal products reported in the present book/article[201]: not reported

incidence: 18/26, conc. range: 0.032–0.132 µg/kg, ∅ conc.: 0.062 µg/kg, sample origin: supermarkets in Seoul (capital), Korea, sample year: 1997, country: Korea[241]

- Co-contamination: not reported
- Further contamination of animal products reported in the present book/article[241]: Milk (cow milk), Milk powder, Yogurt, AFM_1, literature[241]

incidence: 6/9*, conc. range: 0.010–0.049 µg/kg (2 sa.), conc.: ≥0.050 µg/kg (1 sa.), sample origin: supermarkets in Dohar (capital), Qatar, sample year: unknown, country: Qatar/Italy[277], *baby food (milk-based cereals)

- Co-contamination: For detailed information see the article.
- Further contamination of animal products reported in the present book/article[277]: Milk (infant formula), OTA, T-2, and ZEA, literature[277]; Milk, AFM_1, literature[277]

incidence: 20/20, conc. range: 0.01655–0.15414 µg/kg, ∅ conc.: 0.12026 µg/kg, sample origin: Jordanian market, Jordan, sample year: 2014-2015, country: Jordan[278]

- Co-contamination: not reported
- Further contamination of animal products reported in the present book/article[278]: Milk, Milk (camel milk), Milk (cow milk), Milk (goat milk), Milk (sheep milk), Milk powder, AFM_1, literature[278]

incidence: 2/185*, conc. range: 0.0118–0.0153 µg/l, ∅ conc.: 0.01355 µg/l, sample origin: marketed in Italy, sample year: 2006–2007, country: Italy[354], *infant cow's milk-based formula

- Co-contamination: not reported
- Further contamination of animal products reported in the present book/article[354]: Milk (infant formula), OTA, literature[354]

incidence: 7/12*, conc. range: 0.014–0.017 µg/l, ∅ conc.: 0.015 µg/l, sample origin: directly from producer or from supermarkets and drugstores, southern Brazil, sample year: unknown, country: Brazil[407], *infant formula for infants from 0 to 6 months

incidence: 2/14*, conc. range: tr, sample origin: directly from producer or from supermarkets and drugstores, southern Brazil, sample year: unknown, country: Brazil[407], *follow-on formula for infants from 6 months to 1 year

- Co-contamination: not reported
- Further contamination of animal products reported in the present book/article[407]: Dairy products, AFM_1, literature[407]

incidence: 1/7*, conc.: 0.0006 µg/kg, sample origin: pharmacies in Pamplona (city), northern Spain, sample year: March 2007–January 2008, country: Spain[443], *preterm infant formula

incidence: 6/17*, conc. range: <LOD? (5 sa.), 0.003 µg/kg (1 sa.), ∅ conc.: 0.0018 µg/kg, sample origin: pharmacies in Pamplona (city), northern Spain, sample year: March 2007–January 2008, country: Spain[443], *starter infant formula

incidence: 8/19*, conc. range: <LOD? (7 sa.), 0.0042 µg/kg (1 sa.), ∅ conc.: 0.0020 µg/kg, sample origin: pharmacies in Pamplona (city), northern Spain, sample year: March 2007–January 2008, country: Spain[443], *follow-up infant formula

incidence: 11/13*, conc. range: <LOD? (5 sa.), LOD-0.025 µg/kg (6 sa., maximum: 0.0115 µg/kg), ∅ conc.: 0.0049 µg/kg, sample origin: pharmacies in Pamplona (city), northern Spain, sample year: March 2007–January 2008, country: Spain[443], *toddler infant formula

incidence: 0/7*, conc.: no contamination, sample origin: pharmacies in Pamplona (city), northern Spain, sample year: March 2007–January 2008, country: Spain[443], *hypoallergenic infant formula

incidence: 0/6*, conc.: no contamination, sample origin: pharmacies in Pamplona (city), northern Spain, sample year: March 2007–January 2008, country: Spain[443], *lactose-free infant formula

- Co-contamination: not reported
- Further contamination of animal products reported in the present book/article[443]: not reported

incidence: 1/6*, conc.: 0.0216 µg/l, sample origin: local markets and supermarkets, China, sample year: unknown, country: China[448], *milk-based cereal weaning food

- Co-contamination: not reported
- Further contamination of animal products reported in the present book/article[448]: Milk, Milk powder, AFM_1, literature[448]

incidence: 18/18, conc. range: 0.150–0.730 µg/l, sample origin: markets in Goa (state), India, sample year: unknown, country: India[464]

- Co-contamination: not reported
- Further contamination of animal products reported in the present book/article[464]: Milk, AFM_1, literature[464]

incidence: 1/18, conc.: 0.19 µg/kg*, sample origin: local stores, USA/Taiwan, sample year: unknown, country: USA/Taiwan[534], *imported milk-based infant formula sa.

- Co-contamination: not reported
- Further contamination of animal products reported in the present book/article[534]: Cream, Milk, AFM_1, literature[534]; Milk powder, AFB_1 and AFB_2, literature[534]

Aflatoxin/s

incidence: 1/24*, conc.: 70.1 µg/kg**, sample origin: factories, retailers, and stores, Libya, sample year: unknown, country: UK/Libya[394], *cereal with milk, vegetable, and fruits, **AFB_1, AFB_2, AFG_1, and AFG_2

- Co-contamination: not reported
- Further contamination of animal products reported in the present book/article[394]: not reported

Aspergillus and *Penicillium* Toxins

Ochratoxin A

incidence: 25/63*, conc. range: 0.27–4.50 µg/kg, sample origin: pharmacies and supermarkets in Ankara (capital), Turkey, sample year: 2003–2004, country: Turkey[89], *baby food and infant formula

- Co-contamination: not reported
- Further contamination of animal products reported in the present book/article[89]: Milk (infant formula), AFB_1 and AFM_1, literature[89]

For detailed information see the article.

incidence: 0/6*, conc.: no contamination, sample origin: supermarkets and drug stores in Corum (city), Turkey, sample year: January–February 2011, country: Turkey[93], *infant formula (for infants aged up to 4–6 months)

- Co-contamination: not reported
- Further contamination of animal products reported in the present book/article[93]: Milk (infant formula), AFM_1, literature[93]

incidence: 2/36*, conc. range: 0.017–0.029 µg/kg or µg/l, ∅ conc.: 0.023 µg/kg or µg/l, sample origin: supermarkets and drug stores in Corum (city), Turkey, sample year: January–February 2011, country: Turkey[93], *follow-on formula (for infants aged from 6 months)

- Co-contamination: not reported
- Further contamination of animal products reported in the present book/article[93]: Milk (infant formula), AFM_1, literature[93]

incidence: 10/20*, conc. range: 0.024–0.184 µg/kg or µg/l, ∅ conc.: 0.119 µg/kg or µg/l, sample origin: supermarkets and drug stores in Corum (city), Turkey, sample year: January–February 2011, country: Turkey[93], *toddler formula (for young children aged 12–36 months)

- Co-contamination: 1 sa. co-contaminated with AFM_1 and OTA; 9 sa. contaminated solely with OTA
- Further contamination of animal products reported in the present book/article[93]: Milk (infant formula), AFM_1, literature[93]

incidence: 1/9*, conc.: 0.05–0.49 µg/kg (1 sa.), sample origin: supermarkets in Dohar (capital), Qatar, sample year: unknown, country: Qatar/Italy[277], *baby food (milk-based cereals)

- Co-contamination: For detailed information see the article.
- Further contamination of animal products reported in the present book/article[277]: Milk (infant formula), AFM_1, T-2, and ZEA, literature[277]; Milk, AFM_1, literature[277]

incidence: 40/50*, conc. range: ≤0.6895 µg/l, sample origin: marketed in Italy, sample year: 2006–2007, country: Italy[354], *preterm liquid cow's milk-based infant formula incidence: 5/5*, conc. range: 0.08049–0.1273 µg/l, ∅ conc.: 0.09610 µg/l, sample origin: marketed in Italy, sample year: 2006–2007, country: Italy[354], *preterm powder cow's milk-based infant formula

incidence: 40/50*, conc. range: ≤0.2991 µg/l, sample origin: marketed in Italy, sample year: 2006–2007, country: Italy[354], *starter liquid cow's milk-based infant formula

incidence: 48/80*, conc. range: ≤0.3558 µg/l, sample origin: marketed in Italy, sample year: 2006–2007, country: Italy[354], *starter powder cow's milk-based infant formula
- Co-contamination: not reported
- Further contamination of animal products reported in the present book/article[354]: Milk (infant formula), AFM_1, literature[354]

Fusarium Toxins

Fumonisin/s

Fumonisin B_1
incidence: 2/5*, conc. range: 3–13 µg/kg**, sample origin: retail stores in Bari (city), Italy, sample year: unknown, country: Italy[94], *milk-based flours containing maize included milk or skimmed milk, cereals and/or cereal products, and further ingredients, **the range accounts also for 3 other sa.
- Co-contamination: not reported
- Further contamination of cereals and cereal products reported in the present book/article[94]: not reported

For detailed information see the article.

Type A Trichothecenes

T-2 Toxin
incidence: 7/9*, conc. range: 1.00–9.99 µg/kg (7 sa.), sample origin: supermarkets in Dohar (capital), Qatar, sample year: unknown, country: Qatar/Italy[277], *baby food (milk-based cereals)
- Co-contamination: For detailed information see the article.
- Further contamination of animal products reported in the present book/article[277]: Milk (infant formula), AFM_1, OTA, and ZEA, literature[277]; Milk, AFM_1, literature[277]

Other ***Fusarium*** Toxins

Zearalenone
incidence: 17/185*, conc. range: ≤0.76 µg/l, sample origin: marketed in Italy, sample year: 2007–2008, country: Italy[164], *preterm and starter infant formula
- Co-contamination: 3 sa. co-contaminated with ZEA, α-ZEL, and ß-ZEL; 14 sa. contaminated solely with ZEA
- Further contamination of animal products reported in the present book/article[164]: see also Food, chapter: Further Mycotoxins and Microbial Metabolites

For detailed information see the article.

incidence: 1/9*, conc.: 1.00–4.99 µg/kg (1 sa.), sample origin: supermarkets in Dohar (capital), Qatar, sample year: unknown, country: Qatar/Italy[277], *baby food (milk-based cereals)
- Co-contamination: For detailed information see the article.
- Further contamination of animal products reported in the present book/article[277]: Milk (infant formula), AFM_1, OTA, and T-2, literature[277]; Milk, AFM_1, literature[277]

see also **Food** and **Milk powder**

Milk (sheep milk) may contain the following mycotoxins:

Aspergillus Toxins

Aflatoxin M_1
incidence: 12/40, conc. range: 0.004–0.023 µg/l, sample origin: sheep and dairy farms or market in Western Sicily (island, autonomous region), Italy, sample year: November 2001–June 2002, country: Italy[37]
- Co-contamination: not reported
- Further contamination of animal products reported in the present book/article[37]: Cheese (sheep cheese), AFM_1, literature[37]

incidence: 204?/407*, conc. range: 0.005–0.010 µg/kg (114 sa.), >0.010–0.020 µg/kg (36 sa.), >0.020–0.050 µg/kg (11 sa.), >0.050 µg/kg (3 sa., maximum: 0.0839 µg/kg), sample origin: farms in the Castilla-La Mancha region, southeast Spain, sample year: autumn–winter 2007 and 2008, country: Spain[153], *bulk tank ewe milk sa. for Manchego cheese production

incidence: 50?/407*, conc. range: 0.005–0.010 µg/kg (40 sa.), >0.010–0.020 µg/kg (7 sa.), >0.020–0.050 µg/kg (2 sa.), 0.1299 µg/kg (1 sa.), sample origin: farms in the Castilla-La Mancha region, southeast Spain, sample year: autumn–winter 2007 and 2008, country: Spain[153], *silo ewe milk sa. for Manchego cheese production
- Co-contamination: not reported
- Further contamination of animal products reported in the present book/article[153]: Cheese curd, AFM_1, literature[153]

incidence: 8/12*, conc. range: 0.005–0.010 µg/l (3 sa.), 0.011–0.020 µg/l (3 sa.), 0.021–0.050 µg/l (2 sa.), sample origin: different milk producers, Greece, sample year: December 1999–May 2000, country: Greece[195], *raw sheep milk

incidence: 11/15*, conc. range: 0.005–0.010 µg/l (6 sa.), 0.011–0.020 µg/l (3 sa.), 0.021–0.050 µg/l (1 sa.), >0.05 µg/l (1 sa.), sample origin: different milk producers, Greece, sample year: December 2000–May 2001, country: Greece[195], *raw sheep milk
- Co-contamination: not reported
- Further contamination of animal products reported in the present book/article[195]: Milk, Milk (cow milk), Milk (goat milk), Milk (UHT milk), AFM_1, literature[195]

incidence: 0/1*, conc.: no contamination, sample origin: Campania and Calabria (regions), southern Italy, sample year: 2003, country: Italy[225], *raw sheep/goat milk

incidence: 4/7*, conc. range: 0.006–0.009 µg/kg, sample origin: Campania and Calabria (regions), southern Italy, sample year: 2003, country: Italy[225], *raw sheep/goat milk

incidence: 2/8*, conc. range: 0.009–0.027 µg/kg, ∅ conc.: 0.018 µg/kg, sample origin: Campania and Calabria (regions), southern Italy, sample year: 2004, country: Italy[225], *raw sheep/goat milk

incidence: 1/1*, conc.: 0.031 µg/kg, sample origin: Campania and Calabria (regions), southern Italy, sample year: 2005, country: Italy[225], *raw sheep/goat milk
- Co-contamination: not reported
- Further contamination of animal products reported in the present book/article[225]: Milk (buffalo milk), Milk (cow milk), AFM_1, literature[225]

incidence: 19?/51, conc. range: 0.005–0.020 µg/l (8 sa.), 0.021–0.050 µg/l (8 sa.), >0.050 µg/l (2 sa.), ∅ conc.: 0.0281 µg/l, sample origin: different parts of Khuzestan (province), Iran, sample year: November 2007–December 2008, country: Iran[227]

- Co-contamination: not reported
- Further contamination of animal products reported in the present book/article[227]: Milk (buffalo milk), Milk (camel milk), Milk (cow milk), Milk (goat milk), AFM_1, literature[227]

incidence: 20/20*, conc. range: 0.02356–0.13718 µg/kg, Ø conc.: 0.07025 µg/kg, sample origin: Jordanian market, Jordan, sample year: 2014–2015, country: Jordan[278], *fresh sheep milk

- Co-contamination: not reported
- Further contamination of animal products reported in the present book/article[278]: Milk, Milk (camel milk), Milk (cow milk), Milk (goat milk), Milk (infant formula), Milk powder, AFM_1, literature[278]

incidence: 13/23*, conc. range: 0.006–0.634 µg/l, Ø conc.: 0.067 µg/l, sample origin: north, south, and east of Syria, sample year: April 2005–2006, country: Syria[293], *raw sheep milk

- Co-contamination: not reported
- Further contamination of animal products reported in the present book/article[293]: Milk (cow milk), Milk (goat milk), Milk powder, AFM_1, literature[293]

incidence: 13/42, conc. range: 0.008–0.050 µg/l, Ø conc.: 0.0258 µg/l, sample origin: Chaharmahal va Bakhtiari (province), central and southern Iran, sample year: December 2008–2009, country: Iran/USA[311]

- Co-contamination: not reported
- Further contamination of animal products reported in the present book/article[311]: Milk (cow milk), Milk (goat milk), AFM_1, literature[311]

incidence: 27/54, conc. range: ≤0.0182 µg/l, sample origin: plants of traditional manufacture in the west side of Thessaloniki (prefecture), Greece, sample year: March–June 2002, country: Greece[318]

- Co-contamination: not reported
- Further contamination of animal products reported in the present book/article[318]: not reported

incidence: 25/34*, conc. range: 0.0177–0.0954 µg/l, Ø conc.: 0.0619 µg/l, sample origin: rural regions in Hamadan, Ilam, Kermanshah, and Kurdistan (provinces), Iran, sample year: winter and summer seasons 2014, country: Iran[328], *raw sheep milk

- Co-contamination: not reported
- Further contamination of animal products reported in the present book/article[328]: Ayran, Cheese, Kashk, Milk (cow milk), Milk (goat milk), Tarkhineh, Yogurt, AFM_1, literature[328]

incidence: 195/240*, conc. range: 0.001–0.010 µg/l (145 sa.), >0.010–0.050 µg/l (47 sa.), >0.050 µg/l (3 sa., maximum: 0.108 µg/l), Ø conc.: 0.01536 µg/l, sample origin: Enna (province), Sicily (island, autonomous region), Italy, sample year: October–July 2000?, country: Italy[353], *dairy ewe milk

- Co-contamination: not reported
- Further contamination of animal products reported in the present book/article[353]: not reported

incidence: 45/208, conc. range: 0.010–0.050 µg/kg (21 sa.), 0.051–0.100 µg/kg (22 sa.), >0.100 µg/kg (2 sa., maximum: 0.109 µg/kg), sample origin: Yazd (province), Iran, sample year: 2014, country: Iran[437]

- Co-contamination: not reported
- Further contamination of animal products reported in the present book/article[437]: Milk (camel milk), Milk (cow milk), Milk (goat milk), AFM_1, literature[437]

For detailed information see the article.

incidence: 43/72*, conc. range: 0.015–0.102 µg/l, sample origin: dairy ranches, supermarkets, and retail outlets in Esfahan, Shiraz, and Tabriz (cities) as well as Tehran (capital), Iran, sample year: 2008, country: Iran[440], *raw sheep milk

- Co-contamination: not reported
- Further contamination of animal products reported in the present book/article[440]: Ayran, Cheese (Lighvan cheese), Kashk, Milk (cow milk), Milk (goat milk), Yogurt, AFM_1, literature[440]

incidence: 20/34, conc. range: ≤0.003–0.005 µg/l (15 sa.), ≤0.005–0.010 µg/l (4 sa.), 0.016 µg/l (1 sa.), sample origin: different areas on Sicily (island, autonomous region), Italy, sample year: January–June 2012, country: Italy[508]

- Co-contamination: not reported
- Further contamination of animal products reported in the present book/article[508]: Cheese, Cream, Milk (cow milk), Milk (goat milk), AFM_1, literature[508]

For detailed information see the article.

incidence: 8/517, conc. range: ≥0.008–0.020 µg/l (6 sa.)*, >0.020–0.050 µg/l (1 sa.)*, 0.05882 µg/l (1 sa.)*, Ø conc.: 0.01259 µg/l, sample origin: Sardinia (island), Italy, sample year: 2005–2013, country: Italy[557], *in year 2013

- Co-contamination: not reported
- Further contamination of animal products reported in the present book/article[557]: Milk (goat milk), AFM_1, literature[557]

For detailed information see the article.

Fusarium Toxins

Depsipeptides

Enniatin B

incidence: 18/20*, conc. range: 0.0055–0.0121 µg/kg, Ø conc.: 0.0078 µg/kg, sample origin: Bieszczady Mountains, southeastern Poland, sample year: April, country: China[327], *raw sheep milk

- Co-contamination: not reported
- Further contamination of animal products in the present book/article[327]: not reported

see also **Milk**

Milk (UHT milk) may contain the following mycotoxins:

Aspergillus Toxins

Aflatoxin M_1

incidence: 20/20*, conc. range: 0.010–<0.025 µg/kg (8 sa.), 0.025–0.050 µg/kg (9 sa.), 0.051–0.100 µg/kg (4 sa.), 0.101–0.150 µg/kg (3 sa., maximum: 0.080 µg/kg), Ø conc.: 0.035 µg/kg, sample origin: supermarkets in Adana (city), Turkey, sample year: unknown, country: Turkey[12], *cow milk

- Co-contamination: not reported
- Further contamination of animal products reported in the present book/article[12]: Butter, Cheese (Kashar cheese), Cheese (white cheese), AFM_1, literature[12]

incidence: 3*/52**, conc. range: ≤0.0888 µg/kg, sample origin: local sa. purchased in supermarkets in Kuwait City (capital), Kuwait, sample year: January 2005–March 2007, country: Kuwait[41], *above EC limit 0.050 µg/kg, **long-life milk sa.

incidence: 1*/53**, conc.: 0.0626 µg/kg, sample origin: imported sa. purchased in supermarkets in Kuwait City (capital), Kuwait, sample year: January 2005–March 2007, country: Kuwait[41], *above EC limit 0.050 µg/kg, **long-life milk sa.

- Co-contamination: not reported
- Further contamination of human breast milk and animal products reported in the present book/article[41]: Cheese, Milk (cow milk), Milk (human breast milk), Milk powder, AFM_1, literature[41]

incidence: 67/100, conc. range: 0.010–0.025 μg/kg (12 sa.), 0.026–0.050 μg/kg (24 sa.), 0.051–0.100 μg/kg (18 sa.), >0.100 μg/kg (13 sa., maximum: 0.630 μg/kg), sample origin: supermarkets in Edirne, Izmir, Konya, Tekirdag (cities), and Istanbul (capital), Turkey, sample year: September 2007–January 2008, country: Turkey[143]

- Co-contamination: not reported
- Further contamination of animal products reported in the present book/article[143]: Cheese (Kashar cheese), AFM_1, literature[143]

incidence: 84/153, conc. range: 0.005–0.029 μg/l (47 sa.), 0.030–0.049 μg/l (6 sa.), 0.050–0.099 μg/l (17 sa.), 0.100–0.200 μg/l (14 sa.), sample origin: supermarkets in the 25 most populous cities, China, sample year: July–September 2010, country: China[193]

- Co-contamination: not reported
- Further contamination of animal products reported in the present book/article[193]: Milk (cow milk), AFM_1, literature[193]

incidence: 12/76*, conc. range: 0.02–0.04 μg/l, ∅ conc.: 0.0267 μg/l, sample origin: supermarket in Madrid (capital), Spain, sample year: 1993, country: Spain[194], *included 24 semi-skimmed sa. (all not contaminated)

- Co-contamination: not reported
- Further contamination of animal products reported in the present book/article[194]: Milk, AFM_1, literature[194]

incidence: 14/17, conc. range: 0.005–0.010 μg/l (4 sa.), 0.011–0.020 μg/l (6 sa.), 0.021–0.050 μg/l (4 sa.), sample origin: supermarkets, Greece, sample year: December 1999–May 2000, country: Greece[195]

- Co-contamination: not reported
- Further contamination of animal products reported in the present book/article[195]: Milk, Milk (cow milk), Milk (goat milk), Milk (sheep milk), AFM_1, literature[195]

incidence: 60/60, conc. range: >0–0.05 μg/l (7 sa.), >0.05–0.125 μg/l (5 sa.), >0.125–0.25 μg/l (12 sa.), >0.25–0.5 μg/l (29 sa.), >0.5 μg/l (7 sa.), sample origin: locally manufactured and purchased in retail markets, Thailand, sample year: May 1995–January 1996, country: Thailand[196]

- Co-contamination: not reported
- Further contamination of animal products reported in the present book/article[196]: Milk (cow milk), Milk powder, AFM_1, literature[196]

incidence: 5/105, conc. range: 0.0141–0.0349 μg/l, sample origin: produced mostly in the region of southeastern Brazil and purchased in supermarkets and drugstores in São Paulo (city), Brazil, sample year: April 1998–May 1999, country: Brazil[197]

- Co-contamination: not reported
- Further contamination of animal products reported in the present book/article[197]: Milk (cow milk), AFM_1, literature[197]

incidence: 30/41, conc. range: 0.00642–0.07133 μg/l, sample origin: drugstores and supermarkets in various districts of Burdur (city, province), Turkey, sample year: May–December 2011, country: Turkey[202]

- Co-contamination: not reported
- Further contamination of animal products reported in the present book/article[202]: not reported

incidence: 68/72, conc. range: LOD-0.030 μg/kg (68 sa., maximum: 0.01361 μg/kg), ∅ conc.: 0.00969 μg/kg, sample origin: hypermarkets and supermarkets in 12 main cities of Catalonia (autonomous community), Spain, sample year: June–July 2008, country: Spain[216]

- Co-contamination: not reported
- Further contamination of animal products reported in the present book/article[216]: Yoghurt, AFM_1, literature[216]

incidence: 39/43, conc. range: 0.003–0.010 μg/l (30 sa.), 0.011–0.020 μg/l (6 sa.), 0.021–0.030 μg/l (2 sa.), 0.035 μg/l (1 sa.)*, sample origin: supermarkets in Florence (city), Italy, sample year: unknown, country: Italy[217], *partially skimmed UHT milk

- Co-contamination: not reported
- Further contamination of animal products reported in the present book/article[217]: Milk, Yogurt, AFM_1, literature[217]

incidence: 7*/17, conc. range: ≤0.084 μg/kg, sample origin: retailers and traders in Dagoretti (low-income area), Kenya, sample year: November 2013–October 2014, country: Kenya/Sweden[228], *>0.050 μg/kg

incidence: 16*/55**, conc. range: ≤0.470 μg/kg, sample origin: supermarket in Westlands (middle- to high-income area), Nairobi (capital), Kenya, sample year: November 2013–October 2014, country: Kenya/Sweden[228], *>0.050 μg/kg

- Co-contamination: not reported
- Further contamination of animal products reported in the present book/article[228]: Dairy products, Milk (cow milk), Yogurt, AFM_1, literature[228]

incidence: 9/80*, conc. range: 0.02903–0.1028 μg/l, ∅ conc.: 0.073 μg/l, sample origin: supermarkets and general stores in Karachi (city), Pakistan, sample year: December 2002–May 2003, country: Pakistan[229], *UHT-treated buffalo milk

- Co-contamination: not reported
- Further contamination of animal products reported in the present book/article[229]: Milk (buffalo milk), AFM_1, literature[229]

incidence: 17/18*, conc. range: 0.005–0.010 μg/l (2 sa.), 0.011–0.020 μg/l (8 sa.), 0.021–0.050 μg/l (6 sa.), 0.059 μg/l (1 sa.), sample origin: supermarkets in Lisbon (capital), Portugal, sample year: June–September 1999, country: Portugal[259], *UHT whole milk

incidence: 20/22*, conc. range: 0.011–0.020 μg/l (1 sa.), 0.021–0.050 μg/l (18 sa.), 0.061 μg/l (1 sa.), sample origin: supermarkets in Lisbon (capital), Portugal, sample year: June–September 1999, country: Portugal[259], *UHT semi-skimmed milk

incidence: 23/30*, conc. range: 0.005–0.010 μg/l (7 sa.), 0.011–0.020 μg/l (16 sa.), sample origin: supermarkets in Lisbon (capital), Portugal, sample year: June–September 1999, country: Portugal[259], *UHT skimmed milk

- Co-contamination: not reported
- Further contamination of animal products reported in the present book/article[259]: Milk (cow milk), AFM_1, literature[259]

incidence: 9/12, conc. range: 0.010–0.050 μg/l (3 sa.), 0.051–0.100 μg/l (4 sa.), 0.101–0.200 μg/l (2 sa.), ∅ conc.: 0.075 μg/l, sample origin: bakeries and supermarkets in São Paulo (city), Brazil, sample year: July–December 2004, country: Brazil[263]

- Co-contamination: not reported
- Further contamination of animal products reported in the present book/article[263]: Milk (cow milk), AFM_1, and CPA, literature[263]

incidence: 53/60*, conc. range: 0.015–0.050 µg/l (36 sa.), 0.050–0.500 µg/l (17 sa.), sample origin: supermarkets in Ribeirão Preto (municipality), São Paulo (state), Brazil, sample year: 1999–2000, country: Brazil[270], *partially skimmed, skimmed, and whole milk

- Co-contamination: not reported
- Further contamination of animal products reported in the present book/article[270]: Milk (cow milk), AFM_1, literature[270]

incidence: 68/109, conc. range: 0.0056–0.5159 µg/l, Ø conc.: 0.0743 µg/l, sample origin: supermarkets and retail shops in Esfahan and Yazd (provinces), Iran, sample year: December 2008–June 2009, country: Iran[288]

- Co-contamination: not reported
- Further contamination of animal products reported in the present book/article[288]: Milk, AFM_1, literature[288]

incidence: 40/40, conc. range: 0.010–0.020 µg/kg (28 sa.), 0.020–0.050 µg/kg (11 sa.), 0.050–0.500 µg/kg (1 sa.), sample origin: supermarkets in São Paulo (city), Brazil, sample year: September–November 2006, country: Brazil[305]

- Co-contamination: not reported
- Further contamination of animal products reported in the present book/article[305]: Milk, Milk powder, AFM_1, literature[305]

incidence: 34/42, conc. range: >LOD (11 sa.), 0.02–0.05 µg/l (20 sa.), >0.05–<0.5 µg/l (3 sa.), sample origin: supermarkets in Marílía and São Paulo (cities), Brazil, sample year: 2002–2003, country: Brazil[306]

- Co-contamination: not reported
- Further contamination of animal products reported in the present book/article[306]: Milk (cow milk), AFM_1, literature[306]

incidence: 11/19*, conc. range: 0.010–0.020 µg/l (5 sa.), 0.021–0.050 µg/l (5 sa.), 0.0505 µg/l (1 sa.), sample origin: supermarkets in Ankara (capital), Turkey, sample year: unknown, country: Turkey[310], *UHT whole milk

incidence: 3/5*, conc. range: 0.010–0.020 µg/l (2 sa.), 0.021–0.050 µg/l (1 sa.), sample origin: supermarkets in Ankara (capital), Turkey, sample year: unknown, country: Turkey[310], *UHT skimmed milk

- Co-contamination: not reported
- Further contamination of animal products reported in the present book/article[310]: Milk (cow milk), AFM_1, literature[310]

incidence: 10/12*, conc. range: 0.010–0.050 µg/l (4 sa.), 0.051–0.100 µg/l (5 sa.), 0.101–0.200 µg/l (1 sa.), Ø conc.: 0.058 µg/l, sample origin: bakeries and supermarkets in Campinas (city), São Paulo (state), Brazil, sample year: October–December 2004, March–May 2005, country: Brazil[319], *goat UHT milk

- Co-contamination: not reported
- Further contamination of animal products reported in the present book/article[319]: Milk (goat milk), Milk powder, AFM_1, literature[319]

incidence: 4/16, conc. range: tr-0.1 µg/l, sample origin: dairy facility in Léon (province), northwestern Spain, sample year: winter 1985, country: Spain[355]

incidence: 6/11, conc. range: tr-0.08 µg/l, Ø conc.: 0.036 µg/l, sample origin: dairy facility in Léon (province), northwestern Spain, sample year: spring 1985, country: Spain[355]

incidence: 3/10, conc. range: tr-0.04 µg/l, sample origin: dairy facility in Léon (province), northwestern Spain, sample year: summer 1985, country: Spain[355]

incidence: 5/10, conc. range: tr-0.05 µg/l, sample origin: dairy facility in Léon (province), northwestern Spain, sample year: autumn 1985, country: Spain[355]

- Co-contamination: not reported
- Further contamination of animal products reported in the present book/article[355]: not reported

For detailed information see the article.

incidence: 125/161, conc. range: 0.001–0.010 µg/l (112 sa.), >0.010–0.050 µg/l (13 sa., maximum: 0.0235 µg/l), Ø conc.: 0.00628 µg/l, sample origin: drugstores and supermarkets in 4 big Italian cities, Italy, sample year: 1996, country: Italy[356]

- Co-contamination: not reported
- Further contamination of animal products reported in the present book/article[356]: Milk powder, Yogurt, AFM_1, literature[356]

incidence: 50/50*, conc. range: 0.005–0.244 µg/l, Ø conc.: 0.1012 µg/l, sample origin: markets in Bursa (city), Turkey, sample year: April 2004, country: Turkey[357], *commercial whole milk

- Co-contamination: not reported
- Further contamination of animal products reported in the present book/article[357]: not reported

incidence: 75/129*, conc. range: <0.01 µg/l (4 sa.), 0.010–0.049 µg/l (10 sa.), 0.050–0.499 µg/l (57 sa.), ≥0.500 µg/l (4 sa., maximum: 0.54364 µg/l), sample origin: supermarkets in Kayseri, Konya, Nigde, and Sivas (cities), as well as Ankara (capital), Central Anatolia (region), Turkey, sample year: January–February 2005, country: Turkey[358], *commercial whole UHT milk (packaged)

- Co-contamination: not reported
- Further contamination of animal products reported in the present book/article[358]: not reported

incidence: 26/52, conc. range: 0.019–0.217 µg/l, sample origin: supermarkets in Tehran (province), Iran, sample year: May (spring indicator) 2008, country: Iran[359]

incidence: 19/53, conc. range: 0.008–0.160 µg/l, sample origin: supermarkets in Tehran (province), Iran, sample year: August (summer indicator) 2008, country: Iran[359]

incidence: 31/52, conc. range: 0.015–0.194 µg/l, sample origin: supermarkets in Tehran (province), Iran, sample year: November (autumn indicator) 2007, country: Iran[359]

incidence: 40/53, conc. range: 0.012–0.249 µg/l, sample origin: supermarkets in Tehran (province), Iran, sample year: February (winter indicator) 2008, country: Iran[359]

- Co-contamination: not reported
- Further contamination of animal products reported in the present book/article[359]: not reported

incidence: 22/22, conc. range: 0.02240–0.08480 µg/kg, Ø conc.: 0.06550 µg/kg, sample origin: markets in various districts of Tehran (province), Iran, sample year: summer 2004, country: Iran[360]

incidence: 30/30, conc. range: 0.01940–0.09360 µg/kg, Ø conc.: 0.06922 µg/kg, sample origin: markets in various districts of Tehran (province), Iran, sample year: autumn 2004, country: Iran[360]

- Co-contamination: not reported
- Further contamination of animal products reported in the present book/article[360]: not reported

incidence: 49/49, conc. range: <0.050 µg/l (8 sa.), 0.050–0.80 µg/l (14 sa.), >0.80 µg/l (27 sa., maximum: 0.259 µg/l), Ø conc.: 0.11145 µg/l, sample origin: supermarkets in Tabriz (city), Iran, sample year: July–December 2008, country: Iran[368]

- Co-contamination: not reported
- Further contamination of animal products reported in the present book/article[368]: not reported

incidence: 74/94, conc. range: 0.005–0.010 µg/l (16 sa.), 0.011–0.025 µg/l (37 sa.), 0.026–0.050 µg/l (17 sa.), 0.050–0.080 µg/l (4 sa., maximum: 0.6226 µg/l), Ø conc.: 0.0179 µg/l, sample origin: supermarkets in Prishtina (city), Kosovo, sample year: July–December 2013, country: Serbia[401]

- Co-contamination: not reported
- Further contamination of animal products reported in the present book/article[401]: Milk, AFM_1, literature[401]

incidence: 23/75, conc. range: 1.0–2.0 µg/l (13 sa.), 2.1–3.0 µg/l (8 sa.), 3.1–4.2 µg/l (2 sa.), sample origin: cities of Minas Gerais (state), Brazil, sample year: July–November 2019, country: Brazil[414]

- Co-contamination: not reported
- Further contamination of animal products reported in the present book/article[414]: not reported

incidence: 112/233, conc. range: LOD-0.010 µg/kg (18 sa.), 0.010–0.050 µg/kg (46 sa.), 0.050–0.100 µg/kg (48 sa., maximum: 0.09573 µg/kg), Ø conc.: 0.02149 µg/kg, sample origin: retail stores and supermarkets across China, sample year: 2006–2007, country: China[447]

- Co-contamination: not reported
- Further contamination of animal products reported in the present book/article[447]: Yogurt, AFM_1, literature[447]

incidence: 14/17, conc. range: 0.003–0.008 µg/l (1 sa.), 0.009–0.100 µg/l (9 sa.), 0.101–0.250 µg/l (4 sa., maximum: 0.215 µg/l), sample origin: supermarkets in Ribeirão Petro (municipality) in São Paulo (state), Brazil, sample year: 2010, country: Brazil/USA[452]

- Co-contamination: not reported
- Further contamination of animal products reported in the present book/article[452]: Milk (cow milk), Milk powder, AFM_1, literature[452]

incidence: 38/45*, conc. range: 0.051–0.075 µg/l (33 sa.), >0.075 µg/l (5 sa.), sample origin: big retail markets in Kars (city) vicinity, Turkey, sample year: May–July 2006, country: Turkey[462], *raw cow milk

- Co-contamination: not reported
- Further contamination of animal products reported in the present book/article[462]: Milk (cow milk), AFM_1, literature[462]

incidence: 69/69, conc. range: 0.005–0.05 µg/kg (6 sa.), 0.05–0.50 µg/kg (63 sa., maximum: 0.41 µg/kg), Ø conc.: 0.19 µg/kg, sample origin: produced in large dairies and purchased in supermarkets, Serbia, sample year: February–May 2013, country: Serbia[468]

- Co-contamination: not reported
- Further contamination of human breast milk and animal products reported in the present book/article[468]: Milk (cow milk), Milk (donkey milk), Milk (goat milk), Milk (human breast milk), AFM_1, literature[468]

incidence: 27/28*, conc. range: 0.01–0.80 µg/kg, Ø conc.: 0.31 µg/kg, sample origin: produced in Bosnia-Herzegowina (country) and Serbia (country) purchased in Novi Sad (city), northern Serbia, sample year: February/April/May 2013, country: Serbia/Czech Republic[515], *sterilized cow milk with 0.5–3.5% milk fat (1 organic milk sa. without AFM_1 contamination)

- Co-contamination: not reported
- Further contamination of animal products reported in the present book/article[515]: Milk (cow milk), AFM_1, literature[515]

For detailed information see the article.

incidence: 30/45*, conc. range: 0.05–0.1 µg/l (5 sa.), 0.1–0.5 µg/l (14 sa.), >0.5 µg/l (11 sa., maximum: 2.1 µg/l), sample origin: retail shops in Mysore (city), Karnataka (state), India, sample year: April–July 2011, country: India[518], *21 plain milk sa. thereof 21 sa. contaminated and 24 flavored milk sa. thereof 9 sa. contaminated

- Co-contamination: not reported
- Further contamination of animal products reported in the present book/article[518]: Milk (cow milk), AFM_1, literature[518]

For detailed information see the article.

incidence: 12/12, conc. range: 0.01893–0.09127 µg/kg, Ø conc.: 0.0431 µg/kg, sample origin: city markets in Şanlıurfa (city) and locally produced dairy products, Turkey, sample year: January–February 2012, country: Turkey[523]

- Co-contamination: not reported
- Further contamination of animal products reported in the present book/article[523]: Cheese (white cheese), Milk (cow milk), Yogurt, AFM_1, literature[523]

incidence: 1/6*, conc.: 0.147 µg/kg, sample origin: major manufactures, China, sample year: unknown, country: China[531], *5 UHT milks and 1 UHT goat milk (not contaminated)

- Co-contamination: not reported
- Further contamination of animal products reported in the present book/article[531]: Milk (cow milk), Milk powder, AFM_1, literature[531]

incidence: 11/16, conc. range: 0.005–0.042 µg/kg, sample origin: local supermarkets in Rio de Janeiro (city), Brazil, sample year: May–October 2014, country: Brazil[543]

- Co-contamination: 3 sa. co-contaminated with AFM_1 and AFM_2; 8 sa. contaminated solely with AFM_1
- Further contamination of animal products reported in the present book/article[543]: Milk (UHT milk), AFM_2, literature[543]; Milk powder, AFM_1 and AFM_2, literature[543]

incidence: 58/111, conc. range: 0.005–0.0499 µg/l (56 sa.), 0.050–0.0995 µg/l (2 sa., maximum: 0.0725 µg/l), sample origin: Henan, Hubei, and Hunan (provinces), central China, sample year: November 2016–February 2017, country: China[547]

- Co-contamination: not reported
- Further contamination of animal products reported in the present book/article[547]: Milk (cow milk), AFM_1, literature[547]

incidence: 2/69, conc. range: 0.0072–0.0099 µg/l, sample origin: different regions of Kosovo, sample year: April–October 2009 and February–November 2010, country: Serbia/Italy[548]

- Co-contamination: not reported
- Further contamination of animal products reported in the present book/article[548]: Milk, AFM_1, literature[548]

Aflatoxin M_2

incidence: 3/16, conc. range: >0.009 µg/kg, sample origin: local supermarkets in Rio de Janeiro (city), Brazil, sample year: May–October 2014, country: Brazil[543]

- Co-contamination: 3 sa. co-contaminated with AFM_1 and AFM_2
- Further contamination of animal products reported in the present book/article[543]: Milk (UHT milk), AFM_1, literature[543]; Milk powder, AFM_1 and AFM_2, literature[543]

see also **Milk (cow milk)**

Milk confectionery see **Dairy products**

Milk powder may contain the following mycotoxins:

Alternaria Toxins

Tenuazonic Acid

incidence: 3/3*, conc. range: 130–900 µg/kg, Ø conc.: 399 µg/kg, sample origin: retail stores, Germany, sample year: unknown, country: Germany[108], *sorghum (whole meal) and milk powder

incidence: 2/2*, conc. range: 1.3**–8*** µg/kg, Ø conc.: 4.7 µg/kg, sample origin: retail stores, Germany, sample year: unknown, country: Germany[108], *wheat and milk powder** or wheat with milk powder and banana***

- Co-contamination: not reported
- Further contamination of animal products reported in the present book/article[108]: not reported

Aspergillus Toxins

Aflatoxin B_1

incidence: 1/5*, conc.: 0.003 µg/kg, sample origin: pharmacies and supermarkets in Lisbon (capital), Portugal, sample year: May–June 2007, country: Portugal/Netherlands[90], *sa. consisted of milk and different kinds of flours in different concentrations

- Co-contamination: 1 sa. co-contaminated with AFB_1, AFM_1, and OTA
- Further contamination of animal products reported in the present book/article[90]: Milk powder, AFM_1 and OTA, literature[90]

incidence: 0/2*, no contamination, sample origin: organic produce retail outlets (bioshops) in Lisbon (capital), Portugal, sample year: May–June 2007, country: Portugal/Netherlands[90], *sa. consisted of milk and vegetable oils

- Co-contamination: not reported
- Further contamination of animal products reported in the present book/article[90]: Milk powder, AFM_1 and OTA, literature[90]

incidence: 0/4*, no contamination, sample origin: supermarkets in Lisbon (capital), Portugal, sample year: May–June 2007, country: Portugal/Netherlands[90], *sa. consisted of milk and vegetable and/or other ingredients

- Co-contamination: not reported
- Further contamination of animal products reported in the present book/article[90]: Milk powder, AFM_1 and OTA, literature[90]

For detailed information see the article.

incidence: 0/170*, conc.: no contamination, sample origin: commercial, sample year: unknown, country: France/Netherlands[114], *dried milk

incidence: 5/168*, conc. range: tr-3 µg/kg, sample origin: commercial, sample year: unknown, country: France/Netherlands[114], *reconstituted infant milk powders

- Co-contamination: not reported
- Further contamination of animal products reported in the present book/article[114]: Cheese (blue-veined cheese) PA and MA, literature[114]; Cheese (goat cheese), PA and PAT, literature[114]; Cheese (hard cheese), PA, PAT, MA, and STG, literature[114]; Milk powder, AFM_1, literature[114]

incidence: 1/8, conc.: 1.14* µg/kg, sample origin: local stores, USA/Taiwan, sample year: unknown, country: USA/Taiwan[534], *imported milk powder sa.

- Co-contamination: 1 sa. co-contaminated with AFB_1 and AFB_2
- Further contamination of animal products reported in the present book/article[534]: Cream, Milk, Milk (infant formula), AFM_1, literature[534]; Milk powder, AFB_2, literature[534]

Aflatoxin B_2

incidence: 1/8, conc.: 0.20 µg/kg*, sample origin: local stores, USA/Taiwan, sample year: unknown, country: USA/Taiwan[534], *imported milk powder sa.

- Co-contamination: 1 sa. co-contaminated with AFB_1 and AFB_2
- Further contamination of animal products reported in the present book/article[534]: Cream, Milk, Milk (infant formula), AFM_1, literature[534]; Milk powder, AFB_1, literature[534]

Aflatoxin M_1

incidence: 1/16, conc.: tr, sample origin: Division of Dairy Technology, National Dairy Research Institute, Karnal (city), India, sample year: September 1974–February 1975, country: India[27]

- Co-contamination: not reported
- Further contamination of animal products reported in the present book/article[27]: Cheese, AFB_1, AFB_2, and AFG_1, literature[27]; Dairy products, AFB_1 and AFG_1, literature[27]; Milk, Milk (buffalo milk), Milk (cow milk), AFM_1, literature[27]

incidence: 30/41, conc. range: 0.2–1.0 µg/kg (24 sa.), 1.0–2.0 µg/kg (6 sa., maximum: 2.0 µg/kg), Ø conc.: 0.50 µg/kg, sample origin: unknown, sample year: September 1972–December 1974, country: Germany[28]

- Co-contamination: not reported
- Further contamination of animal products reported in the present book/article[28]: Cheese (Camembert cheese), Cheese (fresh cheese), Cheese (hard cheese), Cheese (processed cheese), Milk (cow milk), Yogurt, AFM_1, literature[28]

incidence: 27/27*, conc. range: 0.00204–0.00413 µg/kg, sample origin: supermarkets in Kuwait City (capital), Kuwait, sample year: January 2005–March 2007, country: Kuwait[41], *baby formula powder

- Co-contamination: not reported
- Further contamination of human breast milk and animal products reported in the present book/article[41]: Cheese, Milk (cow milk), Milk (human breast milk), Milk (UHT milk), AFM_1, literature[41]

incidence: 4/15, conc. range: 0.1–0.35 µg/kg, Ø conc.: 0.20 µg/kg, sample origin: small markets in several urban Shanghai (city) areas, China, sample year: April 1988–1989, country: China/USA[46]

- Co-contamination: not reported
- Further contamination of animal products reported in the present book/article[46]: Butter (peanut butter), Pig liver, AFB_1, literature[46]; Milk, AFM_1, literature[46]

incidence: 3/5*, conc. range: 0.017–0.023 µg/kg, Ø conc.: 0.019 µg/kg, sample origin: pharmacies and supermarkets in Lisbon (capital), Portugal, sample year: May–June 2007, country: Portugal/Netherlands[90], *sa. consisted of milk and different kinds of flours in different concentrations

- Co-contamination: 1 sa. co-contaminated with AFB_1, AFM_1, and OTA; 2 sa. co-contaminated with AFM_1 and OTA
- Further contamination of animal products reported in the present book/article[90]: Milk powder, AFB_1 and OTA, literature[90]

incidence: 2/2*, conc. range: 0.007–0.011 µg/kg, ∅ conc.: 0.009 µg/kg, sample origin: organic produce retail outlets (bioshops) in Lisbon (capital), Portugal, sample year: May–June 2007, country: Portugal/Netherlands[90], *sa. consisted of milk and vegetable oils

- Co-contamination: 1 sa. co-contaminated with AFM_1 and OTA; 1 sa. contaminated solely with AFM_1
- Further contamination of animal products reported in the present book/article[90]: Milk powder, AFB_1 and OTA, literature[90]

incidence: 4/4*, conc. range: 0.005–0.041 µg/kg, ∅ conc.: 0.017 µg/kg, sample origin: supermarkets in Lisbon (capital), Portugal, sample year: May–June 2007, country: Portugal/Netherlands[90], *sa. consisted of milk and vegetable and/or other ingredients

- Co-contamination: 1 sa. co-contaminated with AFM_1 and OTA; 3 sa. contaminated solely with AFM_1
- Further contamination of animal products reported in the present book/article[90]: Milk powder, AFB_1 and OTA, literature[90]

For detailed information see the article.

incidence: 1/15, conc.: 15 µg/kg, sample origin: Denmark, imported and purchased in Alexandria (province), Egypt, sample year: unknown, country: Egypt[107]

- Co-contamination: not reported
- Further contamination of animal products reported in the present book/article[107]: not reported

incidence: 6/170*, conc. range: tr, sample origin: commercial, sample year: unknown, country: France/Netherlands[114], *dried milk

incidence: 7/168*, conc. range: tr-2 µg/kg, sample origin: commercial, sample year: unknown, country: France/Netherlands[114], *reconstituted infant milk powders

- Co-contamination: not reported
- Further contamination of animal products reported in the present book/article[114]: Cheese (blue-veined cheese) PA and MA, literature[114]; Cheese (goat cheese), PA and PAT, literature[114]; Cheese (hard cheese), PA, PAT, MA, and STG, literature[114]; Milk powder, AFB_1, literature[114]

incidence: 1/10, conc.: 5 µg/kg, sample origin: Egyptians markets, Egypt, sample year: 1999–2000, country: Egypt[138]

- Co-contamination: not reported
- Further contamination of human breast milk and animal products reported in the present book/article[138]: Cheese (Domiati cheese), Milk (cow milk), AFM_1, literature[138]; Cheese (Ras cheese), AFB_1, AFG_1, and AFM_1, literature[138]; Milk (human breast milk), AFM_1 and OTA, literature[138]

incidence: 5/9*, conc. range: nd-0.046 µg/kg, ∅ conc.: 0.024 µg/kg, sample origin: markets in Londrina (city), Brazil, sample year: March–April 2013, country: Brazil/Japan[161], *national milk powder sa. from southeastern Brazil

incidence: 2/7*, conc. range: nd-0.034 µg/kg, ∅ conc.: 0.024 µg/kg, sample origin: markets in Londrina (city), Brazil, sample year: March–April 2013, country: Brazil/Japan[161], *imported milk powder sa. from Argentina, Germany, and USA

- Co-contamination: not reported
- Further contamination of animal products reported in the present book/article[161]: Milk (human breast milk), AFM_1, literature[161]

incidence: 3/11, conc. range: 0.0035–0.086 µg/kg, ∅ conc.: 0.0395 µg/kg, sample origin: retail shops in Marang and Kuala Terengganu (town and city), Malaysia, sample year: July 2013?, country: Malaysia[163]

- Co-contamination: not reported
- Further contamination of animal products reported in the present book/article[163]: Cheese, Dairy beverages, Milk, Milk (cow milk), Yogurt, AFM_1, literature[163]

incidence: 2/13, conc. range: >0–0.05 µg/l (1 sa.), >0.05–0.125 µg/l (1 sa.), sample origin: imported from Europe and purchased in retail markets, Thailand, sample year: May 1995–January 1996, country: Thailand[196]

incidence: 7/7*, conc. range: >0.125–0.25 µg/l (2 sa.), >0.25–0.5 µg/l (4 sa.), >0.5 µg/l (1 sa.), sample origin: locally manufactured and purchased in retail markets, Thailand, sample year: May 1995–January 1996, country: Thailand[196], *pelleted milk

- Co-contamination: not reported
- Further contamination of animal products reported in the present book/article[196]: Milk (cow milk), Milk (UHT milk), AFM_1, literature[196]

incidence: 81/97, conc. range: 0.001–0.010 µg/kg (24 sa.), >0.010–0.050 µg/kg (47 sa.), >0.050 µg/kg (10 sa., maximum: 0.10125 µg/kg), ∅ conc.: 0.02177 µg/kg, sample origin: imported and domestic, Italy, sample year: 1995, country: Italy[234]

- Co-contamination: not reported
- Further contamination of animal products reported in the present book/article[234]: Milk, Yogurt, AFM_1, literature[234]

incidence: 35/277, conc. range: <0.03 µg/kg? (24 sa.), 0.01–0.02 µg/kg (6 sa.), 0.02–0.04 µg/kg (5 sa.), sample origin: creameries in North England, Midlands, Wales, and West Country, UK, sample year: 1981–1983, country: UK[235]

- Co-contamination: not reported
- Further contamination of animal products reported in the present book/article[235]: Milk (cow milk), AFM_1, literature[235]

incidence: 9/9, conc. range: 0.010–0.024 µg/kg (2 sa.), 0.050–0.099 µg/kg (2 sa.), 0.100–0.300 (5 sa., maximum: 0.280 µg/kg.), ∅ conc.: 0.134 µg/kg, sample origin: dairies in southern Italy, sample year: unknown, country: Italy[236]

- Co-contamination: not reported
- Further contamination of animal products reported in the present book/article[236]: Milk (cow milk), AFM_1, literature[236]

incidence: 17/24, conc. range: 0.083–0.342 µg/kg, ∅ conc.: 0.221 µg/kg, sample origin: supermarkets in Seoul (capital), Korea, sample year: 1997, country: Korea[241]

- Co-contamination: not reported
- Further contamination of animal products reported in the present book/article[241]: Milk (cow milk), Milk (infant formula), Yogurt, AFM_1, literature[241]

incidence: 5/10, conc. range: 0.6–15 µg/kg*, sample origin: retail market in Anantapur (city), Andhra Pradesh (state), southern India, sample year: 2001–2002, country: India/UK[256], *authors pointed out the high levels of AFM_1 in milk (g of dry milk/ml of solution)

- Co-contamination: not reported
- Further contamination of animal products reported in the present book/article[256]: Dairy products, Milk, Milk (buffalo milk), Milk (cow milk), AFM_1, literature[256]

incidence: 4/5, conc. range: 0.010–0.014 µg/l, ∅ conc.: 0.0125 µg/l, sample origin: Argentina, sample year: March–September 1999, country: Argentina[268]

- Co-contamination: not reported
- Further contamination of animal products reported in the present book/article[268]: Milk (cow milk), AFM_1, literature[268]

incidence: 15/15*, conc. range: 0.0180–0.28868 µg/kg, ∅ conc.: 0.10395 µg/kg, sample origin: Jordanian market, Jordan, sample year: 2014–2015, country: Jordan[278], *full cream powdered milk

- Co-contamination: not reported
- Further contamination of animal products reported in the present book/article[278]: Milk, Milk (camel milk), Milk (cow milk), Milk (goat milk), Milk (infant formula), Milk (sheep milk), AFM_1, literature[278]

incidence: 15/15, conc. range: 0.16–0.32 µg/l* (10 sa.), 0.32–0.5 µg/l* (5 sa.), sample origin: supermarkets in Heilongjiang (province), northeast China, sample year: March–May 2008, country: China/Russia/Korea[283], *g of dry milk/ml of solution

- Co-contamination: not reported
- Further contamination of animal products reported in the present book/article[283]: Cheese, Dairy products, Milk (cow milk), AFM_1, literature[283]

incidence: 1/8, conc.: 0.012 µg/l, sample origin: north, south, and east of Syria, sample year: April 2005–2006, country: Syria[293]

- Co-contamination: not reported
- Further contamination of animal products reported in the present book/article[293]: Milk (cow milk), Milk (goat milk), Milk (sheep milk), AFM_1, literature[293]

incidence: 62/65, conc. range: 0.010–0.020 µg/kg (8 sa.), 0.020–0.050 µg/kg (27 sa.), 0.050–0.500 µg/kg (27 sa.), sample origin: municipal day-care centers and elementary schools in São Paulo (city), Brazil, sample year: September–November 2006, country: Brazil[305]

incidence: 10/10, conc. range: 0.010–0.020 µg/kg (1 sa.), 0.020–0.050 µg/kg (4 sa.), 0.050–0.500 µg/kg (5 sa.), sample origin: supermarkets in São Paulo (city), Brazil, sample year: September–November 2006, country: Brazil[305]

- Co-contamination: not reported
- Further contamination of animal products reported in the present book/article[305]: Milk, Milk (UHT milk), AFM_1, literature[305]

incidence: 8/12*, conc. range: 0.010–0.050 µg/l (4 sa.), 0.051–0.100 µg/l (4 sa.), ∅ conc.: 0.056 µg/l, sample origin: bakeries and supermarkets in Campinas (city), São Paulo (state), Brazil, sample year: October–December 2004, March–May 2005, country: Brazil[319], *goat milk powder

- Co-contamination: not reported
- Further contamination of animal products reported in the present book/article[319]: Milk (goat milk), Milk (UHT milk), AFM_1, literature[319]

incidence: 54/125*, conc. range: <0.005 µg/l (19 sa.), 0.005–0.025 µg/l (35 sa., maximum: 0.0218 µg/l), sample origin: stores from different districts of Minoufiya (governorate), Egypt, sample year: March–August 2010, country: Egypt[340], *infant formula milk powder (cans containing imported cows' milk-based formula)

- Co-contamination: not reported
- Further contamination of human breast milk and animal products reported in the present book/article[340]: Milk (human breast milk), AFM_1, literature[340]

incidence: 50/92*, conc. range: 0.001–0.010 µg/kg (9 sa.), >0.010–0.050 µg/kg (37 sa.), >0.050 µg/kg (4 sa., maximum: 0.0796 µg/kg), ∅ conc.: 0.0322 µg/kg, sample origin: supermarkets and drugstores in 4 big Italian cities, Italy, sample year: 1996, country: Italy[356], *sa. partly imported from Denmark, France, Germany, and Switzerland

- Co-contamination: not reported
- Further contamination of animal products reported in the present book/article[356]: Milk (UHT milk), Yogurt, AFM_1, literature[356]

incidence: 3/18, conc. range: 0.040–0.095 µg/kg, ∅ conc.: 0.066 µg/kg, sample origin: manufactures, Italy, sample year: unknown, country: Italy[369]

- Co-contamination: not reported
- Further contamination of animal products reported in the present book/article[369]: not reported

incidence: 47/95, conc. range: 0.1–1.0 µg/kg (46 sa.) 2.55 µg/kg (1 sa.), sample origin: dairies, Germany, sample year: unknown, country: Germany[370]

- Co-contamination: not reported
- Further contamination of animal products reported in the present book/article[370]: not reported

incidence: 21/28, conc. range: tr-0.464 µg/kg, sample origin: China, sample year: 1992–1993, country: Japan/Italy[371]

incidence: 3/12, conc. range: tr, sample origin: Italy, sample year: 1992–1993, country: Japan/Italy[371]

incidence: 0/3, conc.: no contamination, sample origin: New Zealand, sample year: 1992–1993, country: Japan/Italy[371]

incidence: 3/3, conc. range: tr-0.085 µg/kg, sample origin: Poland, sample year: 1992–1993, country: Japan/Italy[371]

incidence: 5/10, conc. range: tr-0.243 µg/kg, sample origin: USA, sample year: 1992–1993, country: Japan/Italy[371]

- Co-contamination: not reported
- Further contamination of animal products reported in the present book/article[371]: not reported

incidence: 469/837, conc. range: 0.04–0.10 µg/kg (351 sa.), 0.11–0.20 µg/kg (105 sa.), >0.20 µg/kg (13 sa., maximum: 0.69 µg/kg), sample origin: Austrian milk powder plants, Austria, sample year: March 1978–February 1979, country: Austria[372]

- Co-contamination: not reported
- Further contamination of animal products reported in the present book/article[372]: not reported

incidence: 4/4, conc. range: 0.00928–0.08160 µg/l, ∅ conc.: 0.04224 µg/l, sample origin: unknown, sample year: unknown, country: USA[373]

- Co-contamination: not reported
- Further contamination of animal products reported in the present book/article[373]: Yogurt, AFM_1, literature[373]

incidence: 8*/166**, conc. range: 0.67–2.0 µg/kg, ∅ conc.: 1.25 µg/kg, sample origin: manufactures, Germany, sample year: unknown, country: Germany[374], *1 milk product for sucklings, 6 spray- and 1 roller-dried milk product(s), **commercial dried milk products with different composition

- Co-contamination: not reported
- Further contamination of animal products reported in the present book/article[374]: not reported

incidence: 33/300*, conc. range: 0.10–0.20 µg/kg (17 sa.), 0.20–0.30 µg/kg (10 sa.), >0.50 µg/kg (6 sa., maximum: 1.00 µg/kg), ∅ conc.: 0.27 µg/kg, sample origin: municipal nurseries, São Paulo (city), Brazil, sample year: October 1992–January 1993, country: Brazil[375], *whole milk powder

- Co-contamination: not reported
- Further contamination of animal products reported in the present book/article[375]: not reported

incidence: 5/5, conc. range: 4.45–5.30 µg/kg, ∅ conc.: 4.85 µg/kg, sample origin: supermarkets in Peoria (city), Illinois (state), USA, sample year: unknown, country: USA[376]

- Co-contamination: not reported
- Further contamination of animal products reported in the present book/article[376]: not reported

incidence: 1/1, conc.: 1.2 µg/kg, sample origin: unknown, sample year: unknown, country: France[377]

- Co-contamination: not reported
- Further contamination of animal products reported in the present book/article[377]: not reported

incidence: 1*/1, conc.: 0.201 µg/kg, sample origin: Thailand, sample year: February 1990, country: Japan[378], *determined by HPLC

incidence: 0*/4, conc.: no contamination, sample origin: Tokyo (capital) Japan, sample year: May 1991, country: Japan[378], *determined by HPLC

incidence: 0*/2, conc.: no contamination, sample origin: Indonesia, sample year: July 1991, country: Japan[378], *determined by HPLC

incidence: 2*/4, conc. range: 0.107–0.128 µg/kg, ∅ conc.: 0.1175 µg/kg, sample origin: France, sample year: August 1992, country: Japan[378], *determined by HPLC

incidence: 1*/10, conc.: 0.216 µg/kg, sample origin: USA, sample year: October 1992, country: Japan[378], *determined by HPLC

- Co-contamination: not reported
- Further contamination of animal products reported in the present book/article[378]: not reported

incidence: 50/80, conc. range: 0.001–0.500 µg/kg (14 sa:), 0.501–0.600 µg/kg (19 sa.), >0.601 µg/kg (17 sa.), sample origin: retail stores of Erzurum, Kars, Konya, and Mersin (cities), as well as Ankara (capital), Turkey, sample year: unknown, country: Turkey[379]

- Co-contamination: not reported
- Further contamination of animal products reported in the present book/article[379]: not reported

incidence: 12/12, conc. range: <0.05 µg/kg (6 sa.), 0.05–0.99 µg/kg (6 sa., maximum: 0.85 µg/kg), ∅ conc.: 0.29 µg/kg, sample origin: imported by milk companies in Khartoum (state), central Sudan, sample year: unknown, country: Sudan[393]

- Co-contamination: not reported
- Further contamination of animal products reported in the present book/article[393]: Milk (cow milk), AFM_1, literature[393]

incidence: 5/14*, conc. range: 0.005–0.025 µg/l** (5 sa., maximum: 0.0165 µg/l**), ∅ conc.: 0.0137 µg/l**, sample origin: Lebanese regions and markets, Lebanon, sample year: March–July 2010, country: Lebanon[404], *cow (13 sa.) and **goat milk powder (1 sa.)

- Co-contamination: not reported
- Further contamination of animal products reported in the present book/article[404]: Milk (cow milk), AFM_1, literature[404]

incidence: 9/14, conc. range: tr-0.103 µg/l, sample origin: imported and purchased in supermarkets around Constantine (city), Algeria, sample year: February–October 2011, country: Algeria/France[406]

- Co-contamination: not reported
- Further contamination of animal products reported in the present book/article[406]: Milk, AFM_1, literature[406]

incidence: 42/45, conc. range: 0.051–0.100 µg/kg (16 sa.), 0.101–0.250 µg/kg (12 sa.), 0.251–0.500 µg/kg (13 sa.), 0.800 µg/kg (1 sa.), sample origin: various districts in Burdur (city), Turkey, sample year: 2008, country: Turkey[418]

- Co-contamination: not reported
- Further contamination of animal products reported in the present book/article[418]: Butter, Cheese (white cheese), Ice cream, Milk (cow milk), Yogurt, AFM_1, literature[418]

incidence: 11/11*, conc. range: 0.10**–0.88 µg/kg, ∅ conc.: 0.32 µg/kg, sample origin: produced in Argentina and purchased in supermarkets and grocery stores, Argentina, sample year: 2012, country: Argentina/Brazil[438], *whole milk powder, **goat whole milk powder

incidence: 9/9*, conc. range: 0.22–0.92 µg/kg, ∅ conc.: 0.50 µg/kg, sample origin: produced in Argentina and purchased in supermarkets and grocery stores, Argentina, sample year: 2012, country: Argentina/Brazil[438], *skimmed milk powder

incidence: 1/1*, conc.: 0.32 µg/kg, sample origin: produced in Argentina and purchased in supermarkets and grocery stores, Argentina, sample year: 2012, country: Argentina/Brazil[438], *infant milk formula

incidence: 7/7*, conc. range: <LOQ-0.74 µg/kg, sample origin: produced in Brazil and purchased in supermarkets and grocery stores, Brazil, sample year: 2012, country: Argentina/Brazil[438], *whole milk powder

incidence: 2/2*, conc. range: 0.27–0.81 µg/kg, ∅ conc.: 0.54 µg/kg, sample origin: produced in Argentina and purchased in supermarkets and grocery stores, Brazil, sample year: 2012, country: Argentina/Brazil[438], *whole milk powder

- Co-contamination: not reported
- Further contamination of animal products reported in the present book/article[438]: not reported

incidence: 7/9, conc. range: 0.0182–0.2534 µg/l, ∅ conc.: 0.0712 µg/l, sample origin: local markets and supermarkets, China, sample year: unknown, country: China[448]

- Co-contamination: not reported
- Further contamination of animal products reported in the present book/article[448]: Milk, Milk (infant formula), AFM_1, literature[448]

incidence: 6/6*, conc. range: 0.009–0.100 µg/kg (2 sa.), 0.101–0.250 µg/kg (1 sa.), 0.251–0.500 µg/kg (3 sa.), sample origin: supermarkets in Ribeirão Petro (municipality) in São Paulo (state), Brazil, sample year: 2010, country: Brazil/USA[452], *powdered whole milk incidence: 6/6*, conc. range: 0.009–0.100 µg/kg (2 sa.), 0.101–0.250 µg/kg (2 sa.), >0.500 µg/kg (2 sa., maximum: 0.760 µg/kg), sample origin: supermarkets in Ribeirão Petro (municipality) in São Paulo (state), Brazil, sample year: 2010, country: Brazil/USA[452], *powdered skim milk

- Co-contamination: not reported
- Further contamination of animal products reported in the present book/article[452]: Milk (cow milk), Milk (UHT milk), AFM_1, literature[452]

incidence: 2/8, conc. range: ≤0.0218 µg/kg, ∅ conc.: 0.0160 µg/kg, sample origin: supermarkets in Beijing (capital), China, sample year: April 2012, country: China[456]

- Co-contamination: For detailed information see the article
- Further contamination of animal products reported in the present book/article[456]: Milk (cow milk), AFM_1, OTA, and ZEA, literature[456]; Milk powder, OTA and ZEA, literature[452]; see also Milk (cow milk), Milk powder, chapter: Further Mycotoxins and Microbial Metabolites

incidence: 2/4, conc. range: 0.50–0.81 µg/kg, ∅ conc.: 0.655 µg/kg, sample origin: households from employees of the University of São Paulo at Pirassununga (municipality), Brazil, sample year: June 2011–March 2012, country: Brazila[458]

- Co-contamination: not reported
- Further contamination of animal products reported in the present book/article[458]: Cheese, Milk (cow milk), AFM_1, literature[458]

incidence: 4/16*, conc. range: 0.077–0.197 µg/kg, ∅ conc.: 0.136 µg/kg, sample origin: major manufactures, China, sample year: unknown, country: China[531], *included 4 whole milk powders (2 contaminated), 4 skim milk powders (1 contaminated), 5 infant milk powders, and 3 goat milk powders (1 contaminated)

- Co-contamination: not reported
- Further contamination of animal products reported in the present book/article[531]: Milk (cow milk), Milk (UHT milk), AFM_1, literature[531]

For detailed information see the article.

incidence: 7/94, conc. range: 0.003–0.018 µg/kg, ∅ conc.: 0.010 µg/kg, sample origin: control points such as import, primary storage, and market, Cyprus, sample year: 2004–2013, country: Cyprus[542]

- Co-contamination: not reported
- Further contamination of animal products reported in the present book/article[542]: Cheese, Cheese (Anari cheese), Cheese (white cheese), Ice cream, Milk, Yogurt, AFM_1, literature[542]

incidence: 53/72*, conc. range: 0.08–1.19 µg/kg, sample origin: local supermarkets in Rio de Janeiro (city), Brazil, sample year: November 2009–May 2012, country: Brazil[543], *whole powdered milk sa.

- Co-contamination: 17 sa. co-contaminated with AFM_1 and AFM_2; 36 sa. contaminated solely with AFM_1
- Further contamination of animal products reported in the present book/article[543]: Milk (UHT milk), AFM_1 and AFM_2, literature[543]; Milk powder, AFM_2, literature[543]

incidence: 30/30*, conc. range: 0.35–1.19 µg/kg, ∅ conc.: 0.60 µg/kg, sample origin: supermarkets and grocery stores in the Caribbean region, Colombia, sample year: December 2017–May 2018, country: Brazil[562], *whole milk powder sa.

incidence: 6/6*, conc. range: 0.46–1.03 µg/kg, ∅ conc.: 0.75 µg/kg, sample origin: supermarkets and grocery stores in the Caribbean region, Colombia, sample year: December 2017–May 2018, country: Brazil[562], *skimmed milk powder sa.

incidence: 9/9*, conc. range: 0.33–1.18 µg/kg, ∅ conc.: 0.70 µg/kg, sample origin: supermarkets and grocery stores in the Caribbean region, Colombia, sample year: December 2017–May 2018, country: Brazil[562], *semi-skimmed lactose-free milk powder sa.

incidence: 6/6*, conc. range: 0.20–0.34 µg/kg, ∅ conc.: 0.25 µg/kg, sample origin: supermarkets and grocery stores in the Caribbean region, Colombia, sample year: December 2017–May 2018, country: Brazil[562], *infant milk formula sa.

- Co-contamination: not reported
- Further contamination of animal products reported in the present book/article[562]: not reported

incidence: 1/7*, conc.: 0.006 µg/kg, sample origin: local supermarkets and drugstores of Castellon (city, province) Spain, sample year: unknown, country: Spain[564]

- Co-contamination: not reported
- Further contamination of animal products reported in the present book/article[564]: not reported[564]

Aflatoxin M_2

incidence: 17/72*, conc. range: >0.08 µg/kg, sample origin: local supermarkets in Rio de Janeiro (city), Brazil, sample year: November 2009–May 2012, country: Brazil[543], *whole powdered milk sa.

- Co-contamination: 17 sa. co-contaminated with AFM_1 and AFM_2
- Further contamination of animal products reported in the present book/article[543]: Milk (UHT milk), AFM_1 and AFM_2, literature[543]; Milk powder, AFM_1, literature[543]

Aflatoxin/s

incidence: 2/2*, conc. range: 0.499–0.849 µg/kg**, ∅ conc.: 0.678 µg/kg**, sample origin: groceries and supermarkets, Turkey, sample year: 2003, country: Turkey[95], *skim milk powder, **AFS total

- Co-contamination: not reported
- Further contamination of cereals and cereal products reported in the present book/article[95]: Cream, AF/S, literature[95]

Aspergillus and *Penicillium* Toxins

Ochratoxin A

incidence: 4/5*, conc. range: 0.010–0.150 µg/kg, ∅ conc.: 0.078 µg/kg, sample origin: pharmacies and supermarkets in Lisbon (capital), Portugal, sample year: May–June 2007, country: Portugal/Netherlands[90], *sa. consisted of milk and different kinds of flours in different concentrations

- Co-contamination: 1 sa. co-contaminated with AFB_1, AFM_1, and OTA; 2 sa. co-contaminated with AFM_1 and OTA; 1 sa. contaminated solely with OTA
- Further contamination of animal products reported in the present book/article[90]: Milk powder, AFB_1 and AFM_1, literature[90]

incidence: 1/2*, conc.: 0.011 µg/kg, sample origin: organic produce retail outlets (bioshops) in Lisbon (capital), Portugal, sample year: May–June 2007, country: Portugal/Netherlands[90], *sa. consisted of milk and vegetable oils

- Co-contamination: 1 sa. co-contaminated with AFM_1 and OTA
- Further contamination of animal products reported in the present book/article[90]: Milk powder, AFB_1 and AFM_1, literature[90]

incidence: 1/4*, conc.: 0.135 µg/kg, sample origin: supermarkets in Lisbon (capital), Portugal, sample year: May–June 2007, country: Portugal/Netherlands[90], *sa. consisted of milk and vegetable and/or other ingredients

- Co-contamination: 1 sa. co-contaminated with AFM_1 and OTA
- Further contamination of animal products reported in the present book/article[90]: Milk powder, AFB_1 and AFM_1, literature[90]

For detailed information see the article.

incidence: 5/8, conc. range: ≤0.0494 µg/kg, Ø conc.: 0.0270 µg/kg, sample origin: supermarkets in Beijing (capital), China, sample year: April 2012, country: China[456]

- Co-contamination: For detailed information see the article
- Further contamination of animal products reported in the present book/article[456]: Milk (cow milk), AFM_1, OTA, and ZEA, literature[456]; Milk powder, AFM_1 and ZEA, literature[456]; see also Milk (cow milk), Milk powder, chapter: Further Mycotoxins and Microbial Metabolites

Fusarium Toxins

Zearalenone

incidence: 6/20, conc. range: 3.1–12.5 µg/kg, Ø conc.: 6.4 µg/kg, sample origin: shops and supermarkets in Alexandria (province), Egypt, sample year: unknown, country: Egypt[141]

- Co-contamination: not reported
- Further contamination of animal products reported in the present book/article[141]: Beef meat, Beefburger, Cheese (Hard Roumy cheese), Cheese (Karish cheese), Meat, Milk, Sausage, ZEA, literature[141]

incidence: 2/8, conc. range: ≤0.0124 µg/kg, Ø conc.: 0.0116 µg/kg, sample origin: supermarkets in Beijing (capital), China, sample year: April 2012, country: China[456]

- Co-contamination: For detailed information see the article
- Further contamination of animal products reported in the present book/article[456]: Milk (cow milk), AFM_1, OTA, and ZEA, literature[456]; Milk powder, AFM_1 and OTA, literature[456]; see also Milk (cow milk), Milk powder, chapter: Further Mycotoxins and Microbial Metabolites

see also **Milk**

Milk products see **Dairy products**

Muhallebi see **Dairy desserts**

Paneer see **Dairy products**

Panga sp. see **Fish**

Pâté may contain the following mycotoxins:

Aspergillus and ***Penicillium*** Toxins

Ochratoxin A

incidence: 1/10*, conc.: <0.4 µg/kg, sample origin: retail outlets and supermarkets, UK, sample year: August 1996, country: UK[361], *pork pâté

- Co-contamination: not reported
- Further contamination of animal products in the present book/article[361]: not reported

incidence: 2/28*, conc. range: ≤0.9 µg/kg, sample origin: small shops and supermarkets, Spain, sample year: unknown, country: Spain[362], *pig-derived liver pâtés

incidence: 1/10*, conc.: 1.77 µg/kg, sample origin: homemade, Spain, sample year: unknown, country: Spain[362], *pig-derived liver pâtés

- Co-contamination: not reported
- Further contamination of animal products in the present book/article[362]: not reported

Petrocephalus sp. see **Fish**

Pig blood may contain the following mycotoxins:

Aspergillus and ***Penicillium*** Toxins

Ochratoxin A

incidence: 36/195, conc. range: 3–270 µg/l, sample origin: slaughterhouse in the central of Poland, sample year: April 1982–March 1983, country: Poland/Sweden[3]

- Co-contamination: not reported
- Further contamination of animal products in the present book/article[3]: Pig kidney, OTA, literature[3]

incidence: 63/105, conc. range: ≤122 µg/l, sample origin: unknown, sample year: 1991–1992, country: Poland[4]

- Co-contamination: not reported
- Further contamination of animal products in the present book/article[4]: Pig kidney, OTA, literature[4]

incidence: 147/255*, conc. range: 0.1–1 µg/l (98 sa.), 1–5 µg/l (44 sa.), 5–20 µg/l (5 sa.), Ø conc.: 1.9 µg/l, sample origin: slaughterhouse, Czechoslovakia, sample year: 1989, country: Czechoslovakia[5], *pig blood serum

- Co-contamination: not reported
- Further contamination of animal products in the present book/article[5]: Pig kidney, OTA, literature[5]

incidence: 179/359*, conc. range: ≥2 µg/l (136 sa.), ≥5 µg/l (29 sa.), ≥10 µg/l (14 sa.), Ø conc.: 8.2 µg/l, sample origin: 8 slaughterhouses, Sweden, sample year: unknown, country: Sweden[6], *collected at slaughter in the beginning of the storage of the grain crop of 1987 feeding

incidence: 94/174*, conc. range: ≥2 µg/l (49 sa.), ≥5 µg/l (29 sa.), ≥10 µg/l (18 sa.), Ø conc.: 13.2 µg/l, sample origin: 8 slaughterhouses, Sweden, sample year: unknown, country: Sweden[6], *collected at slaughter at the end of the storage of the grain crop of 1987 feeding
(400 µg/l OTA in blood was the maximum value)

- Co-contamination: not reported
- Further contamination of animal products in the present book/article[6]: not reported

For detailed information see the article.

incidence: 26/122, conc. range: 2–62 µg/l, Ø conc.: 8.69 µg/l, sample origin: 8 slaughterhouses, Sweden, sample year: August–September 1983, country: Sweden[7]

- Co-contamination: not reported
- Further contamination of animal products in the present book/article[7]: not reported

For detailed information see the article.

incidence: 26/45*, conc. range: 0.3–69.5 μg/l, sample origin: 5 abattoirs and meat processing plants in southern Wielkopolska region, Poland, sample year: February–May 1999, country: Poland[8], *porcine blood serum

- Co-contamination: not reported
- Further contamination of animal products in the present book/article[8]: not reported

incidence: 47/279*, conc. range: 2–187 μg/l, ∅ conc.: 15.74 μg/l, sample origin: 9 slaughterhouses in different parts of Sweden, sample year: May–August 1978, country: Sweden[455], *pig herds (each herd was represented by 2 pigs)

- Co-contamination: not reported
- Further contamination of animal products in the present book/article[455]: not reported

see also **Pig plasma** and **Pig serum**

Pig fat may contain the following mycotoxins:

Fusarium Toxins

Type A Trichothecenes

T-2 Toxin

incidence: 5/10*, conc. range: 0.0240–0.0906 μg/kg, ∅ conc.: 0.0231 μg/kg, sample origin: local markets and supermarkets, Chongqing (city), China, sample year: unknown, country: China[103], *pig back fat

- Co-contamination: not reported
- Further contamination of animal products in the present book/article[103]: Chicken muscle, Pig muscle, T-2, literature[103]; Pig fat, DON, literature[103]

Type B Trichothecenes

Deoxynivalenol

incidence: 3/10*, conc. range: 0.1232–0.4265 μg/kg, ∅ conc.: 0.1048 μg/kg, sample origin: local markets and supermarkets, Chongqing (city), China, sample year: unknown, country: China[103], *pig back fat

- Co-contamination: not reported
- Further contamination of animal products in the present book/article[103]: Chicken muscle, Pig muscle, Pig fat, T-2, literature[103]

Pig kidney may contain the following mycotoxins:

Aspergillus and ***Penicillium*** Toxins

Citrinin

incidence: 9/125, conc. range: 0.5–<1.0 μg/kg (1 sa.), 1.0–<5.0 μg/kg (4 sa.), 5.0–<10.0 μg/kg (2 sa.), >10.0 μg/kg (2 sa.), sample origin: abattoir, south of England, sample year: September, country: UK[169]

- Co-contamination: not reported
- Further contamination of animal products in the present book/article[169]: Pig kidney, OTA, literature[169]

Ochratoxin A

incidence: 27/61, conc. range: ≤9.33 μg/kg, sample origin: CMA, Germany, sample year: unknown, country: Germany[1]

- Co-contamination: not reported
- Further contamination of animal products in the present book/article[1]: Beef meat, Joints, Pig liver, Pork meat, Meat products, Sausage (beef sausage), Sausage (blood sausage), Sausage (Bologna sausage), Sausage (liver sausage), Sausage (poultry sausage), Sausage (raw sausage), OTA, literature[1]

incidence: 27/113*, conc. range: tr-23 μg/kg**, sample origin: slaughterhouse in the central part of Poland, sample year: April 1982–March 1983, country: Poland/Sweden[3], *suspected kidneys taken from 40 pigs in spring and 73 pigs in autumn, **22 contaminated cases in spring and 5 cases in autumn

- Co-contamination: not reported
- Further contamination of animal products in the present book/article[3]: Pig blood, OTA, literature[3]

incidence: 35*/85**, conc. range: ≤3.1 μg/kg, sample origin: unknown, sample year: 1991–1992, country: Poland[4], *suspected kidneys, **slaughtered pigs

- Co-contamination: not reported
- Further contamination of animal products in the present book/article[4]: Pig blood, OTA, literature[4]

incidence: 1/63*, conc.: 2.8 μg/kg, sample origin: slaughterhouse, Czechoslovakia, sample year: 1989, country: Czechoslovakia[5], *kidneys from healthy pigs

incidence: 20/96*, conc. range: 1–5 μg/kg (18 sa.), 5–20 μg/kg (2 sa.), sample origin: slaughterhouse, Czechoslovakia, sample year: 1989, country: Czechoslovakia[5], *kidneys with macroscopic lesions (unfit for human consumption)

- Co-contamination: not reported
- Further contamination of animal products in the present book/article[5]: Pig blood, OTA, literature[5]

incidence: 42/150*, conc. range: 2.41–40.6 μg/kg, ∅ conc.: 21.3 μg/kg, sample origin: markets, slaughterhouses, and supermarkets, Egypt, sample year: unknown, country: Egypt[98], *most probably pig kidneys

- Co-contamination: not reported
- Further contamination of animal products in the present book/article[98]: Chicken muscle, Cow liver, OTA, literature[98]

incidence: 1/12, conc.: 0.46 μg/kg, sample origin: 12 regions, Czech Republic, sample year: 2011–2013, country: Czech Republic[101]

- Co-contamination: not reported
- Further contamination of animal products reported in the present book/article[101]: Chicken meat, Pork meat, OTA, literature[101]

incidence: 42/125, conc. range: 0.1–<0.5 μg/kg (10 sa.), 0.5–<1.0 μg/kg (7 sa.), 1.0–<5.0 μg/kg (17 sa.), 5.0–<10.0 μg/kg (5 sa.), >10.0 μg/kg (3 sa.), sample origin: abattoir, south of England, sample year: September, country: UK[169]

- Co-contamination: not reported
- Further contamination of animal products in the present book/article[169]: Pig kidney, CIT, literature[169]

incidence: 40/300*, conc. range: LOD-LOQ (37 sa.), 0.48–1.40 μg/kg (3 sa.), sample origin: 12 French departments (main producers of pork), France, sample year: 1997, country: France[170], *normal kidneys

incidence: 26/100*, conc. range: LOD-LOQ (20 sa.), 0.16–0.48 μg/kg (6 sa.), sample origin: 12 French departments (main producers of pork), France, sample year: 1997, country: France[170], *nephropathic kidneys with macroscopic lesions and rejected for human consumption

incidence: 238/710*, conc. range: LOD-LOQ (184 sa.), 0.5–5 μg/kg (54 sa.), sample origin: 12 French departments (main producers of pork), France, sample year: 1998, country: France[170], *normal kidneys

- Co-contamination: not reported
- Further contamination of animal products in the present book/article[170]: not reported

incidence: 52*/122**, conc. range: 1–<2 µg/kg (27 sa.), 2–10 µg/kg (25 sa.), sample origin: slaughterhouse in Poznań (city), Poland, sample year: spring and autumn 1983 as well as spring 1984, country: Poland/Sweden[171], *suspected kidneys, **pigs with macroscopical kidney changes

- Co-contamination: not reported
- Further contamination of animal products in the present book/article[171]: Pig serum, OTA, literature[171]

incidence: 718/1561*, conc. range: >25 µg/kg (624 sa.), >150 µg/kg (94 sa.), sample origin: slaughterhouses across Denmark, sample year: first quarter of 1983, country: Denmark[172], *suspected kidneys

incidence: 1526/2878*, conc. range: >25 µg/kg (1468 sa.), >150 µg/kg (58 sa.), sample origin: slaughterhouses across Denmark, sample year: second quarter of 1983, country: Denmark[172], *suspected kidneys

incidence: 1351/2112*, conc. range: >25 µg/kg (1309 sa.), >150 µg/kg (42 sa.), sample origin: slaughterhouses across Denmark, sample year: third quarter of 1983, country: Denmark[172], *suspected kidneys

incidence: 903/1088*, conc. range: >25 µg/kg (892 sa.), >150 µg/kg (11 sa.), sample origin: slaughterhouses across Denmark, sample year: fourth quarter of 1983, country: Denmark[172], *suspected kidneys

- Co-contamination: not reported
- Further contamination of animal products in the present book/article[172]: not reported

incidence: 41/52*, conc. range: ≤3.18 µg/kg, Ø conc.: 0.54 µg/kg, sample origin: Bihor, Mures, and Timis (counties), central and western Romania, sample year: August 1998, country: Romania/Germany[173], *slaughtered pigs

- Co-contamination: not reported
- Further contamination of animal products in the present book/article[173]: Pig liver, Pig muscle, OTA, literature[173]; Pig serum, OTA and ZEA, literature[173]

incidence: 21/71*, conc. range: 1–5 µg/kg (18 sa.), 5–20 µg/kg (3 sa.), sample origin: slaughterhouse, Czechoslovakia, sample year: unknown, country: Czechoslovakia[174], *pig kidneys with macroscopic lesions

- Co-contamination: not reported
- Further contamination of animal products in the present book/article[174]: not reported

incidence: 284/300*, conc. range: 0.02–0.06 µg/kg (54 sa.), 0.06–0.09 µg/kg (27 sa.), 0.09–0.5 µg/kg (140 sa.), 0.50–1.00 µg/kg (39 sa.), >1.00 µg/kg (24 sa., maximum: 15 µg/kg), sample origin: slaughterhouses across Denmark, sample year: March 1999, country: Denmark[175], *healthy slaughtered pigs

- Co-contamination: not reported
- Further contamination of animal products in the present book/article[175]: Pork meat, OTA, literature[175]

incidence: 21/60*, conc. range: 2–68 µg/kg, sample origin: slaughterhouses across Denmark, sample year: 1975, country: Denmark[176], *condemned kidneys

- Co-contamination: not reported
- Further contamination of animal products in the present book/article[176]: not reported

incidence: 112/303*, conc. range: 0.5–<1.0 µg/kg (51 sa.), 1.0–<2.0 µg/kg (39 sa.), 2.0–<5.0 µg/kg (14 sa.), 5.0–<10.0 µg/kg (6 sa.), >10.0 µg/kg (2 sa., maximum: 12.4 µg/kg), sample origin: abattoir, southern England, sample year: September, country: UK[177], *kidneys unsuitable for human consumption

- Co-contamination: not reported
- Further contamination of animal products in the present book/article[177]: not reported

incidence: 20/74*, conc. range: 0.20–0.99 µg/kg (6 sa.), 1.00–4.99 µg/kg (12 sa.), 5.00–9.99 µg/kg (2 sa.), sample origin: Institute of Animal Pathology of the State University of Ghent (city), Belgium, sample year: February 1986–1987?, country: Belgium[178], *macroscopically suspected kidneys

incidence: 8/95*, conc. range: 1.00–4.99 µg/kg (7 sa.), 5.00–9.99 µg/kg (1 sa.), sample origin: Institute of Animal Disease Prevention at Torhout (municipality), Belgium, sample year: February 1986–1987, country: Belgium[178], *macroscopically suspected kidneys

incidence: 6/13*, conc. range: 0.8–1.6 µg/kg, Ø conc.: 1.22 µg/kg, sample origin: unknown, sample year: January 1986, country: Belgium[178], *pale, swollen, and degenerated piglet kidneys

incidence: 3/4*, conc. range: 0.5–1.8 µg/kg, Ø conc.: 1.33 µg/kg, sample origin: unknown, sample year: May 1986, country: Belgium[178], *piglet kidneys

incidence: 68/385*, conc. range: 0.20–0.99 µg/kg (24 sa.), 1.00–4.99 µg/kg (35 sa.), 5.00–9.99 µg/kg (4 sa.), >10 µg/kg (5 sa., maximum: 12.1 µg/kg), sample origin: slaughterhouses in various districts, Belgium, sample year: February–June 1987, country: Belgium[178], *macroscopically suspected kidneys and declared unfit for human consumption

- Co-contamination: not reported
- Further contamination of animal products in the present book/article[178]: Pig serum, OTA, literature[178]

incidence: 32/129*, conc. range: ≥2–<5 µg/kg (25 sa.), ≥5–<10 µg/kg (2 sa.), ≥10 µg/kg (5 sa., maximum: 104 µg/kg), sample origin: southern, southeastern, southwestern, and northern parts of Sweden, sample year: unknown, country: Sweden[179], *most of these kidneys showed histological changes

- Co-contamination: not reported
- Further contamination of animal products in the present book/article[179]: not reported

incidence: 3/160*, conc. range: <0.1–5.0 µg/kg, sample origin: slaughterhouses in different parts of Finland, sample year: September 1981–February 1982, country: Finland[180], *normal kidneys

incidence: 7/33*, conc. range: <0.1–5.0 µg/kg, sample origin: slaughterhouses in different parts of Finland, sample year: September 1981–February 1982, country: Finland[180], *suspected kidneys

- Co-contamination: not reported
- Further contamination of animal products in the present book/article[180]: not reported

incidence: 10/12, conc. range: 4.0–112.7 µg/kg, Ø conc.: 46.1 µg/kg, sample origin: Laboratory of Veterinary Research, Ringsted (city), Denmark, sample year: unknown, country: Belgium/Scotland[181]

- Co-contamination: not reported
- Further contamination of animal products in the present book/article[181]: not reported

incidence: 187/250*, conc. range: 0.2–0.5 µg/kg (151 sa.), 0.51–1.0 µg/kg (29 sa.), 1.01–1.5 µg/kg (4 sa.), 1.51–2.0 µg/kg (2 sa.), 2.3 µg/kg (1 sa.), sample origin: pig slaughterhouses, England, sample year: February 2002, country: UK[182], *condemned kidneys with lesions

- Co-contamination: not reported
- Further contamination of animal products in the present book/article[182]: not reported

incidence: 24/90*, conc. range: >2–88 μg/kg, sample origin: slaughterhouses, Sweden, sample year: January–November 1978, country: Sweden[183], *nephropathic kidneys
- Co-contamination: not reported
- Further contamination of animal products in the present book/article[183]: not reported

For detailed information see the article.

incidence: 8/30, conc. range: 0.01–1 μg/kg (5 sa.), 1–5 μg/kg (2 sa.), 6.5 μg/kg (1 sa.), sample origin: slaughterhouse in Vladimirci (municipality, village), Serbia, sample year: September 2006–February 2007, country: Serbia[184]

incidence: 11/30, conc. range: 0.01–1 μg/kg (5 sa.), 1–5 μg/kg (4 sa.), >5 μg/kg (2 sa., maximum: 17 μg/kg), sample origin: slaughterhouse in Senta (municipality, town), Serbia, sample year: September 2006–February 2007, country: Serbia[184]

incidence: 11?/30, conc. range: 0.01–1 μg/kg (5 sa.), 1–5 μg/kg (4 sa.), 52.5 μg/kg (1 sa.), sample origin: slaughterhouse in Bogatić (municipality, village), Serbia, sample year: September 2006–February 2007, country: Serbia[184]
- Co-contamination: not reported
- Further contamination of animal products in the present book/article[184]: Pig liver, Pig serum, OTA, literature[184]

incidence: 1/1, conc.: 1.9 μg/kg, sample origin: unknown, sample year: unknown, country: Italy[185]
- Co-contamination: not reported
- Further contamination of animal products in the present book/article[185]: not reported

incidence: 42*/54, conc. range: 0.26–3.05 μg/kg, sample origin: local slaughterhouse, Italy, sample year: unknown, country: Italy[186], *9 pig sa. from Belgium, 13 pig sa. from Germany, 9 pig sa. from Italy, and 11 pig sa. from Netherlands
- Co-contamination: not reported
- Further contamination of animal products in the present book/article[186]: not reported

incidence: 5/5*, conc. range: 23.9–27.5 μg/kg, Ø conc.: 25.6 μg/kg, sample origin: Apulia? (region), Italy, sample year: unknown, country: Italy[204], *pigs fed with OTA-contaminated feed
- Co-contamination: not reported
- Further contamination of animal products in the present book/article[204]: Pig liver, OTA, literature[204]

For detailed information see the article.

incidence: 89/104, conc. range: ≤9.30 μg/kg, sample origin: Netherlands, sample year: 1990, country: EU[215]
- Co-contamination: not reported
- Further contamination of animal products reported in the present book/article[215]: Dairy products, Meat, Pig liver, Pork meat, Sausage, OTA, literature[215]

incidence: 315/1092, conc. range: ≥0.2–5 μg/kg (278 sa.), ≥5–10 μg/kg (24 sa.), ≥10 μg/kg (13 sa., maximum: 29.2 μg/kg), sample origin: slaughterhouses, Poland, sample year: 2003–2012, country: Poland[365]
- Co-contamination: not reported
- Further contamination of animal products in the present book/article[365]: not reported

For detailed information see the article.

incidence: 30/90, conc. range: 0.17–52.5 μg/kg, Ø conc.: 4.14 μg/kg, sample origin: slaughterhouses in northern and central Serbia, sample year: 2009–2010, country: Serbia[366]
- Co-contamination: not reported
- Further contamination of animal products in the present book/article[366]: Chicken gizzard, Chicken kidney, Chicken liver, Pig liver, Pig serum, OTA, literature[366]

incidence: 42/60, conc. range: 1.3–22.0 μg/kg, sample origin: slaughterhouses in Backa Topola, Kovij, Sabac, and Senta (regions), Serbia, sample year: unknown, country: Serbia[367]
- Co-contamination: not reported
- Further contamination of animal products in the present book/article[367]: Pig liver, Pig serum, OTA, literature[367]

Pig liver may contain the following mycotoxins:

Aspergillus Toxins

Aflatoxin B_1

incidence: 17/47, conc. range: 0.2–0.87 μg/kg, Ø conc.: 0.43 μg/kg, sample origin: small markets in several urban Shanghai (city) areas, China, sample year: April 1988–1989, country: China/USA[46]
- Co-contamination: not reported
- Further contamination of animal products reported in the present book/article[46]: Butter (peanut butter), AFB_1, literature[46]; Milk, Milk powder, AFM_1, literature[46]

incidence: 1*/43, conc.: 27 μg/kg*, sample origin: farms in Santa Catarina and Rio Grande do Sul (states), Brazil, sample year: unknown, country: Brazil[188], *animal fed with industrial feed as well as farm grains
- Co-contamination: not reported
- Further contamination of animal products in the present book/article[188]: Poultry kidney, AFM_1, literature[188]

incidence: 3/60*, conc. range: 0.04–0.06 μg/kg, Ø conc.: 0.05 μg/kg, sample origin: Iowa and South Carolina (states), USA, sample year: January 1989, country: USA[190], *hogs
- Co-contamination: not reported
- Further contamination of animal products in the present book/article[190]: not reported

incidence: 4/100*, conc. range: 0.04–0.24 μg/kg, Ø conc.: 0.123 μg/kg, sample origin: Iowa, North Carolina, and South Carolina (states), USA, sample year: April 1989, country: USA[190], *hogs
- Co-contamination: 3 sa. co-contaminated with AFB_1 and AFM_1; 1 sa. contaminated solely with AFB_1
- Further contamination of animal products in the present book/article[190]: Pig liver, AFM_1, literature[190]

incidence: 3/50, conc. range: 0.25–0.42 μg/kg, Ø conc.: 0.32 μg/kg, sample origin: slaughterhouses in northern Italy, sample year: May 2004, country: Italy[410]
- Co-contamination: 2 sa. co-contaminated with AFB_1 and AFM_1; 1 sa. contaminated solely with AFB_1
- Further contamination of animal products in the present book/article[410]: Pig liver, AFM_1, literature[410]

incidence: 1/3, conc.: 0.68 μg/kg, sample origin: local supermarket, China, sample year: unknown, country: China[538]
- Co-contamination: 1 sa. co-contaminated with AFB_1, AFB_2, and AFM_1
- Further contamination of animal products reported in the present book/article[538]: Chicken liver, AFB_1, CIT, and OTA,

literature[538]; Chicken muscle, AFB_1 and CIT, literature[538]; Pig liver, AFB_2, AFM_1, CIT, and PAT, literature[538]; Pig muscle, AFB_1, OTA, and STG, literature[538]

Aflatoxin B_2

incidence: 1/3, conc.: 0.52 µg/kg, sample origin: local supermarket, China, sample year: unknown, country: China[538]

- Co-contamination: 1 sa. co-contaminated with AFB_1, AFB_2, and AFM_1
- Further contamination of animal products reported in the present book/article[538]: Chicken liver, AFB_1, CIT, and OTA, literature[538]; Chicken muscle, AFB_1 and CIT, literature[538]; Pig liver, AFB_1, AFM_1, CIT, and PAT, literature[538]; Pig muscle, AFB_1, OTA, and STG, literature[538]

Aflatoxin M_1

incidence: 0/60*, conc.: no contamination, sample origin: Iowa and South Carolina (states), USA, sample year: January 1989, country: USA[190], *hogs

- Co-contamination: not reported
- Further contamination of animal products in the present book/article[190]: not reported

incidence: 4/100*, conc. range: 0.20–0.44 µg/kg, ∅ conc.: 0.285 µg/kg, sample origin: Iowa, North Carolina, and South Carolina (states), USA, sample year: April 1989, country: USA[190], *hogs

- Co-contamination: 3 sa. co-contaminated with AFB_1 and AFM_1; 1 sa. contaminated solely with AFM_1
- Further contamination of animal products in the present book/article[190]: Pig liver, AFB_1, literature[190]

incidence: 3/49, conc. range: 0.10–0.28 µg/kg (2 sa.), 1.05 µg/kg (1 sa.), sample origin: slaughterhouses in northern Italy, sample year: May 2004, country: Italy[410]

- Co-contamination: 2 sa. co-contaminated with AFB_1 and AFM_1; 1 sa. contaminated solely with AFM_1
- Further contamination of animal products in the present book/article[410]: Pig liver, AFB_1, literature[410]

incidence: 2/3, conc. range: 0.63–0.87 µg/kg, ∅ conc.: 0.75 µg/kg, sample origin: local supermarket, China, sample year: unknown, country: China[538]

- Co-contamination: 1 sa. co-contaminated with AFB_1, AFB_2, and AFM_1; 1 sa. co-contaminated with AFM_1 and PAT
- Further contamination of animal products reported in the present book/article[538]: Chicken liver, AFB_1, CIT, and OTA, literature[538]; Chicken muscle, AFB_1 and CIT, literature[538]; Pig liver, AFB_1, AFB_2, CIT, and PAT, literature[538]; Pig muscle, AFB_1, OTA, and STG, literature[538]

Aspergillus and *Penicillium* Toxins

Citrinin

incidence: 1/3, conc.: 1.45 µg/kg, sample origin: local supermarket, China, sample year: unknown, country: China[538]

- Co-contamination: not reported
- Further contamination of animal products reported in the present book/article[538]: Chicken liver, AFB_1, CIT, and OTA, literature[538]; Chicken muscle, AFB_1 and CIT, literature[538]; Pig liver, AFB_1, AFB_2, AFM_1, and PAT, literature[538]; Pig muscle, AFB_1, OTA, and STG, literature[538]

Ochratoxin A

incidence: 10/59, conc. range: ≤2.72 µg/kg, sample origin: CMA, Germany, sample year: unknown, country: Germany[1]

- Co-contamination: not reported
- Further contamination of animal products in the present book/article[1]: Beef meat, Joints, Meat products, Pig kidney, Pork meat, Sausage (beef sausage), Sausage (blood sausage), Sausage (Bologna sausage), Sausage (liver sausage), Sausage (poultry sausage), Sausage (raw sausage), OTA, literature[1]

incidence: 39/52, conc. range: ≤0.61 µg/kg, ∅ conc.: 0.16 µg/kg, sample origin: Bihor, Mures, and Timis (counties), central and western Romania, sample year: August 1998, country: Romania/Germany[173]

- Co-contamination: not reported
- Further contamination of animal products in the present book/article[173]: Pig kidney, Pig muscle, OTA, literature[173]; Pig serum, OTA and ZEA, literature[173]

incidence: 11/30, conc. range: 0.01–1 µg/kg (5 sa.), 1–5 µg/kg (6 sa., maximum: 2.2 µg/kg), sample origin: slaughterhouse in Vladimirci (municipality, village), Serbia, sample year: September 2006–February 2007, country: Serbia[184]

incidence: 4/30, conc. range: 0.01–1 µg/kg (1 sa.), 1–5 µg/kg (2 sa.), 14.5 µg/kg (1 sa.), sample origin: slaughterhouse in Senta (municipality, town), Serbia, sample year: September 2006–February 2007, country: Serbia[184]

incidence: 9/30, conc. range: 0.01–1 µg/kg (2 sa.), 1–5 µg/kg (6 sa.), 5.46 µg/kg (1 sa.), sample origin: slaughterhouse in Bogatić (municipality, village), Serbia, sample year: September 2006–February 2007, country: Serbia[184]

- Co-contamination: not reported
- Further contamination of animal products in the present book/article[184]: Pig kidney, Pig serum, OTA, literature[184]

incidence: 5/5*, conc. range: 3.2–5.3 µg/kg, ∅ conc.: 4.4 µg/kg, sample origin: Apulia? (region), Italy, sample year: unknown, country: Italy[204], *pigs fed with OTA-contaminated feed

- Co-contamination: not reported
- Further contamination of animal products in the present book/article[204]: Pig kidney, OTA, literature[204]

For detailed information see the article.

incidence: 1/10*, conc. range: LOD/LOQ-0.9 µg/kg (1 sa.), sample origin: Netherlands, sample year: 1997, country: EU[215], *pork liver sausage/pate

- Co-contamination: not reported
- Further contamination of animal products reported in the present book/article[215]: Dairy products, Meat, Pig kidney, Pork meat, Sausage, OTA, literature[215]

incidence: 24/90, conc. range: 0.18–14.5 µg/kg, ∅ conc.: 2.14 µg/kg, sample origin: slaughterhouses in northern and central Serbia, sample year: 2009–2010, country: Serbia[366]

- Co-contamination: not reported
- Further contamination of animal products in the present book/article[366]: Chicken gizzard, Chicken kidney, Chicken liver, Pig kidney, Pig serum, OTA, literature[366]

incidence: 39/60, conc. range: 1.2–19.5 µg/kg, sample origin: slaughterhouses in Backa Topola, Kovij, Sabac, and Senta (regions), Serbia, sample year: unknown, country: Serbia[367]

- Co-contamination: not reported
- Further contamination of animal products in the present book/article[367]: Pig kidney, Pig serum, OTA, literature[367]

Patulin

incidence: 1/3, conc.: 0.69 µg/kg, sample origin: local supermarket, China, sample year: unknown, country: China[538]

- Co-contamination: not reported
- Further contamination of animal products reported in the present book/article[538]: Chicken liver, AFB_1, CIT, and OTA, literature[538]; Chicken muscle, AFB_1 and CIT, literature[538]; Pig liver, AFB_1, AFB_2, AFM_1, and CIT, literature[538]; Pig muscle, AFB_1, OTA, and STG, literature[538]

Pig muscle may contain the following mycotoxins:

Aspergillus Toxins

Aflatoxin B_1

incidence: 2/4, conc. range: 0.46–0.74 µg/kg, ∅ conc.: 0.6 µg/kg, sample origin: local supermarket, China, sample year: unknown, country: China[538]

- Co-contamination: 1 sa. co-contaminated with AFB_1 and STG; 1 sa. contaminated solely with AFB_1
- Further contamination of animal products reported in the present book/article[538]: Chicken liver, AFB_1, CIT, and OTA, literature[538]; Chicken muscle, AFB_1 and CIT, literature[538]; Pig liver, AFB_1, AFB_2, AFM_1, CIT, and PAT, literature[538]; Pig muscle, OTA and STG, literature[538]

Sterigmatocystin

incidence: 2/4, conc. range: 0.76–1.23 µg/kg, ∅ conc.: 1.00 µg/kg, sample origin: local supermarket, China, sample year: unknown, country: China[538]

- Co-contamination: 1 sa. co-contaminated with AFB_1 and STG; 1 sa. contaminated solely with STG
- Further contamination of animal products reported in the present book/article[538]: Chicken liver, AFB_1, CIT, and OTA, literature[538]; Chicken muscle, AFB_1 and CIT, literature[538]; Pig liver, AFB_1, AFB_2, AFM_1, CIT, and PAT, literature[538]; Pig muscle, AFB_1 and OTA, literature[538]

Aspergillus and *Penicillium* Toxins

Ochratoxin A

incidence: 2/22*, conc. range: ≤0.04–0.06 µg/kg, ∅ conc.: 0.05 µg/kg, sample origin: industrial plants for ham production in the Emilia region, northern Italy, sample year: 2001–2002, country: Italy[152], *muscle sa. for hams

- Co-contamination: not reported
- Further contamination of animal products reported in the present book/article[152]: Coppa, Ham, Sausage (salami), Sausage (Würstel), OTA, literature[152]

incidence: 6/6*, conc. range: 1.05–4.20 µg/kg, ∅ conc.: 2.10 µg/kg, sample origin: slaughterhouses and manufacturing plants, northern Italy, sample year: 2007–2010, country: Italy[167], *pork ham muscle

- Co-contamination: not reported
- Further contamination of animal products reported in the present book/article[167]: Ham, OTA, literature[167]

incidence: 9/52, conc. range: ≤0.53 µg/kg, ∅ conc.: 0.15 µg/kg, sample origin: Bihor, Mures, and Timis (counties), central and western Romania, sample year: August 1998, country: Romania/Germany[173]

- Co-contamination: not reported
- Further contamination of animal products in the present book/article[173]: Pig kidney, Pig liver, OTA, literature[173]; Pig serum, OTA and ZEA, literature[173]

incidence: 7/13, conc. range: LOD-LOQ (6 sa.), 0.12 µg/kg (1 sa.), ∅ conc.: 0.01 µg/kg, sample origin: supermarkets in Coimbra (city), Portugal, sample year: October 2002–February 2003, country: Spain/Portugal[486]

- Co-contamination: not reported
- Further contamination of animal products in the present book/article[486]: Turkey muscle, OTA, literature[486]

incidence: 1/4, conc.: 0.88 µg/kg, sample origin: local supermarket, China, sample year: unknown, country: China[538]

- Co-contamination: not reported
- Further contamination of animal products reported in the present book/article[538]: Chicken liver, AFB_1, CIT, and OTA, literature[538]; Chicken muscle, AFB_1 and CIT, literature[538]; Pig liver, AFB_1, AFB_2, AFM_1, CIT, and PAT, literature[538]; Pig muscle, AFB_1 and STG, literature[538]

Fusarium Toxins

Type A Trichothecenes

T-2 Toxin

incidence: 6/20*, conc. range: 0.0240–0.4515 µg/kg, ∅ conc.: 0.0987 µg/kg, sample origin: local markets and supermarkets, Chongqing (city), China, sample year: unknown, country: China[103], *pig dorsal muscle

- Co-contamination: not reported
- Further contamination of animal products in the present book/article[103]: Chicken muscle, T-2, literature[103]; Pig fat, DON and T-2, literature[103]

Pig plasma may contain the following mycotoxins:

Aspergillus and *Penicillium* Toxins

Ochratoxin A

incidence: 10/16*, conc. range: LOD/LOQ-4.9 µg/kg (5 sa.), 5.0–9.9 µg/kg (1 sa.), 10.0–24.9 µg/kg (3 sa.), 4.4 µg/kg? (1 sa.), sample origin: Netherlands, sample year: 1985, country: EU[192], *porcine plasma powder

- Co-contamination: not reported
- Further contamination of animal products in the present book/article[192]: Pork meat, Poultry meat, OTA, literature[192]

incidence: 191/216*, conc. range: ≥0.10 µg/l (178 sa.), ≥1.0 µg/l (11 sa.), ≥5.0 µg/l (2 sa., maximum: 12.5 µg/l), ∅ conc.: 0.50 µg/l, sample origin: slaughterhouses in different parts of Norway, sample year: June 1991, country: Norway[384], *in part a combination of plasma and serum was analyzed

- Co-contamination: not reported
- Further contamination of animal products in the present book/article[384]: not reported

For detailed information see the article.

see also **Pig blood** and **Pig serum**

Pig serum may contain the following mycotoxins:

Aspergillus and *Penicillium* Toxins

Citrinin

incidence: 7/10, ∅ conc.: 1.3 µg/l, sample origin: MPN affected farms in Bulgaria, sample year: 2006, country: Bulgaria/South Africa[519], *enlarged and mottled or pale appearance of kidneys

- Co-contamination: not reported
- Further contamination of animal products in the present book/article[519]: Pig serum, DON, OTA, PA, PEA, and ZEA, literature[519]

incidence: 8/10, ∅ conc.: 1.6 µg/l, sample origin: MPN affected farms in Bulgaria, sample year: 2007, country: Bulgaria/South Africa[519], *enlarged and mottled or pale appearance of kidneys
- Co-contamination: not reported
- Further contamination of animal products in the present book/article[519]: Pig serum, OTA, PA, PEA, and ZEA, literature[519]

Ochratoxin A

incidence: 148/388*, conc. range: 1–520 µg/l, sample origin: slaughterhouse in Poznań (city), Poland, sample year: October 1983–July 1984, country: Poland/Sweden[171], *porcine serum sa. not biased for nephropathy
- Co-contamination: not reported
- Further contamination of animal products in the present book/article[171]: Pig kidney, OTA, literature[171]

incidence: 51/52, conc. range: ≤13.4 µg/l, ∅ conc.: 2.43 µg/l, sample origin: Bihor, Mures, and Timis (counties), central and western Romania, sample year: August 1998, country: Romania/Germany[173]
- Co-contamination: not reported
- Further contamination of animal products in the present book/article[173]: Pig kidney, Pig liver, Pig muscle, OTA, literature[173]; Pig serum, ZEA, literature[173]

incidence: 2/4*, conc. range: 3.1–3.7 µg/l, ∅ conc.: 3.4 µg/l, sample origin: unknown, sample year: January 1986, country: Belgium[178], *piglets had pale, swollen, and degenerated piglet kidneys

incidence: 4/4*, conc. range: 2.3–3.7 µg/l, ∅ conc.: 2.95 µg/l, sample origin: unknown, sample year: May 1986, country: Belgium[178], *piglet kidneys
- Co-contamination: not reported
- Further contamination of animal products in the present book/article[178]: Pig kidney, OTA, literature[178]

incidence: 5/30, conc. range: 0.01–1 µg/l (3 sa.), 1–5 µg/l (2 sa., maximum: 2.56 µg/l), sample origin: slaughterhouse in Vladimirci (municipality, village), Serbia, sample year: September 2006–February 2007, country: Serbia[184]

incidence: 13/30, conc. range: 0.01–1 µg/l (6 sa.), 1–5 µg/l (4 sa.), >5 µg/l (3 sa., maximum: 35.7 µg/l), sample origin: slaughterhouse in Senta (municipality, town), Serbia, sample year: September 2006–February 2007, country: Serbia[184]

incidence: 10/30, conc. range: 0.01–1 µg/l (5 sa.), 1–5 µg/l (2 sa.), >5 µg/l (3 sa., maximum: 221 µg/l), sample origin: slaughterhouse in Bogatić (municipality, village), Serbia, sample year: September 2006–February 2007, country: Serbia[184]
- Co-contamination: not reported
- Further contamination of animal products in the present book/article[184]: Pig kidney, Pig liver, OTA, literature[184]

incidence: 28/90, conc. range: 0.24–220.8 µg/kg, ∅ conc.: 11.87 µg/kg, sample origin: slaughterhouses in northern and central Serbia, sample year: 2009–2010, country: Serbia[366]
- Co-contamination: not reported
- Further contamination of animal products in the present book/article[366]: Chicken gizzard, Chicken kidney, Chicken liver, Pig kidney, Pig liver, OTA, literature[366]

incidence: 36/60, conc. range: 2.5–33.3 µg/kg, sample origin: slaughterhouses in Backa Topola, Kovij, Sabac, and Senta (regions), Serbia, sample year: unknown, country: Serbia[367]
- Co-contamination: not reported
- Further contamination of animal products in the present book/article[367]: Pig kidney, Pig liver, OTA, literature[367]

incidence: 52/200, conc. range: 0.1–0.5 µg/l (46 sa.), 0.51–1.0 µg/l (4 sa.), >1.0 µg/l (2 sa., maximum: 1.24 µg/l), sample origin: Styrian slaughterhouses, Austria, sample year: 1995, country: Austria[382]

incidence: 66/287, conc. range: 0.1–0.5 µg/l (58 sa.), 0.51–1.0 µg/l (1 sa.), >1.0 µg/l (7 sa., maximum: 30.36 µg/l), sample origin: Styrian slaughterhouses, Austria, sample year: 1998, country: Austria[382]
- Co-contamination: not reported
- Further contamination of animal products in the present book/article[382]: not reported

incidence: 45/85, conc. range: 0.29–17.6 µg/kg, sample origin: Trier-Saarburg (county), Germany, sample year: September 1988–April 1989, country: Germany[383]
- Co-contamination: not reported
- Further contamination of animal products in the present book/article[383]: not reported

incidence: 3/4, conc. range: 209.4–363.1 µg/l, ∅ conc.: 285.1 µg/l, sample origin: unknown, sample year: unknown, country: Belgium[385]
- Co-contamination: not reported
- Further contamination of animal products in the present book/article[385]: not reported

incidence: 51/62, conc. range: 0.17–2.81 µg/l, sample origin: Tuscany (region), Italy, sample year: unknown, country: Italy[415]
- Co-contamination: not reported
- Further contamination of animal products in the present book/article[415]: not reported

incidence: 14/19, conc. range: 0.3–2.0 µg/l, ∅ conc.: 0.67 µg/l, sample origin: Meat Inspection Station in Ueda (city), Nagano prefecture, Japan, sample year: October 1988–February 1989, country: Japan[431]

incidence: 104/104*, conc. range: 0.003–2.440 µg/l, ∅ conc.: 0.362 µg/l, sample origin: Meat Inspection Station in Iida (city), Nagano prefecture, Japan, sample year: November 1985, country: Japan[431], *6 months old pigs (marketed)

incidence: 8/8*, conc. range: 0.072–0.425 µg/l, ∅ conc.: 0.246 µg/l, sample origin: Meat Inspection Station in Iida (city), Nagano prefecture, Japan, sample year: November 1985, country: Japan[431], *1 year old pigs for breeding

incidence: 7/7*, conc. range: 0.056–0.340 µg/l, ∅ conc.: 0.174 µg/l, sample origin: Meat Inspection Station in Iida (city), Nagano prefecture, Japan, sample year: November 1985, country: Japan[431], *over 1 year old pigs (not marketed)

incidence: 3/4*, conc. range: ≤0.102 µg/l, sample origin: Meat Inspection Station in Iida (city), Nagano prefecture, Japan, sample year: November 1985, country: Japan[431], *pigs with malnutrition (not marketed)

incidence: 18/18, conc. range: 1.930–9.000 µg/l, ∅ conc.: 5.201 µg/l, sample origin: unknown, sample year: October–November 1989, country: Japan[431]
- Co-contamination: not reported
- Further contamination of animal products in the present book/article[431]: not reported

incidence: 205/205*, conc. range: 0.11–0.87 μg/l, sample origin: swine farms in Piedmont (region), northwest Italy, sample year: September 2006–March 2009, country: Italy[498], *conventional

incidence: 80/80*, conc. range: 0.15–6.24 μg/l, sample origin: swine farms in Piedmont (region), northwest Italy, sample year: September 2006–March 2009, country: Italy[498], *organic

- Co-contamination: not reported
- Further contamination of animal products in the present book/article[498]: not reported

incidence: 8/10, ∅ conc.: 28.8 μg/l, sample origin: MPN affected farms in Bulgaria, sample year: 2006, country: Bulgaria/South Africa[519], *enlarged and mottled or pale appearance of kidneys

- Co-contamination: not reported
- Further contamination of animal products in the present book/article[519]: Pig serum, CIT, DON, PA, PEA, and ZEA, literature[519]

incidence: 9/10, ∅ conc.: 6.3 μg/l, sample origin: MPN affected farms in Bulgaria, sample year: 2007, country: Bulgaria/South Africa[519], *enlarged and mottled or pale appearance of kidneys

- Co-contamination: not reported
- Further contamination of animal products in the present book/article[519]: Pig serum, CIT, PA, PEA, and ZEA, literature[519]

Penicillic Acid

incidence: 8/10, ∅ conc.: 23.3 μg/l, sample origin: MPN affected farms in Bulgaria, sample year: 2006, country: Bulgaria/South Africa[519], *enlarged and mottled or pale appearance of kidneys

- Co-contamination: not reported
- Further contamination of animal products in the present book/article[519]: Pig serum, CIT, DON, OTA, PEA, and ZEA, literature[519]

incidence: 9/10, ∅ conc.: 22.9 μg/l, sample origin: MPN affected farms in Bulgaria, sample year: 2007, country: Bulgaria/South Africa[519], *enlarged and mottled or pale appearance of kidneys

- Co-contamination: not reported
- Further contamination of animal products in the present book/article[519]: Pig serum, CIT, OTA, PEA, and ZEA, literature[519]

Fusarium Toxins

Type B Trichothecenes

Deoxynivalenol

incidence: 1/10, conc.: 7.6 μg/l, sample origin: MPN affected farms in Bulgaria, sample year: 2006, country: Bulgaria/South Africa[519], *enlarged and mottled or pale appearance of kidneys

- Co-contamination: not reported
- Further contamination of animal products in the present book/article[519]: Pig serum, CIT, OTA, PA, PEA, and ZEA, literature[519]

incidence: 0/10, conc.: no contamination, sample origin: MPN affected farms in Bulgaria, sample year: 2007, country: Bulgaria/South Africa[519], *enlarged and mottled or pale appearance of kidneys

- Co-contamination: not reported
- Further contamination of animal products in the present book/article[519]: Pig serum, CIT, OTA, PA, PEA, and ZEA, literature[519]

Other *Fusarium* Toxins

Zearalenone

incidence: 9/52, conc. range: ≤0.96 μg/l, ∅ conc.: 0.80 μg/l, sample origin: Bihor, Mures, and Timis (counties), central and western Romania, sample year: August 1998, country: Romania/Germany[173]

- Co-contamination: not reported
- Further contamination of animal products in the present book/article[173]: Pig kidney, Pig liver, Pig muscle, Pig serum, OTA, literature[173]

incidence: 5/10, ∅ conc.: 0.24 μg/l, sample origin: MPN affected farms in Bulgaria, sample year: 2006, country: Bulgaria/South Africa[519], *enlarged and mottled or pale appearance of kidneys

- Co-contamination: not reported
- Further contamination of animal products in the present book/article[519]: Pig serum, CIT, DON, OTA, PA, and PEA, literature[519]

incidence: 5/10, ∅ conc.: 0.33 μg/l, sample origin: MPN affected farms in Bulgaria, sample year: 2007, country: Bulgaria/South Africa[519], *enlarged and mottled or pale appearance of kidneys

- Co-contamination: not reported
- Further contamination of animal products in the present book/article[519]: Pig serum, CIT, OTA, PA, and PEA, literature[519]

Penicillium Toxins

Penitrem A

incidence: 3/10, ∅ conc.: 64.0 μg/l, sample origin: MPN affected farms in Bulgaria, sample year: 2006, country: Bulgaria/South Africa[519], *enlarged and mottled or pale appearance of kidneys

- Co-contamination: not reported
- Further contamination of animal products in the present book/article[519]: Pig serum, CIT, DON, OTA, PA, and ZEA, literature[519]

incidence: 3/10, ∅ conc.: 45.6 μg/l, sample origin: MPN affected farms in Bulgaria, sample year: 2007, country: Bulgaria/South Africa[519], *enlarged and mottled or pale appearance of kidneys

- Co-contamination: not reported
- Further contamination of animal products in the present book/article[519]: Pig serum, CIT, OTA, PA, and ZEA, literature[519]

see also **Pig blood** and **Pig plasma**

Pork meat may contain the following mycotoxins:

Aspergillus and *Penicillium* Toxins

Ochratoxin A

incidence: 10/58, conc. range: ≤0.14 μg/kg, sample origin: CMA, Germany, sample year: unknown, country: Germany[1]

- Co-contamination: not reported
- Further contamination of animal products in the present book/article[1]: Beef meat, Beef, Joints, Meat products, Pig kidney, Pig liver, Sausage (beef sausage), Sausage (blood sausage), Sausage (Bologna sausage), Sausage (liver sausage), Sausage (poultry sausage), Sausage (raw sausage), OTA, literature[1]

incidence: 1/1*, conc.: 0.06 μg/kg, sample origin: supermarkets and fast-food chains in Quebec (city), Canada, sample year: September–October 2008, country: Canada[87], *pork, cured incidence: 1/1*, conc.: 0.20 μg/kg, sample origin: supermarkets and fast-food chains in Calgary (city), Canada, sample year: September–October 2009, country: Canada[87], *pork, cured

incidence: 1/1*, conc.: 0.03 μg/kg, sample origin: supermarkets and fast-food chains in Quebec (city), Canada, sample year: September–October 2008, country: Canada[87], *pork, fresh

incidence: 1/1*, conc.: 0.23 μg/kg, sample origin: supermarkets and fast-food chains in Calgary (city), Canada, sample year: September–October 2009, country: Canada[87], *pork, fresh

- Co-contamination: not reported
- Further contamination of animal products reported in the present book/article[87]: Beefburger, Butter (peanut butter), Chicken nuggets, Chickenburger, Hot-dog, Ice cream, Meat, Meat products, Sausage, OTA, literature[87]

incidence: 1/12*, conc.: 0.20 μg/kg, sample origin: 12 regions, Czech Republic, sample year: 2011–2013, country: Czech Republic[101], *pork, fresh

- Co-contamination: not reported
- Further contamination of animal products reported in the present book/article[101]: Chicken meat, Pig kidney, OTA, literature[101]

incidence: 64/76*, conc. range: LOD-0.99 μg/kg (62 sa.), ≥1.0 μg/kg (2 sa., maximum: 1.3 μg/kg), sample origin: retail shops, Denmark, sample year: 1995, country: Denmark[102], *conventional

incidence: 4/7*, conc. range: LOD-0.99 μg/kg (4 sa., maximum: 0.12 μg/kg), sample origin: retail shops, Denmark, sample year: 1995, country: Denmark[102], *ecological

- Co-contamination: not reported
- Further contamination of animal products reported in the present book/article[102]: Chicken meat, Duck liver, Duck meat, Goose liver, Goose meat, Turkey liver, Turkey meat, OTA, literature[102]

incidence: 228/300*, conc. range: 0.03–0.06 μg/kg (134 sa.), 0.06–0.09 μg/kg (27 sa.), 0.09–0.5 μg/kg (55 sa.), 0.50–1.00 μg/kg (3 sa.), >1.00 μg/kg (9 sa., maximum: 2.9 μg/kg), sample origin: slaughterhouses across Denmark, sample year: March 1999, country: Denmark[175], *healthy slaughtered pigs

- Co-contamination: not reported
- Further contamination of animal products in the present book/article[175]: Pig kidney, OTA, literature[175]

incidence: 64/76, conc. range: LOD/LOQ-4.9 μg/kg (maximum: 1.31 μg/kg), sample origin: Denmark, sample year: 1993–1994, country: EU[192]

- Co-contamination: not reported
- Further contamination of animal products reported in the present book/article[192]: Pig plasma, Poultry meat, OTA, literature[192]

incidence: 103/1011*, conc. range: LOD/LOQ-0.9 μg/kg (83 sa.), 1.0–2.9 μg/kg (18 sa.), 3.0–4.9 (1 sa.), 6.1 μg/kg (1 sa.), sample origin: France, sample year: 1997–1998, country: EU[215], *pork edible offal

incidence: 8/58, conc. range: LOD/LOQ-0.9 μg/kg (8 sa., maximum: 0.1360 μg/kg), sample origin: Germany, sample year: 1995–1998, country: EU[215]

- Co-contamination: not reported
- Further contamination of animal products reported in the present book/article[215]: Dairy products, Meat, Pig kidney, Pig liver, Sausage, OTA, literature[215]

incidence: 3/94, conc. range: 0.9–1.8 μg/kg, sample origin: retail markets across USA, sample year: 2012–2014, country: USA[276]

- Co-contamination: not reported
- Further contamination of animal products reported in the present book/article[276]: not reported

incidence: 1/12*, conc.: 5 μg/kg, sample origin: Slavonski Brod (city), Yugoslavia (now Croatia), sample year: 1971–1975, country: Denmark/Yugoslavia[363], *sa. collected from BEN area

- Co-contamination: not reported
- Further contamination of animal products reported in the present book/article[363]: not reported

incidence: 5/20, conc. range: ≤0.578 μg/kg, ∅ conc.: 0.405 μg/kg, sample origin: region of Porto (city), Portugal, sample year: unknown, country: Portugal[364]

- Co-contamination: not reported
- Further contamination of animal products reported in the present book/article[364]: not reported

Fusarium Toxins

Zearalenone

incidence: 7/11*, conc. range: ≤4.31 μg/kg, sample origin: pig farms in northern and western Croatia, sample year: 2014, country: Croatia[505], *meat from male pigs

incidence: 10/19*, conc. range: ≤1.09 μg/kg, sample origin: pig farms in northern and western Croatia, sample year: 2014, country: Croatia[505], *meat from female pigs

- Co-contamination: not reported
- Further contamination of animal products reported in the present book/article[505]: not reported

Poultry kidney may contain the following mycotoxins:

Aspergillus Toxins

Aflatoxin M_1

incidence: 1/43, conc.: <0.1 μg/kg, sample origin: farms in Santa Catarina and Rio Grande do Sul (states), Brazil, sample year: unknown, country: Brazil[188]

- Co-contamination: not reported
- Further contamination of animal products in the present book/article[188]: Pig liver AFB_1, literature[188]

Aspergillus and ***Penicillium*** Toxins

Ochratoxin A

incidence: 5/14*, conc. range: 4.3–29.2 μg/kg, sample origin: poultry slaughterhouse, Danpo (company), Vamdrup (municipality), Jutland (peninsula), Denmark, sample year: unknown, country: Denmark[187], *condemned birds with enlarged pale kidneys

- Co-contamination: not reported
- Further contamination of animal products in the present book/article[187]: not reported

Poultry meat may contain the following mycotoxins:

Aspergillus and ***Penicillium*** Toxins

Ochratoxin A

incidence: 62/113, conc. range: LOD/LOQ-4.9 μg/kg (maximum: 0.18 μg/kg), sample origin: Denmark, sample year: 1993–1994, country: EU[192]

- Co-contamination: not reported
- Further contamination of animal products reported in the present book/article[192]: Pig plasma, Pork meat, OTA, literature[192]

incidence: 17/51*, conc. range: ≤0.67 μg/kg, ∅ conc.: 0.070 μg/kg, sample origin: supermarkets and fresh provision shops, Hong Kong (special administrative region), China, sample year: August–September 2005, country: China[214], *poultry meat and their products

- Co-contamination: not reported
- Further contamination of animal products reported in the present book/article[214]: not reported

Salami see Sausage (salami)

Sausage may contain the following mycotoxins:

Aspergillus Toxins

Aflatoxin B_1

incidence: 1/25, conc.: 7 μg/kg, sample origin: local companies in Cairo (capital), Egypt, sample year: unknown, country: Egypt[2]

- Co-contamination: 1 sa. co-contaminated with AFB1 and AFB_2
- Further contamination of animal products in the present book/article[2]: Beefburger, AFB_1, literature[2]; Hot-dog, Kubeba, Meat, AFB_1 and AFB_2, literature[2]; Sausage, AFB_2, literature[2]

incidence: 1/15*, conc.: 1.5 μg/kg, sample origin: individual producer or Croatian meat industrial facilities, Croatia, sample year: unknown, country: Croatia[481], *game sausages included rabbit, wild boar, deer, roe deer, and mixed sausages (wild boar, deer, and pork)

- Co-contamination: 1 sa. co-contaminated with AFB_1, CIT, and OTA
- Further contamination of animal products in the present book/article[481]: Meat products, AFB_1, CIT, and OTA, literature[481]; Sausage, CIT and OTA, literature[481]

incidence: 2/25*, conc. range: 1.0–1.1 μg/kg, ∅ conc.: 1.05 μg/kg, sample origin: individual producer or Croatian meat industrial facilities, Croatia, sample year: unknown, country: Croatia[481], *semidry sausages included grill, Kranjska, Slavonian, Zagorska, and homemade garlic sausages

- Co-contamination: not reported
- Further contamination of animal products in the present book/article[481]: Meat products, AFB_1, CIT, and OTA, literature[481]; Sausage, CIT and OTA, literature[481]

incidence: 1/28*, conc. range: 1.07 μg/kg, sample origin: markets, fairs, or directly from the producing household, Primorje, Zagorje, Medimurje, Slavonia, and Baranja (regions), Croatia, sample year: 2011–2014, country: Croatia/Italy[500], *dry-fermented sausage (Istrian type)

incidence: 4/76*, conc. range: 1.12–1.29 μg/kg, sample origin: markets, fairs, or directly from the producing household, Primorje, Zagorje, Medimurje, Slavonia, and Baranja (regions), Croatia, sample year: 2011–2014, country: Croatia/Italy[500], *dry-fermented sausage (Slavonian type)

incidence: 3/69*, conc. range: 0.96–1.25 μg/kg, sample origin: markets, fairs, or directly from the producing household, Primorje, Zagorje, Medimurje, Slavonia, and Baranja (regions), Croatia, sample year: 2011–2014, country: Croatia/Italy[500], *dry-fermented sausage (northern type)

- Co-contamination: not reported
- Further contamination of animal products reported in the present book/article[500]: Bacon, Sausage, OTA, literature[500]; Ham, Sausage (liver sausage), AFB_1 and OTA literature[500]

Aflatoxin B_2

incidence: 1/25, conc.: 3 μg/kg, sample origin: local companies in Cairo (capital), Egypt, sample year: unknown, country: Egypt[2]

- Co-contamination: 1 sa. co-contaminated with AFB1 and AFB_2
- Further contamination of animal products in the present book/article[2]: Beefburger, Sausage, AFB_1, literature[2]; Hot-dog, Kubeba, Meat, AFB_1 and AFB_2, literature[2]

Aspergillus and ***Penicillium*** Toxins

Citrinin

incidence: 1/15*, conc.: 1.0 μg/kg, sample origin: individual producer or Croatian meat industrial facilities, Croatia, sample year: unknown, country: Croatia[481], *game sausages included rabbit, wild boar, deer, roe deer, and mixed sausages (wild boar, deer, and pork)

- Co-contamination: For detailed information see the article.
- Further contamination of animal products in the present book/article[481]: Meat products, AFB_1, CIT, and OTA, literature[481]; Sausage, AFB_1 and OTA, literature[481]

incidence: 1/25*, conc.: LOQ, sample origin: individual producer or Croatian meat industrial facilities, Croatia, sample year: unknown, country: Croatia[481], *semidry sausages included grill, Kranjska, Slavonian, Zagorska, and homemade garlic sausages

- Co-contamination: 1 sa. co-contaminated with CIT and OTA
- Further contamination of animal products in the present book/article[481]: Meat products, AFB_1, CIT, and OTA, literature[481]; Sausage, AFB_1 and OTA, literature[481]

Ochratoxin A

incidence: 1/1*, conc.: 0.12 μg/kg, sample origin: supermarkets and fast-food chains in Quebec (city), Canada, sample year: September–October 2008, country: Canada[87], *wieners and sausages

incidence: 1/1*, conc.: 0.06 μg/kg, sample origin: supermarkets and fast-food chains in Calgary (city), Canada, sample year: September–October 2009, country: Canada[87], *wieners and sausages

- Co-contamination: not reported
- Further contamination of animal products reported in the present book/article[87]: Beefburger, Butter (peanut butter), Chicken nuggets, Chickenburger, Hot-dog, Ice cream, Meat, Meat products, Pork meat, OTA, literature[87]

incidence: 4/32*, conc. range: ≤1.80 μg/kg, sample origin: Netherlands, sample year: 1990, country: EU[215], *black pudding

- Co-contamination: not reported
- Further contamination of animal products reported in the present book/article[215]: Dairy products, Meat, Pig kidney, Pig liver, Pork meat, OTA, literature[215]

incidence: 72/160*, conc. range: 3–18 μg/kg**, sample origin: plants in northern Italy, sample year: unknown, country: Italy/Croatia[380], *artisanal and industrial sausages (including Bergamasco, Brianza, Cremonese, Mantovano, Milano, Napoli, Ungherese, and Varzi sausages), **in casings of various types of sausages

- Co-contamination: not reported
- Further contamination of animal products reported in the present book/article[380]: not reported

For detailed information see the article.

incidence: 14/15*, conc. range: ≤3.07 µg/kg, sample origin: individual producer or Croatian meat industrial facilities, Croatia, sample year: unknown, country: Croatia[481], *game sausages included rabbit, wild boar, deer, roe deer, and mixed sausages (wild boar, deer, and pork)

- Co-contamination: 1 sa. co-contaminated with AFB_1, CIT, and OTA; 1 sa. co-contaminated with CIT and OTA; 12 sa. contaminated solely with OTA
- Further contamination of animal products in the present book/article[481]: Meat products, AFB_1, CIT, and OTA, literature[481]; Sausage, AFB_1 and CIT, literature[481]

incidence: 21/25*, conc. range: ≤3.28 µg/kg**, sample origin: individual producer or Croatian meat industrial facilities, Croatia, sample year: unknown, country: Croatia[481], *semidry sausages included grill, Kranjska**, Slavonian, Zagorska, and homemade garlic sausages

- Co-contamination: 1 sa. co-contaminated with CIT and OTA; 20 sa. contaminated solely with OTA
- Further contamination of animal products in the present book/article[481]: Meat products, AFB_1, CIT, and OTA, literature[481]; Sausage, AFB_1 and CIT, literature[481]

incidence: 2/35*, conc. range: 0.95–1.52 µg/kg, ∅ conc.: 1.24 µg/kg, sample origin: markets, fairs, or directly from the producing household, Primorje, Zagorje, Medimurje, Slavonia, and Baranja (regions), Croatia, sample year: 2011–2014, country: Croatia/Italy[500], *dry-fermented sausage (Slavonski Kulen)

incidence: 2/28*, conc. range: 1.15–2.64 µg/kg, ∅ conc.: 1.9 µg/kg, sample origin: markets, fairs, or directly from the producing household, Primorje, Zagorje, Medimurje, Slavonia, and Baranja (regions), Croatia, sample year: 2011–2014, country: Croatia/Italy[500], *dry-fermented sausage (Istrian type)

incidence: 5/76*, conc. range: 1.07–2.35 µg/kg, sample origin: markets, fairs, or directly from the producing household, Primorje, Zagorje, Medimurje, Slavonia, and Baranja (regions), Croatia, sample year: 2011–2014, country: Croatia/Italy[500], *dry-fermented sausage (Slavonian type)

incidence: 5/69*, conc. range: 0.99–5.10 µg/kg, sample origin: markets, fairs, or directly from the producing household, Primorje, Zagorje, Medimurje, Slavonia, and Baranja (regions), Croatia, sample year: 2011–2014, country: Croatia/Italy[500], *dry-fermented sausage (northern type)

- Co-contamination: not reported
- Further contamination of animal products reported in the present book/article[500]: Bacon, OTA, literature[500]; Ham, Sausage (liver sausage), AFB_1 and OTA, literature[500]; Sausage, AFB_1, literature[500]

Fusarium Toxins

Zearalenone

incidence: 5/20, conc. range: 2.1–8.9 µg/kg, ∅ conc.: 6.3 µg/kg, sample origin: shops and supermarkets in Alexandria (province), Egypt, sample year: unknown, country: Egypt[141]

- Co-contamination: not reported
- Further contamination of animal products reported in the present book/article[141]: Beef meat, Beefburger, Cheese (Hard Roumy cheese), Cheese (Karish cheese), Meat, Milk, Milk powder, ZEA, literature[141]

Sausage (beef sausage) may contain the following mycotoxins:

Aspergillus and ***Penicillium*** Toxins

Ochratoxin A

incidence: 5/31, conc. range: ≤0.19 µg/kg, sample origin: CMA, Germany, sample year: unknown, country: Germany[1]

- Co-contamination: not reported
- Further contamination of animal products in the present book/article[1]: Beef meat, Joints, Meat products, Pig kidney, Pig liver, Pork meat, Sausage (blood sausage), Sausage (Bologna sausage), Sausage (liver sausage), Sausage (poultry sausage), Sausage (raw sausage), OTA, literature[1]

Sausage (Bergamasco sausage) see **Sausage**

Sausage (blood sausage) may contain the following mycotoxins:

Aspergillus and ***Penicillium*** Toxins

Ochratoxin A

incidence: 44/57, conc. range: ≤3.16 µg/kg, sample origin: CMA, Germany, sample year: unknown, country: Germany[1]

- Co-contamination: not reported
- Further contamination of animal products in the present book/article[1]: Beef meat, Joints, Meat products, Pig kidney, Pig liver, Pork meat, Sausage (beef sausage), Sausage (Bologna sausage), Sausage (liver sausage), Sausage (poultry sausage), Sausage (raw sausage), OTA, literature[1]

Sausage (Bologna sausage) may contain the following mycotoxins:

Aspergillus and ***Penicillium*** Toxins

Ochratoxin A

incidence: 21/45*, conc. range: ≤0.38 µg/kg, sample origin: CMA, Germany, sample year: unknown, country: Germany[1], *Bologna-type sausage

- Co-contamination: not reported
- Further contamination of animal products in the present book/article[1]: Beef meat, Joints, Meat products, Pig kidney, Pig liver, Pork meat, Sausage (beef sausage), Sausage (blood sausage), Sausage (liver sausage), Sausage (poultry sausage), Sausage (raw sausage), OTA, literature[1]

Sausage (Brianza sausage) see **Sausage**

Sausage (Calabrian salami) see **Sausage (salami)**

Sausage (Kranjska sausage) see **Sausage**

Sausage (liver sausage) may contain the following mycotoxins:

Aspergillus Toxins

Aflatoxin B_1

incidence: 3/27, conc. range: 1.18–1.69 µg/kg, sample origin: markets, fairs, or directly from the producing household, Primorje, Zagorje, Medimurje, Slavonia, and Baranja (regions), Croatia, sample year: 2011–2014, country: Croatia/Italy[500]

- Co-contamination: not reported
- Further contamination of animal products reported in the present book/article[500]: Ham, Sausage, AFB_1 and OTA, literature[500]; Bacon, Sausage (liver sausage), OTA, literature[500]

Aspergillus **and** ***Penicillium*** **Toxins**

Ochratoxin A

incidence: 36/53*, conc. range: ≤4.56 µg/kg, sample origin: CMA, Germany, sample year: unknown, country: Germany[1], *liver-type sausage

- Co-contamination: not reported
- Further contamination of animal products in the present book/article[1]: Beef meat, Joints, Meat products, Pig kidney, Pig liver, Pork meat, Sausage (beef sausage), Sausage (blood sausage), Sausage (Bologna sausage), Sausage (poultry sausage), Sausage (raw sausage), OTA, literature[1]

incidence: 3/27, conc. range: 1.12–3.13 µg/kg, sample origin: markets, fairs, or directly from the producing household, Primorje, Zagorje, Medimurje, Slavonia, and Baranja (regions), Croatia, sample year: 2011–2014, country: Croatia/Italy[500]

- Co-contamination: not reported
- Further contamination of animal products reported in the present book/article[500]: Bacon, OTA, literature[500]; Ham, Sausage, AFB_1 and OTA, literature[500]; Sausage (liver sausage), AFB_1, literature[500]

Sausage (Mantovano sausage) see **Sausage**

Sausage (Milano sausage) see **Sausage**

Sausage (Napoli sausage) see **Sausage**

Sausage (Piedmontese salami) see **Sausage (salami)**

Sausage (poultry sausage) may contain the following mycotoxins:

Aspergillus **and** ***Penicillium*** **Toxins**

Ochratoxin A

incidence: 7/40, conc. range: ≤0.03 µg/kg, sample origin: CMA, Germany, sample year: unknown, country: Germany[1]

- Co-contamination: not reported
- Further contamination of animal products in the present book/article[1]: Beef meat, Joints, Meat products, Pig kidney, Pig liver, Pork meat, Sausage (beef sausage), Sausage (blood sausage), Sausage (Bologna sausage), Sausage (liver sausage), Sausage (raw sausage), OTA, literature[1]

Sausage (raw sausage) may contain the following mycotoxins:

Aspergillus **and** ***Penicillium*** **Toxins**

Ochratoxin A

incidence: 28/56, conc. range: ≤0.27 µg/kg, sample origin: CMA, Germany, sample year: unknown, country: Germany[1]

- Co-contamination: not reported
- Further contamination of animal products in the present book/article[1]: Beef meat, Joints, Meat products, Pig kidney, Pig liver, Pork meat, Sausage (beef sausage), Sausage (blood sausage), Sausage (Bologna sausage), Sausage (liver sausage), Sausage (poultry sausage), OTA, literature[1]

Sausage (salami) may contain the following mycotoxins:

Aspergillus **and** ***Penicillium*** **Toxins**

Ochratoxin A

incidence: 4/12, conc. range: ≤0.08 µg/kg, Ø conc.: 0.06 µg/kg, sample origin: retail outlets in the Emilia region, northern Italy, sample year: 2001–2002, country: Italy[152]

- Co-contamination: not reported
- Further contamination of animal products reported in the present book/article[152]: Coppa, Ham, Pig muscle, Sausage (Würstel), OTA, literature[152]

incidence: 14/30, conc. range: 0.006–0.06 µg/kg (9 sa.), 0.06–1 µg/kg (5 sa., maximum: 0.4 µg/kg), sample origin: local retailers in southern Italy, sample year: unknown, country: Italy[381]

- Co-contamination: not reported
- Further contamination of animal products reported in the present book/article[381]: not reported

incidence: 5/50*, conc. range: 0.06–103.69 µg/kg, Ø conc.: 20.91 µg/kg, sample origin: family-run farms, holiday farms, and mountain huts ("malghe") in Veneto (region), Treviso and Vicenza (provinces), Italy, sample year: September–October 2013, country: Italy[399], *artisan salami

- Co-contamination: not reported
- Further contamination of animal products reported in the present book/article[399]: not reported

incidence: 4/52*, conc. range: 0.013–5.66 µg/kg, Ø conc.: 1.86 µg/kg, sample origin: farms and small salami factories in Piedmont (region), Italy, sample year: January 2015–May 2016, country: Italy[423], *Piedmontese salami

incidence: 0/16*, conc.: no contamination, sample origin: farms and small salami factories in Veneto (region), Italy, sample year: January 2015–May 2016, country: Italy[423], *Venetian salami

incidence: 13/50*, conc. range: 0.007–0.62 µg/kg, Ø conc.: 0.18 µg/kg, sample origin: farms and small salami factories in Calabria (region), Italy, sample year: January 2015–May 2016, country: Italy[423], *Calabrian salami

incidence: 5/54*, conc. range: 0.007–1.3 µg/kg, Ø conc.: 0.29 µg/kg, sample origin: farms and small salami factories in Sicily (region), Italy, sample year: January 2015–May 2016, country: Italy[423], *Sicilian salami

- Co-contamination: not reported
- Further contamination of animal products reported in the present book/article[423]: not reported

incidence: 13/133*, conc. range: 1.14–691 µg/kg, Ø conc.: 61.1 µg/kg, sample origin: small-scale and industrial producers in Lombardy and Emilia Romagna (regions), northern Italy, sample year: 2013–2015, country: Italy[473], *traditional salami

- Co-contamination: not reported
- Further contamination of animal products reported in the present book/article[473]: not reported

Sausage (Sicilian salami) see **Sausage (salami)**

Sausage (Slavonian sausage) see **Sausage**

Sausage (Slavonski Kulen sausage) may contain the following mycotoxins:

Aspergillus **Toxins**

Aflatoxin B_1

incidence: 0/9*, conc.: no contamination**, sample origin: small rural household, Croatia, sample year: unknown, country: Croatia[499], *0 months ripening time, **external, interior, and central of the sausage

incidence: 0/9*, conc.: no contamination**, sample origin: small rural household, Croatia, sample year: unknown, country: Croatia[499], *3 months ripening time, **external, interior, and central of the sausage

incidence: 0/9*, conc.: no contamination**, sample origin: small rural household, Croatia, sample year: unknown, country: Croatia[499], *6 months ripening time, **external, interior, and central of the sausage

incidence: 3/9*, conc. range: 1.62–2.78 μg/kg**, sample origin: small rural household, Croatia, sample year: unknown, country: Croatia[499], *9 months ripening time, **external, interior, and central of the sausage

incidence: 7/9*, conc. range: 1.84–14.46 μg/kg**, sample origin: small rural household, Croatia, sample year: unknown, country: Croatia[499], *12 months ripening time, **external, interior, and central of the sausage

- Co-contamination: not reported
- Further contamination of animal products reported in the present book/article[499]: Sausage (Slavonski Kulen sausage), OTA, literature[499]

For detailed information see the article.

Aspergillus **and** ***Penicillium*** **Toxins**

Ochratoxin A

incidence: 0/9*, conc.: no contamination**, sample origin: small rural household, Croatia, sample year: unknown, country: Croatia[499], *0 months ripening time, **external, interior, and central of the sausage

incidence: 0/9*, conc.: no contamination**, sample origin: small rural household, Croatia, sample year: unknown, country: Croatia[499], *3 months ripening time, **external, interior, and central of the sausage

incidence: 2/9*, conc. range: 1.91–2.43 μg/kg**, sample origin: small rural household, Croatia, sample year: unknown, country: Croatia[499], *6 months ripening time, **external, interior, and central of the sausage

incidence: 5/9*, conc. range: 1.37–3.91 μg/kg**, sample origin: small rural household, Croatia, sample year: unknown, country: Croatia[499], *9 months ripening time, **external, interior, and central of the sausage

incidence: 9/9*, conc. range: 1.97–19.84 μg/kg**, sample origin: small rural household, Croatia, sample year: unknown, country: Croatia[499], *12 months ripening time, **external, interior, and central of the sausage

- Co-contamination: not reported
- Further contamination of animal products reported in the present book/article[499]: Sausage (Slavonski Kulen sausage), AFB_1, literature[499]

For detailed information see the article.

incidence: 3/9*, conc. range: 1.82–2.14 μg/kg**, ∅ conc.: 1.98 μg/kg**, sample origin: small rural household, Croatia, sample year: unknown, country: Croatia[501], *30 days of production, **intact casing

incidence: 3/9*, conc. range: 1.84–2.25 μg/kg**, ∅ conc.: 2.03 μg/kg**, sample origin: small rural household, Croatia, sample year: unknown, country: Croatia[501], *60 days of production, **intact casing

incidence: 4/9*, conc. range: 1.90–2.24 μg/kg**, ∅ conc.: 2.04 μg/kg**, sample origin: small rural household, Croatia, sample year: unknown, country: Croatia[501], *90 days of production, **intact casing

incidence: 4/9*, conc. range: 2.10–2.61 μg/kg**, ∅ conc.: 2.36 μg/kg**, sample origin: small rural household, Croatia, sample year: unknown, country: Croatia[501], *120 days of production, **intact casing

incidence: 4/9*, conc. range: 1.83–2.59 μg/kg**, ∅ conc.: 2.17 μg/kg**, sample origin: small rural household, Croatia, sample year: unknown, country: Croatia[501], *150 days of production, **intact casing

incidence: 3/9*, conc. range: 2.14–3.18 μg/kg**, ∅ conc.: 2.54 μg/kg**, sample origin: small rural household, Croatia, sample year: unknown, country: Croatia[501], *180 days of production, **intact casing

- Co-contamination: not reported
- Further contamination of animal products reported in the present book/article[501]: not reported

For detailed information see the article.

Sausage (Ungherese sausage) see Sausage

Sausage (Varzi sausage) see Sausage

Sausage (Venetian salami) see Sausage (salami)

Sausage (Wiener sausage) see Sausage

Sausage (Würstel) may contain the following mycotoxins:

Aspergillus **and** ***Penicillium*** **Toxins**

Ochratoxin A

incidence: 1/12, conc.: ≤0.06 μg/kg, sample origin: retail outlets in the Emilia region, northern Italy, sample year: 2001–2002, country: Italy[152]

- Co-contamination: not reported
- Further contamination of animal products reported in the present book/article[152]: Coppa, Ham, Pig muscle, Sausage (salami), OTA, literature[152]

Sausage (Zagorska sausage) see Sausage

Seafood may contain the following mycotoxins:

Aspergillus **and** ***Penicillium*** **Toxins**

Ochratoxin A

incidence: 1/8*, conc.: 1.9 μg/kg, sample origin: supermarkets in Shanghai (city), China, sample year: unknown, country: China/Belgium[522], *dried seafoods

- Co-contamination: not reported

- Further contamination of animal products reported in the present book/article[522]: Fish, AFB_2, OTA, and ZEA, literature[522]; Seafood, ZEA, literature[522]

Fusarium Toxins

Zearalenone

incidence: 5/8*, conc. range: 3.5–317.3 µg/kg, ∅ conc.: 93.98 µg/kg, sample origin: supermarkets in Shanghai (city), China, sample year: unknown, country: China/Belgium[522], *dried seafoods

- Co-contamination: not reported
- Further contamination of animal products reported in the present book/article[522]: Fish, AFB_2, OTA, and ZEA, literature[522]; Seafood, OTA, literature[522]

Sour cream see Cream (sour)

Sparus aurata see Fish

Sutlac see Dairy desserts

Swine see Pig

Tarkhineh may contain the following mycotoxins:

Aspergillus Toxins

Aflatoxin M_1

incidence: 4/7, conc. range: 0.0223–0.0319 µg/kg, ∅ conc.: 0.0255 µg/kg, sample origin: rural regions in Hamadan, Ilam, Kermanshah, and Kurdistan (provinces), Iran, sample year: winter and summer seasons 2014, country: Iran[328]

- Co-contamination: not reported
- Further contamination of animal products reported in the present book/article[328]: Ayran, Cheese, Kashk, Milk (cow milk), Milk (goat milk), Milk (sheep milk), Yogurt, AFM_1, literature[328]

Tarkhineh (Tarhana) is a fermented mixture of grain and fermented milk or yogurt which is heated and sun-dried (up to 3 days).

Tilapia sp. see Fish

Toddler formula see Baby food

Turkey liver may contain the following mycotoxins:

Aspergillus and *Penicillium* Toxins

Ochratoxin A

incidence: 3/17, conc. range: LOD-0.99 µg/kg (3 sa., maximum: 0.28 µg/kg), sample origin: retail shops, Denmark, sample year: 1995, country: Denmark[102]

- Co-contamination: not reported
- Further contamination of animal products reported in the present book/article[102]: Chicken meat, Duck liver, Duck meat, Goose liver, Goose meat, Pork meat, Turkey meat, OTA, literature[102]

Turkey meat may contain the following mycotoxins:

Aspergillus and *Penicillium* Toxins

Ochratoxin A

incidence: 10/17, conc. range: LOD-0.99 µg/kg (10 sa., maximum: 0.11 µg/kg), sample origin: retail shops, Denmark, sample year: 1995, country: Denmark[102]

- Co-contamination: not reported
- Further contamination of animal products reported in the present book/article[102]: Chicken meat, Duck liver, Duck meat, Goose liver, Goose meat, Pork meat, Turkey liver, OTA, literature[102]

Turkey muscle may contain the following mycotoxins:

Aspergillus and *Penicillium* Toxins

Ochratoxin A

incidence: 9/13, conc. range: LOD-LOQ (5 sa.), >LOQ-≤0.04 µg/kg (4 sa.), ∅ conc.: 0.02 µg/kg, sample origin: supermarkets in Coimbra (city), Portugal, sample year: October 2002–February 2003, country: Spain/Portugal[486]

- Co-contamination: not reported
- Further contamination of animal products in the present book/article[486]: Pig muscle, OTA, literature[486]

Whey see Cheese whey

Wild boar kidney may contain the following mycotoxins:

Aspergillus and *Penicillium* Toxins

Ochratoxin A

incidence: 23/23*, conc. range: 0.1–3.9 µg/kg, ∅ conc.: 1.1 µg/kg, sample origin: Calabria region, southern Italy, sample year: 2009–2010, country: Italy[408], *hunted wild boars

- Co-contamination: not reported
- Further contamination of animal products reported in the present book/article[408]: Wild boar liver, Wild boar muscle, OTA, literature[408]

incidence: 27/39*, conc. range: ≤969 µg/kg, ∅ conc.: 6.64 µg/kg, sample origin: regions in the western Kujawsko-Pomorskie Province, northwestern Poland, sample year: November–December 2006, country: Poland[430], *hunted wild boars

incidence: 16/62*, conc. range: ≤907 µg/kg, ∅ conc.: 2.64 µg/kg, sample origin: regions in the western Kujawsko-Pomorskie Province, northwestern Poland, sample year: 2007, country: Poland[430], *hunted wild boars

- Co-contamination: not reported
- Further contamination of animal products reported in the present book/article[430]: Wild boar serum, OTA, literature[430]

For detailed information see the article.

incidence: 12/12*, conc. range: 0.24–1.90 µg/kg, ∅ conc.: 1.1 µg/kg, sample origin: MSRM, Tyrrhenian coast in the region of Pisa and Lucca (provinces), Italy, sample year: 2014, country: Italy[478], *female wild boars

incidence: 14/14*, conc. range: 0.19–3.23 µg/kg, ∅ conc.: 0.63 µg/kg, sample origin: MSRM, Tyrrhenian coast in the region of Pisa and Lucca (provinces), Italy, sample year: 2014, country: Italy[478], *male wild boars

incidence: 11/11*, conc. range: 0.07–1.72 µg/kg, ∅ conc.: 0.45 µg/kg, sample origin: MSRM, Tyrrhenian coast in the region of Pisa and Lucca (provinces), Italy, sample year: 2015, country: Italy[478], *female wild boars

incidence: 11/11*, conc. range: 0.09–1.15 µg/kg, ∅ conc.: 0.25 µg/kg, sample origin: MSRM, Tyrrhenian coast in the region of Pisa and Lucca (provinces), Italy, sample year: 2015, country: Italy[478], *male wild boars

- Co-contamination: not reported
- Further contamination of animal products reported in the present book/article[478]: Wild boar kidney, Wild boar muscle, OTA, literature[478]

Wild boar liver may contain the following mycotoxins:

Aspergillus and ***Penicillium*** Toxins

Ochratoxin A

incidence: 23/23*, conc. range: 0.1–2.0 µg/kg, ∅ conc.: 0.6 µg/kg, sample origin: Calabria region, southern Italy, sample year: 2009–2010, country: Italy[408], *hunted wild boars

- Co-contamination: not reported
- Further contamination of animal products reported in the present book/article[408]: Wild boar kidney, Wild boar muscle, OTA, literature[408]

incidence: 12/12*, conc. range: 0.06–1.93 µg/kg, ∅ conc.: 0.15 µg/kg, sample origin: MSRM, Tyrrhenian coast in the region of Pisa and Lucca (provinces), Italy, sample year: 2014, country: Italy[478], *female wild boars

incidence: 14/14*, conc. range: 0.04–1.76 µg/kg, ∅ conc.: 0.15 µg/kg, sample origin: MSRM, Tyrrhenian coast in the region of Pisa and Lucca (provinces), Italy, sample year: 2014, country: Italy[478], *male wild boars

incidence: 11/11*, conc. range: 0.05–1.31 µg/kg, ∅ conc.: 0.23 µg/kg, sample origin: MSRM, Tyrrhenian coast in the region of Pisa and Lucca (provinces), Italy, sample year: 2015, country: Italy[478], *female wild boars

incidence: 11/11*, conc. range: 0.02–0.91 µg/kg, ∅ conc.: 0.20 µg/kg, sample origin: MSRM, Tyrrhenian coast in the region of Pisa and Lucca (provinces), Italy, sample year: 2015, country: Italy[478], *male wild boars

- Co-contamination: not reported
- Further contamination of animal products reported in the present book/article[478]: Wild boar kidney, Wild boar muscle, OTA, literature[478]

Wild boar muscle may contain the following mycotoxins:

Aspergillus and ***Penicillium*** Toxins

Ochratoxin A

incidence: 23/23*, conc. range: 0.1–1.3 µg/kg, ∅ conc.: 0.3 µg/kg, sample origin: Calabria region, southern Italy, sample year: 2009–2010, country: Italy[408], *hunted wild boars

- Co-contamination: not reported
- Further contamination of animal products reported in the present book/article[408]: Wild boar kidney, Wild boar liver, OTA, literature[408]

incidence: ?/12*, conc. range: <LOD-0.77 µg/kg, sample origin: MSRM, Tyrrhenian coast in the region of Pisa and Lucca (provinces), Italy, sample year: 2014, country: Italy[478], *female wild boars

incidence: ?/14*, conc. range: <LOD-0.77 µg/kg, sample origin: MSRM, Tyrrhenian coast in the region of Pisa and Lucca (provinces), Italy, sample year: 2014, country: Italy[478], *male wild boars

incidence: 11/11*, conc. range: 0.03–0.38 µg/kg, ∅ conc.: 0.18 µg/kg, sample origin: MSRM, Tyrrhenian coast in the region of Pisa and Lucca (provinces), Italy, sample year: 2015, country: Italy[478], *female wild boars

incidence: 11/11*, conc. range: 0.06–0.50 µg/kg, ∅ conc.: 0.12 µg/kg, sample origin: MSRM, Tyrrhenian coast in the region of Pisa and Lucca (provinces), Italy, sample year: 2015, country: Italy[478], *male wild boars

- Co-contamination: not reported
- Further contamination of animal products reported in the present book/article[478]: Wild boar kidney, Wild boar liver, OTA, literature[478]

Wild boar serum may contain the following mycotoxins:

Aspergillus and ***Penicillium*** Toxins

Ochratoxin A

incidence: 28/39*, conc. range: ≤11.700 µg/l, ∅ conc.: 58.9 µg/l, sample origin: regions in the western Kujawsko-Pomorskie Province, northwestern Poland, sample year: November-December 2006, country: Poland[430], *hunted wild boars

incidence: 30/62*, conc. range: ≤783 µg/l, ∅ conc.: 5.91 µg/l, sample origin: regions in the western Kujawsko-Pomorskie Province, northwestern Poland, sample year: 2007, country: Poland[430], *hunted wild boars

- Co-contamination: not reported
- Further contamination of animal products reported in the present book/article[430]: Wild boar kidney, OTA, literature[430]

For detailed information see the article.

Yogurt may contain the following mycotoxins:

Aspergillus Toxins

Aflatoxin M_1

incidence: 44/54, conc. range: 0.05–0.20 µg/kg (36 sa.), 0.21–0.50 µg/kg (8 sa., maximum: 0.47 µg/kg), ∅ conc.: 0.20 µg/kg, sample origin: unknown, sample year: September 1972–December 1974, country: Germany[28]

- Co-contamination: not reported
- Further contamination of animal products reported in the present book/article[28]: Cheese (Camembert cheese), Cheese (fresh cheese), Cheese (hard cheese), Cheese (processed cheese), Milk (cow milk), Milk powder, AFM_1, literature[28]

incidence: 32/40, conc. range: 0.061–0.090 µg/kg (18 sa.), 0.091–0.120 µg/kg (8 sa.), >0.120 µg/kg (6 sa., maximum: 0.36564 µg/kg), sample origin: Ankara (capital), Turkey, sample year: 2004, country: Turkey[38]

- Co-contamination: not reported
- Further contamination of animal products reported in the present book/article[38]: Cheese, AFM_1, literature[38]

incidence: 6/6, conc. range: >0.003–0.010 µg/kg (2 sa.), 0.011–0.050 µg/kg (4 sa.), sample origin: Ribeirão Preto (municipality), São Paulo (state), Brazil, sample year: unknown, country: Brazil/USA[44]

- Co-contamination: not reported
- Further contamination of animal products reported in the present book/article[44]: Cheese (Minas cheese), Cheese (Prato cheese), Dairy beverages, AFM_1, literature[44]

incidence: 1/1*, conc.: 0.19 µg/kg, sample origin: homemade in Lattakia? (city), Syria, sample year: January 1989, country: Syria[106], *Kashk (made from drained yogurt), it is a secondary dairy product

- Co-contamination: not reported

- Further contamination of animal products reported in the present book/article[106]: not reported

incidence: 4/13*, conc. range: 0.02–0.05 µg/kg (4 sa., maximum: 0.04 µg/kg), sample origin: retail outlets, England and Wales, sample year: unknown, country: UK[131], *plain yogurt

incidence: 2/17*, conc. range: 0.02–0.05 µg/kg (2 sa., maximum: 0.03 µg/kg), sample origin: retail outlets, England and Wales, sample year: unknown, country: UK[131], *fruit yogurt

- Co-contamination: not reported
- Further contamination of animal products reported in the present book/article[131]: Cheese (Cheddar cheese), Cheese (Cheshire cheese), Cheese (Double Gloucester cheese), Cheese (Lancashire cheese), Cheese (Leicester cheese), Cheese (Wensleydale cheese), Milk (cow milk), Milk (infant formula), AFM_1, literature[131]

incidence: 18/25*, conc. range: 0.0025–0.0695 µg/kg, sample origin: Kayseri (city), Turkey, sample year: January–March 2010, country: Turkey/Kyrgyzstan[144], *whole-fat yogurt sa.

incidence: 10/25*, conc. range: 0.0036–0.078 µg/kg, sample origin: Kayseri (city), Turkey, sample year: January–March 2010, country: Turkey/Kyrgyzstan[144], *semi-fat yogurt sa.

- Co-contamination: not reported
- Further contamination of animal products reported in the present book/article[144]: Cheese (Kashar cheese), Cheese (Tulum cheese), Cheese (white cheese), Dairy desserts, Milk, AFM_1, literature[144]

incidence: 2/5, conc. range: 0.0075–0.0254 µg/kg, ∅ conc.: 0.0165 µg/kg, sample origin: retail shops in Marang and Kuala Terengganu (town and city), Malaysia, sample year: July 2013?, country: Malaysia[163]

- Co-contamination: not reported
- Further contamination of animal products reported in the present book/article[163]: Cheese, Dairy beverages, Milk, Milk (cow milk), Milk powder, AFM_1, literature[163]

incidence: 2/72*, conc. range: 0.0251–0.05158 µg/kg, ∅ conc.: 0.03834 µg/kg, sample origin: hypermarkets and supermarkets in 12 main cities of Catalonia (autonomous community), Spain, sample year: June–July 2008, country: Spain[216], *natural yogurt

- Co-contamination: not reported
- Further contamination of animal products reported in the present book/article[216]: Milk (UHT milk), AFM_1, literature[216]

incidence: 64/66*, conc. range: 0.003–0.010 µg/l (56 sa.), 0.011–0.020 µg/l (7 sa.), 0.038 µg/l (1 sa.), sample origin: supermarkets in Florence (city), Italy, sample year: unknown, country: Italy[217], *yogurt made from pasteurized milk

- Co-contamination: not reported
- Further contamination of animal products reported in the present book/article[217]: Milk, Milk (UHT milk), AFM_1, literature[217]

incidence: 13*/17, conc. range: ≤0.270 µg/kg, sample origin: retailers and traders in Dagoretti (low income area), Kenya, sample year: November 2013–October 2014, country: Kenya/Sweden[228], *>0.050 µg/kg

incidence: 12*/21**, conc. range: ≤1.100 µg/kg, sample origin: supermarket in Westlands (middle- to high-income area), Nairobi (capital), Kenya, sample year: November 2013–October 2014, country: Kenya/Sweden[228], *>0.050 µg/kg

- Co-contamination: not reported
- Further contamination of animal products reported in the present book/article[228]: Dairy products, Milk (cow milk), Milk (UHT milk), AFM_1, literature[228]

incidence: 91/114, conc. range: 0.001–0.010 µg/l (61 sa.), >0.010–0.050 µg/l (28 sa.), >0.050 µg/l (2 sa., maximum: 0.49647 µg/l), ∅ conc.: 0.01808 µg/l, sample origin: imported and domestic, Italy, sample year: 1995, country: Italy[234]

- Co-contamination: not reported
- Further contamination of animal products reported in the present book/article[234]: Milk, Milk powder, AFM_1, literature[234]

incidence: 31/60, conc. range: 0.017–0.124 µg/kg, ∅ conc.: 0.045 µg/kg, sample origin: supermarkets, Seoul (capital), Korea, sample year: 1997, country: Korea[241]

- Co-contamination: not reported
- Further contamination of animal products reported in the present book/article[241]: Milk (cow milk), Milk (infant formula), Milk powder, AFM_1, literature[241]

incidence: 3/24*, conc. range: 0.007–0.009 µg/l (2 sa.), 0.044 µg/l (1 sa.), ∅ conc.: 0.02 µg/l, sample origin: supermarkets, convenience stores, and drug stores in counties, Taiwan, sample year: June–August 2002, country: Taiwan[291], *drinking yogurt

- Co-contamination: not reported
- Further contamination of animal products reported in the present book/article[291]: Milk (cow milk), AFM_1, literature[291]

incidence: 6/10, conc. range: 0.0171–0.0358 µg/kg, ∅ conc.: 0.0273 µg/kg, sample origin: rural regions in Hamadan, Ilam, Kermanshah, and Kurdistan (provinces), Iran, sample year: winter and summer seasons 2014, country: Iran[328]

- Co-contamination: not reported
- Further contamination of animal products reported in the present book/article[328]: Ayran, Cheese, Kashk, Milk (cow milk), Milk (goat milk), Milk (sheep milk), Tarkhineh, AFM_1, literature[328]

incidence: 73/120*, conc. range: 0.001–0.010 µg/kg (49 sa.), >0.010–0.050 µg/kg (24 sa., maximum: 0.0321 µg/kg), ∅ conc.: 0.00906 µg/kg, sample origin: drugstores and supermarkets in 4 big Italian cities, Italy, sample year: 1996, country: Italy[356], *sa. partly imported from Belgium and France

- Co-contamination: not reported
- Further contamination of animal products reported in the present book/article[356]: Milk (UHT milk), Milk powder, AFM_1, literature[356]

incidence: 3/4, conc. range: 0.00971–0.06375 µg/kg, ∅ conc.: 0.03392 µg/kg, sample origin: unknown, sample year: unknown, country: USA[373]

- Co-contamination: not reported
- Further contamination of animal products reported in the present book/article[373]: Milk powder, AFM_1, literature[373]

incidence: 2/48*, conc. range: 0.043–0.045 µg/kg, ∅ conc.: 0.044 µg/kg, sample origin: supermarkets in Lisbon (capital), Portugal, sample year: 2001, country: Portugal[386], *commercial natural yogurt

incidence: 16/48*, conc. range: 0.011–0.035 µg/kg (6 sa.), 0.036–0.050 µg/kg (4 sa.), 0.051–0.065 µg/kg (2 sa.), >0.065 µg/kg (4 sa., maximum: 0.098 µg/kg), ∅ conc.: 0.05112 µg/kg, sample origin: supermarkets in Lisbon (capital), Portugal, sample year: 2001, country: Portugal[386], *commercial yogurt with pieces of strawberries

- Co-contamination: not reported
- Further contamination of animal products reported in the present book/article[386]: not reported

incidence: 68/104*, conc. range: 0.001–0.030 μg/kg (40 sa.), 0.031–0.050 μg/kg (16 sa.), 0.051–0.100 μg/kg (12 sa.), sample origin: groceries, markets, and supermarkets in Afyonkarahisar (city), Turkey, sample year: March–July 2005 (sa. analyzation), country: Turkey[387], *ordinary yogurt

incidence: 7/21*, conc. range: 0.001–0.030 μg/kg (3 sa.), 0.031–0.050 μg/kg (2 sa.), 0.051–0.100 μg/kg (2 sa.), sample origin: groceries, markets, and supermarkets in Afyonkarahisar (city), Turkey, sample year: March–July 2005 (sa. analyzation), country: Turkey[387], *fruit yogurt

incidence: 29/52*, conc. range: 0.001–0.030 μg/kg (12 sa.), 0.031–0.050 μg/kg (6 sa.), 0.051–0.100 μg/kg (6 sa.), 0.101–0.500 μg/kg (5 sa.), sample origin: groceries, markets, and supermarkets in Afyonkarahisar (city), Turkey, sample year: March–July 2005 (sa. analyzation), country: Turkey[387], *strained yogurt (Laban = similar to quark)

- Co-contamination: not reported
- Further contamination of animal products reported in the present book/article[387]: not reported

incidence: 5/40*, conc. range: >0.025 μg/l (4 sa.), 0.0617 μg/l (1 sa.), sample origin: Mazandaran (province), northern Iran, sample year: autumn 2009, country: Iran[388], *pasteurized yogurt

incidence: 4/10*, conc. range: >0.025 μg/l (3 sa.), 0.053 μg/l (1 sa.), sample origin: Mazandaran (province), northern Iran, sample year: autumn 2009, country: Iran[388], *local yogurt

- Co-contamination: not reported
- Further contamination of animal products reported in the present book/article[388]: not reported

incidence: 70?/80, conc. range: 0.005–0.025 μg/kg (21 sa.), 0.026–0.050 μg/kg (32 sa.), 0.051–0.100 μg/kg (6 sa.), >0.100 μg/kg (10 sa., maximum: 0.475 μg/kg), ∅ conc.: 0.0661 μg/kg, sample origin: markets in Erzurum (province), Turkey, sample year: September 2007–2008, country: Turkey[389]

incidence: 72/80*, conc. range: 0.005–0.025 μg/kg (28 sa.), 0.026–0.050 μg/kg (33 sa.), 0.051–0.100 μg/kg (8 sa.), >0.100 μg/kg (3 sa., maximum: 0.264 μg/kg), ∅ conc.: 0.0365 μg/kg, sample origin: markets in Erzurum (province), Turkey, sample year: September 2007–2008, country: Turkey[389], *ayran: yogurt drink

- Co-contamination: not reported
- Further contamination of animal products reported in the present book/article[389]: not reported

incidence: 15/18, conc. range: 0.0078–0.0121 μg/kg, ∅ conc.: 0.01025 μg/kg, sample origin: Sistan and Baluchestan (provinces), south-east of Iran, sample year: summer and winter 2015, country: Iran[417]

- Co-contamination: not reported
- Further contamination of animal products reported in the present book/article[417]: Butter, Cheese (white cheese), Milk, AFM_1, literature[417]

For detailed information see the article.

incidence: 20/45, conc. range: 0.051–0.100 μg/kg (6 sa.), 0.101–0.250 μg/kg (10 sa.), 0.251–0.500 μg/kg (4 sa., maximum: 0.360 μg/kg), sample origin: various districts in Burdur (city), Turkey, sample year: 2008, country: Turkey[418]

- Co-contamination: not reported
- Further contamination of animal products reported in the present book/article[418]: Butter, Cheese (white cheese), Ice cream, Milk (cow milk), Milk powder, AFM_1, literature[418]

incidence: 48/60, conc. range: 0.0197–0.3194 μg/l, sample origin: retail markets in Isfahan (province), Iran, sample year: February 2011–2012, country: Iran[424]

- Co-contamination: not reported
- Further contamination of animal products reported in the present book/article[424]: Cheese, Ice cream, AFM_1, literature[424]

For detailed information see the article.

incidence: 21/64, conc. range: 0.005–0.050 μg/l (17 sa.), >0.050 μg/l (4 sa.), sample origin: supermarkets, Lebanon, sample year: unknown, country: Lebanon[432]

- Co-contamination: not reported
- Further contamination of animal products reported in the present book/article[432]: Milk, AFM_1, literature[432]

incidence: 45/68, conc. range: 0.015–0.119 μg/kg, sample origin: retail stores and supermarkets in Esfahan, Shiraz, and Yazd (cities), as well as Tehran (capital), Iran, sample year: winter and summer 2009, country: Iran[435]

- Co-contamination: not reported
- Further contamination of animal products reported in the present book/article[435]: Butter, Cheese (white cheese), Ice cream, Milk, AFM_1, literature[435]

For detailed information see the article.

incidence: 30/61*, conc. range: 0.015–0.102 μg/kg, sample origin: dairy ranches, supermarkets, and retail outlets in Esfahan, Shiraz, and Tabriz (cities) as well as Tehran (capital), Iran, sample year: 2008, country: Iran[440], *industrial yogurt

incidence: 14/60*, conc. range: 0.015–0.036 μg/kg, sample origin: dairy ranches, supermarkets, and retail outlets in Esfahan, Shiraz, and Tabriz (cities) as well as Tehran (capital), Iran, sample year: 2008, country: Iran[440], *traditional yogurt

- Co-contamination: not reported
- Further contamination of animal products reported in the present book/article[440]: Ayran, Cheese (Lighvan cheese), Kashk, Milk (cow milk), Milk (goat milk), Milk (sheep milk), AFM_1, literature[440]

incidence: 8/178, conc. range: 0.050–0.100 μg/kg (8 sa., maximum: 0.08259), ∅ conc.: 0.02710 μg/kg, sample origin: retail stores and supermarkets across China, sample year: 2006–2007, country: China[447]

- Co-contamination: not reported
- Further contamination of animal products reported in the present book/article[447]: Milk (UHT milk), AFM_1, literature[447]

incidence: 59/96, conc. range: <0.050 μg/kg (31 sa.), >0.050–0.150 μg/kg (18 sa.), >0.151 μg/kg (10 sa., maximum: 0.6158 μg/kg), ∅ conc.: 0.1471 μg/kg, sample origin: main districts of Punjab (province), Pakistan, sample year: November 2010–April 2011, country: Pakistan[451]

- Co-contamination: not reported
- Further contamination of animal products reported in the present book/article[451]: Butter, Cheese (cream cheese), Cheese (white cheese), Milk, AFM_1, literature[451]

incidence: 7/9*, conc. range: 0.011–0.050 µg/kg (5 sa.), 0.051–0.100 µg/kg (1 sa.), <0.530 µg/kg (1 sa.), sample origin: grocery stores in Ribeirão Preto (municipality), São Paulo (state), Brazil, sample year: 2010, country: Brazil/USA[453], *whole yogurt

incidence: 15/21*, conc. range: 0.011–0.050 µg/kg (12 sa.), 0.051–0.100 µg/kg (2 sa.), <0.530 µg/kg (1 sa.), sample origin: grocery stores in Ribeirão Preto (municipality), São Paulo (state), Brazil, sample year: 2010, country: Brazil/USA[453], *semi-skimmed yogurt

incidence: 15/23*, conc. range: 0.011–0.050 µg/kg (13 sa.), 0.051–0.100 µg/kg (2 sa.), sample origin: grocery stores in Ribeirão Preto (municipality), São Paulo (state), Brazil, sample year: 2010, country: Brazil/USA[453], *skimmed yogurt

- Co-contamination: not reported
- Further contamination of animal products reported in the present book/article[453]: Cheese (Minas cheese), Dairy beverages, AFM_1, literature[453]

incidence: 6/10, conc. range: 0.0105–0.0130 µg/l, ∅ conc.: 0.0111 µg/l, sample origin: dairy farms, sale points, bazaars, and markets in Faisalabad (city), Pakistan, sample year: unknown, country: Pakistan[482]

- Co-contamination: not reported
- Further contamination of animal products reported in the present book/article[482]: Butter, Milk (cow milk), AFM_1, literature[482]

For detailed information see the article.

incidence: 2/60, conc. range: 0.024–0.028 µg/kg, ∅ conc.: 0.026 µg/kg, sample origin: Corum (province), Turkey, sample year: October 2012–March 2015, country: Turkey[510]

- Co-contamination: not reported
- Further contamination of animal products reported in the present book/article[510]: Ayran, Milk (cow milk), AFM_1, literature[510]

incidence: 80/80*, conc. range: 0.0042–0.010 µg/kg (20 sa.), 0.010–0.025 µg/kg (38 sa.), 0.025–0.050 µg/kg (14 sa.), >0.050 µg/kg (8 sa., maximum: 0.0647 µg/kg), ∅ conc.: 0.0261 µg/kg, sample origin: dairy plants in Guilan (province), northern Iran, sample year: 2009, country: Iran[520], *industrial yogurt

incidence: 40/40*, conc. range: 0.007–0.010 µg/kg (6 sa.), 0.010–0.025 µg/kg (16 sa.), 0.025–0.050 µg/kg (10 sa.), >0.050 µg/kg (8 sa., maximum: 0.0789 µg/kg), ∅ conc.: 0.0329 µg/kg, sample origin: local people in Guilan (province), northern Iran, sample year: 2009, country: Iran[520], *traditional yogurt

- Co-contamination: not reported
- Further contamination of animal products reported in the present book/article[520]: not reported

incidence: 50/50, conc. range: 0.04062–0.07204 µg/kg, ∅ conc.: 0.05528 µg/kg, sample origin: city markets in Şanlıurfa (city) and locally produced dairy products, Turkey, sample year: January–February 2012, country: Turkey[523]

- Co-contamination: not reported
- Further contamination of animal products reported in the present book/article[523]: Cheese (white cheese), Milk (cow milk), Milk (UHT milk), AFM_1, literature[523]

incidence: 22/25, conc. range: 0.0211–0.1376 µg/kg, sample origin: dairy factories in Tehran (capital), Iran, sample year: winter 2008, country: Iran/India[525]

incidence: 13/25, conc. range: 0.022–0.0651 µg/kg, sample origin: dairy factories in Tehran (capital), Iran, sample year: summer 2009, country: Iran/India[525]

- Co-contamination: not reported
- Further contamination of animal products reported in the present book/article[525]: not reported

incidence: 35/162, conc. range: 0.002–0.072 µg/kg, ∅ conc.: 0.017 µg/kg, sample origin: control points such as import, primary storage, and market, Cyprus, sample year: 2004–2013, country: Cyprus[542]

- Co-contamination: not reported
- Further contamination of animal products reported in the present book/article[542]: Cheese, Cheese (Anari cheese), Cheese (white cheese), Ice cream, Milk, Milk powder, AFM_1, literature[542]

incidence: 9/27*, conc. range: 0.02–0.07 µg/kg (8 sa.), 0.07–0.15 µg/kg (1 sa.), ∅ conc.: 0.039 µg/kg, sample origin: supermarkets, South Korea, sample year: October 2013–March 2014, country: South Korea[561], *liquid yogurt sa.

incidence: 6/28*, conc. range: 0.02–0.07 µg/kg (4 sa.), 0.07–0.15 µg/kg (2 sa.), ∅ conc.: 0.062 µg/kg, sample origin: supermarkets, South Korea, sample year: October 2013–March 2014, country: South Korea[561], *plain yogurt sa.

- Co-contamination: not reported
- Further contamination of animal products reported in the present book/article[561]: Cheese, Milk (cow milk), AFM_1, literature[561]

see also **Ayran**, **Cheese**, and **Kashk**

Milk (human breast milk) may contain the following mycotoxins:

Aspergillus Toxins

Aflatoxin B_1

incidence: 7/78, conc. range: 0.056–0.291 µg/l, ∅ conc.: 0.147 µg/l, sample origin: Nabon (canton), Ecuador, sample year: November 2012–January 2013, country: Belgium/Ecuador[280]

- Co-contamination: not reported
- Further contamination of human breast milk reported in the present book/article[280]: Milk (human breast milk), AFM_1, literature[280]

incidence: 41/800, conc. range: 0.150–55.792 µg/l, sample origin: hospitals in Khartoum (capital), Sudan; Embu (town), Kenya; Accra (capital), Ghana, sample year: unknown, country: UK[320]

- Co-contamination: not reported
- Further contamination of human breast milk reported in the present book/article[320]: Milk (human breast milk), AFB_2, AFG_1, AFG_2, AFM_1, and AFM_2, literature[320]; see also Milk (human breast milk), chapter: Further Mycotoxins and Microbial Metabolites

incidence: 20/113, conc. range: 0.05–372 µg/l, sample origin: Njala (town, University Health Center) and Bo (city, Government Hospital) in Southern Province, Sierra Leone, sample year: unknown, country: Sierra Leone/UK[321]

- Co-contamination: 7 sa. co-contaminated with AFB_1 and OTA; 3 sa. co-contaminated with AFB_1, AFM_2, and OTA; 3 sa. co-contaminated with AFB_1, AFM_1, and OTA; 3 sa. co-contaminated with AFB_1, AFL, and OTA; 1 sa. contaminated solely with AFB_1; no further information available
- Further contamination of human breast milk reported in the present book/article[321]: Milk (human breast milk), AFG_1, AFG_2, AFM_1, AFM_2, and OTA, literature[321]; see also Milk (human breast milk), chapter: Further Mycotoxins and Microbial Metabolites

For detailed information see the article.

incidence: 1/231, conc.: 0.0114 µg/l, sample origin: hospitals in Lombardy (administrative region), Italy, sample year: 1999, country: Italy[322]

- Co-contamination: 1 sa. co-contaminated with AFB_1 and AFM_1; no further information available
- Further contamination of human breast milk reported in the present book/article[322]: Milk (human breast milk), AFM_1 and OTA, literature[322]

incidence: 75/75, conc. range: 0.094–0.129 µg/l (17 sa.), 0.130–0.149 µg/l (15 sa.), 0.150–0.199 µg/l (25 sa.), 0.200–0.300 µg/l (10 sa.), >0.300 µg/l (8 sa., maximum: 4.12380 µg/l), sample origin: Hacettepe University Faculty of Medicine in Ankara (capital), Turkey, sample year: October 2007–March 2008, country: Turkey[323]

- Co-contamination: not reported
- Further contamination of human breast milk reported in the present book/article[323]: Milk (human breast milk), AFM_1, literature[323]

incidence: 17/264, conc. range: 0.130–8.218 µg/l, sample origin: Accra (capital), Ghana, sample year: mid-dry season-onset of wet season, country: UK/Ghana/Nigeria[472]

- Co-contamination: not reported
- Further contamination of human breast milk reported in the present book/article[472]: Milk (human breast milk), AFB_2, AFM_1, and AFM_2, literature[472]; see also Milk (human breast milk), chapter: Further Mycotoxins and Microbial Metabolites

For detailed information see the article.

Aflatoxin B_2

incidence: 10/800, conc. range: 0.049–0.623 µg/l, sample origin: hospitals in Khartoum (capital), Sudan; Embu (town), Kenya; Accra (capital), Ghana, sample year: unknown, country: UK[320]

- Co-contamination: not reported
- Further contamination of human breast milk reported in the present book/article[320]: Milk (human breast milk), AFB_1, AFG_1, AFG_2, AFM_1, and AFM_2, literature[320]; see also Milk (human breast milk), chapter: Further Mycotoxins and Microbial Metabolites

incidence: 2/224, conc. range: 0.005 µg/l, ∅ conc.: 0.005 µg/l, sample origin: Federal District, Brazil, sample year: May 2011–February 2012, country: Brazil[395]

- Co-contamination: not reported
- Further contamination of human breast milk reported in the present book/article[395]: not reported

incidence: 2/264, conc. range: 0.049–0.050 µg/l, ∅ conc.: 0.0495 µg/l, sample origin: Accra (capital), Ghana, sample year: mid-dry season-onset of wet season, country: UK/Ghana/Nigeria[472]

- Co-contamination: not reported
- Further contamination of human breast milk reported in the present book/article[472]: Milk (human breast milk), AFB_1, AFM_1, and AFM_2, literature[472]; see also Milk (human breast milk), chapter: Further Mycotoxins and Microbial Metabolites

For detailed information see the article.

Aflatoxin G_1

incidence: 4/800, conc. range: 1.890–5.180 µg/l, sample origin: hospitals in Khartoum (capital), Sudan; Embu (town), Kenya; Accra (capital), Ghana, sample year: unknown, country: UK[320]

- Co-contamination: not reported
- Further contamination of human breast milk reported in the present book/article[320]: Milk (human breast milk), AFB_1, AFB_2, AFG_2, AFM_1, and AFM_2, literature[320]; see also Milk (human breast milk), chapter: Further Mycotoxins and Microbial Metabolites

incidence: 22/113, conc. range: 0.005–139 µg/l, sample origin: Njala (town, University Health Center) and Bo (city, Government Hospital) in Southern Province, Sierra Leone, sample year: unknown, country: Sierra Leone/UK[321]

- Co-contamination: 6 sa. co-contaminated with AFG_1 and OTA; no further information available
- Further contamination of human breast milk reported in the present book/article[321]: Milk (human breast milk), AFB_1, AFG_2, AFM_1, AFM_2, and OTA, literature[321]; see also Milk (human breast milk), chapter: Further Mycotoxins and Microbial Metabolites

For detailed information see the article.

incidence: 3/5, conc. range: pr*, sample origin: Gambia, sample year: unknown, country: USA/France/UK[324], *18, 70, and 114 pg AFG_1

- Co-contamination: 3 sa. co-contaminated with AFG_1 and AFM_1
- Further contamination of human breast milk reported in the present book/article[324]: Milk (human breast milk), AFM_1, literature[324]

Aflatoxin G_2

incidence: 3/800, conc. range: 0.010–0.087 µg/l, sample origin: hospitals in Khartoum (capital), Sudan; Embu (town), Kenya; Accra (capital), Ghana, sample year: unknown, country: UK[320]

- Co-contamination: not reported
- Further contamination of human breast milk reported in the present book/article[320]: Milk (human breast milk), AFB_1, AFB_2, AFG_1, AFM_1, and AFM_2, literature[320]; see also Milk (human breast milk), chapter: Further Mycotoxins and Microbial Metabolites

incidence: 25/113, conc. range: 0.003–366 µg/l, sample origin: Njala (town, University Health Center) and Bo (city, Government Hospital) in Southern Province, Sierra Leone, sample year: unknown, country: Sierra Leone/UK[321]

- Co-contamination: 7 sa. co-contaminated with AFG_2 and OTA; 1 sa. contaminated solely with AFG_2; no further information available
- Further contamination of human breast milk reported in the present book/article[321]: Milk (human breast milk), AFB_1, AFG_1, AFM_1, AFM_2, and OTA, literature[321]; see also Milk (human breast milk), chapter: Further Mycotoxins and Microbial Metabolites

For detailed information see the article.

Aflatoxin M_1

incidence: 6/12*, conc. range: 0.00883–0.0152 µg/kg, sample origin: Kuwait, sample year: January 2005– March 2007?, country: Kuwait[41]

- Co-contamination: not reported
- Further contamination of human breast milk and animal products reported in the present book/article[41]: Cheese, Milk (cow milk), Milk (UHT milk), Milk powder, AFM_1, literature[41]

incidence: 2/10, conc. range: 0.5–5 µg/l, ∅ conc.: 2.75 µg/l, sample origin: Egypt, sample year: 1999–2000, country: Egypt[138]

- Co-contamination: not reported
- Further contamination of human breast milk and animal products reported in the present book/article[138]: Cheese (Domiati cheese), Milk (cow milk), Milk powder, AFM_1, literature[138]; Cheese (Ras cheese), AFB_1, AFG_1, and AFM_1, literature[138]; Milk (human breast milk), OTA, literature[138]

incidence: 3/62, conc. range: <0.05 µg/l (2 sa.), 0.625 µg/l (1 sa.), sample origin: hospitals in Yaounde (capital), Cameroon, sample year: 1991–1995, country: Cameroon[156]

- Co-contamination: not reported
- Further contamination of human breast milk and animal products reported in the present book/article[156]: Eggs, AF/S, literature[156]; Milk (cow milk), AFM_1, literature[156]

incidence: 43/178*, conc. range: 0.003–0.067 µg/l, ∅ conc.: 0.009 µg/l, sample origin: Salesi Children's Hospital in Ancona (city) and Umberto I University Hospital in Rome (capital), Italy, sample year: 2011–2013, country: Italy[160], *healthy breastfeeding women (control)

incidence: 104/275*, conc. range: 0.003–0.340 µg/l, ∅ conc.: 0.012 µg/l, sample origin: Salesi Children's Hospital in Ancona (city) and Umberto I University Hospital in Rome (capital), Italy, sample year: 2011–2013, country: Italy[160], *breastfeeding women with celiac disease

- Co-contamination: not reported
- Further contamination of human breast milk reported in the present book/article[160]: Milk (human breast milk), OTA and ZEA, literature[160]

incidence: 5/94, conc. range: 0.013–0.025 µg/kg, ∅ conc.: 0.018 µg/kg, sample origin: hospitals in Londrina (city), Brazil, sample year: June–August 2013, country: Brazil/Japan[161]

- Co-contamination: not reported
- Further contamination of human breast milk reported in the present book/article[161]: Milk powder, AFM_1, literature[161]

incidence: 10/64*, conc. range: 0.3–1.3 µg/l, ∅ conc.: 0.77 µg/l, sample origin: Corniche Maternity Hospital and Al-Nahyan Clinic for Maternity and Childhood in Abu Dhabi (capital), UAE, sample year: unknown, country: UAE/UK[230], TLC tested

incidence: 15/15*, conc. range: 0.007–0.023 µg/l, sample origin: Corniche Maternity Hospital and Al-Nahyan Clinic for Maternity and Childhood in Abu Dhabi (capital), UAE, sample year: unknown, country: UAE/UK[230], HPLC tested

- Co-contamination: not reported
- Further contamination of human breast milk and animal products reported in the present book/article[230]: Milk (camel milk), AFM_1, literature[230]

For detailed information see the article.

incidence: 11/78, conc. range: LOD-LOQ (1 sa.), 0.053–0.458 µg/l (10 sa.), ∅ conc.: 0.216 µg/l, sample origin: Nabon (canton), Ecuador, sample year: November 2012–January 2013, country: Belgium/Ecuador[280]

- Co-contamination: not reported
- Further contamination of human breast milk reported in the present book/article[280]: Milk (human breast milk), AFB_1, literature[280]

incidence: 51/94, conc. range: ≤0.05 µg/kg (15 sa.), ≤0.1 µg/kg (4 sa.), ≤0.25 µg/kg (9 sa.), ≤0.5 µg/kg (10 sa.), ≤1.0 µg/kg (7 sa.), ≤2.0 µg/kg (4 sa.), >2 µg/kg (2 sa., maximum: 2.561 µg/kg), sample origin: pediatric departments of Omdurman and Bahry Teaching Hospital, Sudan, sample year: unknown, country: Sudan[287]

- Co-contamination: not reported
- Further contamination of human breast milk reported in the present book/article[287]: not reported

incidence: 8/61*, conc. range: 0.00510–0.00690 µg/l, ∅ conc.: 0.00568 µg/l, sample origin: Istanbul (capital), Turkey, sample year: 2006–2007, country: Turkey[299]

- Co-contamination: not reported
- Further contamination of human breast milk and animal products reported in the present book/article[299]: Milk (cow milk), AFM_1, literature[299]

incidence: 80/80*, conc. range: 0.00917–0.13718 µg/kg, ∅ conc.: 0.06778 µg/kg, sample origin: Jordan, sample year: February–July 2011, country: Jordan[313]

- Co-contamination: not reported
- Further contamination of human breast milk and animal products reported in the present book/article[313]: Milk (buttermilk), Milk (cow milk), AFM_1, literature[313]

incidence: 121/800, conc. range: 0.005–1.379 µg/l, sample origin: hospitals in Khartoum (capital), Sudan; Embu (town), Kenya; Accra (capital), Ghana, sample year: unknown, country: UK[320]

- Co-contamination: not reported

- Further contamination of human breast milk reported in the present book/article[320]: Milk (human breast milk), AFB_1, AFB_2, AFG_1, AFG_2, and AFM_2, literature[320]; see also Milk (human breast milk), chapter: Further Mycotoxins and Microbial Metabolites

incidence: 35/113, conc. range: 0.2–99 µg/l, sample origin: Njala (town, University Health Center) and Bo (city, Government Hospital) in Southern Province, Sierra Leone, sample year: unknown, country: Sierra Leone/UK[321]

- Co-contamination: 10 sa. co-contaminated with AFM_1 and OTA; 5 sa. co-contaminated with AFL, AFM_1, and OTA; 5 sa. co-contaminated with AFM_1, AFM_2, and OTA; 3 sa. co-contaminated with AFB_1, AFM_1, and OTA; no further information available
- Further contamination of human breast milk reported in the present book/article[321]: Milk (human breast milk), AFB_1, AFG_1, AFG_2, AFM_2, and OTA, literature[321]; see also Milk (human breast milk), chapter: Further Mycotoxins and Microbial Metabolites

For detailed information see the article.

incidence: 1/231, conc.: 0.194 µg/l, sample origin: hospitals in Lombardy (administrative region), Italy, sample year: 1999, country: Italy[322]

- Co-contamination: 1 sa. co-contaminated with AFB_1 and AFM_1; no further information available
- Further contamination of human breast milk reported in the present book/article[322]: Milk (human breast milk), AFB_1 and OTA, literature[322]

incidence: 75/75, conc. range: 0.060–0.079 µg/l (13 sa.), 0.080–0.099 µg/l (24 sa.), 0.100–>0.150 µg/l (38 sa., maximum: 0.29999 µg/l), sample origin: Hacettepe University Faculty of Medicine in Ankara (capital), Turkey, sample year: October 2007–March 2008, country: Turkey[323]

- Co-contamination: not reported
- Further contamination of human breast milk reported in the present book/article[323]: Milk (human breast milk), AFB_1, literature[323]

incidence: 5/5, conc. range: ≤0.0014 µg/l, sample origin: Gambia, sample year: unknown, country: USA/France/UK[324]

- Co-contamination: 3 sa. co-contaminated with AFG_1 and AFM_1; 2 sa. contaminated solely with AFM_1
- Further contamination of human breast milk reported in the present book/article[324]: Milk (human breast milk), AFG_1, literature[324]

incidence: 4/15, conc. range: 0.028–0.078 µg/l, ∅ conc.: 0.0575 µg/l, sample origin: Nursing Mothers' Association of Australia, sample year: June–August 1991, country: Australia/UK[325]

incidence: 3/14, conc. range: 0.048–0.063 µg/l, ∅ conc.: 0.0563 µg/l, sample origin: Middle Eastern ethnic women, sample year: June–August 1991, country: Australia/UK[325]

incidence: 7/44, conc. range: <0.10–1.031 µg/l, sample origin: Royal Women's Hospital in Melbourne (city), Victoria (state), Australia, sample year: November–December 1992, country: Australia/UK[325]

incidence: 5/11, conc. range: 0.039–1.736 µg/l, ∅ conc.: 0.776 µg/l, sample origin: Thai mothers, sample year: 1991, country: Australia/UK[325]

- Co-contamination: not reported
- Further contamination of human breast milk reported in the present book/article[325]: not reported

incidence: 248/443, conc. range: 0.0042–0.889 µg/l, sample origin: New El-Qalyub Hospital, Qalyubiyah (governorate), Egypt, sample year: 2003, country: Finland/UK/Egypt[329]

- Co-contamination: not reported
- Further contamination of human breast milk reported in the present book/article[329]: not reported

For detailed information see the article.

incidence: 6/64, conc. range: 0.0141–0.0505 µg/l, ∅ conc.: 0.03365 µg/l, sample origin: agricultural region in northern Zimbabwe, sample year: March–April (rainy season), country: France/Zimbabwe[330]

- Co-contamination: not reported
- Further contamination of human breast milk reported in the present book/article[330]: not reported

incidence: 443/445, conc. range: 0.002–3.000 µg/l, sample origin: donors from different countries at the Corniche Hospital or the Al-Nahyan Clinic for Maternity and Childhood in Abu Dhabi, UAE, sample year: November 1989–June 1990, country: UAE/UK[331]

- Co-contamination: not reported
- Further contamination of human breast milk reported in the present book/article[331]: not reported

For detailed information see the article.

incidence: 138/388, conc. range: 0.0056–5.131 µg/l, sample origin: New El-Qalyub Hospital in Qalyubiyah (governorate), Egypt, sample year: May–September 2003, country: Finland/UK/Egypt[332]

- Co-contamination: not reported
- Further contamination of human breast milk reported in the present book/article[332]: not reported

For detailed information see the article.

incidence: 1/22, conc.: 0.02 µg/l, sample origin: Human Milk Bank of the Southern Regional Hospital in São Paulo (city), Brazil, sample year: winter 2001–2002, country: Brazil[333]

incidence: 0/28, conc.: no contamination, sample origin: Human Milk Bank of the Southern Regional Hospital in São Paulo (city), Brazil, sample year: summer 2002, country: Brazil[333]

- Co-contamination: not reported
- Further contamination of human breast milk reported in the present book/article[333]: Milk (human breast milk), OTA, literature[333]

incidence: 129/140*, conc. range: ≤3.400 µg/l, sample origin: Al Ain Hospital, UAE, sample year: January 1999–December 2000, country: UAE[334], *lactating mothers from UAE and other countries

- Co-contamination: not reported
- Further contamination of human breast milk reported in the present book/article[334]: not reported

incidence: 66/120, conc. range: 0.2–2.09 µg/l, sample origin: Abo El-Rish Hospital—Cairo (capital) University, Egypt, sample year: April 2000–May 2002, country: Egypt[335]

- Co-contamination: not reported
- Further contamination of human breast milk reported in the present book/article[335]: Milk (human breast milk), OTA, literature[335]

For detailed information see the article.

incidence: 26/99, conc. range: 0.005–0.064 µg/l, ∅ conc.: 0.019 µg/l, sample origin: Children's Emergency Hospital, Khartoum (capital), Sudan, sample year: unknown, country: UK/Sudan[336]

- Co-contamination: 13 sa. co-contaminated with AFM_1 and AFM_2; 13 sa. contaminated solely with AFM_1
- Further contamination of human breast milk reported in the present book/article[336]: Milk (human breast milk), AFM_2, literature[336]

incidence: 4/82, conc. range: 0.005–0.010 µg/l (1 sa.), >0.010–0.050 µg/l (2 sa.), 0.140 µg/l (1 sa.), ∅ conc.: 0.05535 µg/l, sample origin: Italian hospitals, Italy, sample year: January–December 2006, country: Italy[337]

- Co-contamination: not reported
- Further contamination of human breast milk reported in the present book/article[337]: Milk (human breast milk), OTA, literature[337]

incidence: 157/160, conc. range: 0.0003–0.0267 µg/kg, sample origin: clinics in Tehran (capital), Iran, sample year: May–September 2006, country: Iran[339]

- Co-contamination: not reported
- Further contamination of human breast milk reported in the present book/article[339]: not reported

incidence: 87/125, conc. range: 0.005–0.025 µg/l (22 sa.), >0.025–0.050 µg/l (18 sa.), >0.050–0.100 µg/l (26 sa.), >0.100–0.200 µg/l (17 sa.), >0.200 µg/l (4 sa., maximum: 0.3286 µg/l), sample origin: Minoufiya (governorate), Egypt, sample year: March–August 2010, country: Egypt[340]

- Co-contamination: not reported
- Further contamination of human breast milk and animal products reported in the present book/article[340]: Milk powder, AFM_1, literature[340]

incidence: 20*/182, conc. range: 0.0051–0.0081 µg/l, ∅ conc.: 0.00696 µg/l, sample origin: urban (91 sa.) and rural (91 sa.) areas of Tabriz (city), Iran, sample year: March–April 2007, country: Iran[341], *only breast milk sa. from rural areas contaminated

- Co-contamination: not reported
- Further contamination of human breast milk reported in the present book/article[341]: not reported

incidence: 1/136, conc.: 0.01358 µg/l, sample origin: hospitals in Sari (city), Iran, sample year: May–August 2011, country: Iran[390]

- Co-contamination: not reported
- Further contamination of human breast milk reported in the present book/article[390]: Milk (human breast milk), OTA, literature[390]

incidence: 15/15, conc. range: 0.00465–0.010 µg/l (1 sa.), 0.011–0.025 µg/l (6 sa.), >0.025 µg/l (8 sa., maximum: 0.09214 µg/l), ∅ conc.: 0.03500 µg/l, sample origin: Ogun Central (senatorial district), Nigeria, sample year: unknown, country: Nigeria/Italy/Ireland[391]

incidence: 12/18, conc. range: 0.001–0.010 µg/l (12 sa., maximum: 0.01858 µg/l?), sample origin: Ogun East (senatorial district), Nigeria, sample year: unknown, country: Nigeria/Italy/Ireland[391]

incidence: 14/17, conc. range: 0.001–0.010 µg/l (14 sa., maximum: 0.00540 µg/l), sample origin: Ogun West (senatorial district), Nigeria, sample year: unknown, country: Nigeria/Italy/Ireland[391]

- Co-contamination: not reported
- Further contamination of human breast milk reported in the present book/article[391]: not reported

incidence: 1/75, conc.: <LOQ, sample origin: Ilishan Remo (town), Ogun State, Nigeria, sample year: January–February 2016, country: Austria/Nigeria/Cameroon/Germany/USA[403]

- Co-contamination: not reported
- Further contamination of human breast milk reported in the present book/article[403]: Milk (human breast milk), BEA, ENB, and OTA, literature[403]

incidence: 6/29*, conc. range: 0.0020–0.0045 µg/l, ∅ conc.: 0.00306 µg/l, sample origin: Erzurum (city), Turkey, sample year: December 2008–April 2009, country: Turkey[405], *sa. from mothers with no habit of moldy cheese consumption

incidence: 12/44*, conc. range: 0.0013–0.0060 µg/l, ∅ conc.: 0.00297 µg/l, sample origin: Erzurum (city), Turkey, sample year: December 2008–April 2009, country: Turkey[405], *sa. from mothers with the habit of moldy cheese consumption

- Co-contamination: not reported
- Further contamination of human breast milk reported in the present book/article[405]: not reported

incidence: 20/20, conc. range: 0.00375–0.02085 µg/l, ∅ conc.: 0.01278 µg/l, sample origin: Human Milk Bank of the Hospital General "Dr. Maximiliano Ruiz Castañeda in Naucalpan de Juarez (city), Estado de México (state), México, sample year: January and February (winter) 2014, country: Mexico[409]

incidence: 35/42, conc. range: 0.00301–0.03424 µg/l, ∅ conc.: 0.01209 µg/l, sample origin: Human Milk Bank of the Hospital General "Dr. Maximiliano Ruiz Castañeda" in Naucalpan de Juarez (city), Estado de México (state), México, sample year: March, April, and May (spring) 2014, country: Mexico[409]

incidence: 45/50, conc. range: 0.00321–0.01890 µg/l, ∅ conc.: 0.00791 µg/l, sample origin: Human Milk Bank of the Hospital General "Dr. Maximiliano Ruiz Castañeda" in Naucalpan de Juarez (city), Estado de México (state), México, sample year: June, July, and August (summer) 2014, country: Mexico[409]

- Co-contamination: not reported
- Further contamination of human breast milk reported in the present book/article[409]: not reported

For detailed information see the article.

incidence: 45/50, conc. range: 0.0009–0.0185 µg/l, ∅ conc.: 0.0052 µg/l, sample origin: The Mercy Hospital Foundation, Colombia, sample year: May–September 2013, country: Colombia[413]

- Co-contamination: not reported
- Further contamination of human breast milk reported in the present book/article[413]: not reported

For detailed information see the article.

incidence: 1/80, conc.: 0.0068 µg/l, sample origin: Isfahan (city), Iran, sample year: January–February 2011, country: Iran[420]

- Co-contamination: not reported
- Further contamination of human breast milk reported in the present book/article[420]: not reported

incidence: 2/100, conc. range: 0.003–0.008 µg/l (2 sa.), ∅ conc.: 0.0055 µg/l, sample origin: Human Milk Bank of Hospital das Clínicas—USP in Ribeirão Preto (city), São Paulo (state), Brazil, sample year: June 2011–August 2012, country: Brazil/USA[454]

- Co-contamination: not reported
- Further contamination of human breast milk reported in the present book/article[454]: Milk (Human breast milk), OTA, literature[454]

incidence: 4/160, conc. range: 0.01419–0.02389 µg/l, ∅ conc.: 0.01996 µg/l, sample origin: urban area, Iran, sample year: winter 2016, country: Iran[459]

incidence: 35/90, conc. range: 0.01107–0.03925 µg/l, ∅ conc.: 0.02112 µg/l, sample origin: rural area, Iran, sample year: winter 2016, country: Iran[459]

- Co-contamination: not reported
- Further contamination of human breast milk reported in the present book/article[459]: not reported

For detailed information see the article.

incidence: 6/10, conc. range: 0.005–0.05 µg/kg (6 sa., maximum: 0.022 µg/kg), ∅ conc.: 0.01 µg/kg, sample origin: Serbia, sample year: February–May 2013, country: Serbia[468]

- Co-contamination: not reported
- Further contamination of human breast milk and animal products reported in the present book/article[468]: Milk (cow milk), Milk (donkey milk), Milk (goat milk), Milk (UHT milk), AFM_1, literature[468]

incidence: 35/40, conc. range: 0.0113–0.0278 µg/l, ∅ conc.: 0.0158 µg/l, sample origin: Department of Pediatrics, Unit of Neonatology in Harran University Hospital in Şanliurfa (city), Turkey, sample year: December 2014, country: Turkey[470]

incidence: 31/34, conc. range: 0.0096–0.080 µg/l, ∅ conc.: 0.0229 µg/l, sample origin: Department of Pediatrics, Unit of Neonatology in Harran University Hospital in Şanliurfa (city), Turkey, sample year: June 2015, country: Turkey[470]

- Co-contamination: not reported
- Further contamination of human breast milk reported in the present book/article[470]: not reported

incidence: 59/264, conc. range: 0.002–1.816 µg/l, sample origin: Accra (capital), Ghana, sample year: mid-dry season-onset of wet season, country: UK/Ghana/Nigeria[472]

- Co-contamination: not reported
- Further contamination of human breast milk reported in the present book/article[472]: Milk (human breast milk), AFB_1, AFB_2, and AFM_2, literature[472]; see also Milk (human breast milk), chapter: Further Mycotoxins and Microbial Metabolites

For detailed information see the article.

incidence: 143/143*, conc. range: 0.01–0.55 µg/l, sample origin: Kikelelwa and Mbomai (villages) in Rombo (district), northern Tanzania, sample year: November 2011–February 2012, country: Belgium/Tanzania[479], *lactation stage: month 1

incidence: 121/121*, conc. range: 0.01–0.47 µg/l, sample origin: Kikelelwa and Mbomai (villages) in Rombo (district), northern Tanzania, sample year: November 2011–February 2012, country: Belgium/Tanzania[479], *lactation stage: month 3

incidence: 118/118*, conc. range: 0.01–0.34 µg/l, sample origin: Kikelelwa and Mbomai (villages) in Rombo (district), northern Tanzania, sample year: November 2011–February 2012, country: Belgium/Tanzania[479], *lactation stage: month 5

- Co-contamination: not reported
- Further contamination of human breast milk reported in the present book/article[479]: not reported

For detailed information see the article.

incidence: 85/85, conc. range: 0.002–0.010 µg/l, ∅ conc.: 0.00591 µg/l, sample origin: Ilam (province), western Iran, sample year: April–October 2014, country: Iran[485]

- Co-contamination: not reported
- Further contamination of human breast milk reported in the present book/article[485]: not reported

incidence: 24/87, conc. range: 0.00013–0.00491 µg/l, ∅ conc.: 0.00056 µg/l, sample origin: rural health centers in Khorrambid (city), Fars (province), Iran, sample year: June–July 2011, country: Iran[503]

- Co-contamination: not reported
- Further contamination of human breast milk reported in the present book/article[503]: not reported

incidence: 41/43*, conc. range: 0.00022–0.00789 µg/l, ∅ conc.: 0.0041 µg/l, sample origin: different areas in Lebanon, sample year: November 2015–December 2016, country: Lebanon[539], *sa. collected in fall and winter

incidence: 63/68*, conc. range: 0.00023–0.00779 µg/l, ∅ conc.: 0.0050 µg/l, sample origin: different areas in Lebanon, sample year: November 2015–December 2016, country: Lebanon[539], *sa. collected in spring and summer

- Co-contamination: not reported
- Further contamination of human breast milk reported in the present book/article[539]: not reported

incidence: 27/34*, conc. range: 0.00536–0.02844 µg/l, ∅ conc.: 0.00836 µg/l, sample origin: rural and urban areas of the Famagusta District, Cyprus, sample year: March 2015–May 2015, country: Turkey[540], *sa. of mothers with no habit of moldy food consumption

incidence: 13/16*, conc. range: 0.00601–0.01055 µg/l, ∅ conc.: 0.00834 µg/l, sample origin: rural and urban areas of the Famagusta District, Cyprus, sample year: March 2015–May 2015, country: Turkey[540], *sa. of mothers with habits of moldy food consumption

incidence: 35/44*, conc. range: 0.00537–0.02844 µg/l, ∅ conc.: 0.00842 µg/l, sample origin: rural and urban areas of the Famagusta District, Cyprus, sample year: March 2015–May 2015, country: Turkey[540], *nonsmoker

incidence: 4/5*, conc. range: 0.00711–0.00973 µg/l, ∅ conc.: 0.00849 µg/l, sample origin: rural and urban areas of the Famagusta District, Cyprus, sample year: March 2015–May 2015, country: Turkey[540], *smoker

- Co-contamination: not reported
- Further contamination of human breast milk reported in the present book/article[540]: not reported

incidence: 98/150, conc. range: 0.2–19.0 µg/l, ∅ conc.: 7.1 µg/l, sample origin: lactation clinic in the Cairo University Children's Hospital, Egypt, sample year: January 2008–July 2009, country: Egypt[541]

- Co-contamination: not reported
- Further contamination of human breast milk reported in the present book/article[541]: not reported

Aflatoxin M_2

incidence: 103/800, conc. range: 0.003–6.368 µg/l, sample origin: hospitals in Khartoum (capital), Sudan, Embu (town), Kenya, Accra (capital) Ghana, sample year: unknown, country: UK[320]

- Co-contamination: not reported
- Further contamination of human breast milk reported in the present book/article[320]: Milk (human breast milk), AFB_1, AFB_2, AFG_1, AFG_2, and AFM_1, literature[320]; see also Milk (human breast milk), chapter: Further Mycotoxins and Microbial Metabolites

incidence: 70/113, conc. range: 0.07–77.5 µg/l, sample origin: Njala (town, University Health Center) and Bo (city, Government Hospital) in Southern Province, Sierra Leone, sample year: unknown, country: Sierra Leone/UK[321]

- Co-contamination: 25 sa. co-contaminated with AFM_2 and OTA; 9 sa. co-contaminated with AFL, AFM_2, and OTA; 5 sa. co-contaminated with AFM_1, AFM_2, and OTA; 3 sa. co-contaminated with AFB_1, AFM_2, and OTA; 9 sa. contaminated solely with AFM_2; no further information available
- Further contamination of human breast milk reported in the present book/article[321]: Milk (human breast milk), AFB_1, AFG_1, AFG_2, AFM_1, and OTA, literature[321]; see also Milk (human breast milk), chapter: Further Mycotoxins and Microbial Metabolites

For detailed information see the article.

incidence: 24/99, conc. range: 0.003–0.020 µg/l, ∅ conc.: 0.0122 µg/l, sample origin: Children's Emergency Hospital, Khartoum (capital), Sudan, sample year: unknown, country: UK/Sudan[336]

- Co-contamination: 13 sa. co-contaminated with AFM_1 and AFM_2; 11 sa. contaminated solely with AFM_2
- Further contamination of human breast milk reported in the present book/article[336]: Milk (human breast milk), AFM_1, literature[336]

incidence: 18/264, conc. range: 0.016–2.075 µg/l, sample origin: Accra (capital), Ghana, sample year: mid-dry season-onset of wet season, country: UK/Ghana/Nigeria[472]

- Co-contamination: not reported
- Further contamination of human breast milk reported in the present book/article[472]: Milk (human breast milk), AFB_1, AFB_2, and AFM_1, literature[472]; see also Milk (human breast milk), chapter: Further Mycotoxins and Microbial Metabolites

For detailed information see the article.

Aspergillus and ***Penicillium*** Toxins

Ochratoxin A

incidence: 4/40, conc. range: 0.005–0.014* µg/kg, ∅ conc.: 0.0725 µg/kg, sample origin: north of the Alps, sample year: 1992–1993, country: Switzerland[104], *diabetic woman

- Co-contamination: not reported
- Further contamination of human breast milk reported in the present book/article[104]: not reported

incidence: 3/10, conc.: 3–15 µg/l, ∅ conc.: 8.87 µg/kg, sample origin: Egypt, sample year: 1999–2000, country: Egypt[138]

- Co-contamination: not reported
- Further contamination of human breast milk and animal products reported in the present book/article[138]: Cheese (Domiati cheese), Milk (cow milk), Milk (human breast milk), Milk powder, AFM_1, literature[138]; Cheese (Ras cheese), AFB_1, AFG_1, and AFM_1, literature[138]

incidence: 1/178*, conc.: 0.017–0.056 µg/l, sample origin: Salesi Children's Hospital in Ancona (city) and Umberto I University Hospital in Rome (capital), Italy, sample year: 2011–2013, country: Italy[160], *healthy breastfeeding women (control)

incidence: 6/275*, conc. range: 0.017–0.123 µg/l, sample origin: Salesi Children's Hospital in Ancona (city) and Umberto I University Hospital in Rome (capital), Italy, sample year: 2011–2013, country: Italy[160], *breastfeeding women with celiac disease

- Co-contamination: not reported
- Further contamination of human breast milk reported in the present book/article[160]: Milk (human breast milk), AFM_1 and ZEA, literature[160]

incidence: 23/40, conc. range: 0.010–0.040 µg/l, sample origin: Västerbotten (county), middle north of Sweden, sample year: late 1990–early 1991, country: Sweden[315]

- Co-contamination: 4 sa. co-contaminated with OTA and OT-α; 36 sa. contaminated solely with OTA
- Further contamination of human breast milk and animal products reported in the present book/article[315]: Milk (cow milk), OTA, literature[315]; see also Milk (human breast milk), chapter: Further Mycotoxins and Microbial Metabolites

incidence: 40/113, conc. range: 0.2–337 µg/l, sample origin: Njala (town, University Health Center) and Bo (city, Government Hospital) in Southern Province, Sierra Leone, sample year: unknown, country: Sierra Leone/UK[321]

- Co-contamination: 25 sa. co-contaminated with AFM_2 and OTA; 16 sa. co-contaminated with AFL and OTA; 10 sa. co-contaminated with AFM_1 and OTA; 9 sa. co-contaminated with AFL, AFM_2, and OTA; 7 sa. co-contaminated with AFB_1 and OTA; 6 sa. co-contaminated with AFG_1 and OTA; 5 sa. co-contaminated with AFL, AFM_1, and OTA; 5 sa. co-contaminated with AFM_1, AFM_2, and OTA; 3 sa. co-contaminated with AFB_1, AFL, and OTA; 3 sa. co-contaminated with AFB_1, AFM_1, and OTA; 3 sa. co-contaminated with AFB_1, AFM_2, and OTA; 4 sa. contaminated solely with OTA; no further information available
- Further contamination of human breast milk reported in the present book/article[321]: Milk (human breast milk), AFB_1, AFG_1, AFG_2, AFM_1, and AFM_2, literature[321]; see also Milk (human breast milk), chapter: Further Mycotoxins and Microbial Metabolites

For detailed information see the article.

incidence: 198/231, conc. range: 0.001–0.057 µg/l, sample origin: hospitals in Lombardy (administrative region), Italy, sample year: 1999, country: Italy[322]

- Co-contamination: not reported
- Further contamination of human breast milk reported in the present book/article[322]: Milk (human breast milk), AFB_1 and AFM_1, literature[322]

incidence: 0/22, conc.: no contamination, sample origin: Human Milk Bank of the Southern Regional Hospital in São Paulo (city), Brazil, sample year: winter 2001–2002, country: Brazil[333]

incidence: 2/28, conc. range: 0.01–0.02 µg/l, ∅ conc.: 0.015 µg/l, sample origin: Human Milk Bank of the Southern Regional Hospital in São Paulo (city), Brazil, sample year: summer 2002, country: Brazil[333]

- Co-contamination: not reported
- Further contamination of human breast milk reported in the present book/article[333]: Milk (human breast milk), AFM_1, literature[333]

incidence: 43/120, conc. range: 5.07–45.01 µg/l, sample origin: Abo El-Rish Hospital—Cairo (capital) University, Egypt, sample year: April 2000–May 2002, country: Egypt[335]

- Co-contamination: not reported
- Further contamination of human breast milk reported in the present book/article[335]: Milk (human breast milk), AFM_1, literature[335]

For detailed information see the article.

incidence: 61/82, conc. range: 0.005–0.010 µg/l (28 sa.), >0.010–0.050 µg/l (27 sa.), >0.050 µg/l (6 sa., maximum: 0.405 µg/l), ∅ conc.: 0.03043 µg/l, sample origin: Italian hospitals, Italy, sample year: January–December 2006, country: Italy[337]

- Co-contamination: not reported
- Further contamination of human breast milk reported in the present book/article[337]: Milk (human breast milk), AFM_1, literature[337]

incidence: 7/48, conc. range: 0.010–0.040 µg/l (6 sa.), 0.056 µg/l (1 sa.), ∅ conc.: 0.0274 µg/l, sample origin: Bodø (town and municipality, north coast), Norway, sample year: May–August 1994, country: Norway[342]

incidence: 11/19, conc. range: 0.010–0.040 µg/l (6 sa.), >0.040 µg/l (5 sa., maximum: 0.102 µg/l), ∅ conc.: 0.0490 µg/l, sample origin: Trondheim (city and municipality, middle coast), Norway, sample year: May–August 1994, country: Norway[342]

incidence: 20/48, conc. range: 0.010–0.040 µg/l (12 sa.), >0.040 µg/l (8 sa., maximum: 0.130 µg/l), ∅ conc.: 0.0432 µg/l, sample origin: Elverum (city and municipality, south-east inland), Norway, sample year: May–August 1994, country: Norway[342]

- Co-contamination: not reported
- Further contamination of human breast milk reported in the present book/article[342]: not reported

incidence: 2/100, conc. range: 3.0–3.6 µg/l, ∅ conc.: 3.3 µg/l, sample origin: Australia, sample year: unknown, country: Australia[343]

- Co-contamination: not reported
- Further contamination of human breast milk reported in the present book/article[343]: not reported

incidence: 2/13, conc. range: 0.024–0.030 µg/l, ∅ conc.: 0.027 µg/l, sample origin: Munich (city), Germany, sample year: 1986, country: Germany[344]

incidence: 2/23, conc. range: 0.017–0.024 µg/l, ∅ conc.: 0.0205 µg/l, sample origin: Bayreuth (city), Germany, sample year: 1986, country: Germany[344]

- Co-contamination: not reported
- Further contamination of human breast milk reported in the present book/article[344]: not reported

incidence: 17/80, conc. range: 0.010–0.182 µg/l, ∅ conc.: 0.030 µg/l, sample origin: Oslo area (capital), Norway, sample year: July 1995–September 1996, country: Norway[345]

- Co-contamination: not reported
- Further contamination of human breast milk reported in the present book/article[345]: not reported

For detailed information see the article

incidence: 41/52*, conc. range: >0.001–0.002 µg/l (4 sa.), >0.002–0.004 µg/l (17 sa.), >0.004–0.005 µg/l (3 sa.), >0.005–0.010 µg/l (6 sa.), >0.010–0.020 µg/l (5 sa.), >0.020–0.030 µg/l (4 sa.), >0.030 µg/l (2 sa., maximum: 0.0751 µg/l), ∅ conc.: 0.010 µg/l, sample origin: Department of Obstetrics and Gynaecology of the "G. da Saliceto" Hospital in Piacenza (city), Italy, sample year: January–June 2007, country: Italy[346], *breast milk sa. from Italian and non-Italian women

- Co-contamination: not reported
- Further contamination of human breast milk reported in the present book/article[346]: not reported

For detailed information see the article.

incidence: 75/75, conc. range: 0.600–1.499 µg/l (28 sa.), 1.500–2.499 µg/l (31 sa.), 2.500–2.999 µg/l (3 sa.), 3.000–3.499 µg/l (3 sa.), >3.500 µg/l (10 sa., maximum: 13.11130 µg/l), sample origin: Department of Pediatrics, Hacettepe University Faculty of Medicine, Ankara (capital), Turkey, sample year: October 2007–March 2008, country: Turkey[347]

- Co-contamination: not reported
- Further contamination of human breast milk reported in the present book/article[347]: not reported

incidence: 23/76, conc. range: 0.0048–0.0144 µg/l (14 sa.), ≤0.0603 µg/l (9 sa.), sample origin: Clinic of Children and Adolescents in Martin (city), Slovakia, sample year: March–August 2007, country: Slovakia[349]

- Co-contamination: not reported
- Further contamination of human breast milk reported in the present book/article[349]: not reported

incidence: 38/92, conc. range: 0.22–1 µg/l (13 sa.), 1–2 µg/l (12 sa.), 2–3 µg/l (8 sa.), 3–5 µg/l (3 sa.), >5 µg/l (2 sa., maximum: 7.63 µg/l), sample origin: Maternity Ward of the Hospital of Kaposvár (city), Hungary, sample year: August-October 1992, country: Hungary[350]

- Co-contamination: not reported
- Further contamination of human breast milk reported in the present book/article[350]: not reported

incidence: 5/13, conc. range: 0.0053–0.017 µg/l, ∅ conc.: 0.01026 µg/l, sample origin: Mother and Child Institute, Warsaw (capital), Poland, sample year: October 1998–April 1999, country: Poland[351]

- Co-contamination: not reported
- Further contamination of human breast milk reported in the present book/article[351]: not reported

incidence: 9/9, conc. range: 0.071–0.184 µg/l*, ∅ conc.: 0.117 µg/l, sample origin: Higueras Hospital in Talcahuano (city), Chile, sample year: October 2008–January 2009, country: Chile/Germany[352], *values of the first day of lactation

- Co-contamination: not reported
- Further contamination of human breast milk reported in the present book/article[352]: not reported

For detailed information see the article.

incidence: 2/136, conc.: 0.0090–0.140 µg/l, ∅ conc.: 0.115 µg/l, sample origin: hospitals in Sari (city), Iran, sample year: May–August 2011, country: Iran[390]

- Co-contamination: not reported
- Further contamination of human breast milk reported in the present book/article[390]: Milk (human breast milk), AFM_1, literature[390]

incidence: 11/75, conc.: <LOQ, sample origin: Ilishan Remo (town), Ogun State, Nigeria, sample year: January–February 2016, country: Austria/Nigeria/Cameroon/Germany/USA[403]

- Co-contamination: not reported
- Further contamination of human breast milk reported in the present book/article[403]: Milk (human breast milk), AFM_1, BEA, and ENB, literature[403]

incidence: 84/84, conc.: <1 µg/l (17 sa.), 1–2 µg/l (36 sa.), 2–3 µg/l (17 sa.), 3–4 µg/l (7 sa.), 4–5 µg/l (2 sa.), >5 µg/l (5 sa., maximum:

7.34 μg/l), sample origin: clinic in Jiroft city, Kerman (province), south-east of Iran, sample year: April 2016–January 2017, country: Iran[416]

- Co-contamination: not reported
- Further contamination of human breast milk reported in the present book/article[416]: not reported

incidence: 84/87, conc.: 0.0016–0.060 μg/l, ∅ conc.: 0.02457 μg/l, sample origin: health centers of Khorrambid Town, Fars (province), Iran, sample year: June–July 2011, country: Iran[426]

- Co-contamination: not reported
- Further contamination of human breast milk reported in the present book/article[426]: not reported

incidence: 40/78, conc.: >0.005–0.0187* μg/l, sample origin: Poland, sample year: unknown, country: Poland[429], *colostrum sa.

- Co-contamination: not reported
- Further contamination of human breast milk reported in the present book/article[429]: not reported

incidence: 34/100, conc. range: LOD-<0.0003 μg/l (2 sa.), 0.0003–<0.0008 μg/l (3 sa.), 0.0008–0.005 μg/l (27 sa.), 0.011–0.015 μg/l (1 sa.), 0.021 μg/l (1 sa.), ∅ conc.: 0.004 μg/l, sample origin: Human Milk Bank of Hospital das Clínicas—USP in Ribeirão Preto (city), São Paulo (state), Brazil, sample year: June 2011–August 2012, country: Brazil/USA[454]

- Co-contamination: not reported
- Further contamination of human breast milk reported in the present book/article[454]: Milk (Human breast milk), AFM_1, literature[454]

incidence: 40/50*, conc. range: LOD-0.186 μg/l, sample origin: Talcahuano (city) in the Bio-Bio region, Chile, sample year: November 2008–February 2010, country: Germany/Chile[491], *colostrum and mature breast milk sa.

- Co-contamination: not reported
- Further contamination of human breast milk reported in the present book/article[491]: not reported

For detailed information see the article.

Fusarium Toxins

Depsipeptides

Beauvericin

incidence: 42/75, conc. range: <LOQ-0.019 μg/l, ∅ conc.: 0.010 μg/l, sample origin: Ilishan Remo (town), Ogun State, Nigeria, sample year: January–February 2016, country: Austria/Nigeria/Cameroon/Germany/USA[403]

- Co-contamination: not reported
- Further contamination of human breast milk reported in the present book/article[403]: Milk (human breast milk), AFM_1, ENB, and OTA, literature[403]

Enniatin A

incidence: 2/35, conc. range: 20.1–25.2 μg/l, ∅ conc.: 22.65 μg/l, sample origin: València (city), Spain, sample year: January–December 2012, country: Spain/Czech Republic/France[513]

- Co-contamination: 1 sa. co-contaminated with ENA, ENA_1, ENB, ENB_1, and NEO; 1 sa. co-contaminated with ENA, ENA_1, ENB, and ENB_1
- Further contamination of human breast milk reported in the present book/article[513]: Milk (human breast milk), ENA_1, ENB, ENB_1, HT-2, NEO, NIV, and ZEA, literature[513]; see also Milk (human breast milk), chapter: Further Mycotoxins and Microbial Metabolites

Enniatin A_1

incidence: 2/35, conc. range: 42.1–51.1 μg/l, ∅ conc.: 46.6 μg/l, sample origin: València (city), Spain, sample year: January–December 2012, country: Spain/Czech Republic/France[513]

- Co-contamination: 1 sa. co-contaminated with ENA, ENA_1, ENB, ENB_1, and NEO; 1 sa. co-contaminated with ENA, ENA_1, ENB, and ENB_1
- Further contamination of human breast milk reported in the present book/article[513]: Milk (human breast milk), ENA, ENB, ENB_1, HT-2, NEO, NIV, and ZEA, literature[513]; see also Milk (human breast milk), chapter: Further Mycotoxins and Microbial Metabolites

Enniatin B

incidence: 7/75, conc. range: <LOQ-0.009 μg/l, ∅ conc.: 0.005 μg/l, sample origin: Ilishan Remo (town), Ogun State, Nigeria, sample year: January–February 2016, country: Austria/Nigeria/Cameroon/Germany/USA[403]

- Co-contamination: not reported
- Further contamination of human breast milk reported in the present book/article[403]: Milk (human breast milk), AFM_1, BEA, and OTA, literature[403]

incidence: 2/35, conc. range: 99.8–110.3 μg/l, ∅ conc.: 105.1 μg/l, sample origin: València (city), Spain, sample year: January–December 2012, country: Spain/Czech Republic/France[513]

- Co-contamination: 1 sa. co-contaminated with ENA, ENA_1, ENB, ENB_1, and NEO; 1 sa. co-contaminated with ENA, ENA_1, ENB, and ENB_1
- Further contamination of human breast milk reported in the present book/article[513]: Milk (human breast milk), ENA, ENA_1, ENB_1, HT-2, NEO, NIV, and ZEA, literature[513]; see also Milk (human breast milk), chapter: Further Mycotoxins and Microbial Metabolites

Enniatin B_1

incidence: 2/35, conc. range: 90.7–101.1 μg/l, ∅ conc.: 95.9 μg/l, sample origin: València (city), Spain, sample year: January–December 2012, country: Spain/Czech Republic/France[513]

- Co-contamination: 1 sa. co-contaminated with ENA, ENA_1, ENB, ENB_1, and NEO; 1 sa. co-contaminated with ENA, ENA_1, ENB, and ENB_1
- Further contamination of human breast milk reported in the present book/article[513]: Milk (human breast milk), ENA, ENA_1, ENB, HT-2, NEO, NIV, and ZEA, literature[513]; see also Milk (human breast milk), chapter: Further Mycotoxins and Microbial Metabolites

Fumonisin/s

Fumonisin B_1

incidence: 58/131, conc. range: 0.00657–0.47105 μg/l, sample origin: Kikelewa and Mbomai (villages) in Rombo (district), northern Tanzania, sample year: unknown, country: Belgium/Tanzania[477]

- Co-contamination: not reported
- Further contamination of human breast milk reported in the present book/article[477]: not reported

Type A Trichothecenes

HT-2 Toxin

incidence: 10/35, conc. range: 12.2–62.5 µg/l, ∅ conc.: 36.49 µg/l, sample origin: València (city), Spain, sample year: January–December 2012, country: Spain/Czech Republic/France[513]

- Co-contamination: 4 sa. co-contaminated with HT-2, NEO, and ZEN; 2 sa. co-contaminated with HT-2 and ZEN; 1 sa. co-contaminated with HT-2 and NEO; 3 sa. contaminated solely with HT-2
- Further contamination of human breast milk reported in the present book/article[513]: Milk (human breast milk), ENA, ENA_1, ENB, ENB_1, NEO, NIV, and ZEA, literature[513]; see also Milk (human breast milk), chapter: Further Mycotoxins and Microbial Metabolites

Neosolaniol

incidence: 7/35, conc. range: 10.3–36.9 µg/l, ∅ conc.: 17.86 µg/l, sample origin: València (city), Spain, sample year: January–December 2012, country: Spain/Czech Republic/France[513]

- Co-contamination: 4 sa. co-contaminated with HT-2, NEO, and ZEN; 1 sa. co-contaminated with ENA, ENA_1, ENB, ENB_1, and NEO; 1 sa. co-contaminated with HT-2 and NEO; 1 sa. co-contaminated with NEO, NIV, α-ZEL, and ß-ZEL
- Further contamination of human breast milk reported in the present book/article[513]: Milk (human breast milk), ENA, ENA_1, ENB, ENB_1, HT-2, NIV, and ZEA, literature[513]; see also Milk (human breast milk), chapter: Further Mycotoxins and Microbial Metabolites

Type B Trichothecenes

Nivalenol

incidence: 3/35, conc. range: 53.1–69.7 µg/l, ∅ conc.: 63.3 µg/l, sample origin: València (city), Spain, sample year: January–December 2012, country: Spain/Czech Republic/France[513]

- Co-contamination: 1 sa. co-contaminated with NEO, NIV, α-ZEL, and ß-ZEL; 1 sa. co-contaminated with NIV and ZEA; 1 sa. contaminated solely with NIV
- Further contamination of human breast milk reported in the present book/article[513]: Milk (human breast milk), ENA, ENA_1, ENB, ENB_1, HT-2, NEO, and ZEA, literature[513]; see also Milk (human breast milk), chapter: Further Mycotoxins and Microbial Metabolites

Other *Fusarium* Toxins

Zearalenone

incidence: 15/178*, conc. range: 2.0–22 µg/l, ∅ conc.: 2.7 µg/l, sample origin: Salesi Children's Hospital in Ancona (city) and Umberto I University Hospital in Rome (capital), Italy, sample year: 2011–2013, country: Italy[160], *healthy breastfeeding women (control)

incidence: 12/275*, conc. range: 2.0–17 µg/l, ∅ conc.: 2.1 µg/l, sample origin: Salesi Children's Hospital in Ancona (city) and Umberto I University Hospital in Rome (capital), Italy, sample year: 2011–2013, country: Italy[160], *breastfeeding women with celiac disease

- Co-contamination: not reported
- Further contamination of human breast milk reported in the present book/article[160]: Milk (human breast milk), AFM_1 and OTA, literature[160]

incidence: 47/47, conc. range: 0.26–1.78 µg/l, ∅ conc.: 1.13 µg/l, sample origin: Caucasian mothers living in the Naples (city) countryside area, Italy, sample year: unknown, country: Italy[274]

- Co-contamination: not reported
- Further contamination of human breast milk reported in the present book/article[274]: not reported

incidence: 13/35, conc. range: 2.1–14.3 µg/l, ∅ conc.: 9.36 µg/l, sample origin: València (city), Spain, sample year: January–December 2012, country: Spain/Czech Republic/France[513]

- Co-contamination: 4 sa. co-contaminated with HT-2, NEO, and ZEA; 2 sa. co-contaminated with HT-2 and ZEA; 1 sa. co-contaminated with NIV and ZEA; 6 sa. contaminated solely with ZEA
- Further contamination of human breast milk reported in the present book/article[513]: Milk (human breast milk), ENA, ENA_1, ENB, ENB_1, HT-2, NEO, and NIV, literature[513]; see also Milk (human breast milk), chapter: Further Mycotoxins and Microbial Metabolites

Further Mycotoxins and Microbial Metabolites

Aquatic food (literature[536]) contained further mycotoxins/microbial metabolites:
Tentoxin

Chicken gizzard (literature[326]) contained further mycotoxins/microbial metabolites:
α-Zearalanol, ß-zearalanol, and α-zearalenol

Chicken heart (literature[326]) contained further mycotoxins/microbial metabolites:
α-Zearalanol, ß-zearalanol, and α-zearalenol

Chicken liver (literature[326]) contained further mycotoxins/microbial metabolites:
α-Zearalanol, ß-zearalanol, and α-zearalenol

Dairy products (literature[457]) contained further mycotoxins/microbial metabolites:
Ochratoxin-α

Eggs (literature[326]) contained further mycotoxins/microbial metabolites:
α-Zearalanol, ß-zearalanol, α-zearalenol, and ß-zearalenol

Eggs (literature[535]) contained further mycotoxins/microbial metabolites:
ß-Zearalenol

Eggs (literature[536]) contained further mycotoxins/microbial metabolites:
Tentoxin

Milk (cow milk) (literature[231]) contained further mycotoxins/microbial metabolites:
Aflatoxicol

Milk (cow milk) (literature[456]) contained further mycotoxins/microbial metabolites:
α-Zearalenol

Milk (human breast milk) (literature[315]) contained further mycotoxins/microbial metabolites:
Ochratoxin-α

Milk (human breast milk) (literature[320]) contained further mycotoxins/microbial metabolites:
Aflatoxicol

Milk (human breast milk) (literature[321]) contained further mycotoxins/microbial metabolites:
Aflatoxicol

Milk (human breast milk) (literature[472]) contained further mycotoxins/microbial metabolites:
Aflatoxicol

Milk (human breast milk) (literature[513]) contained further mycotoxins/microbial metabolites:
α-Zearalenol and ß-zearalenol

Milk (infant formula) (literature[164]) contained further mycotoxins/microbial metabolites:
α-Zearalenol and ß-zearalenol

Milk powder (literature[456]) contained further mycotoxins/microbial metabolites:
α-Zearalenol

For incidence and concentration as well as possible co-contamination of these compounds see the corresponding article.

M. Weidenbörner, *Mycotoxins in Animal Products*, https://doi.org/10.1007/978-3-030-30919-0_2

Tables

Mycotoxins and Their Animal Product Spectrum as well as Human Breast Milk

Table 1. ***Alternaria*** **toxins in animal products**

Altenuene	
	Aquatic food
Alternariol	
	Insects
Alternariol methyl ether	
	Aquatic food; eggs; insects
Tenuazonic acid	
	Aquatic food; eggs; milk; milk powder

Table 2. ***Aspergillus*** **toxins in animal products**

Aflatoxin B_1	
	Beefburger; butter (cocoa butter); butter (nut butter); butter (peanut butter); cheese; cheese (Bhutanese cheese); cheese (cottage cheese); cheese (Ras cheese); chicken gizzard; chicken kidney; chicken liver; chicken muscle; dairy products; eggs; fish; fish oil; ham; hot-dog; infant food; kashk; kubeba; meat; meat products; milk (cow milk); milk (human breast milk); milk (infant formula); milk powder; pig liver; pig muscle; sausage; sausage (liver sausage); sausage (Slavonski Kulen sausage)
Aflatoxin B_2	
	Butter (cocoa butter); butter (nut butter); butter (peanut butter); cheese; cheese (Bhutanese cheese); cheese (cottage cheese); eggs; fish; hot-dog; infant food; kubeba; meat; milk (human breast milk); milk powder; pig liver; sausage
Aflatoxin G_1	
	Butter (nut butter); butter (peanut butter); cheese; cheese (Ras cheese); dairy products; eggs; fish; infant food; meat; milk (human breast milk)
Aflatoxin G_2	
	Butter (nut butter); butter (peanut butter); eggs; fish; infant food; milk (human breast milk)
Aflatoxin M_1	
	Ayran; butter; butter (peanut butter); cheese; cheese (Anari cheese); cheese (blue-veined cheese); cheese (Brie cheese); cheese (Butter cheese); cheese (Camembert cheese); cheese (Cecil cheese); cheese (Cheddar cheese); cheese (Cheshire cheese); cheese (Civil cheese); cheese (cow cheese); cheese (cream cheese); cheese (Domiati cheese); cheese (Double Gloucester cheese); cheese (Edam cheese); cheese (Feta cheese); cheese (fondue cheese); cheese (fresh cheese); cheese (goat cheese); cheese (goat-sheep cheese); cheese (Gouda cheese); cheese (Grana Padano cheese); cheese (Graviera cheese); cheese (Gruyere cheese); cheese (hard cheese); cheese (Havarti cheese); cheese (Kashar cheese); cheese (Lancashire cheese); cheese (Leicester cheese); cheese (Lighvan cheese); cheese (Lor cheese); cheese (Maribo cheese); cheese (Minas cheese); cheese (Mozzarella cheese); cheese (Münster cheese); cheese (Oaxaca cheese); cheese (Parmesan cheese); cheese (Prato cheese); cheese (processed cheese); cheese (Ras cheese); cheese (Samsoe cheese); cheese (semihard cheese); cheese (sheep cheese); cheese (soft cheese); cheese (Surk cheese); cheese (Tulum cheese); cheese (Urfa cheese); cheese (Van otlu cheese); cheese (Wensleydale cheese); cheese (white cheese); cheese curd; cheese whey; cream; dairy beverages; dairy desserts; dairy products; food; ice cream; kashk; milk; milk (buffalo milk); milk (buttermilk); milk (camel milk); milk (cow milk); milk (donkey milk); milk (goat milk); milk (human breast milk); milk (infant formula); milk (sheep milk); milk (UHT milk); milk powder; pig liver; poultry kidney; tarkhineh; yogurt

(continued)

M. Weidenbörner, *Mycotoxins in Animal Products*, https://doi.org/10.1007/978-3-030-30919-0_3

Table 2. (continued)

Aflatoxin M_2
Butter (peanut butter); cheese; cheese (Oaxaca cheese); cream (sour); milk (cow milk); milk (human breast milk); milk (UHT milk); milk powder

Aflatoxins (M_1, M_2)
Milk (cow milk)

Aflatoxin/s
Beef meat; beefburger; butter (peanut butter); butter (sesame butter); chicken kidney; chicken liver; chicken meat; chicken muscle; cream; custard powder; eggs; fish; fish products; fish/shrimp; insects; meat; meat products; milk; milk (infant formula); milk powder

Sterigmatocystin
Cheese; cheese (hard cheese); cheese (Ras cheese); eggs; pig muscle

Table 3. ***Aspergillus*** **and** ***Penicillium*** **toxins in animal products**

Citrinin
Cheese; cheese (Bhutanese cheese); chicken muscle; meat products; pig kidney; pig liver; pig serum; sausage

Cyclopiazonic acid
Cheese (Camembert cheese); milk (cow milk)

Ochratoxin A
Bacon; beef meat; beefburger; butter (cocoa butter); butter (nut butter); butter (peanut butter); cheese; cheese (Bhutanese cheese); cheese (fresh cheese); cheese (Gorgonzola cheese); cheese (hard cheese); cheese (Roquefort cheese); cheese (semihard cheese); chicken gizzard; chicken kidney; chicken liver; chicken meat; chicken muscle; chicken nuggets; chickenburger; coppa; cow liver; dairy products; duck liver; duck meat; eggs; fish; goose liver; goose meat; ham; hot-dog; ice cream; joints; meat; meat products; milk; milk (cow milk); milk (human breast milk); milk (infant formula); milk powder; pâté; pig blood; pig kidney; pig liver; pig muscle; pig plasma; pig serum; pork meat; poultry kidney; poultry meat; sausage; sausage (beef sausage); sausage (blood sausage); sausage (Bologna sausage); sausage (liver sausage); sausage (poultry sausage); sausage (raw sausage); sausage (salami sausage); sausage (Slavonski Kulen sausage); sausage (Würstel); seafood; turkey liver; turkey meat; turkey muscle; wild boar kidney; wild boar liver; wild boar muscle; wild boar serum

Ochratoxins (A, B)
Cheese (cottage cheese)

Patulin
Cheese (goat cheese); cheese (hard cheese); cheese (semihard cheese); pig liver

Penicillic acid
Cheese (blue-veined cheese); cheese (goat cheese); cheese (hard cheese); pig serum

Table 4. ***Fusarium*** **toxins in animal products**

Depsipeptides
Beauvericin
Aquatic food; eggs; milk; milk (human breast milk)

Enniatin A
Eggs; milk (human breast milk)

Enniatin A_1
Eggs; fish; milk (human breast milk)

Enniatin B
Aquatic food; eggs; fish; meat; milk; milk (human breast milk); milk (sheep milk)

Enniatin B_1
Aquatic food; eggs; fish; milk (human breast milk)

Fumonisin/s
Fumonisin B_1
Dairy products; milk (cow milk); milk (human breast milk); milk (infant formula)

Type A trichothecenes
HT-2 toxin
Insects; milk (human breast milk)

Neosolaniol
Milk (human breast milk)

T-2 toxin
Chicken muscle; milk (infant formula); pig fat; pig muscle

Type B trichothecenes
Deoxynivalenol
Custard powder; eggs; pig fat; pig serum

(continued)

Table 4. (continued)

3-Acetyldeoxynivalenol
Eggs
15-Acetyldeoxynivalenol
Eggs
Fusarenon-X (4-acetylnivalenol)
Fish
Nivalenol
Insects; milk (human breast milk)
Other *Fusarium* Toxins
Zearalenone
Beef meat; beefburger; cheese (Hard Roumy cheese); cheese (Karish cheese); chicken gizzard; chicken heart; chicken liver; chicken meat; eggs; fish; insects; meat; milk; milk (cow milk); milk (human breast milk); milk (infant formula); milk powder; pig serum; pork meat; sausage; seafood

Table 5. ***Penicillium*** **toxins in animal products**

Isofumigaclavine A
Cheese (blue-veined cheese); cheese (Stilton cheese)
Isofumigaclavine B
Cheese (blue-veined cheese)
Mycophenolic acid
Cheese; cheese (Bleu des Causses cheese); cheese (blue-veined cheese); cheese (Gorgonzola cheese); cheese (hard cheese); cheese (mold-ripened); cheese (Roquefort cheese)
Penitrem A
Cheese (cream cheese); pig serum
Roquefortine C
Cheese (blue-veined cheese); cheese (Danablu cheese); cheese (Edelpilzkäse); cheese (Gorgonzola cheese); cheese (mold ripened); cheese (Roquefort cheese); cheese (Stilton cheese); insects

Animal Products as well as Human Breast Milk and Their Mycotoxins

Aquatic food

Alternaria toxins
Altenuene, alternariol methyl ether, tenuazonic acid

Fusarium toxins
Depsipeptides: beauvericin, enniatin B, enniatin B_1

Ayran

Aspergillus toxins
Aflatoxin/s: aflatoxin M_1

Bacon

Aspergillus and ***Penicillium*** toxins
Ochratoxin A

Beef meat

Aspergillus toxins
Aflatoxin/s

Aspergillus and ***Penicillium*** toxins
Ochratoxin A

Fusarium toxins
Zearalenone

Beefburger

Aspergillus toxins
Aflatoxin/s: aflatoxin B_1, aflatoxin/s

Aspergillus and ***Penicillium*** toxins
Ochratoxin A

Fusarium toxins
Zearalenone

Butter

Aspergillus toxins
Aflatoxin/s: aflatoxin M_1

Butter (cocoa butter)

Aspergillus toxins
Aflatoxin/s: aflatoxin B_1, aflatoxin B_2

Aspergillus and ***Penicillium*** toxins
Ochratoxin A

Butter (nut butter)

Aspergillus toxins
Aflatoxin/s: aflatoxin B_1, aflatoxin B_2, aflatoxin G_1, aflatoxin G_2

Aspergillus and ***Penicillium*** toxins
Ochratoxin A

Butter (peanut butter)

Aspergillus toxins
Aflatoxin/s: aflatoxin B_1, aflatoxin B_2, aflatoxin G_1, aflatoxin G_2, aflatoxin M_1, aflatoxin M_2, aflatoxin/s

Aspergillus and ***Penicillium*** toxins
Ochratoxin A

Butter (sesame butter)

Aspergillus toxins
Aflatoxin/s

Cheese

Aspergillus toxins
Aflatoxin/s: aflatoxin B_1, aflatoxin B_2, aflatoxin G_1, aflatoxin M_1, aflatoxin M_2
Other *Aspergillus* toxins: sterigmatocystin

Aspergillus and ***Penicillium*** toxins
Citrinin, ochratoxin A

Penicillium toxins
Mycophenolic acid

Cheese (Anari cheese)

Aspergillus toxins
Aflatoxin/s: aflatoxin M_1

Cheese (Bhutanese cheese)

Aspergillus toxins
Aflatoxin/s: aflatoxin B_1, aflatoxin B_2

Aspergillus and ***Penicillium*** toxins
Citrinin, ochratoxin A

Cheese (Bleu des Causses cheese)

Penicillium toxins
Mycophenolic acid

Cheese (blue-veined cheese)

Aspergillus toxins
Aflatoxin/s: aflatoxin M_1

Aspergillus and ***Penicillium*** toxins
Penicillic acid

Penicillium toxins
Isofumigaclavine A, isofumigaclavine B, mycophenolic acid, roquefortine C

M. Weidenbörner, *Mycotoxins in Animal Products*, https://doi.org/10.1007/978-3-030-30919-0_4

Cheese (Brie cheese)

Aspergillus toxins
Aflatoxin/s: aflatoxin M_1

Cheese (butter cheese)

Aspergillus toxins
Aflatoxin/s: aflatoxin M_1

Cheese (Camembert cheese)

Aspergillus toxins
Aflatoxin/s: aflatoxin M_1

Aspergillus and ***Penicillium*** toxins
Cyclopiazonic acid

Cheese (Cecil cheese)

Aspergillus toxins
Aflatoxin/s: aflatoxin M_1

Cheese (Cheddar cheese)

Aspergillus toxins
Aflatoxin/s: aflatoxin M_1

Cheese (Cheshire cheese)

Aspergillus toxins
Aflatoxin/s: aflatoxin M_1

Cheese (Civil cheese)

Aspergillus toxins
Aflatoxin/s: aflatoxin M_1

Cheese (cottage cheese)

Aspergillus toxins
Aflatoxin/s: aflatoxin B_1, aflatoxin B_2

Aspergillus and ***Penicillium*** toxins
Ochratoxins (A, B)

Cheese (cow cheese)

Aspergillus toxins
Aflatoxin/s: aflatoxin M_1

Cheese (cream cheese)

Aspergillus toxins
Aflatoxin/s: aflatoxin M_1

Penicillium toxins
Penitrem A

Cheese (Danablu cheese)

Penicillium toxins
Roquefortine C

Cheese (Domiati cheese)

Aspergillus toxins
Aflatoxin/s: aflatoxin M_1

Cheese (Double Gloucester cheese)

Aspergillus toxins
Aflatoxin/s: aflatoxin M_1

Cheese (Edam cheese)

Aspergillus toxins
Aflatoxin/s: aflatoxin M_1

Cheese (Edelpilzkäse)

Penicillium toxins
Roquefortine C

Cheese (Feta cheese)

Aspergillus toxins
Aflatoxin/s: aflatoxin M_1

Cheese (fondue cheese)

Aspergillus toxins
Aflatoxin/s: aflatoxin M_1

Cheese (fresh cheese)

Aspergillus toxins
Aflatoxin/s: aflatoxin M_1

Aspergillus and ***Penicillium*** toxins
Ochratoxin A

Cheese (goat cheese)

Aspergillus toxins
Aflatoxin/s: aflatoxin M_1

Aspergillus and ***Penicillium*** toxins
Patulin, penicillic acid

Cheese (goat-sheep cheese)

Aspergillus toxins
Aflatoxin/s: aflatoxin M_1

Cheese (Gorgonzola cheese)

Aspergillus and ***Penicillium*** toxins
Ochratoxin A

Penicillium toxins
Mycophenolic acid, roquefortine C

Cheese (Gouda cheese)

Aspergillus toxins
Aflatoxin/s: aflatoxin M_1

Cheese (Grana Padano cheese)

Aspergillus toxins
Aflatoxin/s: aflatoxin M_1

Cheese (Graviera cheese)

Aspergillus toxins
Aflatoxin/s: aflatoxin M_1

Cheese (Gruyere cheese)

Aspergillus toxins
Aflatoxin/s: aflatoxin M_1

Cheese (hard cheese)

Aspergillus toxins
Aflatoxin/s: aflatoxin M_1
Other *Aspergillus* toxins: sterigmatocystin

Aspergillus and ***Penicillium*** toxins
Ochratoxin A, patulin, penicillic acid

Penicillium toxins
Mycophenolic acid

Cheese (Hard Roumy cheese)

Fusarium toxins
Zearalenone

Cheese (Havarti cheese)

Aspergillus toxins
Aflatoxin/s: aflatoxin M_1

Cheese (Karish cheese)

Fusarium toxins
Zearalenone

Cheese (Kashar cheese)

Aspergillus toxins
Aflatoxin/s: aflatoxin M_1

Cheese (Lancashire cheese)

Aspergillus toxins
Aflatoxin/s: aflatoxin M_1

Cheese (Leicester cheese)

Aspergillus toxins
Aflatoxin/s: aflatoxin M_1

Cheese (Lighvan cheese)

Aspergillus toxins
Aflatoxin/s: aflatoxin M_1

Cheese (Lor cheese)

Aspergillus toxins
Aflatoxin/s: aflatoxin M_1

Cheese (Maribo cheese)

Aspergillus toxins
Aflatoxin/s: aflatoxin M_1

Cheese (Minas cheese)

Aspergillus toxins
Aflatoxin/s: aflatoxin M_1

Cheese (mold-ripened)

Penicillium toxins
Mycophenolic acid, roquefortine C

Cheese (Mozzarella cheese)

Aspergillus toxins
Aflatoxin/s: aflatoxin M_1

Cheese (Münster cheese)

Aspergillus toxins
Aflatoxin/s: aflatoxin M_1

Cheese (Oaxaca cheese)

Aspergillus toxins
Aflatoxin/s: aflatoxin M_1, aflatoxin M_2

Cheese (Parmesan cheese)

Aspergillus toxins
Aflatoxin/s: aflatoxin M_1

Cheese (Prato cheese)

Aspergillus toxins
Aflatoxin/s: aflatoxin M_1

Cheese (processed cheese)

Aspergillus toxins
Aflatoxin/s: aflatoxin M_1

Cheese (Ras cheese)

Aspergillus toxins
Aflatoxin/s: aflatoxin B_1, aflatoxin G_1, aflatoxin M_1
Other *Aspergillus* toxins: sterigmatocystin

Cheese (Roquefort cheese)

Aspergillus and ***Penicillium*** toxins
Ochratoxin A

Penicillium toxins
Mycophenolic acid, roquefortine C

Cheese (Samsoe cheese)

Aspergillus toxins
Aflatoxin/s: aflatoxin M_1

Cheese (semihard cheese)

Aspergillus toxins
Aflatoxin/s: aflatoxin M_1

Aspergillus and ***Penicillium*** toxins
Ochratoxin A, patulin

Cheese (sheep cheese)

Aspergillus toxins
Aflatoxin/s: aflatoxin M_1

Cheese (soft cheese)

Aspergillus toxins
Aflatoxin/s: aflatoxin M_1

Cheese (Stilton cheese)

Penicillium toxins
Isofumigaclavine A, roquefortine C

Cheese (Surk cheese)

Aspergillus toxins
Aflatoxin/s: aflatoxin M_1

Cheese (Tulum cheese)

Aspergillus toxins
Aflatoxin/s: aflatoxin M_1

Cheese (Urfa cheese)

Aspergillus toxins
Aflatoxin/s: aflatoxin M_1

Cheese (Van otlu cheese)

Aspergillus toxins
Aflatoxin/s: aflatoxin M_1

Cheese (Wensleydale cheese)

Aspergillus toxins
Aflatoxin/s: aflatoxin M_1

Cheese (white cheese)

Aspergillus toxins
Aflatoxin/s: aflatoxin M_1

Cheese curd

Aspergillus toxins
Aflatoxin/s: aflatoxin M_1

Cheese whey

Aspergillus toxins
Aflatoxin/s: aflatoxin M_1

Chicken gizzard

Aspergillus toxins
Aflatoxin/s: aflatoxin B_1

Aspergillus and ***Penicillium*** toxins
Ochratoxin A

Fusarium toxins
Zearalenone

Chicken heart

Fusarium toxins
Zearalenone

Chicken kidney

Aspergillus toxins
Aflatoxin/s: aflatoxin B_1, aflatoxin/s

Aspergillus and ***Penicillium*** toxins
Ochratoxin A

Chicken liver

Aspergillus toxins
Aflatoxin/s: aflatoxin B_1, aflatoxin/s

Aspergillus and ***Penicillium*** toxins
Citrinin, ochratoxin A

Fusarium toxins
Zearalenone

Chicken meat

Aspergillus toxins
Aflatoxin/s

Aspergillus and ***Penicillium*** toxins
Ochratoxin A

Fusarium toxins
Zearalenone

Chicken muscle

Aspergillus toxins
Aflatoxin/s: aflatoxin B_1, aflatoxin/s

Aspergillus and ***Penicillium*** toxins
Citrinin; ochratoxin A

Fusarium toxins
Type A trichothecenes: T-2 toxin

Chicken nuggets

Aspergillus and ***Penicillium*** toxins
Ochratoxin A

Chickenburger

Aspergillus and ***Penicillium*** toxins
Ochratoxin A

Coppa

Aspergillus and ***Penicillium*** toxins
Ochratoxin A

Cow liver

Aspergillus and ***Penicillium*** toxins
Ochratoxin A

Cream

Aspergillus toxins
Aflatoxin/s: aflatoxin M_1, aflatoxins

Cream (sour)

Aspergillus toxins
Aflatoxin/s: aflatoxin M_2

Custard powder

Aspergillus toxins
Aflatoxin/s

Fusarium toxins
Type B trichothecenes: deoxynivalenol

Dairy beverages

Aspergillus toxins
Aflatoxin/s: aflatoxin M_1

Dairy desserts

Aspergillus toxins
Aflatoxin/s: aflatoxin M_1

Dairy products

Aspergillus toxins
Aflatoxin/s: aflatoxin B_1, aflatoxin G_1, aflatoxin M_1

Aspergillus and ***Penicillium*** toxins
Ochratoxin A

Fusarium toxins
Fumonisin/s: fumonisin B_1

Duck liver

Aspergillus and ***Penicillium*** toxins
Ochratoxin A

Duck meat

Aspergillus and ***Penicillium*** toxins
Ochratoxin A

Eggs

Alternaria toxins
Alternariol methyl ether, tenuazonic acid

Aspergillus toxins
Aflatoxin/s: aflatoxin B_1, aflatoxin B_2, aflatoxin G_1, aflatoxin G_2, aflatoxin/s
Other *Aspergillus* toxins: sterigmatocystin

Aspergillus and ***Penicillium*** toxins
Ochratoxin A

Fusarium toxins
Depsipeptides: beauvericin, enniatin A, enniatin A_1, enniatin B, enniatin B_1
Type B trichothecenes: deoxynivalenol, 3-acetyldeoxynivalenol, 15-acetyldeoxynivalenol
Other *Fusarium* toxins: zearalenone

Fish

Aspergillus toxins
Aflatoxin/s: aflatoxin B_1, aflatoxin B_2, aflatoxin G_1, aflatoxin G_2, aflatoxin/s

Aspergillus and ***Penicillium*** toxins
Ochratoxin A

Fusarium toxins
Depsipeptides: enniatin A_1, enniatin B, enniatin B_1
Type B trichothecenes: fusarenon-X (4-acetylnivalenol)
Other *Fusarium* toxins: zearalenone

Fish oil

Aspergillus toxins
Aflatoxin/s: aflatoxin B_1

Fish products

Aspergillus toxins
Aflatoxin/s

Fish/shrimp

Aspergillus toxins
Aflatoxin/s

Food

Aspergillus toxins
Aflatoxin/s: aflatoxin M_1

Goose liver

Aspergillus and ***Penicillium*** toxins
Ochratoxin A

Goose meat

Aspergillus and ***Penicillium*** toxins
Ochratoxin A

Ham

Aspergillus toxins
Aflatoxin/s: aflatoxin B_1

Aspergillus and ***Penicillium*** toxins
Ochratoxin A

Hot-dog

Aspergillus toxins
Aflatoxin/s: aflatoxin B_1, aflatoxin B_2

Aspergillus and ***Penicillium*** toxins
Ochratoxin A

Ice cream

Aspergillus toxins
Aflatoxin/s: aflatoxin M_1

Aspergillus and ***Penicillium*** toxins
Ochratoxin A

Infant food

Aspergillus toxins
Aflatoxin/s: aflatoxin B_1, aflatoxin B_2, aflatoxin G_1, aflatoxin G_2

Insects

Alternaria toxins
Alternariol, alternariol methyl ether

Aspergillus toxins
Aflatoxin/s

Fusarium toxins
Type A trichothecenes: HT-2 toxin
Type B trichothecenes: nivalenol
Other *Fusarium* toxins: zearalenone

Penicillium toxins
Roquefortine C

Joints

Aspergillus and ***Penicillium*** toxins
Ochratoxin A

Kashk

Aspergillus toxins
Aflatoxin/s: aflatoxin B_1, aflatoxin M_1

Kubeba

Aspergillus toxins
Aflatoxin/s: aflatoxin B_1, aflatoxin B_2

Meat

Aspergillus toxins
Aflatoxin/s: aflatoxin B_1, aflatoxin B_2, aflatoxin G_1, aflatoxin/s

Aspergillus and ***Penicillium*** toxins
Ochratoxin A

Fusarium toxins
Depsipeptides: enniatin B
Other *Fusarium* toxins: zearalenone

Meat products

Aspergillus toxins
Aflatoxin/s: aflatoxin B_1, aflatoxin/s

Aspergillus and ***Penicillium*** toxins
Citrinin, ochratoxin A

Milk

Alternaria toxins
Tenuazonic acid

Aspergillus toxins
Aflatoxin/s: aflatoxin M_1, aflatoxin/s

Aspergillus and ***Penicillium*** toxins
Ochratoxin A

Fusarium toxins
Depsipeptides: beauvericin, enniatin B
Other *Fusarium* toxins: zearalenone

Milk (buffalo milk)

Aspergillus toxins
Aflatoxin/s: aflatoxin M_1

Milk (buttermilk)

Aspergillus toxins
Aflatoxin/s: aflatoxin M_1

Milk (camel milk)

Aspergillus toxins
Aflatoxin/s: aflatoxin M_1

Milk (cow milk)

Aspergillus toxins
Aflatoxin/s: aflatoxin B_1, aflatoxin M_1, aflatoxin M_2, aflatoxins (M_1, M_2)

Aspergillus and ***Penicillium*** toxins
Cyclopiazonic, ochratoxin A

Fusarium toxins
Fumonisin/s: fumonisin B_1
Other *Fusarium* toxins: zearalenone

Milk (donkey milk)

Aspergillus toxins
Aflatoxin/s: aflatoxin M_1

Milk (goat milk)

Aspergillus toxins
Aflatoxin/s: aflatoxin M_1

Milk (human breast milk)

Aspergillus toxins
Aflatoxin/s: aflatoxin B_1, aflatoxin B_2, aflatoxin G_1, aflatoxin G_2, aflatoxin M_1, aflatoxin M_2

Aspergillus and ***Penicillium*** toxins
Ochratoxin A

Fusarium toxins
Depsipeptides: beauvericin, enniatin A, enniatin A_1, enniatin B, enniatin B_1
Fumonisin/s: fumonisin B_1
Type A trichothecenes: HT-2 toxin; neosolaniol
Type B trichothecenes: nivalenol
Other *Fusarium* toxins: zearalenone

Milk (infant formula)

Aspergillus toxins
Aflatoxin/s: aflatoxin B_1, aflatoxin M_1, aflatoxins

Aspergillus and ***Penicillium*** toxins
Ochratoxin A

Fusarium toxins
Fumonisin/s: fumonisin B_1
Type A trichothecenes: T-2 toxin
Other *Fusarium* toxins: zearalenone

Milk (sheep milk)

Aspergillus toxins
Aflatoxin/s: aflatoxin M_1

Fusarium toxins
Depsipeptides: enniatin B

Milk (UHT milk)

Aspergillus toxins
Aflatoxin/s: aflatoxin M_1, aflatoxin M_2

Milk powder

Alternaria toxins
Tenuazonic acid

Aspergillus toxins
Aflatoxin/s: aflatoxin B_1, aflatoxin B_2, aflatoxin M_1, aflatoxin M_2, aflatoxin/s

Aspergillus and ***Penicillium*** toxins
Ochratoxin A

Fusarium toxins
Zearalenone

Pâté

Aspergillus and ***Penicillium*** toxins
Ochratoxin A

Pig blood

Aspergillus and ***Penicillium*** toxins
Ochratoxin A

Pig fat

Fusarium toxins
Type A trichothecenes: T-2 toxin
Type B trichothecenes: deoxynivalenol

Pig kidney

Aspergillus and ***Penicillium*** toxins
Citrinin, ochratoxin A

Pig liver

Aspergillus toxins
Aflatoxin/s: aflatoxin B_1, aflatoxin B_2, aflatoxin M_1

Aspergillus and ***Penicillium*** toxins
Citrinin, ochratoxin A, patulin

Pig muscle

Aspergillus toxins
Aflatoxin/s: aflatoxin B_1
Other *Aspergillus* toxins: sterigmatocystin

Aspergillus and ***Penicillium*** toxins
Ochratoxin A

Fusarium toxins
Type A trichothecenes: T-2 toxin

Pig plasma

Aspergillus and ***Penicillium*** toxins
Ochratoxin A

Pig serum

Aspergillus and ***Penicillium*** toxins
Citrinin, ochratoxin A, penicillic acid

Fusarium toxins
Type B trichothecenes: deoxynivalenol
Other *Fusarium* toxins: zearalenone

Penicillium toxins
Penitrem A

Pork meat

Aspergillus and ***Penicillium*** toxins
Ochratoxin A

Fusarium toxins
Zearalenone

Poultry kidney

Aspergillus toxins
Aflatoxin/s: aflatoxin M_1

Aspergillus and ***Penicillium*** toxins
Ochratoxin A

Poultry meat

Aspergillus and ***Penicillium*** toxins
Ochratoxin A

Sausage

Aspergillus toxins
Aflatoxin/s: aflatoxin B_1, aflatoxin B_2

Aspergillus and ***Penicillium*** toxins
Citrinin, ochratoxin A

Fusarium toxins
Zearalenone

Sausage (beef sausage)

Aspergillus and ***Penicillium*** toxins
Ochratoxin A

Sausage (blood sausage)

Aspergillus and ***Penicillium*** toxins
Ochratoxin A

Sausage (Bologna sausage)

Aspergillus and ***Penicillium*** toxins
Ochratoxin A

Sausage (liver sausage)

Aspergillus toxins
Aflatoxin/s: aflatoxin B_1

Aspergillus and ***Penicillium*** toxins
Ochratoxin A

Sausage (poultry sausage)

Aspergillus and ***Penicillium*** toxins
Ochratoxin A

Sausage (raw sausage)

Aspergillus and ***Penicillium*** toxins
Ochratoxin A

Sausage (salami sausage)

Aspergillus and ***Penicillium*** toxins
Ochratoxin A

Sausage (Slavonski Kulen sausage)

Aspergillus toxins
Aflatoxin/s: aflatoxin B_1

Aspergillus and ***Penicillium*** toxins
Ochratoxin A

Sausage (Würstel)

Aspergillus and ***Penicillium*** toxins
Ochratoxin A

Seafood

Aspergillus and ***Penicillium*** toxins
Ochratoxin A

Fusarium toxins
Zearalenone

Tarkhineh

Aspergillus toxins
Aflatoxin/s: aflatoxin M_1

Turkey liver

Aspergillus and ***Penicillium*** toxins
Ochratoxin A

Turkey meat

Aspergillus and ***Penicillium*** toxins
Ochratoxin A

Turkey muscle

Aspergillus and ***Penicillium*** toxins
Ochratoxin A

Wild boar kidney

Aspergillus and ***Penicillium*** toxins
Ochratoxin A

Wild boar liver

Aspergillus and ***Penicillium*** toxins
Ochratoxin A

Wild boar muscle

Aspergillus and ***Penicillium*** toxins
Ochratoxin A

Wild boar serum

Aspergillus and ***Penicillium*** toxins
Ochratoxin A

Yogurt

Aspergillus toxins
Aflatoxin/s: aflatoxin M_1

Mycotoxin Contamination in Conventional and Organic Animal Products

Alvito, P.C., Sizoo, E.A., Almeida, C.M.M., Van Egmond, H.P., 2010: Occurrence of aflatoxins and ochratoxin A in baby foods in Portugal. Food Analytical Methods 3, 22-30. doi: https://doi.org/10.1007/s12161-008-9064-x (90).

Armorini, S., Altafini, A., Zaghini, A., Roncada, P., 2016: Occurrence of aflatoxin M_1 in conventional and organic milk offered for sale in Italy. Mycotoxin Research 32, 237-246. doi: https://doi.org/10.1007/s12550-016-0256-8 (558).

Food Standards Agency, 2001: Survey of milk for mycotoxins. Food-Survey-Information-Sheet No. 17/01 (208).

Food Standards Agency, 2002: Survey of nuts, nut products and dried tree fruits for mycotoxins. Food-Survey-Information-Sheet No. 21/02 (9).

Gazzotti, T., Lugoboni, B., Zironi, E., Barbarossa, A., Serraino, A., Pagliuca, G., 2009: Determination of fumonisin B_1 in bovine milk by LC-MS/MS. Food Control 20, 1171-1174. doi: https://doi.org/10.1016/j.foodcont.2009.02.009 (442).

Ghiasian, S.A., Maghsood, A.H., Neyestani, T.R., Mirhendi, S.H., 2007: Occurrence of aflatoxin M_1 in raw milk during the summer and winter seasons in Hamedan, Iran. Journal of Food Safety 27, 188-198 (294).

Ghidini, S., Zanardi, E., Battaglia, A., Varisco, G., Ferretti, E., Campanini, G., Chizzolini, R., 2005: Comparison of contaminant and residue levels in organic and conventional milk and meat products from northern Italy. Food Additives and Contaminants 22, 9-14. doi: https://doi.org/10.1080/02652030400027995 (295).

Gilbert, J., Shepherd, M.J., 1985: A survey of aflatoxins in peanut butter, nuts and nut confectionery products by HPLC with fluorescence detection. Food Additives and Contaminants 2, 171–183 (18).

Gutiérrez, R., Rosell, P., Vega, S., Pérez, J., Ramírez, A., Coronado, M., 2013: Self and foreign substances in organic and conventional milk produced in the eastern region of Mexico. Food and Nutrition Sciences 4, 586-593. doi: https://doi.org/10.4236/fns.2013.45076 (421).

Jørgensen, K., 1998: Survey of pork, poultry, coffee, beer and pulses for ochratoxin A. Food Additives and Contaminants 15, 550-554 (102).

Pattono, D., Gallo, P.F., Civera, T., 2011: Detection and quantification of ochratoxin A in milk produced in organic farms. Food Chemistry 127, 374-377. doi: https://doi.org/10.1016/j.foodchem.2010.12.051 (224).

Pozzo, L., Cavallarin, L:, Nucera, D., Antoniazzi, S., Schiavone, A., 2010: A survey of ochratoxin A contamination in feeds and sera from organic and standard swine farms in northwest Italy. Journal of the Science of Food and Agriculture 90, 1467-1472. doi: https://doi.org/10.1002/jsfa.3965 (498).

Skaug, M.A., 1999: Analysis of Norwegian milk and infant formulas for ochratoxin A. Food Additives and Contaminants 16, 75-78 (316).

Virdis, S., Corgiolu, G., Scarano, C., Pilo, A.L., De Santis, E.P.L., 2008: Occurrence of aflatoxin M_1 in tank bulk goat milk and ripened goat cheese. Food Control 19, 44-49. doi: https://doi.org/10.1016/j.foodcont.2007.02.001 (31) (intensive versus extensive husbandry system).

Yoon, B.R., Hong, S.-Y., Cho, S.M., Lee, K.R., Kim, M., Chung, S.H., 2016: Aflatoxin M_1 levels in dairy products from South Korea determined by high performance liquid chromatography with fluorescence detection. Journal of Food and Nutrition Research 55, 171-180 (561).

Mycotoxin Contamination in Organic Animal Products

Kos, J., Levic, J., Duragic, O., Kokic, B., Miladinovic, I., 2014: Occurrence and estimation of aflatoxin M_1 exposure in milk in Serbia. Food Control 38, 41-46 doi: https://doi.org/10.1016/j.foodcont.2013.09.060 (468).

Mortimer, D.N., Shepherd, M.J., Gilbert, J., Morgan, M.R.A., 1987: A survey of the occurrence of aflatoxin B_1 in peanut butters by enzyme-linked immunosorbent assay. Food Additives and Contaminants 5, 127-132 (52).

Turcotte, A.-M., Scott, P.M., Tague, B., 2013: Analysis of cocoa products for ochratoxin A and aflatoxins. Mycotoxin Research 29, 193-201. doi: https://doi.org/10.1007/s12550-013-0167-x (16).

M. Weidenbörner, *Mycotoxins in Animal Products*, https://doi.org/10.1007/978-3-030-30919-0_5

Literature

Numerical Bibliography

1. Gareis, M., Scheuer, R., 2000: Ochratoxin A in meat and meat products. Archiv für Lebensmittelhygiene 51, 102-104.
2. Aziz, N.H., Youssef, Y.A., 1991: Occurrence of aflatoxins and aflatoxin-producing moulds in fresh and processed meat in Egypt. Food Additives and Contaminants 8, 321-331.
3. Goliński, P., Hult, K., Grabarkiewicz-Szczesna, J., Chelkowski, J., Kneblewski, P., Szebiotko, K., 1984: Mycotoxic porcine nephropathy and spontaneous occurrence of ochratoxin A residues in kidneys and blood of Polish swine. Applied and Environmental Microbiology 47, 1210-1212.
4. Kotowski, K., Kostecki, M., Grabarkiewicz-Szczęsna, J., Golinski, P., 1993: Ochratoxin A residue in kidney and blood of pigs. Medycyna Weterynaryjna 49, 554-556.
5. Fukal, L., 1991: Spontaneous occurrence of ochratoxin A residues in Czechoslovak slaughter pigs determined by immunoassay. Deutsche Lebensmittel-Rundschau 87, 316-319.
6. Holmberg, T., Breitholtz, A., Bengtsson, A., Hult, K., 1990: Ochratoxin A in swine blood in relation to moisture content in feeding barley at harvest. Acta Agriculturæ Scandinavica 40, 201-204.
7. Hult, K., Rutqvist, L., Holmberg, T., Thafvelin, B., Gatenbeck, S., 1984: Ochratoxin A in blood of slaughter pigs. Nordisk Veterinaermedicin 36, 314-316.
8. Kotowski, K., Grabarkiewicz-Szczesna, J., Waskiewicz, A., Kostecki, M., Golinski, P., 2000: Ochratoxin A in porcine blood and in consumed feed samples. Mycotoxin Research 16, 66-72.
9. Food Standards Agency, 2002: Survey of nuts, nut products and dried tree fruits for mycotoxins. Food-Survey-Information-Sheet No. 21/02.
10. Aycicek, H., Yarsan, E., Sarimehmetoglu, B., Cakmak, O., 2002: Aflatoxin M_1 in white cheese and butter consumed in Istanbul, Turkey. Veterinary and Human Toxicology 44, 295-296.
11. Aycicek, H., Aksoy, A., Saygi, S., 2005: Determination of aflatoxin levels in some dairy and food products which consumed in Ankara, Turkey. Food Control 16, 263-266. doi: https://doi.org/10.1016/j.foodcont.2004.03.004.
12. Var, I., Kabak, B., 2008: Detection of aflatoxin M_1 in milk and dairy products consumed in Adana, Turkey. International Journal of Dairy Technology 62, 15-18. doi: https://doi.org/10.1111/j.1471-0307.2008.00440.x.
13. Tekinşen, K.K., Uçar, G., 2008: Aflatoxin M_1 levels in butter and cream cheese consumed in Turkey. Food Control 19, 27-30. doi: https://doi.org/10.1016/j.foodcont.2007.01.003.
14. Copetti, M.V., Iamanaka, B.T., Pereira, J.L., Lemes, D.P., Nakano, F., Taniwaki, M.H., 2012: Determination of aflatoxins in by-products of industrial processing of cocoa beans. Food Additives and Contaminants: Part A 29, 972-978. doi: https://doi.org/10.1080/19440049.2012.660657.
15. Copetti, M.V., Iamanaka, B.T., Nester, M.A., Efraim, P., Taniwaki, M.H., 2013: Occurrence of ochratoxin A in cocoa by-products and determination of its reduction during chocolate manufacture. Food Chemistry 136, 100-104. doi: https://doi.org/10.1016/j.foodchem.2012.07.093.
16. Turcotte, A.-M., Scott, P.M., Tague, B., 2013: Analysis of cocoa products for ochratoxin A and aflatoxins. Mycotoxin Research 29, 193-201. doi: https://doi.org/10.1007/s12550-013-0167-x.
17. Mounjouenpou, P., Mbang, J.A.A., Guyot, B., Guiraud, J.-P., 2012: Traditional procedures of cocoa processing and occurrence of ochratoxin - A in the derived products. Journal of Chemical and Pharmaceutical Research 4, 1332-1339.
18. Gilbert, J., Shepherd, M.J., 1985: A survey of aflatoxins in peanut butters, nuts and nut confectionery products by HPLC with fluorescence detection. Food Additives and Contaminants 2, 171-183.
19. Herry, M.-P., Lemetayer, N., 1992: Aflatoxin B_1 contamination in oilseeds, dried fruits and spices. Microbiologie – Aliments – Nutrition 10, 261-266.
20. Jang, M.-R., Lee, C.-H., Cho, S.-H., Park, J.-S., Kwon, E.-Y., Lee, E.-J., Kim, S.-H., Kim, D.-B., 2007: A survey of total aflatoxins in food using high performance liquid chromatography-fluorescence detector (HPLC-FLD) and liquid chromatography tandem mass spectrometry (LC-MS/MS). Korean Journal of Food Science and Technology 39, 488-493.
21. Ok, H.E., Kim, H.J., Shim, W.B., Lee, H., Bae, D.-H., Chung, D.-H., Chun, H.S., 2007: Natural occurrence of aflatoxin B_1 in marketed foods and risk estimates of dietary exposure in Koreans. Journal of Food Protection 70, 2824-2828.

M. Weidenbörner, *Mycotoxins in Animal Products*, https://doi.org/10.1007/978-3-030-30919-0

22. Fritz, W., Engst, R., 1981: Survey of selected mycotoxins in food. Journal of Environmental Science and Health B16, 193-210.
23. Koirala, P., Kumar, S., Yadav, B.K., Premarajan, K.C., 2005: Occurrence of aflatoxin in some of the food and feed in Nepal. Indian Journal of Medical Sciences 59, 331-336.
24. Tabata, S., Kamimura, H., Tamura, Y., Yasuda, K., Ushiyama, H., Hashimoto, H., Nishijima, M., Nishima, T., 1987: Investigation of aflatoxins contamination in foods and foodstuffs. Journal of Food Hygienic Society of Japan 28, 395-401.
25. Noviandi, C.T., Razzazi, E., Agus, A., Böhm, J., Hulan, H.W., Wedhastri, S., Maryudhani, Y.B., Nuryono, Sardjono, Leibetseder, J., 2001: Natural occurrence of aflatoxin B_1 in some Indonesian food and feed products in Yogyakarta in year 1998-1999. Mycotoxin Research 17A, 174-178.
26. Boutrif, E., Jemmali, M., Campbell, A.D., Pohland, A.E., 1977: Aflatoxin in Tunisian foods and foodstuffs. Annales de la Nutrition et de l'Alimentation 31, 431-434.
27. Paul, R., Kalra, M.S., Singh, A., 1976: Incidence of aflatoxins in milk and milk products. Indian Journal of Dairy Science 29, 318-321.
28. Polzhofer, K., 1977: Determination of aflatoxins in milk and milk products. Zeitschrift für Lebensmittel-Untersuchung und -Forschung 163, 175-177.
29. Saito, K., Nishijima, M., Yasuda, K., Kamimura, H., Ibe, A., Nagayama, T., Ushiyama, H., Naoi, Y., 1980: Natural occurrence of aflatoxins in commercial cheese. Journal of the Hygienic Society of Japan, Tokyo 21, 472-475.
30. Trucksess, M.W., Page, S.W., 1986: Examination of imported cheeses for aflatoxin M_1. Journal of Food Protection 49, 632-633.
31. Virdis, S., Corgiolu, G., Scarano, C., Pilo, A.L., De Santis, E.P.L., 2008: Occurrence of aflatoxin M_1 in tank bulk goat milk and ripened goat cheese. Food Control 19, 44-49. doi: https://doi.org/10.1016/j.foodcont.2007.02.001.
32. Piva, G., Pietri, A., Galazzi, L., Curto, O., 1987: Aflatoxin M_1 occurrence in dairy products marketed in Italy. Food Additives and Contaminants 5, 133-139.
33. Oruç, H.H., Sonal, S., 2001: Determination of aflatoxin M_1 levels in cheese and milk consumed in Bursa, Turkey. Veterinary and Human Toxicology 43, 292-293.
34. Barrios, M.J., Gualda, M.J., Cabanas, J.M., Medina, L.M., Jordano, R., 1996: Occurrence of aflatoxin M_1 in cheeses from the south of Spain. Journal of Food Protection 59, 898-900.
35. Sarımehmetoglu, B., Kuplulu, O., Celik, T.H., 2004: Detection of aflatoxin M_1 in cheese samples by ELISA. Food Control 15, 45-49. doi: https://doi.org/10.1016/S0956-7135(03)00006-9.
36. Finoli, C., Vecchio, A., 1997: Aflatoxin M_1 in goat dairy products. Microbiologie - Aliments - Nutrition 15, 47-52.
37. Finoli, C., Vecchio, A., 2003: Occurrence of aflatoxins in feedstuff, sheep milk and dairy products in Western Sicily. Italian Journal of Animal Science 2, 191-196.
38. Gürbay, A., Engin, A.B., Çağlayan, A., Şahin, G., 2006: Aflatoxin M_1 levels in commonly consumed cheese and yogurt samples in Ankara, Turkey. Ecology of Food and Nutrition 45, 449-459. doi: https://doi.org/10.1080/03670240600985274.
39. Rahimi, E., Karim, G., Shakerian, A., 2009: Occurrence of aflatoxin M_1 in traditional cheese consumed in Esfahan, Iran. World Mycotoxin Journal 2, 91-94. doi: https://doi.org/10.3920/WMJ2008.1082.
40. Godič Torkar, K.G., Vengušt, A., 2008: The presence of yeasts, moulds and aflatoxin M_1 in raw milk and cheese in Slovenia. Food Control 19, 570-577. doi: https://doi.org/10.1016/j.foodcont.2007.06.008.
41. Dashti, B., Al-Hamli, S., Alomirah, H., Al-Zenki, S., Abbas, A.B., Sawaya, W., 2009: Levels of aflatoxin M_1 in milk, cheese consumed in Kuwait and occurrence of total aflatoxin in local and imported animal feed. Food Control 20, 686-690. doi: https://doi.org/10.1016/j.foodcont.2009.01.001.
42. Goto, T., Manabe, M., Matsuura, S., 1982: Analysis of aflatoxins in milk and milk products by high-performance liquid chromatography. Agricultural and Biological Chemistry 46, 801-802.
43. Montagna, M.T., Napoli, C., De Giglio, O., Iatta, R., Barbuti, G., 2008: Occurrence of aflatoxin M_1 in dairy products in southern Italy. International Journal of Molecular Sciences 9, 2614-2621. doi: https://doi.org/10.3390/ijms9122614.
44. Iha, M.H., Barbosa, C.B., Favaro, R.M.D., Trucksess, M.W., 2011: Chromatographic method for the determination of aflatoxin M_1 in cheese, yogurt, and dairy beverages. Journal of the Association of Official Analytical Chemists International 94, 1513-1518. doi: https://doi.org/10.5740/jaoac.int.10-432.
45. Lohiya, G., Nichols, L., Hsieh, D., Lohiya, S., Nguyen, H., 1987: Aflatoxin content of foods served to a population with a high incidence of hepatocellular carcinoma. Hepatology 7, 750-752.
46. Qian, G.-S., Ross, R.K., Yu, M.C., Yuan, J.-M., Gao, Y.-T., Henderson, B.E., Wogan, G.N., Groopman, J.D., 1994: A follow-up study of urinary markers of aflatoxin exposure and liver cancer risk in Shanghai, People's Republic of China. Cancer Epidemiology, Biomarkers & Prevention 3, 3-10.
47. Kumagai, S., Nakajima, M., Tabata, S., Ishikuro, E., Tanaka, T., Norizuki, H., Itoh, Y., Aoyama, K., Fujita, K., Kai, S., Sato, T., Saito, S., Yoshiike, N., Sugita-Konishi, Y., 2008: Aflatoxin and ochratoxin A contamination of retail foods and intake of these mycotoxins in Japan. Food Additives and Contaminants: Part A 25, 1101-1106. doi: https://doi.org/10.1080/02652030802226187.
48. Taguchi, S., Fukushima, S., Sumimoto, T., Yoshida, S., Nishimune, T., 1995: Aflatoxins in foods collected in Osaka, Japan, from 1988 to 1992. Journal of the Association of Official Analytical Chemists International 78, 325-327.
49. Cervino, C., Asam, S., Knopp, D., Rychlik, M., Niessner, R., 2008: Use of isotope-labeled aflatoxins for LC-MS/MS stable isotope dilution analysis of foods. Journal of Agricultural and Food Chemistry 56, 1873-1879. doi: https://doi.org/10.1021/jf073231z.
50. Ioannou-Kakouri, E., Aletrari, M., Christou, E., Hadjioannou-Ralli, A., Koliou, A., Akkelidou, D., 1999: Surveillance and control of aflatoxins B_1, B_2, G_1, G_2, and M_1 in foodstuffs in the Republic of Cyprus: 1992-1996. Journal of the Association of Official Analytical Chemists International 82, 883-892.
51. Wei, D.-L., Wei, R.-D., 1980: High pressure liquid chromatographic determination of aflatoxins in peanut and peanut products of Taiwan. Proceedings of the National Science Council, Republic of China 4, 152-155.
52. Mortimer, D.N., Shepherd, M.J., Gilbert, J., Morgan, M.R.A., 1987: A survey of the occurrence of aflatoxin B_1 in peanut butters by enzyme-linked immunosorbent assay. Food Additives and Contaminants 5, 127-132.
53. Ram, B.P., Hart, L.P., Cole, R.J., Pestka, J.J., 1986: Application of ELISA to retail survey of aflatoxin B_1 in peanut butter. Journal of Food Protection 49, 792-795.
54. Tung, T.C., Ling, K.H., 1968: Study on aflatoxin of foodstuffs in Taiwan. The Journal of Vitaminology 14, 48-52.

55. Vahl, M., Jørgensen, K., 1998: Determination of aflatoxins in food using LC/MS/MS. Zeitschrift für Lebensmittel-Untersuchung und -Forschung 206, 243-245.
56. Siame, B.A., Mpuchane, S.F., Gashe, B.A., Allotey, J., Teffera, G., 1998: Occurrence of aflatoxins, fumonisin B_1, and zearalenone in foods and feeds in Botswana. Journal of Food Protection 61, 1670-1673.
57. Elshafie, S.Z.B., ElMubarak, A., El-Nagerabi, S.A.F., Elshafie, A.E., 2011: Aflatoxin B_1 contamination of traditionally processed peanuts butter for human consumption in Sudan. Mycopathologia 171, 435-439. doi: https://doi.org/10.1007/s11046-010-9378-2.
58. Chun, H.S., Kim, H.J., Ok, H.E., Hwang, J.-B., Chung, D.-H., 2007: Determination of aflatoxin levels in nuts and their products consumed in South Korea. Food Chemistry 102, 385-391. doi: https://doi.org/10.1016/j.foodchem.2006.05.031.
59. Yentür, G., Er, B., Özkan, M.G., Öktem, A.B., 2006: Determination of aflatoxins in peanut butter and sesame samples using high-performance liquid chromatography method. European Food Research and Technology 224, 167-170. doi: https://doi.org/10.1007/s00217-006-0310-4.
60. Francis, O.J., Jr., Lipinski, L.J., Gaul, J.A., Campbell, A.D., 1982: High pressure liquid chromatographic determination of aflatoxins in peanut butter using a silicia gel-packed flowcell for fluorescence detection. Journal of the Association of Official Analytical Chemists 65, 672-676.
61. Li, F.-Q., Li, Y.-W., Wang, Y.-R., Luo, X.-Y., 2009: Natural occurrence of aflatoxins in Chinese peanut butter and sesame paste. Journal of Agricultural and Food Chemistry 57, 3519-3524. doi: https://doi.org/10.1021/jf804055n.
62. Hendrickse, R.G., Coulter, J.B.S., Lamplugh, S.M., MacFarlane, S.B.J., Williams, T.E., Omer, M.I.A., Suliman, G.I., 1982: Aflatoxins and kwashiorkor: a study in Sudanese children. British Medical Journal 285, 843-846.
63. Leong, Y.-H., Ismail, N., Latif, A.A., Ahmad, R., 2010: Aflatoxin occurrence in nuts and commercial nutty products in Malaysia. Food Control 21, 334-338. doi: https://doi.org/10.1016/j.foodcont.2009.06.002.
64. Razzazi-Fazeli, E., Noviandi, C.T., Porasuphatana, S., Agus, A., Böhm, J., 2004: A survey of aflatoxin B_1 and total aflatoxin contamination in baby food, peanut and corn products sold at retail in Indonesia analysed by ELISA and HPLC. Mycotoxin Research 20, 51-58.
65. Huang, B., Han, Z., Cai, Z., Wu, Y., Ren, Y., 2010: Simultaneous determination of aflatoxin B_1, B_2, G_1, G_2, M_1 and M_2 in peanuts and their derivative products by ultra-high-performance liquid chromatography-tandem mass spectrometry. Analytica Chimica Acta 662, 62-68. doi: https://doi.org/10.1016/j.aca.2010.01.002.
66. Cressey, P., Pearson, A., 2014: Risk Profile: Mycotoxins in the New Zealand Food Supply. ESR Client Report FW 14005. ESR, Christchurch, New Zealand.
67. Elzupir, A.O., Salih, A.O.A., Suliman, S.A., Adam, A.A., Elhussein, A.M., 2011: Aflatoxins in peanut butter in Khartoum State, Sudan. Mycotoxin Research 27, 183-186. doi: https://doi.org/10.1007/s12550-011-0094-7.
68. Matumba, L., Monjerezi, M., Biswick, T., Mwatseteza, J., Makumba, W., Kamangira, D., Mtukuso, A., 2014 : A survey of the incidence and level of aflatoxin contamination in a range of locally and imported processed foods on Malawian retail market. Food Control 39, 87-91. doi: https://doi.org/10.1016/j.foodcont.2013.09.068.
69. Mupunga, I., Lebelo, S.L., Mngqawa, P., Rheeder, J.P., Katerere, D.R., 2014: Natural occurrence of aflatoxins in peanuts and peanut butter from Bulawayo, Zimbabwe. Journal of Food Protection 77, 1814-1818. doi: https://doi.org/10.4315/0362-028X.JPF-14-129.
70. Bulatao-Jayme, J., Almero, E.M., Castro Ma, C.A., Jardeleza Ma, T.R., Salamat, L.A., 1982: A case-control dietary study of primary liver cancer risk from aflatoxin exposure. International Journal of Epidemiology 11, 112-119.
71. Gelda, C.S., Luyt, L.J., 1977: Survey of total aflatoxin content in peanuts, peanut butter, and other foodstuffs. Annales de la Nutrition et de l'Alimentation 31, 477-483.
72. Wood, G.E., 1989: Aflatoxins in domestic and imported foods and feeds. Journal of the Association of Official Analytical Chemists 72, 543-548.
73. Zhou, J., Xu, J.-J., Cong, J.-M., Cai, Z.-X., Zhang, J.-S., Wang, J.-L., Ren, Y.-P., 2018: Optimization for quick, easy, cheap, effective, rugged and safe extraction of mycotoxins and veterinary drugs by response surface methodology for application to egg and milk. Journal of Chromatography A 1532, 20-29. doi: https://doi.org/10.1016/j.chroma.2017.11.050.
74. MAFF - UK, 1996: Aflatoxin surveillance of retail and imported nuts, nut products, dried figs and fig products. Food-Surveillance-Information-Sheet No. 81.
75. Morgan, M.R.A., Kang, A.S., Chan, H.W.-S., 1986: Aflatoxin determination in peanut butter by enzyme-linked immunosorbent assay. Journal of the Science of Food and Agriculture 37, 908-914.
76. Abdulkadar, A.H.W., Al-Ali, A.A., Al-Kildi, A.M., Al-Jedah, J.H., 2004: Mycotoxins in food products available in Qatar. Food Control 15, 543-548. doi: https://doi.org/10.1016/j.foodcont.2003.08.008.
77. Younis, Y.M.H., Malik, K.M., 2003: TLC and HPLC assays of aflatoxin contamination in Sudanese peanuts and peanut products. Kuwait Journal of Science & Engineering 30, 79-93.
78. Nasir, M.S., Jolley, M.E., 2002: Development of a fluorescence polarization assay for the determination of aflatoxins in grains. Journal of Agricultural and Food Chemistry 50, 3116-3121. doi: https://doi.org/10.1021/jf011638c.
79. Gagliardi, S.J., Cheatle, T.F., Mooney, R.L., Llewellyn, G.C., O'Rear, C.E., Llewellyn, G.C., 1991: The occurrence of aflatoxin in peanut butter from 1982 to 1989. Journal of Food Protection 54, 627-631.
80. Ali, N., 2000: Aflatoxins in Malaysian food. Mycotoxins 50, 31-35.
81. Schwartzbord, J.R., Brown, D.L., 2015: Aflatoxin contamination in Haitian peanut products and maize and the safety of oil processed from contaminated peanuts. Food Control 56, 114-118. doi: https://doi.org/10.1016/j.foodcont.2015.03.014.
82. Chen, Y.-C., Liao, C.-D., Lin, H.-Y., Chiueh, L.-C., Shih, D.Y.-C., 2013: Survey of aflatoxin contamination in peanut products in Taiwan from 1997 to 2011. Journal of Food and Drug Analysis 21, 247-252. doi: https://doi.org/10.1016/j.jfda.2013.07.001.
83. Iqbal, S.Z., Asi, M.R., Zuber, M., Akram, N., Batool, N., 2013: Aflatoxins contamination in peanut and peanut products commercially available in retail markets of Punjab, Pakistan. Food Control 32, 83-86. doi: https://doi.org/10.1016/j.foodcont.2012.11.024.
84. Cressey, P., Thomson, B., 2006: Risk Profile: Mycotoxins in the New Zealand Food Supply. ESR Client Report FW 0617. ESR, Christchurch, New Zealand.
85. Cressey, P., Jones, S., 2010: Mycotoxin surveillance programme 2009-2010. Aflatoxins in nuts and nut products. ESR Client Report FW 10036. ESR, Christchurch, New Zealand.

86. Matumba, L., Van Poucke, C., Monjerezi, M., Njumbe Ediage, E., De Saeger, S., 2015: Concentrating aflatoxins on the domestic market through groundnut export: a focus on Malawian groundnut value and supply chain. Food Control 51, 236-239. doi: https://doi.org/10.1016/j.foodcont.2014.11.035.
87. Tam, J., Pantazopoulos, P., Scott, P.M., Moisey, J., Dabeka, R.W., Richard, I.D.K., 2011: Application of isotope dilution mass spectrometry: determination of ochratoxin A in the Canadian total diet study. Food Additives and Contaminants: Part A 28, 754-761. doi: https://doi.org/10.1080/19440049.2010.504750.
88. Nilüfer, D., Boyacioğlu, D., 2002: Comparative study of three different methods for the determination of aflatoxins in Tahini. Journal of Agricultural and Food Chemistry 50, 3375-3379. doi: https://doi.org/10.1021/jf020005a.
89. Baydar, T., Erkekoglu, P., Sipahi, H., Sahin, G., 2007: Aflatoxin B_1, M_1 and ochratoxin A levels in infant formulae and baby foods marketed in Ankara, Turkey. Journal of Food and Drug Analysis 15, 89-92.
90. Alvito, P.C., Sizoo, E.A., Almeida, C.M.M., Van Egmond, H.P., 2010: Occurrence of aflatoxins and ochratoxin A in baby foods in Portugal. Food Analytical Methods 3, 22-30. doi: https://doi.org/10.1007/s12161-008-9064-x.
91. Rastogi, S., Dwivedi, P.D., Khanna, S.K., Das, M., 2004: Detection of aflatoxin M_1 contamination in milk and infant milk products from Indian markets by ELISA. Food Control 15, 287-290. doi: https://doi.org/10.1016/S0956-7135(03)00078-1.
92. Oveisi, M.-R., Jannat, B., Sadeghi, N., Hajimahmoodi, M., Nikzad, A., 2007: Presence of aflatoxin M_1 in milk and infant milk products in Tehran, Iran. Food Control, 18, 1216-1218. doi: https://doi.org/10.1016/j.foodcont.2006.07.021.
93. Kabak, B., 2012: Aflatoxin M_1 and ochratoxin A in baby formulae in Turkey: occurrence and safety evaluation. Food Control 26, 182-187. doi: https://doi.org/10.1016/j.foodcont.2012.01.032.
94. De Girolamo, A., Pereboom-De Fauw, D., Sizoo, E., Van Egmond, H.P., Gambacorta, L., Bouten, K., Stroka, J., Visconti, A., Solfrizzo, M., 2010: Determination of fumonisins B_1 and B_2 in maize-based baby food products by HPLC with fluorimetric detection after immunoaffinity column clean-up. World Mycotoxin Journal 3, 135-146. doi: https://doi.org/10.3920/WMJ2010.1213.
95. Dogan, M., Liman, B.C., Sagdic, O., 2006: Total aflatoxin levels of instant foods in Turkey. Archiv für Lebensmittelhygiene 57, 55-58.
96. Hernández-Martínez, R., Navarro-Blasco, I., 2010: Aflatoxin levels and exposure assessment of Spanish infant cereals. Food Additives and Contaminants: Part B 3, 275-288. doi: https://doi.org/10.1080/19393210.2010.531402.
97. Mardani, M., Rezapour, S., Rezapour, P., 2011: Survey of aflatoxins in Kashkineh: a traditional Iranian food. Iranian Journal of Microbiology 3, 147-151.
98. El-Sayed Abd Alla, A.M., 1996: Natural occurrence of ochratoxin A and citrinin in food stuffs in Egypt. Mycotoxin Research 12, 41-44.
99. Mühlemann, M., Lüthy, J., Hübner, P., 1997: Mycotoxin contamination of food in Ecuador. A: aflatoxins. Mitteilungen aus dem Gebiete der Lebensmitteluntersuchung und Hygiene 88, 474-496.
100. Engel, G., 2000: Ochratoxin A in sweets, oil seeds and dairy products. Archiv für Lebensmittelhygiene 51, 98-101.
101. Ostry, V., Malir, F., Dofkova, M., Skarkova, J., Pfohl-Leszkowicz, A., Ruprich, J., 2015: Ochratoxin A dietary exposure of ten population groups in the Czech Republic: comparison with data over the world. Toxins (Basel) 7 (9), 3608-3635. doi: https://doi.org/10.3390/toxins7093608.
102. Jørgensen, K., 1998: Survey of pork, poultry, coffee, beer and pulses for ochratoxin A. Food Additives and Contaminants 15, 550-554.
103. Zou, Z., He, Z., Li, H., Han, P., Tang, J., Xi, C., Li, Y., Zhang, L., Li, X., 2012: Development and application of a method for the analysis of two trichothecenes: deoxynivalenol and T-2 toxin in meat in China by HPLC-MS/MS. Meat Science 90, 613-617. doi: https://doi.org/10.1016/j.meatsci.2011.10.002.
104. Zimmerli, B., Dick, R., 1995: Determination of ochratoxin A at the ppt level in human blood, serum, milk and some foodstuffs by high-performance liquid chromatography with enhanced fluorescence detection and immunoaffinity column cleanup: methodology and Swiss data. Journal of Chromatography B 666, 85-99.
105. Maaroufi, K., Achour, A., Betbeder, A.M., Hammami, M., Ellouz, F., Creppy, E.E., Bacha, H., 1995: Foodstuffs and human blood contamination by the mycotoxin ochratoxin A: correlation with chronic interstitial nephropathy in Tunisia. Archives of Toxicology 69, 552-558.
106. Haydar, M., Benelli, L., Brera, C., 1990: Occurrence of aflatoxins in Syrian foods and foodstuffs: a preliminary study. Food Chemistry 37, 261-268.
107. El-Gohary, A.H., 1995: Study on aflatoxins in some foodstuffs with special reference to public health hazard in Egypt. Asian-Australasian Journal of Animal Science 8, 571-575.
108. Asam, S., Rychlik, M., 2013: Potential health hazards due to the occurrence of the mycotoxin tenuazonic acid in infant food. European Food Research and Technology 236, 491-497. doi: https://doi.org/10.1007/s00217-012-1901-x.
109. Yang, L.-X., Liu, Y.-P., Miao, H., Dong, B., Yang, N.-J., Chang, F.-Q., Sun, J.-B., 2011: Determination of aflatoxins in edible oil from markets in Hebei Province of China by liquid chromatography-tandem mass spectrometry. Food Additives and Contaminants: Part B 4, 244-247. doi: https://doi.org/10.1080/19393210.2011.632694.
110. Bijl, J.P., Van Peteghem, C.H., Dekeyser, D.A., 1987: Fluorimetric determination of aflatoxin M_1 in cheese. Journal of the Association of Official Analytical Chemists 70, 472-475.
111. Veršilovskis, A., Van Peteghem, C., De Saeger, S., 2009: Determination of sterigmatocystin in cheese by high-performance liquid chromatography-tandem mass spectrometry. Food Additives and Contaminants: Part A 26, 127-133. doi: https://doi.org/10.1080/02652030802342497.
112. Bartoš, J., Matyáš, Z., 1982: Examination of cheeses for sterigmatocystine. Veterinární Medicína 27, 747-752.
113. Northolt, M.D., Van Egmond, H.P., Soentoro, P., Deijll, E., 1980: Fungal growth and the presence of sterigmatocystin in hard cheese. Journal of the Association of Official Analytical Chemists 63, 115-119.
114. Lafont, P., Siriwardana, M.G., DeBoer, E., 1990: Contamination of dairy products by fungal metabolites. Journal of Environmental Pathology, Toxicology and Oncology 10, 99-102.
115. Jarvis, B., 1983: Mould and mycotoxins in mouldy cheeses. Microbiologie – Aliments – Nutrition 1, 187-191.
116. Le Bars, J., 1979: Cyclopiazonic acid production by *Penicillium camemberti* Thom and natural occurrence of this mycotoxin in cheese. Applied and Environmental Microbiology 38, 1052-1055.
117. Zambonin, C.G., Monaci, L., Aresta, A., 2001: Determination of cyclopiazonic acid in cheese samples using solid-phase microextraction and high performance liquid chromatography. Food Chemistry 75, 249-254.

118. Lafont, P., Siriwardana, M.G., Combemale, I., Lafont, J., 1979: Mycophenolic acid in marketed cheeses. Food and Cosmetics Toxicology 17, 147-149.
119. Sinha, A.K., Ranjan, K.S., 1991: A report of mycotoxin contamination in Bhutanese cheese. Journal of Food Science and Technology, India 28, 398-399.
120. Hisada, K., Yamamoto, K., Tsubouchi, H., Sakabe, Y., 1984: Natural occurrence of aflatoxin M_1 in imported and domestic cheese. (Studies on mycotoxins in foods. VI). Journal of Food Hygienic Society of Japan, Tokyo 25, 543-548.
121. Dall'Asta, C., De Dea Lindner, J., Galaverna, G., Dossena, A., Neviani, E., Marchelli, R., 2008: The occurrence of ochratoxin A in blue cheese. Food Chemistry 106, 729-734. doi: https://doi.org/10.1016/j.foodchem.2007.06.049.
122. Scott, P.M., Kennedy, B.P.C., 1976: Analysis of blue cheese for roquefortine and other alkaloids from *Penicillium roqueforti*. Journal of Agricultural and Food Chemistry 24, 865-868.
123. Kokkonen, M., Jestoi, M., Rizzo, A., 2005: Determination of selected mycotoxins in mould cheeses with liquid chromatography coupled to tandem with mass spectrometry. Food Additives and Contaminants 22, 449-456. doi: https://doi.org/10.1080/02652030500089861.
124. Finoli, C., Vecchio, A., Galli, A., Dragoni, I., 2001: Roquefortine C occurrence in blue cheese. Journal of Food Protection 64, 246-251.
125. Ware, G.M., Thorpe, C.W., Pohland, A.E., 1980: Determination of roquefortine in blue cheese and blue cheese dressing by high pressure liquid chromatography with ultraviolet and electrochemical detectors. Journal of the Association of Official Analytical Chemists 63, 637-641.
126. Noroozian, E., Lagerwerf, F., Lingeman, H., Brinkman, U.A.Th., Kerkhoff, M.A.T., 1999: Determination of roquefortine C in blue cheese using on-line column-switching liquid chromatography. Journal of Pharmaceutical and Biomedical Analysis 20, 611-619.
127. Corbion, B., Fremy, J.M., 1978: Recherche des aflatoxines B_1 et M_1 dans les fromages de type "Camembert". Le Lait 58, 133-140.
128. Hisada, K., Terada, H., Yamamoto, K., Tsubouchi, H., Sakabe, Y., 1984: Reverse phase liquid chromatographic determination and confirmation of aflatoxin M_1 in cheese. Journal of the Association of Official Analytical Chemists 67, 601-606.
129. Dragacci, S., Gleizes, E., Fremy, J.M., Candlish, A.A.G., 1995: Use of immunoaffinity chromatography as a purification step for the determination of aflatoxin M_1 in cheeses. Food Additives and Contaminants 12, 59-65.
130. Kamber, U., 2005: Aflatoxin M_1 contamination of some commercial Turkish cheeses from markets in Kars, Turkey. Fresenius Environmental Bulletin 14, 1046-1049.
131. MAFF - UK, 1995: Survey of aflatoxin M_1 in retail milk and milk products. Food-Surveillance-Information-Sheet No. 64.
132. Yapar, K., Elmali, M., Kart, A., Yaman, H., 2008: Aflatoxin M_1 levels in different type of cheese products produced in Turkey. Medycyna Weterynaryjna 64, 53-55.
133. Gürses, M., Erdoğan, A., Çetin, B., 2004: Occurrence of aflatoxin M_1 in some cheese types sold in Erzurum, Turkey. Turkish Journal of Veterinary and Animal Sciences 28, 527-530.
134. El-Sawi, N.M., El-Maghraby, O.M.O., Mohran, H.S., Abo-Gharbia, M.A., 1994: Abnormal contamination of cottage cheese in Egypt. Journal of Applied Animal Research 6, 81-90.
135. Yaroglu, T., Oruc, H.H., Tayar, M., 2005: Aflatoxin M_1 levels in cheese samples from some provinces of Turkey. Food Control 16, 883-885. doi: https://doi.org/10.1016/j.foodcont.2004.08.001.
136. Fallah, A.A., Jafari, T., Fallah, A., Rahnama, M., 2009: Determination of aflatoxin M_1 levels in Iranian white and cream cheese. Food and Chemical Toxicology 47, 1872-1875. doi: https://doi.org/10.1016/j.fct.2009.04.042.
137. Richard, J.L., Arp, L.H., 1979: Natural occurrence of the mycotoxin penitrem A in moldy cream cheese. Mycopathologia 67, 107-109.
138. El-Sayed Abd Alla, A.M., Neamat-Allah, A.A., Soher, E.A., 2000: Situation of mycotoxins in milk, dairy products and human milk in Egypt. Mycotoxin Research 16, 91-100.
139. Kamkar, A., 2006: A study on the occurrence of aflatoxin M_1 in Iranian Feta cheese. Food Control 17, 768-775. doi:https://doi.org/10.1016/j.foodcont.2005.04.018.
140. Pietri, A., Bertuzzi, T., Bertuzzi, P., Piva, G., 1997: Aflatoxin M_1 occurrence in samples of Grana Padano cheese. Food Additives and Contaminants 14, 341-344.
141. El-Hoshy, S.M., 1999: Occurrence of zearalenone in milk, meat and their products with emphasis on influence of heat treatments on its level. Archiv für Lebensmittelhygiene 50, 140-143.
142. Colak, H., Hampikyan, H., Ulusoy, B., Ergun, O., 2006: Comparison of a competitive ELISA with an HPLC method for the determination of aflatoxin M_1 in Turkish White, Kasar and Tulum cheeses. European Food Research and Technology 223, 719-723. doi: https://doi.org/10.1007/s00217-006-0258-4.
143. Tekinşen, K.K., Eken, H.S., 2008: Aflatoxin M_1 levels in UHT milk and kashar cheese consumed in Turkey. Food and Chemical Toxicology 46, 3287-3289. doi: https://doi.org/10.1016/j.fct.2008.07.014.
144. Ertas, N., Gonulalan, Z., Yildirim, Y., Karadal, F., 2011: A survey of concentration of aflatoxin M_1 in dairy products marketed in Turkey. Food Control 22, 1956-1959. doi: https://doi.org/10.1016/j.foodcont.2011.05.009.
145. Prado, G., Oliveira, M.S., Pereira, M.L., Abrantes, F.M., Santos, L.G., Veloso, T., 2000: Aflatoxin M_1 in samples of "Minas" cheese commercialized in the city of Belo Horizonte - Minas Gerais/Brazil. Ciência e Tecnologia de Alimentos 20, 398-400. doi: https://doi.org/10.1590/S0101-20612000000300020.
146. Barbieri, G., Bergamini, C., Ori, E., Resca, P., 1994: Aflatoxin M_1 in Parmesan cheese: HPLC determination. Journal of Food Science 59, 1313, 1331.
147. Abd Alla, E.A.M., Metwally, M.M., Mehriz, A.M., Abu Sree, Y.H., 1996: Sterigmatocystin: incidence, fate and production by *Aspergillus versicolor* in Ras cheese. Nahrung 40, 310-313.
148. Aygun, O., Essiz, D., Durmaz, H., Yarsan, E., Altinas, L., 2009: Aflatoxin M_1 levels in Surk samples, a traditional Turkish cheese from southern Turkey. Bulletin of Environmental Contamination and Toxicology 83, 164-167. doi: https://doi.org/10.1007/s00128-009-9765-x.
149. Tekinşen, K.K., Tekinşen, O.C., 2005: Aflatoxin M_1 in white pickle and Van otlu (herb) cheeses consumed in southeastern Turkey. Food Control 16, 565-568. doi: https://doi.org/10.1016/j.foodcont.2004.02.006.
150. Elgerbi, A.M., Aidoo, K.E., Candlish, A.A.G., Tester, R.F., 2004: Occurrence of aflatoxin M_1 in randomly selected North African milk and cheese samples. Food Additives and Contaminants 21, 592-597. doi: https://doi.org/10.1080/02652030410001687690.

151. Ardic, M., Karakaya, Y., Atasever, M., Adiguzel, G., 2009: Aflatoxin M_1 levels of Turkish white brined cheese. Food Control 20, 196-199. doi: https://doi.org/10.1016/j.foodcont.2008.04.003.
152. Pietri, A,. Bertuzzi, T., Gualla, A., Piva, G., 2006: Occurrence of ochratoxin A in raw ham muscles and in pork products from northern Italy. Italian Journal of Food Science 18, 99-106.
153. Rubio, R., Licón, C.C., Berruga, M.I., Molina, M.P., Molina, A., 2011: Short communication: occurrence of aflatoxin M_1 in the Manchego cheese supply chain. Journal of Dairy Science 94, 2775-2778. doi: https://doi.org/10.3168/jds.2010-4017.
154. Mutlu, A.G., Kursun, O., Kasimoglu, A., Dukel, M., 2010: Determination of aflatoxin M_1 levels and antibiotic residues in the traditional Turkish desserts and ice creams consumed in Burdur City Center. Journal of Animal and Veterinary Advances 9, 2035-2037.
155. Frenich, A.G., Romero-Gonzáles, R., Gómez-Pérez, M.L., Vidal, J.L.M., 2011: Multi-mycotoxin analysis in eggs using a QuEChERS-based extraction procedure and ultra-high-pressure liquid chromatography coupled to triple quadrupole mass spectrometry. Journal of Chromatography A 1218, 4349-4356. doi: https://doi.org/10.1016/j.chroma.2011.05.005.
156. Tchana, A.N., Moundipa, P.F., Tchouanguep, F.M., 2010: Aflatoxin contamination in food and body fluids in relation to malnutrition and cancer status in Cameroon. International Journal of Environmental Research and Public Health 7, 178-188. doi: https://doi.org/10.3390/ijerph7010178.
157. Herzallah, S.M., 2009: Determination of aflatoxins in eggs, milk, meat and meat products using HPLC fluorescent and UV detectors. Food Chemistry 114, 1141-1146. doi: https://doi.org/10.1016/j.foodchem.2008.10.077.
158. Jestoi, M., Rokka, M., Järvenpää, E., Peltonen, K., 2009: Determination of *Fusarium* mycotoxins beauvericin and enniatins (A, A_1, B, B_1) in eggs of laying hens using liquid chromatography-tandem mass spectrometry (LC-MS/MS). Food Chemistry 115, 1120-1127. doi: https://doi.org/10.1016/j.foodchem.2008.12.105.
159. Jonsyn, F.E., Lahai, G.P., 1992: Mycotoxic flora and mycotoxins in smoke-dried fish from Sierra Leone. Nahrung 36, 485-489.
160. Valitutti, F., De Santis, B., Trovato, C.M., Montuori, M., Gatti, S., Oliva, S., Brera, C., Catassi, C., 2018: Assessment of mycotoxin exposure in breastfeeding mothers with celiac disease. Nutrients 10, 336. doi: https://doi.org/10.3390/nu10030336.
161. Ishikawa, A.T., Takabayashi-Yamashita, C.R., Ono, E.Y.S., Bagatin, A.K., Rigobello, F.F., Kawamura, O., Hirooka, E.Y., Itano, E.N., 2016: Exposure assessment of infants to aflatoxin M_1 through consumption of breast milk and infant powdered milk in Brazil. Toxins (Basel) 8, 246. doi: https://doi.org/10.3390/toxins8090246.
162. Shank, R.C., Wogan, G.N., Gibson, J.B., Nondasuta, A., 1972: Dietary aflatoxins and human liver cancer. II. Aflatoxins in marketed foods and foodstuffs of Thailand and Hong Kong. Food Cosmetics and Toxicology 10, 61-69.
163. Farah Nadira, A., Rosita, J., Norhaizan, M.E., Mohd Redzwan, S., 2017: Screening of aflatoxin M_1 occurrence in selected milk and dairy products in Terengganu, Malaysia. Food Control 73, 209-214. doi: https://doi.org/10.1016/j.foodcont.2016.08.004.
164. Meucci, V., Soldani, G., Razzuoli, E., Saggese, G., Massart, F., 2011: Mycoestrogen pollution of Italian infant food. Journal of Pediatrics. 159, 278-283. doi: https://doi.org/10.1016/j.peds.2011.01.028.
165. Chiavaro, E., Lepiani, A., Colla, F., Bettoni, P., Pari, E., Spotti, E., 2002: Ochratoxin A determination in ham by immunoaffinity clean-up and a quick fluorometric method. Food Additives and Contaminants 19, 575-581. doi: https://doi.org/10.1080/02652030210123869.
166. Toscani, T., Moseriti, A., Dossena, A., Dall'Asta, C., Simoncini, N., Virgili, R., 2007: Determination of ochratoxin A in dry-cured meat products by a HPLC-FLD quantitative method. Journal of Chromatography B 855, 242-248. doi: https://doi.org/10.1016/j.jchromb.2007.05.010.
167. Pietri, A., Gualla, A., Rastelli, S., Bertuzzi, T., 2011: Enzyme-assisted extraction for the HPLC determination of ochratoxin A in pork and dry-cured ham. Food Additives and Contaminants: Part A 28, 1717-1723. doi: https://doi.org/10.1080/19440049.2011.609490.
168. Rodríguez, A., Rodríguez, M., Martín, A., Delgado, J., Córdoba, J.J., 2012: Presence of ochratoxin A on the surface of dry-cured Iberian ham after initial fungal growth in the drying stage. Meat Science 92, 728-734. doi: https://doi.org/10.1016/j.meatsci.2012.06.029.
169. Anderson, P.H., Wells, G.A.H., Jackman, R., Morgan, M.R.A., 1984: Ochratoxicosis and ochratoxin A residues in adult pig's kidneys – a pilot survey. In: Moss, M.O., Frank, M. (eds): Proceedings of the 5th Meeting of Mycotoxins in Animal and Human Health. Surrey University Press, Guildford, UK, pp. 23-29.
170. Dragacci, S., Grosso, F., Bire, R., Fremy, J.M., Coulon, S., 1999: A French monitoring programme for determining ochratoxin A occurrence in pig kidneys. Natural Toxins 7, 167-173.
171. Goliński, P., Hult, K., Grabarkiewicz-Szczęsna, J., Chelkowski, J., Szebiotko, K., 1985: Spontaneous occurrence of ochratoxin A residues in porcine kidney and serum samples in Poland. Applied and Environmental Microbiology 49, 1014-1015.
172. Büchmann, N.B., Hald, B., 1985: Analysis, occurrence and control of ochratoxin A residues in Danish pig kidneys. Food Additives and Contaminants 2, 193-199.
173. Curtui, V.G., Gareis, M., Usleber, E., Märtlbauer, E., 2001: Survey of Romanian slaughtered pigs for the occurrence of mycotoxins ochratoxins A and B, and zearalenone. Food Additives and Contaminants 18, 730-738. doi: https://doi.org/10.1080/02652030110035101.
174. Fukal, L., 1990: A survey of cereals, cereal products, feedstuffs and porcine kidneys for ochratoxin A by radioimmunoassay. Food Additives and Contaminants 7, 253-258.
175. Jørgensen, K., Petersen, A., 2002: Content of ochratoxin A in paired kidney and meat samples from healthy Danish slaughter pigs. Food Additives and Contaminants 19, 562-567. doi: https://doi.org/10.1080/02652030110113807.
176. Krogh, P., 1977: Ochratoxin A residues in tissues of slaughter pigs with nephropathy. Nordic Veterinary Medicine A 29, 402-405.
177. Morgan, M.R.A., McNerney, R., Chan, H.W.-S., Anderson, P.H., 1986: Ochratoxin A in pig kidney determined by enzyme-linked immunosorbent assay (ELISA). Journal of the Science of Food and Agriculture 37, 475-480.
178. Rousseau, D.M., Van Peteghem, C.H., 1989: Spontaneous occurrence of ochratoxin A residues in porcine kidneys in Belgium. Bulletin of Environmental Contamination and Toxicology 42, 181-186.
179. Rutqvist, L., Björklund, N.-E., Hult, K., Gatenbeck, S., 1977: Spontaneous occurrence of ochratoxin residues in kidneys of fattening pigs. Zentralblatt für Veterinär Medizin A 24, 402-408.
180. Tyllinen, M., Hintikka, E.-L., 1982: Occurrence of ochratoxin A in swine kidneys and feed in Finland. Nordisk Jordbruksforskning 64, 298-299.
181. Rousseau, D.M., Candlish, A.A.G., Slegers, G.A., Van Peteghem, C.H., Stimson, W.H., Smith, J.E. 1987: Detection of ochratoxin A in porcine kidneys by a monoclonal antibody-based radioimmunoassay. Applied and Environmental Microbiology 53, 514-518.

182. Gresham, A., Done, S., Livesey, C., MacDonald, S., Chan, D., Sayers, R., Clark, C., Kemp, P., 2006: Survey of pigs' kidneys with lesions consistent with PMWS and PDNS and ochratoxicosis. Part 1: concentrations and prevalence of ochratoxin A. Veterinary Record 159, 737-742.
183. Josefsson, E., 1979: Study of ochratoxin A in pig kidneys. Vår Föda 31, 415-420.
184. Milićević, D., Jurić, V., Stefanović, S., Jovanović, M., Janković, S., 2008: Survey of slaughtered pigs for occurrence of ochratoxin A and porcine nephropathy in Serbia. International Journal of Molecular Sciences 9, 2169-2183. doi: https://doi.org/10.3390/ijms9112169.
185. Losito, I., Monaci, L., Palmisano, F., Tantillo, G., 2004: Determination of ochratoxin A in meat products by high-performance liquid chromatography coupled to electrospray ionisation sequential mass spectrometry. Rapid Communications in Mass Spectrometry 18, 1965-1971. doi: https://doi.org/10.1002/rcm.1577.
186. Monaci, L., Tantillo, G., Palmisano, F., 2004: Determination of ochratoxin A in pig tissues by liquid-liquid extraction and clean-up and high-performance liquid chromatography. Analytical and Bioanalytical Chemistry 378, 1777-1782. doi: https://doi.org/10.1007/s00216-004-2497-1.
187. Elling, F., Hald, B., Jacobsen, Chr., Krogh, P., 1975: Spontaneous toxic nephropathy in poultry associated with ochratoxin A. Acta Pathologica et Microbiologica Scandinavica Section A 83, 739-741.
188. Sabino, M., Purchio, A., Milanez, T.V., 1996: Survey of aflatoxins B_1, M_1 and aflatoxicol in poultry and swine tissues from farms located in the states of Rio Grande do Sul and Santa Catarina, Brazil. Revista de Microbiologia 27, 189-191.
189. Bintvihok, A., Davitiyananda, D., Kositcharoenkul, S., Panichkriangkrai, W., Jamratchai, O., 1998: Residues of aflatoxins and their metabolites in chicken tissues in Thailand. The Journal of Toxicological Sciences 23 (Suppl II), 389.
190. Honstead, J.P., Dreesen, D.W., Stubblefield, R.D., Shotwell, O.L., 1992: Aflatoxins in swine tissues during drought conditions: an epidemiologic study. Journal of Food Protection 55, 182-186.
191. Ismail, M.A., Zaky, Z.M., 1999: Evaluation of the mycological status of luncheon meat with special reference to aflatoxigenic moulds and aflatoxin residues. Mycopathologia 146, 147-154.
192. Commission of the European Communities, 1997. Reports on tasks for scientific cooperation. Report of experts participating in task 3.2.2: assessment of dietary intake of ochratoxin A by the population of EU member states. Report EUR 17523 EN.
193. Zheng, N., Sun, P., Wang, J.Q., Zhen, Y.P., Han, R.W., Xu, X.M., 2013: Occurrence of aflatoxin M_1 in UHT milk and pasteurized milk in China market. Food Control 29, 198-201. doi: https://doi.org/10.1016/j.foodcont.2012.06.020.
194. Díaz, S., Domínguez, L., Prieta, J., Blanco, J.L., Moreno, M.A., 1995: Application of a diphasic membrane procedure for surveying occurrence of aflatoxin M_1 in commercial milk. Journal of Agricultural and Food Chemistry 43, 2678-2680.
195. Roussi, V., Govaris, A., Varagouli, A., Botsoglou, N.A., 2002: Occurrence of aflatoxin M_1 in raw and market milk commercialized in Greece. Food Additives and Contaminants 19, 863-868. doi: https://doi.org/10.1080/02652030210146864.
196. Saitanu, K., 1997: Incidence of aflatoxin M_1 in Thai milk products. Journal of Food Protection 60, 1010-1012.
197. Taveira, J.A., Midio, A.F., 2001: Incidence of aflatoxin M_1 in milk marketed in São Paulo, Brazil. Italian Journal of Food Science 13, 443-447.
198. Choudhary, P.L., Sharma, R.S., Borkhatriya, V.N., Murthi, T.N., Wadodkar, U.R., 1997: Survey on the levels of aflatoxin M_1 in raw and market milk in and around Anand town. Indian Journal of Dairy Science 50, 156-158.
199. De Sylos, C.M., Rodriguez-Amaya, D.B., Carvalho, P.R.N., 1996: Occurrence of aflatoxin M_1 in milk and dairy products commercialized in Campinas, Brazil. Food Additives and Contaminants 13, 169-172.
200. Burdaspal, P.A., Legarda, T.M., Pinilla, I., 1983: Note. Occurrence of aflatoxin M_1 contamination in milk. Revista de Agroquímica y Tecnologia de Alímentos 23, 287-290.
201. Er, B., Demirhan, B., Yentür, G., 2014: Short communication: investigation of aflatoxin M_1 levels in infant follow-on milks and infant formulas sold in the markets of Ankara, Turkey. Journal of Dairy Science 97, 3328-3331. doi: https://doi.org/10.3168/jds.2013-7831.
202. Sahindokuyucu Kocasari, F., 2014: Occurrence of aflatoxin M_1 in UHT milk and infant formula samples consumed in Burdur, Turkey. Environmental Monitoring and Assessment 186, 6363-6368. doi: https://doi.org/10.1007/s10661-014-3860-0.
203. Picinin, L.C.A., Oliveira Pinho Cerqueira, M.M., Azevedo Vargas, E., Quintão Lana, A.M., Toaldo, I.M., Bordignon-Luiz, M.T., 2013: Influence of climate conditions on aflatoxin M_1 contamination in raw milk from Minas Gerais State, Brazil. Food Control 31, 419-424. doi: https://doi.org/10.1016/j.foodcont.2012.10.024.
204. Ceci, E., Bozzo, G., Bonerba, E., Di Pinto, A., Tantillo, M.G., 2007: Ochratoxin A detection by HPLC in target tissues of swine and cytological and histological analysis. Food Chemistry 105, 364-368. doi: https://doi.org/10.1016/j.foodchem.2006.12.019.
205. Srivastava, V.P., Bu-Abbas, A., Alaa-Basuny, W., Al-Johar, W., Al-Mufti, S., Siddiqui, M.K.J., 2001: Aflatoxin M_1 contamination in commercial samples of milk and dairy products in Kuwait. Food Additives and Contaminants 18, 993-997. doi: https://doi.org/10.1080/02652030110050357.
206. Piñeiro, M., Dawson, R., Costarrica, M.L., 1996: Monitoring program for mycotoxin contamination in Uruguayan food and feeds. Natural Toxins 4, 242-245.
207. MAFF – UK, 1996: Survey of aflatoxin M_1 in farm gate milk. Food-Surveillance-Information-Sheet No. 78.
208. Food Standards Agency, 2001: Survey of milk for mycotoxins. Food-Survey-Information-Sheet No. 17/01.
209. Zinedine, A., González-Osnaya, L., Soriano, J.M., Moltó, J.C., Idrissi, L., Mañes, J., 2007: Presence of aflatoxin M_1 in pasteurized milk from Morocco. International Journal of Food Microbiology 114, 25-29. doi: https://doi.org/10.1016/j.ijfoodmicro.2006.11.001.
210. Alborzi, S., Pourabbas, B., Rashidi, M., Astaneh, B., 2006: Aflatoxin M_1 contamination in pasteurized milk in Shiraz (south of Iran). Food Control 17, 582-584. doi: https://doi.org/10.1016/j.foodcont.2005.03.009.
211. Alonso, V.A., Monge, M.P., Larriestra, A., Dalcero, A.M., Cavaglieri, L.R., Chiachiera, S.M., 2010: Naturally occurring aflatoxin M_1 in raw bulk milk from farm cooling tanks in Argentina. Food Additives and Contaminants: Part A 27, 373-379. doi: https://doi.org/10.1080/19440040903403362.
212. Boudra, H., Barnouin, J., Dragacci, S., Morgavi, D.P., 2007: Aflatoxin M_1 and ochratoxin A in raw bulk milk from French dairy herds. Journal of Dairy Science 90, 3197-3201. doi: https://doi.org/10.3168/jds.2006-565.
213. Nuryono, N., Agus, A., Wedhastri, S., Maryudani, Y.B., Sigit Setyabudi, F.M.C., Böhm, J., Razzazi-Fazeli, E., 2009: A limited survey of aflatoxin M_1 in milk from Indonesia by ELISA. Food Control 20, 721-724. doi: https://doi.org/10.1016/j.foodcont.2008.09.005.

214. Chung, S. W.-C., Kwong, K.P., Tang, A.S.P., Yeung, S.T.K., 2009: Ochratoxin A levels in foodstuffs marketed in Hong Kong. Journal of Food Composition and Analysis. 22, 756-761. doi: https://doi.org/10.1016/j.jfca.2009.02.014.
215. Commission of the European Communities, January 2002. Reports on tasks for scientific cooperation. Report of experts participating in task 3.2.7: assessment of dietary intake of ochratoxin A by the population of EU member states.
216. Cano-Sancho, G., Marin, S., Ramos, A.J., Peris-Vicente, J., Sanchis, V., 2010: Occurrence of aflatoxin M_1 and exposure assessment in Catalonia (Spain). Revista Iberoamericana de Micología 37, 130-135. doi: https://doi.org/10.1016/j.riam.2010.05.003.
217. Capei, R., Neri, P., 2002: Occurrence of aflatoxin M_1 in milk and yoghurt offered for sale in Florence (Italy). Annali di Igiene 14, 313-319.
218. Gholampour Azizi, I., Khoushnevis, S.H., Hashemi, S.J., 2007: Aflatoxin M_1 level in pasteurized and sterilized milk of Babol city. Tehran University Medical Journal 65, Supplement 1, 20-24.
219. Martins, H.M., Guerra, M.M., Bernardo, F., 2005: A six year survey (1999-2004) of the occurrence of aflatoxin M_1 in dairy products produced in Portugal. Mycotoxin Research 21, 192-195.
220. Patel, P.M., Netke, S.P., Gupta, B.S., Dabadghao, A.K., 1981: Note on the survey of consumer milk supplied to Jabalpur city for the incidence of aflatoxin M_1 and M_2. Indian Journal of Animal Science 51, 906.
221. Souza, S.V.C., Vargas, E.A., Junqueira, R.G., 1999: Efficiency of an ELISA kit on the detection and quantification of aflatoxin M_1 in milk and it's occurrence in the state of Minas Gerais. Ciência e Tecnologia de Alimentos 19, 401-405. doi: https://doi.org/10.1590/S0101-20611999000300019.
222. Winterlin, W., Hall, G., Hsieh, D.P.H., 1979: On-column chromatographic extraction of aflatoxin M_1 from milk and determination by reversed phase high performance liquid chromatography. Analytical Chemistry 51, 1873-1874.
223. Losito, I., Monaci, L., Aresta, A., Zambonin, C.G., 2002: LC-ion trap electrospray MS-MS for the determination of cyclopiazonic acid in milk samples. Analyst 127, 499-502. doi: https://doi.org/10.1039/b200394p.
224. Pattono, D., Gallo, P.F., Civera, T., 2011: Detection and quantification of ochratoxin A in milk produced in organic farms. Food Chemistry 127, 374-377. doi: https://doi.org/10.1016/j.foodchem.2010.12.051.
225. Gallo, P., Salzillo, A., Rossini, C., Urbani, V., Serpe, L., 2006: Aflatoxin M_1 determination in milk: method validation and contamination levels in samples from southern Italy. Italian Journal of Food Science 18, 251-259.
226. Motawee, M.M., Bauer, J., McMahon, D.J., 2009: Survey of aflatoxin M_1 in cow, goat, buffalo and camel milks in Ismailia-Egypt. Bulletin of Environmental Contamination and Toxicology 83, 766-769. doi: https://doi.org/10.1007/s00128-009-9840-3.
227. Rahimi, E., Bonyadian, M., Rafei, M., Kazemeini, H.R., 2010: Occurrence of aflatoxin M_1 in raw milk of five dairy species in Ahzav, Iran. Food and Chemical Toxicology 48, 129-131. doi: https://doi.org/10.1016/j.fct.2009.09.028.
228. Lindahl, J.F., Kagera, I.N., Grace, D., 2018: Aflatoxin M_1 levels in different marketed milk products in Nairobi, Kenya. Mycotoxin Research 34, 1-7. doi: https://doi.org/10.1007/s12550-018-0323-4.
229. Raza, R., 2006: Occurrence of aflatoxin M_1 in the milk marketed in the city of Karachi, Pakistan. Journal of the Chemical Society of Pakistan 28, 155-157.
230. Saad, A.M., Abdelgadir, A.M., Moss, M.O., 1989: Aflatoxin in human and camel milk in Abu Dhabi, United Arab Emirates. Mycotoxin Research 5, 57-60.
231. Carvajal, M., Rojo, F., Méndez, I., Bolaños, A., 2003: Aflatoxin B_1 and its interconverting metabolite aflatoxicol in milk: the situation in Mexico. Food Additives and Contaminants 20, 1077-1086. doi: https://doi.org/10.1080/02652030310001594478.
232. Fukal, L., Březina, P., Marek, M., 1990: Immunochemical monitoring of aflatoxin M_1 occurrence in milk produced in Czechoslovakia. Deutsche Lebensmittel-Rundschau 86, 289-291.
233. Gajek, O., 1983: Aflatoxins in protein food for animals and milk. Roczniki Państwowego Zakładu Higieny 33, 415-420.
234. Galvano, F., Galofaro, V., De Angelis, A., Galvano, M., Bognanno, M., Galvano, G., 1998: Survey of the occurrence of aflatoxin M_1 in dairy products marketed in Italy. Journal of Food Protection 61, 738-741.
235. Gilbert, J., Shepherd, M.J., Wallwork, M.A., Knowles, M.E., 1984: A survey of the occurrence of aflatoxin M_1 in UK-produced milk for the period 1981-1983. Food Additives and Contaminants 1, 23-28.
236. Visconti, A., Bottalico, A., Solfrizzo, M., 1985: Aflatoxin M_1 in milk, in southern Italy. Mycotoxin Research 1, 71-75.
237. Dragacci, S., Fremy, J.-M., 1993: Occurrence of aflatoxin M_1 in milk. Fifteen years of sanitary control. Sciences des Aliments 13, 711-722.
238. Karaioannoglou, P., Mantis, A., Koufidis, D., Koidis, P., Triantafillou, J., 1989: Occurrence of aflatoxin M_1 in raw and pasteurized milk and in Feta and Telme cheese samples. Milchwissenschaft 44, 746-748.
239. Kiermeier, F., 1973: Aflatoxin residues in fluid milk. Pure and Applied Microbiology 35, 271-273.
240. Kiermeier, F., Weiss, G., Behringer, G., Miller, M., Ranfft, K., 1977: On the presence and the content of aflatoxin M_1 in milk shipped to a dairy plant. Zeitschrift für Lebensmittel-Untersuchung und –Forschung 163, 171-174.
241. Kim, E.K., Shon, D. H., Ryu, D., Park, J.W., Hwang, H.J., Kim, Y. B., 2000: Occurrence of aflatoxin M_1 in Korean dairy products determined by ELISA and HPLC. Food Additives and Contaminants 17, 59-64.
242. Cathey, C.G., Huang, Z.G., Sarr, A.B., Clement, B.A., Phillips, T.D., 1994: Development and evaluation of a minicolumn assay for the detection of aflatoxin M_1 in milk. Journal of Dairy Science 77, 1223-1231.
243. Veselý, D., Veselá, D., 1983: Determination of M_1 aflatoxin in milk and of its toxic effect on chick embryo. Veterinární Medicína 28, 57-61.
244. Schuddeboom, L.J., 1983: Development of legislation concerning mycotoxins in dairy products in The Netherlands. Microbiologie – Aliments – Nutrition 1, 179-185.
245. Markaki, P., Melissari, E., 1997: Occurrence of aflatoxin M_1 in commercial pasteurized milk determined with ELISA and HPLC. Food Additives and Contaminants 14, 451-456.
246. Nieuwenhof, F.F.J., Hoolwerf, J.D., Van Den Bedem, J.W., 1990: Evaluation of an enzyme immunoassay for the determination of aflatoxin M_1 in milk using antibody-coated polystyrene beads. Milchwissenschaft 45, 584-588.
247. Palermo, D., Palermo, C., Rotunno, T., 2001: Survey of aflatoxin M_1 level in cow milk from Puglia Italy. Italian Journal of Food Science 13, 435-442.
248. Pettersson, H., Holmberg, T., Larsson, K., Kaspersson, A., 1989: Aflatoxins in acid-treated grain in Sweden and occurrence of aflatoxin M_1 in milk. Journal of the Science of Food and Agriculture 48, 411-420.
249. Purchase, I.F.H., Vorster, L.J., 1968: Aflatoxin in commercial milk samples. South African Medical Journal 42, 219.

250. Rajan, A., Ismail, P.K., Radhakrishnan, V., 1995: Survey of milk samples for aflatoxin M_1 in Thrissur, Kerala. Indian Journal of Dairy Science 48, 302-305.
251. Sabino, M., Purchio, A., Zorzetto, M.A.P., 1989: Variations in the levels of aflatoxin in cows milk consumed in the city of São Paulo, Brazil. Food Additives and Contaminants 6, 321-326.
252. Salem, D.A., 2002: Natural occurrence of aflatoxins in feedstuffs and milk of dairy farms in Assiut Province, Egypt. Wiener Tierärztliche Monatsschrift 89, 86-91.
253. Möller, T., Andersson, S., 1983: Aflatoxin M_1 in Swedish milk. Vår Föda 35, 461-465.
254. Bartoš, J., Matyáš, Z., 1981: Research on aflatoxin M_1 in fresh milk. Veterinární Medicína 26, 419-423.
255. Stoloff, L., 1980: Aflatoxin M in perspective. Journal of Food Protection 43, 226-230.
256. Waliyar, F., Reddy, S.V., Subramaniam, K., Reddy, T.Y., Rama Devi, K., Craufurd, P.Q., Wheeler, T.R., 2003: Importance of mycotoxins in food and feed in India. Aspects of Applied Biology 68, 147-154.
257. Shon, D.-H., Lim, S.-H., Lee, Y.-W., 1996: Detection of aflatoxin in cow's milk by an enzyme-linked immunosorbent assay. Korean Journal of Applied Microbiology and Biotechnology 24, 630-635.
258. Bakirci, I., 2001: A study on the occurrence of aflatoxin M_1 in milk and milk products produced in Van Province of Turkey. Food Control 12, 47-51.
259. Martins, M.L., Martins, H.M., 2000: Aflatoxin M_1 in raw and ultra high temperature-treated milk commercialized in Portugal. Food Additives and Contaminants 17, 871-874.
260. Sassahara, M., Pontes Netto, D., Yanaka, E.K., 2005: Aflatoxin occurrence in foodstuff supplied to dairy cattle and aflatoxin M_1 in raw milk in the North of Paraná State. Food and Chemical Toxicology 43, 981-984. doi: https://doi.org/10.1016/j.fct.2005.02.003.
261. Amra, H.A., 1998: Survey of aflatoxin M_1 in Egyptian raw milk by enzyme-linked immunosorbent assay. Revue de Médecine Vétérinaire 149, 695.
262. Domagala, J., Kisza, J., Blüthgen, A., Heeschen, W., 1997: Contamination of milk with aflatoxin M_1 in Poland. Milchwissenschaft 52, 631-633.
263. Oliveira, C.A., Rosmaninho, J., Rosim, R., 2006: Aflatoxin M_1 and cyclopiazonic acid in fluid milk traded in São Paulo, Brazil. Food Additives and Contaminants 23, 196-201. doi: https://doi.org/10.1080/02652030500398379.
264. Awashti, V., Bahman, S., Thakur, L.K., Singh, S.K., Dua, A., Ganguly, S., 2012: Contaminants in milk and impact of heating: an assessment study. Indian Journal of Public Health 56, 95-99. doi: https://doi.org/10.4103/0019-557X.96985.
265. Meerarani, S., Ramadass, P., Padmanaban, V.D., Nachimuthu, K., 1997: Incidence of aflatoxin M_1 in milk samples around Chennai (Madras) City. Journal of Food Science and Technology 34, 506-508.
266. Boccia, A., Micco, C., Miraglia, M., Scioli, M., 1986: A study on milk contamination by aflatoxin M_1 in a restricted area in central Italy. Microbiologie – Aliments – Nutrition 4, 293-298.
267. Diaz, G.J., Espitia, E., 2006: Occurrence of aflatoxin M_1 in retail milk samples from Bogotá, Colombia. Food Additives and Contaminants 23, 811-815. doi: https://doi.org/10.1080/02652030600681617.
268. López, C.E., Ramos, L.L., Ramadán, S.S., Bulacio, L.C., 2003: Presence of aflatoxin M_1 in milk for human consumption in Argentina. Food Control 14, 31-34.
269. Nakajima, M., Tabata, S., Akiyama, H., Itoh, Y., Tanaka, T., Sunagawa, H., Tyonan, T., Yoshizawa, T., Kumagai, S., 2004: Occurrence of aflatoxin M_1 in domestic milk in Japan during the winter season. Food Additives and Contaminants 21, 472-478. doi: https://doi.org/10.1080/02652030410001677817.
270. Garrido, N.S., Iha, M.H., Santos Ortolani, M.R., Duarte Fávaro, R.M., 2003: Occurrence of aflatoxins M_1 and M_2 in milk commercialized in Ribeirão Preto-SP, Brazil. Food Additives and Contaminants 20, 70-73. doi: https://doi.org/10.1080/0265203021000035371.
271. Yurdun, T., Özmenteşe, N., 2001: Presence of aflatoxin M_1 in milk and dairy products in Turkey. Toxicology Letters 123 (Suppl), 39-40.
272. Rodríguez Velasco, M.L., Calonge Delso, M.M., Ortónez Escudero, D., 2003: ELISA and HPLC determination of the occurrence of aflatoxin M_1 in raw cow's milk. Food Additives and Contaminants 20, 276-280. doi: https://doi.org/10.1080/0265203021000045208.
273. Prado, G., Silva de Oliveira, M., Souza Lima, A., Aprigio Moreira, A.P., 2008: Occurrence of aflatoxin M_1 in Parmesan cheese consumed in Minas Gerais, Brazil. Ciência e Agrotecnologia 32, 1906-1911.
274. Massart, F., Micillo, F., Rivezzi, G., Perrone, L., Baggiani, A., Miccoli, M., Meucci, V., 2016: Zearalenone screening of human breast milk from the Naples area. Toxicological and Environmental Chemistry 98, 128-136. doi: https://doi.org/10.1080/02772248.2015.1101112.
275. López, P., De Rijk, T., Sprong, R.C., Mengelers, M.J.B., Castenmiller, J.J.M., Alewijn, M. 2016: A mycotoxin-dedicated total diet study in the Netherlands in 2013: part II - occurrence. World Mycotoxin Journal 9, 89-108. doi: https://doi.org/10.3920/WMJ2015.1906.
276. Mitchell, N.J., Chen, C., Palumbo, J.D., Bianchini, A., Cappozzo, J., Stratton, J., Ryu, D., Wu, F., 2017: A risk assessment of dietary ochratoxin A in the United States. Food and Chemical Toxicology 100, 265-273. doi: https://doi.org/10.1016/j.fct.2016.12.037.
277. Ul Hassan, Z., Al Thani, R., Atia, F.A., Al Meer, S., Migheli, Q., Jaoua, S., 2018: Co-occurrence of mycotoxins in commercial formula milk and cereal-based baby food on the Qatar market. Food Additives and Contaminants: Part B 11, 191-197. doi: https://doi.org/10.1080/19393210.2018.1437785.
278. Omar, S.S., 2016: Aflatoxin M_1 levels in raw milk, pasteurised milk and infant formula. Italian Journal of Food Safety 5:5788, 158-160. doi: https://doi.org/10.4081/ijfs.2016.5788.
279. Mottaghianpour, E., Nazari, F., Reza Mehrasbi, M., Hosseini, M.-J, 2017: Occurrence of aflatoxin B_1 in baby foods marketed in Iran. Journal of the Science of Food and Agriculture. 97, 2690-2694. doi: https://doi.org/10.1002/jsfa.
280. Ortiz, J., Jacxsens, L., Astudillo, G., Ballesteros, A., Donoso, S., Huybregts, L., De Meulenaer, B., 2018: Multiple mycotoxin exposure of infants and young children via breastfeeding and complementary/weaning foods consumption in Ecuadorian highlands. Food and Chemical Toxicology 118, 541-548. doi: https://doi.org/10.1016/j.fct.2018.06.008.
281. Carvajal, M., Bolaños, A., Rojo, F., Méndez, I., 2003: Aflatoxin M_1 in pasteurized and ultrapasteurized milk with different fat content in Mexico. Journal of Food Protection 66, 1885-1892.
282. Çelik, T.H., Sarımehmetoğlu, B., Küplülü, Ö., 2005: Aflatoxin M_1 contamination in pasteurised milk. Veterinarski Arhiv 75, 57-65.
283. Pei, S.C., Zhang, Y.Y., Eremin, S.A., Lee, W.J., 2009: Detection of aflatoxin M_1 in milk products from China by ELISA using monoclonal antibodies. Food Control 20, 1080-1085. doi: https://doi.org/10.1016/j.foodcont.2009.02.004.

284. Corbett, W.T., Brownie, C.F., Hagler, S.B., Hagler, W.M., Jr., 1988: An epidemiological investigation associating aflatoxin M_1 with milk production in dairy cattle. Veterinary and Human Toxicology 30, 5-8.
285. Elzupir, A.O., Elhussein, A.M., 2010: Determination of aflatoxin M_1 in dairy cattle milk in Khartoum State, Sudan. Food Control 21, 945-946. doi: https://doi.org/10.1016/j.foodcont.2009.11.013.
286. Elzupir, A.O., Makawi, S.Z.A., Elhussein, A.M., 2009: Determination of aflatoxins and ochratoxin A in dairy cattle feed and milk in Wad Medani, Sudan. Journal of Animal and Veterinary Advances 8, 2508-2511.
287. Elzupir, A.O., Abas, A.R.A., Hemmat Fadul, M., Modwi, A.K., Ali, N.M.I., Jadian, A.F.F., Ahmed, N.A.A., Adam, S.Y.A., Ahmed, N.A.M., Khairy A.A.A., Khalil, E.A.G., 2012: Aflatoxin M_1 in breast milk of nursing Sudanese mothers. Mycotoxin Research 28, 131-134. doi: https://doi.org/10.1007/s12550-012-0127-x.
288. Fallah, A.A., 2010: Assessment of aflatoxin M_1 contamination in pasteurized and UHT milk marketed in central part of Iran. Food and Chemical Toxicology 48, 988-991. doi: https://doi.org/10.1016/j.fct.2010.01.014.
289. Ayar, A., Sert, D., Çon, A.I., 2007: A study on the occurrence of aflatoxin in raw milk due to feeds. Journal of Food Safety 27, 199-207.
290. Movassagh Ghazani, M.H., 2009: Aflatoxin M_1 contamination in pasteurized milk in Tabriz (northwest of Iran). Food and Chemical Toxicology 47, 1624-1625. doi: https://doi.org/10.1016/j.fct.2009.04.011.
291. Lin, L.-C., Liu, F.-M., Fu, Y.-M., Shih, Y.-C., 2004: Survey of aflatoxin M_1 contamination of dairy products in Taiwan. Journal of Food and Drug Analysis 12, 154-160.
292. Bento, H., Fernandes, A.M., Barbosa, M., 1989: Determination of the aflatoxin M_1 in pasteurized milk and in milk powder. Revista Portuguesa de Ciencias Veterinarias 84, 161-171.
293. Ghanem, I., Orfi, M., 2009: Aflatoxin M_1 in raw, pasteurized and powdered milk available in the Syrian market. Food Control 20, 603-605. doi: https://doi.org/10.1016/j.foodcont.2008.08.018.
294. Ghiasian, S.A., Maghsood, A.H., Neyestani, T.R., Mirhendi, S.H., 2007: Occurrence of aflatoxin M_1 in raw milk during the summer and winter seasons in Hamedan, Iran. Journal of Food Safety 27, 188-198.
295. Ghidini, S., Zanardi, E., Battaglia, A., Varisco, G., Ferretti, E., Campanini, G., Chizzolini, R., 2005: Comparison of contaminant and residue levels in organic and conventional milk and meat products from northern Italy. Food Additives and Contaminants 22, 9-14. doi: https://doi.org/10.1080/02652030400027995.
296. Hussain, I., Anwar, J., 2008: A study on contamination of aflatoxin M_1 in raw milk in the Punjab province of Pakistan. Food Control 19, 393-395. doi: https://doi.org/10.1016/j.foodcont.2007.04.019.
297. Wang, H., Zhou, X.-J., Liu, Y.-Q., Yang, H.-M., Guo, Q.-L., 2011: Simultaneous determination of chloramphenicol and aflatoxin M_1 residues in milk by triple quadrupole liquid chromatography-tandem mass spectrometry. Journal of Agricultural and Food Chemistry 59, 3532-3538. doi: https://doi.org/10.1021/jf2006062.
298. Kamkar, A., 2005: A study on the occurrence of aflatoxin M_1 in raw milk produced in Sarab city of Iran. Food Control 16, 593-599. doi: https://doi.org/10.1016/j.foodcont.2004.06.021.
299. Keskin, Y., Başkaya, R., Karsli, S., Yurdun, T., Özyaral, O., 2009: Detection of aflatoxin M_1 in human breast milk and raw cow's milk in Istanbul, Turkey. Journal of Food Protection 72, 885-889.
300. Peng, K.-Y., Chen, C.-Y., 2009: Prevalence of aflatoxin M_1 in milk and its potential liver cancer risk in Taiwan. Journal of Food Protection 72, 1025-1029.
301. Pietri, A., Bertuzzi, T., Moschini, M., Piva, G., 2003: Aflatoxin M_1 occurrence in milk samples destined for Parmigiano Reggiano cheese production. Italian Journal of Food Science 15, 301-306.
302. Ruangwises, N., Ruangwises, S., 2010: Aflatoxin M_1 contamination in raw milk within the central region of Thailand. Bulletin of Environmental Contamination and Toxicology 85, 195-198. doi: https://doi.org/10.1007/s00128-010-0056-3.
303. Ruangwises, S., Ruangwises, N., 2009: Occurrence of aflatoxin M_1 in pasteurized milk of the school milk project in Thailand. Journal of Food Protection 72, 1761-1763.
304. Sani, A.M., Nikpooyan, H., Moshiri, R., 2010: Aflatoxin M_1 contamination and antibiotic residue in milk in Khorasan province, Iran. Food and Chemical Toxicology 48, 2130-2132. doi: https://doi.org/10.1016/j.fct.2010.05.015.
305. Shundo, L., Navas, S.A., Lamardo, L.C.A., Ruvieri, V., Sabino, M., 2009: Estimate of aflatoxin M_1 exposure in milk and occurrence in Brazil. Food Control 20, 655-657. doi: https://doi.org/10.1016/j.foodcont.2008.09.019.
306. Shundo, L., Sabino, M., 2006: Aflatoxin M_1 in milk by immunoaffinity column cleanup with TLC/HPLC determination. Brazilian Journal of Microbiology 37, 164-167. doi: https://doi.org/10.1590/S1517-83822006000200013.
307. Tajik, H., Rohani, S.M.R., Moradi, M., 2007: Detection of aflatoxin M_1 in raw and commercial pasteurized milk in Urmia, Iran. Pakistan Journal of Biological Sciences 10, 4103-4107.
308. Tajkarimi, M., Aliabadi, F.S., Nejad, M.S., Pursoltani, H., Motallebi, A.A., Mahdavi, H., 2007: Seasonal study of aflatoxin M_1 contamination in milk in five regions in Iran. International Journal of Food Microbiology 116, 346-349. doi: https://doi.org/10.1016/j.ijfoodmicro.2007.02.008.
309. Nemati, M., Mehran, M.A., Hamed, P.K., Masoud, A., 2010: A survey on the occurrence of aflatoxin M_1 in milk samples in Ardabil, Iran. Food Control 21, 1022-1024. doi: https://doi.org/10.1016/j.foodcont.2009.12.021.
310. Gürbay, A., Aydın, S., Girgin, G., Engin, A.B., Şahin, G., 2006: Assessment of aflatoxin M_1 levels in milk in Ankara, Turkey. Food Control 17, 1-4. doi: https://doi.org/10.1016/j.foodcont.2004.07.008.
311. Rahimi, E., Ameri, M., 2012: A survey of aflatoxin M_1 contamination in bulk milk samples from dairy bovine, ovine, and caprine herds in Iran. Bulletin of Environmental Contamination and Toxicology 89, 158-160. doi: https://doi.org/10.1007/s00128-012-0616-9.
312. El Marnissi, B., Belkhou, R., Morgavi, D.P., Bennani, L., Boudra, H., 2012: Occurrence of aflatoxin M_1 in raw milk collected from traditional dairies in Morocco. Food and Chemical Toxicology 50, 2819-2821. doi: https://doi.org/10.1016/j.fct.2012.05.031.
313. Omar, S.S., 2012: Incidence of aflatoxin M_1 in human and animal milk in Jordan. Journal of Toxicology and Environmental Health Part A 75, 1404-1409. doi: https://doi.org/10.1080/15287394.2012.721174.
314. Ferguson-Foos, J., Warren, J.D., 1984: Improved cleanup for liquid chromatographic analysis and fluorescence detection of aflatoxins M_1 and M_2 in fluid milk products. Journal of the Association of Official Analytical Chemists 67, 1111-1114.
315. Breitholtz-Emanuelsson, A., Olsen, M., Oskarsson, A., Palminger, I., Hult, K., 1993: Ochratoxin A in cow's milk and in human milk with corresponding human blood samples. Journal of the Association of Official Analytical Chemists International 76, 842-846.

316. Skaug, M.A., 1999: Analysis of Norwegian milk and infant formulas for ochratoxin A. Food Additives and Contaminants 16, 75-78.
317. Maragos, C.M., Richard, J.L., 1994: Quantitation and stability of fumonisins B_1 and B_2 in milk. Journal of the Association of Official Analytical Chemists International 77, 1162-1167.
318. Kaniou-Grigoriadu, I., Eleftheriadou, A., Mouratidou, T., Katikou, P., 2005: Determination of aflatoxin M_1 in ewe's milk samples and the produced curd and Feta cheese. Food Control 16, 257-261. doi: https://doi.org/10.1016/j.foodcont.2004.03.003.
319. Oliveira, C.A.F., Ferraz, J.C.O., 2007: Occurrence of aflatoxin M_1 in pasteurised, UHT milk and milk powder from goat origin. Food Control 18, 375-378. doi: https://doi.org/10.1016/j.foodcont.2005.11.003.
320. Maxwell, S.M., Apeagyei, F., De Vries, H.R., Mwanmut, D.D., Hendrickse, R.G., 1989: Aflatoxins in breast milk, neonatal cord blood and sera of pregnant women. Journal of Toxicology - Toxin Reviews 8, 19-29.
321. Jonsyn, F.E., Maxwell, S.M., Hendrickse, R.G., 1995: Ochratoxin A and aflatoxins in breast milk samples from Sierra Leone. Mycopathologia 131, 121-126.
322. Turconi, G., Guarcello, M., Liveri, C., Comizzoli, S., Maccarini, L., Castellazzi, A.M., Pietri, A., Piva, G., Roggi, C., 2004: Evaluation of xenobiotics in human milk and ingestion by the newborn. An epidemiological survey in Lombardy (northern Italy). European Journal of Nutrition 43, 191-197. doi: https://doi.org/10.1007/s00394-004-0458-2.
323. Gürbay, A., Atasayar Sabuncuoğlu, S., Girgin, G., Şahin, G., Yiğit, Ş., Yurdakök, M., Tekinalp, G., 2010: Exposure of newborne to aflatoxin M_1 and B_1 from mothers' breast milk in Ankara, Turkey. Food and Chemical Toxicology 48, 314-319. doi: https://doi.org/10.1016/j.fct.2009.10.016.
324. Zarba, A., Wild, C.P., Hall, A.J., Montesano, R., Hudson, G.J., Groopman, J.D., 1992: Aflatoxin M_1 in human breast milk from The Gambia, West Africa, quantified by combined monoclonal antibody immunoaffinity chromatography and HPLC. Carcinogenesis 13, 891-894.
325. El-Nezami, H.S., Nicoletti, G., Neal, G.E., Donohue, D.C., Ahokas, J.T., 1995: Aflatoxin M_1 in human breast milk samples from Victoria, Australia and Thailand. Food and Chemical Toxicology 33, 173-179.
326. Wang, L., Zhang, Q., Yan, Z., Tan, Y., Zhu, R., Yu, D., Yang, H., Wu, A., 2018: Occurrence and quantitative risk assessment of twelve mycotoxins in eggs and chicken tissues in China. Toxins (Basel) 10, 477. doi: https://doi.org/10.3390/toxins10110477.
327. Piątkowska, M., Sulyok, M., Pietruszka, K., Panasiuk, Ł., 2018: Pilot study for the presence of fungal metabolites in sheep milk from first spring milking. Journal of Veterinary Research 62, 167-172. doi: https://doi.org/10.2478/jvetres-2018-0026.
328. Bahrami, R., Shahbazi, Y., Nikousefat, Z., 2016: Aflatoxin M_1 in milk and traditional dairy products from west part of Iran: occurrence and seasonal variation with an emphasis on risk assessment of human exposure. Food Control 62, 250-256. doi: https://doi.org/10.1016/j.foodcont.2015.10.039.
329. Polychronaki, N., West, R.M., Turner, P.C., Amra, H., Abdel-Wahhab, M., Mykkänen, H., El-Nezami, H., 2007: A longitudinal assessment of aflatoxin M_1 excretion in breast milk of selected Egyptian mothers. Food and Chemical Toxicology 45, 1210-1215. doi: https://doi.org/10.1016/j.fct.2007.01.001.
330. Wild, C.P., Pionneau, F.A., Montesano, R., Mutiro, C.F., Chetsanga, C.J., 1987: Aflatoxin detected in human breast milk by immunoassay. International Journal of Cancer 40, 328-333.
331. Saad, A.M., Abdelgadir, A.M., Moss, M.O., 1995: Exposure of infants to aflatoxin M_1 from mothers' breast milk in Abu Dhabi, UAE. Food Additives and Contaminants 12, 255-261.
332. Polychronaki, N., Turner, P.C., Mykkänen, H., Gong, Y., Amra, H., Abdel-Wahhab, M., El-Nezami, H., 2006: Determinants of aflatoxin M_1 in breast milk in a selected group of Egyptian mothers. Food Additives and Contaminants 23, 700-708. doi: https://doi.org/10.1080/02652030600627222.
333. Navas, S.A., Sabino, M., Rodriguez-Amaya, D.B., 2005: Aflatoxin M_1 and ochratoxin A in a human milk bank in the city of São Paulo, Brazil. Food Additives and Contaminants 22, 457-462. doi: https://doi.org/10.1080/02652030500110550.
334. Abdulrazzaq, Y.M., Osman, N., Yousif, Z.M., Al-Falahi, S., 2003: Aflatoxin M_1 in breast-milk of UAE women. Annals of Tropical Paediatrics 23, 173-179. doi: https://doi.org/10.1179/027249303225007671.
335. El-Sayed Abd Alla, A.M., Soher, E.A., Neamat-Allah, A.A., 2002: Human exposure to mycotoxins in Egypt. Mycotoxin Research 18, 23-30.
336. Coulter, J.B.S., Lamplugh, S.M., Suliman, G.I., Omer, M.I.A., Hendrickse, R.G., 1984: Aflatoxins in human breast milk. Annals of Tropical Paediatrics 4, 61-66.
337. Galvano, F., Pietri, A., Bertuzzi, T., Gagliardi, L., Ciotti, S., Luisi, S., Bognanno, M., La Fauci, L., Iacopino, A.M., Nigro, F., Li Volti, G., Vanella, L., Giammanco, G., Tina, G.L., Gazzolo, D., 2008: Maternal dietary habits and mycotoxin occurrence in human mature milk. Molecular Nutrition & Food Research 52, 496-501. doi: https://doi.org/10.1002/mnfr.200700266.
338. Kachapulula, P.W., Akello, J., Bandyopadhyay, R., Cotty, P.J., 2018: Aflatoxin contamination of dried insects and fish in Zambia. Journal of Food Protection 81, 1508-1518. doi: https://doi.org/10.4315/0362-028X.JFP-17-527.
339. Sadeghi, N., Oveisi, M.R., Jannat, B., Hajimahmoodi, M., Bonyani, H., Jannat, F., 2009: Incidence of aflatoxin M_1 in human breast milk in Tehran, Iran. Food Control 20, 75-78. doi: https://doi.org/10.1016/j.foodcont.2008.02.005.
340. El-Tras, W.F., El-Kady, N.N., Tayel, A.A., 2011: Infants exposure to aflatoxin M_1 as a novel foodborne zoonosis. Food and Chemical Toxicology 49, 2816-2819. doi: https://doi.org/10.1016/j.fct.2011.08.008.
341. Mahdavi, R., Nikniaz, L., Arefhosseini, S.R., Vahed Jabbari, M., 2010: Determination of aflatoxin M_1 in breast milk samples in Tabriz-Iran. Maternal and Child Health Journal 14, 141-145. doi: https://doi.org/10.1007/s10995-008-0439-9 (ehem = MyFo II 1639).
342. Skaug, M.A., Størmer, F.C., Saugstad, O.D., 1998: Ochratoxin A: a naturally occurring mycotoxin found in human milk samples from Norway. Acta Pædiatrica 87, 1275-1278.
343. Apostolou, E., El-Nezami, H.S., Ahokas, J.T., Donohue, D.C., 1998: The evaluation of ochratoxin A in breast milk in Victoria (Australia). Revue de Médecine Véterinaire 149, 709.
344. Gareis, M., Märtlbauer, E., Bauer, J., Gedek, B., 1988: Determination of ochratoxin A in human milk. Zeitschrift für Lebensmittel-Untersuchung und -Forschung 186, 114-117.
345. Skaug, M.A., Helland, I., Solvoll, K., Saugstad, O.D., 2001: Presence of ochratoxin A in human milk in relation to dietary intake. Food Additives and Contaminants 18, 321-327. doi: https://doi.org/10.1080/02652030010021477.
346. Biasucci, G., Calabrese, G., Di Giuseppe, R., Carrara, G., Colombo, F., Mandelli, B., Maj, M., Bertuzzi, T., Pietri, A., Rossi, F., 2011: The presence of ochratoxin A in cord serum and in human milk and its correspondence with maternal dietary habits. European Journal of Nutrition 50, 211-218. doi: https://doi.org/10.1007/s00394-010-0130-y.

347. Gürbay, A., Girgin, G., Atasayar Sabuncuoğlu, S., Şahin, G., Yurdakök, M., Yiğit, Ş., Tekinalp, G., 2010: Ochratoxin A: is it present in breast milk samples obtained from mothers from Ankara, Turkey? Journal of Applied Toxicology 30, 329-333. doi: https://doi.org/10.1002/jat.1499.
348. Gul, O., Dervisoglu, M., 2014: Occurrence of aflatoxin M_1 in vacuum packed kashar cheeses in Turkey. International Journal of Food Properties 17, 273-282. doi: https://doi.org/10.1080/10942912.2011.631247.
349. Dostal, A., Jakusova, L., Cajdova, J., Hudeckova, H., 2008: Results of the first studies of occurrence of ochratoxin A in human milk in Slovakia. Bratislavslec Lekarske Listy 109, 276-278.
350. Kovács, F., Sándor, G., Ványi, A., Domány, S., Zomborszky-Kovács, M., 1995: Detection of ochratoxin A in human blood and colostrum. Acta Veterinaria Hungarica 43, 393-400.
351. Postupolski, J., Karłowski, K., Kubik, P., 2006: Ochratoxin A in maternal and foetal blood and in maternal milk. Roczniki Państwowego Zakładu Higieny 57, 23-30.
352. Muñoz, K., Campos, V., Blaszkewicz, M., Vega, M., Alvarez, A., Neira, J., Degen, H.G., 2010: Exposure of neonates to ochratoxin A: first biomonitoring results in human milk (colostrum) from Chile. Mycotoxin Research 26, 59-67. doi: https://doi.org/10.1007/s12550-009-0040-0.
353. Bognanno, M., La Fauci, L., Ritieni, A., Tafuri, A., De Lorenzo, A., Micari, P., Di Renzo, L., Ciappellano, S., Sarullo, V., Galvano, F., 2006: Survey of the occurrence of aflatoxin M_1 in ovine milk by HPLC and its confirmation by MS. Molecular Nutrition & Food Research 50, 300-305. doi: https://doi.org/10.1002/mnfr.200500224.
354. Meucci, V., Razzuoli, E., Soldani, G., Massart, F., 2010: Mycotoxin detection in infant formula milks in Italy. Food Additives and Contaminants: Part A 27, 64-71. doi: https://doi.org/10.1080/02652030903207201.
355. Blanco, J.L., Domínguez, L., Gómez-Lucía, E., Garayzabal, J.F.F., García, J.A., Suárez, G., 1988: Presence of aflatoxin M_1 in commercial ultra-high-temperature-treated milk. Applied and Environmental Microbiology 54, 1622-1623.
356. Galvano, F., Galofaro, V., Ritieni, A., Bognanno, M., De Angelis, A., Galvano, G., 2001: Survey of the occurrence of aflatoxin M_1 in dairy products marketed in Italy: second year of observation. Food Additives and Contaminants 18, 644-646. doi: https://doi.org/10.1080/02652030110035381.
357. Gündinç, U., Filazi, A., 2009: Detection of aflatoxin M_1 concentrations in UHT milk consumed in Turkey markets by ELISA. Pakistan Journal of Biological Sciences 12, 653-656.
358. Unusan, N., 2006: Occurrence of aflatoxin M_1 in UHT milk in Turkey. Food and Chemical Toxicology 44, 1897-1900. doi: https://doi.org/10.1016/j.fct.2006.06.010.
359. Heshmati, A., Milani, J.M., 2010: Contamination of UHT milk by aflatoxin M_1 in Iran. Food Control 21, 19-22. doi: https://doi.org/10.1016/j.foodcont.2009.03.013.
360. Kamkar, A., 2008: The study of aflatoxin M_1 in UHT milk samples by ELISA. Journal of Veterinary Research 63, 7-12.
361. MAFF - UK, 1997: Survey of aflatoxins and ochratoxin A in cereals and retail products. Food-Surveillance-Information-Sheet No. 130.
362. Jiménez, A.M., López De Cerain, A., Gonzalez-Peñas, E., Bello, J., 2001: Determination of ochratoxin A in pig liver-derived pâtés by high-performance liquid chromatography. Food Additives and Contaminants 18, 559-563. doi: https://doi.org/10.1080/02652030010025392.
363. Krogh, P., Hald, B., Plěstina, R., Ćeović, S., 1977: Balkan (Endemic) Nephropathy and foodborne ochratoxin A: preliminary results of a survey of foodstuffs. Acta Pathologica Microbiologica Scandinavica Section B 85, 238-240.
364. Duarte, S.C., Lino, C.M., Pena, A., 2013: Novel IAC-LC-ESI-MS^2 analytical set-up for ochratoxin A determination in pork. Food Chemistry 138, 1055-1061. doi: https://doi.org/10.1016/j.foodchem.2012.11.071.
365. Wiśniewska-Dmytrow, H., Żmudzki, J., Burek, O., Pietruszka, K., 2013: Official control of ochratoxin A in food of animal origin in Poland between 2003 and 2012. Bulletin of the Veterinary Institute in Pulawy 57, 519-523. doi: https://doi.org/10.2478/bvip-2013-0090.
366. Milićević, D., Stefanović, S., Janković, S., Radičević, T., 2012: Risk analysis and exposure assessment of ochratoxin A in Serbia. Veterinary World 5, 412-416. doi: https://doi.org/10.5455/vetworld.2012.412-416.
367. Milićević, D., Jurić, V., Vuković, D.Z., Mandić, M.M., Baltić, T.M., 2009: Residue of ochratoxin A in swine tissues - risk assessment. Archive of Oncology 17, 56-60. doi: https://doi.org/10.2298/AOO0904056M.
368. Movassagh, M.H., 2011: Presence of aflatoxin M_1 in UHT milk in Tabriz (northwest of Iran). Journal of Food Safety 31, 238-241. doi: https://doi.org/10.1111/j.1745-4565.2010.00291.x.
369. Finoli, C., Rondinini, G., 1989: Evaluation of infant formula contamination in Italy. Food Chemistry 32, 1-8.
370. Ranfft, K., 1983: Aflatoxin M_1 in milkpowder – A fast method for determination and a stock-taking of occurrence. Milchwissenschaft 38, 328-329.
371. Kawamura, O., Wang, D.-S., Liang, Y.-X., Hasegawa, A., Saga, C., Visconti, A., Ueno, Y., 1994: Further survey of aflatoxin M_1 in milk powders by ELISA. Food & Agricultural Immunology 6, 465-467.
372. Pfleger, R., Brandl, E., 1980: Aflatoxin residues in Austrian milk powder. Wiener Tierärztliche Monatsschrift 67, 101-106.
373. Fremy, J.M., Chu, F.S., 1984: Direct enzyme-linked immunosorbent assay for determining aflatoxin M_1 at picogram levels in dairy products. Journal of the Association of Official Analytical Chemists 67, 1098-1101.
374. Neumann-Kleinpaul, A., Terplan, G., 1972: Zum Vorkommen von Aflatoxin M_1 in Trockenmilchprodukten. Archiv für Lebensmittelhygiene 6, 128-132.
375. Oliveira, C.A.F., Germano, P.M.L., Bird, C., Pinto, C.A., 1997: Immunochemical assessment of aflatoxin M_1 in milk powder consumed by infants in São Paulo, Brazil. Food Additives and Contaminants 14, 7-1.
376. Stubblefield, R.D., 1979: The rapid determination of aflatoxin M_1 in dairy products. The Journal of the American Oil Chemist's Society 56, 800-802.
377. Fremy, J.-M., Boursier, B., 1981: Rapid determination of aflatoxin M_1 in dairy products by reversed-phase high-performance liquid chromatography. Journal of Chromatography 219, 156-161.
378. Okumura, H., Okimoto, J., Kishimoto, S., Hasegawa, A., Kawamura, O., Nakajima, M., Miyabe, M., Ueno, Y., 1993: An improved indirect competitive ELISA for aflatoxin M_1 in milk powders using novel monoclonal antibodies. Food & Agricultural Immunology 5, 75-84.
379. Elmali, M., Yapar, K., Kart, A., Yaman, H., 2008: Aflatoxin M_1 levels in milk powder consumed in Turkey. Journal of Animal and Veterinary Advances 7, 643-645.

380. Iacumin, L., Chiesa, L., Boscolo, D., Manzano, M., Cantoni, C., Orlic, S., Comi, G., 2009: Moulds and ochratoxin A on surfaces of artisanal and industrial dry sausages. Food Microbiology 26, 65-70. doi: https://doi.org/10.1016/j.fm.2008.07.006.
381. Monaci, L., Palmisano, F., Matrella, R., Tantillo, G., 2005: Determination of ochratoxin A at part-per-trillion level in Italian salami by immunoaffinity clean-up high-performance liquid chromatography with fluorescence detection. Journal of Chromatography A 1090, 184-187. doi: https://doi.org/10.1016/j.chroma.2005.07.020.
382. Dieber, F., Köfer, J., 1999: Ochratoxin A-Nachweis im Serum steirischer Schlachtschweine. Deutsche Lebensmittel-Rundschau 95, 327-329.
383. Majerus, P., Otteneder, H., Hower, C., 1989: Beitrag zum Vorkommen von Ochratoxin A in Schweineblutserum. Deutsche Lebensmittel-Rundschau 85, 307-313.
384. Langseth, W., Nymoen, U., Bergsjø, B., 1993: Ochratoxin A in plasma of Norwegian swine determined by an HPLC column-switching method. Natural Toxins 1, 216-221.
385. Rousseau, D.M., Slegers, G.A., Van Peteghem, C.H., 1986: Solid-phase radioimmunoassay of ochratoxin A in serum. Journal of Agricultural and Food Chemistry 34, 862-865.
386. Martins, M.L., Martins, H.M., 2004: Aflatoxin M_1 in yoghurts in Portugal. International Journal of Food Microbiology 91, 315-317. doi: https://doi.org/10.1016/S0168-1605(02)00363-X.
387. Akkaya, L., Birdane, Y.O., Oguz, H., Cemek, M., 2006: Occurrence of aflatoxin M_1 in yogurt samples from Afyonkarahisar, Turkey. Bulletin of the Veterinary Institute in Pulawy 50, 517-519.
388. Barjesteh, M.H., Azizi, G., Noshfar, E., 2010: Occurrence of aflatoxin M_1 in pasteurized and local yogurt in Mazandaran province (northern Iran) using ELISA. Global Veterinaria 4, 459-462.
389. Atasever, M.A., Atasever, M., Özturan, K., 2011: Aflatoxin M_1 levels in retail yogurt and ayran in Erzurum in Turkey. Turkish Journal of Veterinary and Animal Sciences 35, 59-62. doi: https://doi.org/10.3906/vet-0906-96.
390. Afshar, P., Shokrzadeh, M., Kalhori, S., Babaee, Z., Saeedi Saravi, S.S., 2013: Occurrence of ochratoxin A and aflatoxin M_1 in human breast milk in Sari, Iran. Food Control 31, 525-529. doi: https://doi.org/10.1016/j.foodcont.2012.12.009.
391. Adejumo, O., Atanda, O., Raiola, A., Somorin, Y., Bandyopadhyay, R., Ritieni, A., 2013: Correlation between aflatoxin M_1 content of breast milk, dietary exposure to aflatoxin B_1 and socioeconomic stastus of lactating mothers in Ogun State, Nigeria. Food and Chemical Toxicology 56, 171-177. doi: https://doi.org/10.1016/j.fct.2013.02.027.
392. Abd-Elghany, S.M., Sallam, K.I., 2015: Rapid determination of total aflatoxins and ochratoxins A in meat products by immuno-affinity fluorimetry. Food Chemistry 179, 253-256. doi: https://doi.org/10.1016/j.foodchem.2015.01.140.
393. Ali, M.A.I., El Zubeir, I.E.M., Fadel Elseed, A.M.A., 2014: Aflatoxin M_1 in raw and imported powdered milk sold in Khartoum state, Sudan. Food Additives and Contaminants 7, 208-212. doi: https://doi.org/10.1080/19393210.2014.887149.
394. Aidoo, K.E., Mohamed, S.M., Candlish, A.A., Tester, R.F., Elgerbi, A.M., 2011: Occurrence of fungi and mycotoxins in some commercial baby foods in North Africa. Food Nutrition Sciences 2, 751-758. doi: https://doi.org/10.4236/fns.2011.27103.
395. Andrade, P.D., Gomes Da Silva, J.L., Caldas, E.D., 2013: Simultaneous analysis of aflatoxins B_1, B_2, G_1, G_2, M_1 and ochratoxin A in breast milk by high-performance liquid chromatography/fluorescence after liquid-liquid extraction with low temperature purification (LLE-LTP). Journal of Chromatography A 1304, 61-68. doi: https://doi.org/10.1016/j.chroma.2013.06.049.
396. Amer, A.A., Ibrahim, M.A.E., 2010: Determination of aflatoxin M_1 in raw milk and traditional cheeses retailed in Egyptian markets. Journal of Toxicology and Environmental Health Sciences 2, 50-53.
397. Amirkhizi, B., Arefhosseini, S.R., Ansarin, M., Nemati, M., 2015: Aflatoxin B_1 in eggs and chicken livers by dispersive liquid-liquid micro-extraction and HPLC. Food Additives and Contaminants: Part B 8, 245-249. doi: https://doi.org/10.1080/19393210.2015.1067649.
398. Aslam, N., Tipu, M.Y., Ishaq, M., Cowling, A., McGill, D., Warriach, H.M., Wynn, P., 2016: Higher levels of aflatoxin M_1 contamination and poorer composition of milk supplied by informal milk marketing chains in Pakistan. Toxins (Basel) 8, 347. doi: https://doi.org/10.3390/toxins8120347 (and personal communication).
399. Armorini, S., Altafini, A., Zaghini, A., Roncada, P., 2016: Ochratoxin A in artisan salami in Veneto (Italy). Food Additives and Contaminants: Part B 9, 9-14. doi: https://doi.org/10.1080/19393210.2015.1098735.
400. Anfossi, L., Baggiani, C., Giovannoli, C., D'Arco, G., Passini, C., Giraudi, C., 2012: Occurrence of aflatoxin M_1 in Italian cheese: results of a survey conducted in 2010 and correlation with manufacturing, production season, milking animals, and maturation of cheese. Food Control 25, 125-130. doi: https://doi.org/10.1016/j.foodcont.2011.10.027.
401. Rama, A., Latifi, F., Bajraktari, D., Ramadani, N., 2015: Assessment of aflatoxin M_1 levels in pasteurized and UHT milk consumed in Prishtina, Kosovo. Food Control 57, 351-354. doi: https://doi.org/10.1016/j.foodcont.2015.04.021.
402. Ismail, A., Riaz, M., Levin, R.E., Akhtar, S., Gong, Y.Y., Hameed, A., 2016: Seasonal prevalence level of aflatoxin M_1 and its estimated daily intake in Pakistan. Food Control 60, 461-465. doi: https://doi.org/10.1016/j.foodcont.2015.08.025.
403. Braun, D., Ezekiel, C.N., Abia, W.A., Wisgrill, L., Degen, G.H., Turner, P.C., Marko, D., Warth, B., 2018: Monitoring early life mycotoxin exposures via LC-MS/MS breast milk analysis. Analytical Chemistry 90, 14569-14577. doi: https://doi.org/10.1021/acs.analchem.8b04576.
404. Assem, E., Mohamad, A., Oula, E.A., 2011: A survey on the occurrence of aflatoxin M_1 in raw and processed milk samples marketed in Lebanon. Food Control 22, 1856-1858 doi: https://doi.org/10.1016/j.foodcont.2011.04.026.
405. Atasever, M., Yildirim, Y., Atasever, M., Tastekin, A., 2014: Assessment of aflatoxin M_1 in maternal breast milk in eastern Turkey. Food and Chemical Toxicology 66, 147-149. doi: https://doi.org/10.1016/j.fct.2014.01.037.
406. Redouane-Salah, S., Morgavi, D.P., Arhab, R., Messai, A., Boudra, H., 2015: Presence of aflatoxin M_1 in raw, reconstituted, and powdered milk samples collected in Algeria. Environmental Monitoring and Assessment 187: 375. doi: https://doi.org/10.1007/s10661-015-4627-y.
407. Tonon, K.M., Savi, G.D., Scussel, V.M., 2018: Application of a LC-MS/MS method for multi-mycotoxin analysis in infant formula and milk-based products for young children commercialized in southern Brazil. Journal of Environmental Science and Health: Part B 2, 1-7. doi: https://doi.org/10.1080/03601234.2018.1474560.
408. Bozzo, G., Ceci, E., Bonerba, E., Di Pinto, A., Tantillo, G., De Giglio, E., 2012 : Occurrence of ochratoxin A in the wild boar (*Sus scrofa*): chemical and histological analysis. Toxins (Basel) 4, 1440-1450. doi: https://doi.org/10.3390/toxins4121440.
409. Cantú-Cornelio, F., Aguilar-Toalá, J.E., de León-Rodríguez, C.I., Esparza-Romero, J., Vallejo-Cordoba, B., Gonzáles-Córdova, A.F., García, H.S., Hernández-Mendoza, A., 2016: Occurrence and factors associated with the presence of aflatoxin M_1 in breast milk samples of nursing mothers in central Mexico. Food Control 62, 16-22. doi: https://doi.org/10.1016/j.foodcont.2015.10.004.

410. Chiavaro, E., Cacchioli, C., Berni, E., Spotti, E., 2005: Immunoaffinity clean-up and direct fluorescence measurement of aflatoxins B_1 and M_1 in pig liver: comparison with high-performance liquid chromatography determination. Food Additives and Contaminants 22, 1154-1161. doi: https://doi.org/10.1080/02652030500307115.
411. Chavarría, G., Granados-Chinchilla, F., Alfaro-Cascante, M., Molina, A., 2015: Detection of aflatoxin M_1 in milk, cheese and sour cream samples from Costa Rica using enzyme-assisted extraction and HPLC. Food Additives and Contaminants: Part B 8, 128-135. doi: https://doi.org/10.1080/19393210.2015.1015176.
412. Dimitrieska-Stojković, E., Stojanovska-Dimzoska, B., Ilievska, G., Uzunov, R., Stojković, G., Hajrulai-Musliu, Z., Jankuloski, D., 2016: Assessment of aflatoxin contamination in raw milk and feed in Macedonia during 2013. Food Control 59, 201-206. doi: https://doi.org/10.1016/j.foodcont.2015.05.019.
413. Diaz, G.J., Sánchez, M.P., 2015: Determination of aflatoxin M_1 in breast milk as a biomarker of maternal and infant exposure in Colombia. Food Additives and Contaminants: Part A 32, 1192-1198. doi: https://doi.org/10.1080/19440049.2015.1049563.
414. De Oliveira, C.P., De Fátima Ferreira Soares, N., Veloso De Oliveira, T., Baffa Júnior, J.C., Azevedo De Silva, W., 2013: Aflatoxin M_1 occurrence in ultra high temperature (UHT) treated fluid milk from Minas Gerais/Brazil. Food Control 30, 90-92. doi: https://doi.org/10.1016/j.foodcont.2012.07.026.
415. Meucci, V., Costa, E., Razzuoli, E., Mengozzi, G., Soldani, G., 2005: Occurrence of ochratoxin A in blood of Italian slaughtered pigs. Toxicology Letters, 158S, S116.
416. Kamali, A., Mehni, S., Kamali, M., Taheri Sarvtin, M., 2017: Detection of ochratoxin A in human breast milk in Jiroft city, south of Iran. Current Medical Mycology 3, 1-4. doi: https://doi.org/10.29252/cmm.3.3.1.
417. Sohrabi, N., Gharahkoli, H., 2016: A seasonal study for determination of aflatoxin M_1 level in dairy products in Iranshahr, Iran. Current Medical Mycology 2, 27-31. doi: https://doi.org/10.18869/acadpub.cmm.2.3.27.
418. Kocasari, F.S., Tasci, F., Mor, F., 2012: Survey of aflatoxin M_1 in milk and dairy products consumed in Burdur, Turkey. International Journal of Dairy Technology 65, 365-371. doi: https://doi.org/10.1111/j.1471-0307.2012.00841.x.
419. Khoshnevis, S.H., Azizi, I.G., Shateri, S., Mousavizadeh, M., 2012: Determination of the aflatoxin M_1 in ice cream in Babol City (northern, Iran). Global Veterinaria 8, 205-208.
420. Jafarian-Dehkordi, A., Pourradi, N., 2013: Aflatoxin M_1 contamination of human breast milk in Isfahan, Iran. Advanced Biomedical Research 2, 86. doi: https://doi.org/10.4103/2277-9175.122503.
421. Gutiérrez, R., Rosell, P., Vega, S., Pérez, J., Ramírez, A., Coronado, M., 2013: Self and foreign substances in organic and conventional milk produced in the eastern region of Mexico. Food and Nutrition Sciences 4, 586-593. doi: https://doi.org/10.4236/fns.2013.45076.
422. Garmakhany, A.D., Zighamian, H., Sarhangpour R., Rasti, M., Aghajani, N., 2011: Occurrence of aflatoxin M_1 in raw and pasteurized milk in Esfahan province of Iran. Minerva Biotecnologica 23, 53-57.
423. Altafani, A., Fedrizzi, G., Roncada, P., 2018: Occurrence of ochratoxin A in typical salami produced in different regions of Italy. Mycotoxin Research 22, 77. doi: https://doi.org/10.1007/s12550-018-0338-x.
424. Rahimi, E., 2014: Survey of the occurrence of aflatoxin M_1 in dairy products marketed in Iran. Toxicology and Industrial Health 30, 750-754. doi: https://doi.org/10.1177/0748233712462476.
425. Mohammadi, H., Shokrzadeh, M., Aliabadi, Z., Riahi-Zamjani, B., 2016: Occurrence of aflatoxin M_1 in commercial pasteurized milk samples in Sari, Mazandaran province, Iran. Mycotoxin Research 32, 85-87. doi: https://doi.org/10.1007/s12550-016-0243-0.
426. Dehghan, P., Pakshir, K., Rafiei, H., Chadeganipour, M., Akbari, M., 2014: Prevalence of ochratoxin A in human milk in the Khorrambid town, Fars province, south of Iran. Jundishapur Journal of Microbiology 7 (7), e11220. doi: https://doi.org/10.5812/jjm.11220.
427. Dutton, M.F., Mwanza, M., De Kock, S., Khilosia, L.D., 2012: Mycotoxins in South African foods: a case study on aflatoxin M_1 in milk. Mycotoxin Research 28, 17-23. doi: https://doi.org/10.1007/s12550-011-0112-9.
428. Duarte, S.C., Almeida, A.M., Teixeira, A.S., Pereira, A.L., Falcão, A.C., Pena, A., Lino, C.M., 2013: Aflatoxin M_1 in marketed milk in Portugal: assessment of human and animal exposure. Food Control 30, 411-417. doi: https://doi.org/10.1016/j.foodcont.2012.08.002.
429. Karwowska, W., Pierzynowska, J., Janicki, A., Waskiewicz-Robak, B., Przybylska, A., 2004: Qualitative and quantitative analysis of filamentous fungi in air, food and ochratoxin A in human milk. Polish Journal of Food and Nutrition Sciences 13/54, SI 2, 41-44.
430. Grajewski, J., Twarużek, M., Kosicki, R., 2012: High levels of ochratoxin A in blood serum and kidneys of wild boars *Sus scrofa* in Poland. Wildlife Biology 18, 272-279. doi: https://doi.org/10.2981/11-059 (and personal communication).
431. Kawamura, O., Sato, S., Nagura, M., Kishimoto, S., Ueno, I., Sato, S., Uda, T., Ito, Y., Ueno, Y., 1990: Enzyme-linked immunosorbent assay for detection and survey of ochratoxin A in livestock sera and mixed feeds. Food & Agricultural Immunology 2, 135-143.
432. El Khoury A., Atoui, A., Yaghi, J., 2011: Analysis of aflatoxin M_1 in milk and yogurt and AFM_1 reduction by lactic acid bacteria used in Lebanese industry. Food Control 22, 1695-1699. doi: https://doi.org/10.1016/j.foodcont.2011.04.001.
433. Elkak, A., El Atat, O., Habib, J., Abbas, M., 2012: Occurrence of aflatoxin M_1 in cheese processed and marketed in Lebanon. Food Control 25, 140-143. doi: https://doi.org/10.1016/j.foodcont.2011.10.033.
434. Engel, G., von Milczewski, K.E., Prokopek, D., Teuber, M., 1982: Strain-specific synthesis of mycophenolic acid by *Penicillium roqueforti* in blue-veined cheese. Applied and Environmental Microbiology 43, 1034-1040.
435. Fallah, A.A., 2010: Aflatoxin M_1 contamination in dairy products marketed in Iran during winter and summer. Food Control 21, 1478-1481. doi: https://doi.org/10.1016/j.foodcont.2010.04.017.
436. Fallah, A.A., Barani, A., Nasiri, Z., 2015: Aflatoxin M_1 in raw milk in Qazvin province, Iran: a seasonal study. Food Additives and Contaminants: Part B 8, 195-198. doi: https://doi.org/10.1080/19393210.2015.1046193.
437. Fallah, A.A., Fazlollahi, R., Emami, A., 2016: Seasonal study of aflatoxin M_1 contamination in milk of four dairy species in Yazd, Iran. Food Control 68, 77-82. doi: https://doi.org/10.1016/j.foodcont.2016.03.018.
438. García Londoño, V.A., Boasso A.C., De Paula, M.C.Z., Garcia, L.P., Scussel, V.M., Resnik, S., Pacín, A., 2013: Aflatoxin M_1 survey on randomly collected milk powder commercialized in Argentina and Brazil. Food Control 34, 752-755. doi: https://doi.org/10.1016/j.foodcont.2013.06.030.
439. Fontaine, K., Passeró, E., Vallone, L., Hymery, N., Coton, M., Jany, J.-L., Mounier, J., Coton, E., 2015: Occurrence of roquefortine C, mycophenolic acid and aflatoxin M_1 mycotoxins in blue-veined cheeses. Food Control 47, 634-640. doi: https://doi.org/10.1016/j.foodcont.2014.07.046.

440. Fallah, A.A., Rahnama, M., Jafari, T., Saei-Dehkordi, S.S., 2011: Seasonal variation of aflatoxin M_1 contamination in industrial and traditional Iranian dairy products. Food Control 22, 1653-1656. doi: https://doi.org/10.1016/j.foodcont.2011.03.024.
441. Ghazani, M.H.M., 2009: Aflatoxin M_1 contamination in pasteurized milk in Tabriz (northwest Iran). Food and Chemical Toxicology 47, 1624-1625. doi: https://doi.org/10.1016/j.fct.2009.04.011.
442. Gazzotti, T., Lugoboni, B., Zironi, E., Barbarossa, A., Serraino, A., Pagliuca, G., 2009: Determination of fumonisin B_1 in bovine milk by LC-MS/MS. Food Control 20, 1171-1174. doi: https://doi.org/10.1016/j.foodcont.2009.02.009.
443. Gómez-Arranz, E., Navarro-Blasco, I., 2010: Aflatoxin M_1 in Spanish infant formulae: occurrence and dietary intake regarding type, protein-base and physical state. Food Additives and Contaminants: Part B 3, 193-199. doi: https://doi.org/10.1080/19393210.2010.503353.
444. Golge, O., 2014: A survey on the occurrence of aflatoxin M_1 in raw milk produced in Adana province of Turkey. Food Control 45, 150-155. doi: https://doi.org/10.1016/j.foodcont.2014.04.039.
445. Zheng, N., Wang, J.-Q., Han, R.-W., Zhen, Y.P., Xu, X.-M., Sun, P., 2013: Survey of aflatoxin M_1 in raw milk in the five provinces of China. Food Additives and Contaminants: Part B 6, 110-115 doi: https://doi.org/10.1080/19393210.2012.763191.
446. Hampikyan, H., Bingol, E.B., Cetin, O., Colak, H., 2010: Determination of aflatoxin M_1 levels in Turkish white, kashar and tulum cheeses. Journal of Food, Agriculture and Environment 8, 13-15.
447. Guo, Y., Yuan, Y., Yue, T., 2013: Aflatoxin M_1 in milk products in China and dietary risk assessment. Journal of Food Protection 76, 849-853. doi: https://doi.org/10.4315/0362-028X.JFP-12-419.
448. Guan, D., Li, P., Zhang, Q., Zhang, W., Zhang, D., Jiang, J., 2011: An ultra-sensitive monoclonal antibody-based competitive enzyme immunoassay for aflatoxin M_1 in milk and infant milk products. Food Chemistry 125, 1359-1364. doi: https://doi.org/10.1016/j.foodchem.2010.10.006.
449. Han, R.W., Zheng, N., Wang, J.Q., Zhen, Y.P., Xu, X.M., Li, S.L., 2013: Survey of aflatoxin in dairy cow feed and raw milk in China. Food Control 34, 35-39. doi: https://doi.org/10.1016/j.foodcont.2013.04.008.
450. Iqbal, S.Z., Asi, M.R., Ariño, A., 2011: Aflatoxin M_1 contamination in cow and buffalo milk samples from the North West Frontier Province (NWFP) and Punjab provinces of Pakistan. Food Additives and Contaminants: Part B 4, 282-288. doi: https://doi.org/10.1080/19393210.2011.637237.
451. Iqbal, S.Z., Asi, M.R., 2013: Assessment of aflatoxin M_1 in milk and milk products from Punjab, Pakistan. Food Control 30, 235-239. doi: https://doi.org/10.1016/j.foodcont.2012.06.026.
452. Iha, M.H., Barbosa, C.B., Okada, I.A., Trucksess, M.W., 2013: Aflatoxin M_1 in milk and distribution of aflatoxin M_1 during production and storage of yoghurt and cheese. Food Control 29, 1-6. doi: https://doi.org/10.1016/j.foodcont.2012.05.058.
453. Iha, M.H., Barbosa, C.B., Okada, I.A., Trucksess, M.W., 2011: Occurrence of aflatoxin M_1 in dairy products in Brazil. Food Control 22, 1971-1974. doi: https://doi.org/10.1016/j.foodcont.2011.05.013.
454. Iha, M.H., Barbosa, C.B., Heck, A.R., Trucksess, M.W., 2014: Aflatoxin M_1 and ochratoxin A in human milk in Ribeirão Preto-SP, Brazil. Food Control 40, 310-313. doi: https://doi.org/10.1016/j.foodcont.2013.12.014.
455. Hult, K., Hökby, E., Gatenbeck, S., Rutqvist, L., 1980: Ochratoxin A in blood from slaughter pigs in Sweden: use in evaluation of toxin content of consumed feed. Applied and Environmental Microbiology 39, 828-830.
456. Huang, L.C., Zheng, N., Zheng, B.Q., Wen, F., Cheng, J.B., Han, R.W., Xu, X.M., Li, S.L., Wang, J.Q., 2014: Simultaneous determination of aflatoxin M_1, ochratoxin A, zearalenone and α-zearalenol by UHPLC-MS/MS. Food Chemistry 146, 242-249. doi: https://doi.org/10.1016/j.foodchem.2013.09.047.
457. Jia, W., Chu, X., Ling, Y., Huang, J., Chang, J., 2014: Multi-mycotoxin analysis in dairy products by liquid chromatography coupled to quadrupole orbitrap mass spectrometry. Journal of Chromatography A 1345, 107-114. doi: https://doi.org/10.1016/j.chroma.2014.04.021.
458. Jager, A.V., Tedesco, M.P., Souto, P.C.M.C., Oliveira, C.A.F., 2013: Assessment of aflatoxin intake in São Paulo, Brazil. Food Control 33, 87-92. doi: https://doi.org/10.1016/j.foodcont.2013.02.016.
459. Jafari, T., Fallah, A.A., Kheiri, S., Fadaei, A., Amini, S.A., 2017: Aflatoxin M_1 in human breast milk in Shahrekord, Iran and association with dietary factors. Food Additives and Contaminants: Part B 10, 128-136. doi: https://doi.org/10.1080/19393210.2017.1282545.
460. Iqbal, S.Z., Nisar, S., Asi, M.R., Jinap, S., 2014: Natural incidence of aflatoxins, ochratoxin A and zearalenone in chicken meat and eggs. Food Control 43, 98-103. doi: https://doi.org/10.1016/j.foodcont.2014.02.046.
461. Kav, K., Col, R., Tekinsen, K.K., 2011: Detection of aflatoxin M_1 levels by ELISA in white-brined Urfa cheese consumed in Turkey. Food Control 22, 1883-1886 doi: https://doi.org/10.1016/j.foodcont.2011.04.030.
462. Kart, A., Elmali, M., Yapar, K., Yaman, H., 2009: Occurrence of AFM_1 determined by ELISA in UHT (sterilized) and raw milk samples produced in Turkey. Asian Journal of Chemistry 21, 2047-2051.
463. Kara, R., Ince, S., 2014: Aflatoxin M_1 in buffalo and cow milk in Afyonkarahisar, Turkey. Food Additives and Contaminants: Part B 7, 7-10. doi: https://doi.org/10.1080/19393210.2013.825646.
464. Kanungo, L., Bhand, S., 2014: A survey of aflatoxin M_1 in some commercial milk samples and infant formula milk in Goa, India. Food and Agricultural Immunology 25, 467-476. doi: https://doi.org/10.1080/09540105.2013.849051.
465. Kamkar, A., Yazdankhah, S., Nafchi, A.M., Nejad, A.S.M., 2014: Aflatoxin M_1 in raw cow milk and buffalo milk in Shush city of Iran. Food Additives and Contaminants: Part B 7, 21-24. doi: https://doi.org/10.1080/19393210.2013.830277.
466. Kiermeier, F., Weiß, G., Behringer, G., Miller, M., 1977: On the presence and the content of aflatoxin M_1 in commercial cheese samples. Zeitschrift für Lebensmittel-Untersuchung und –Forschung 163, 268-271.
467. Khan, M.Z., Hameed, M.R., Hussain, T., Khan, A., Javed, I., Ahmad, I., Hussain, A., Saleemi, M.K., Islam, N.U., 2013: Aflatoxin residues in tissues of healthy and sick broiler birds at market age in Pakistan: a one year study. Pakistan Veterinary Journal 33, 423-427.
468. Kos, J., Levic, J., Duragic, O., Kokic, B., Miladinovic, I., 2014: Occurrence and estimation of aflatoxin M_1 exposure in milk in Serbia. Food Control 38, 41-46. doi: https://doi.org/10.1016/j.foodcont.2013.09.060 (and personal communication).
469. Kim, H.J., Lee, J.E., Kwak, B.-M., Ahn, J.-H., Jeong, S.-H., 2010: Occurrence of aflatoxin M_1 in raw milk from South Korea winter seasons using an immunoaffinity column and high performance liquid chromatography. Journal of Food Safety 30, 804-813.
470. Kiliç Altun, S., Gürbüz, S., Ayağ, E., 2017: Aflatoxin M_1 in human breast milk in southeastern Turkey. Mycotoxin Research 33, 103-107. doi: https://doi.org/10.1007/s12550-016-0268-4.

471. Laleh rokhi, M., Darsanaki, R., Mohammadi, M., Kolavani, M.H., Issazadeh, K., Aliabadi, M.A., 2013: Determination of aflatoxin M_1 levels in raw milk samples in Gilan, Iran. Advanced Studies in Biology 5, 151-156.
472. Lamplugh, S.M., Hendrickse, R.G., Apeagyei, F., Mwammut, D.D., 1988: Aflatoxins in breast milk, neonatal cord blood, and serum of pregnant women. British Medical Journal 296, 968.
473. Merla, C., Andreoli, G., Garino, C., Vicari, N., Tosi, G., Guglielminetti, M.L., Moretti, A., Biancardi, A., Alorio, M., Fabbi, M., 2018: Monitoring of ochratoxin A and ochratoxin-producing fungi in traditional salami manufactured in northern Italy. Mycotoxin Research 34, 107-116. doi: https://doi.org/10.1007/s12550-017-0305-y (and personal communication).
474. Biancardi, A., Piro, R., Galaverna, G., Dall'Asta, C., 2013: A simple and reliable liquid chromatography-tandem mass spectrometry method for determination of ochratoxin A in hard cheese. International Journal of Food Sciences and Nutrition 64, 632-640. doi: https://doi.org/10.3109/09637486.2013.763911.
475. Lee, J.E., Kwak, B.-M., Ahn, J.-H., Jeon, T.-H., 2009: Occurrence of aflatoxin M_1 in raw milk in South Korea using an immunoaffinity column and liquid chromatography. Food Control 20, 136-138. doi: https://doi.org/10.1016/j.foodcont.2008.03.002.
476. Mahmoudi, R., Norian, R., 2015: Aflatoxin B_1 and M_1 contamination in cow feeds and milk from Iran. Food and Agricultural Immunology 26, 131-137. doi: https://doi.org/10.1080/09540105.2013.876977.
477. Magoha, H., De Meulenaer, B., Kimanya, M., Hipolite, D., Lachat, C., Kolsteren, P., 2014: Fumonisin B_1 contamination in breast milk and its exposure in infants under 6 months of age in Rombo, northern Tanzania. Food and Chemical Toxicology 74, 112-116. doi: https://doi.org/10.1016/j.fct.2014.09.008.
478. Luci, G., Intorre, L., Ferruzzi, G., Mani, D., Giuliotti, L., Pretti, C., Tognetti, R., Bertini, S., Meucci, V., 2018: Determination of ochratoxin A in tissues of wild boar (*Sus scrofa* L.) by enzymatic digestion (ED) coupled to high-performance liquid chromatography with a fluorescence detector (HPLC-FLD). Mycotoxin Research 34, 1-8. doi: https://doi.org/10.1007/s12550-017-0292-z.
479. Magoha, H., Kimanya, M., De Meulenaer, B., Roberfroid, D., Lachat, C., Kolsteren, P., 2014: Association between aflatoxin M_1 exposure through breast milk and growth impairment in infants from northern Tanzania. World Mycotoxin Journal 7, 277-284. doi: https://doi.org/10.3920/WMJ2014.1705.
480. Liman, B.C., Şeybeck, N., 2001: Quantitative analysis of aflatoxin M_1 in raw milk contaminated at low levels. Toxicology Letters, 123, 42.
481. Markov, K., Pleadin, J., Bevardi, M., Vahčić, N., Sokolić-Mihalak, D., Frece, J., 2013: Natural occurrence of aflatoxin B_1, ochratoxin and citrinin in Croatian fermented meat products. Food Control 34, 312-317. doi: https://doi.org/10.1016/j.foodcont.2013.05.002.
482. Maqbool, U., Anwar-Ul-Haq, Ahmad, M., 2009: ELISA determination of aflatoxin M_1 in milk and dairy products in Pakistan. Toxicological and Environmental Chemistry 91, 241-249. doi: https://doi.org/10.1080/02772240802144562.
483. Mao, J., Lei, S., Liu, Y., Xiao, D., Fu, C., Zhong, L., Ouyang, H., 2015: Quantification of aflatoxin M_1 in raw milk by a core-shell column on a conventional HPLC with large volume injection and step gradient elution. Food Control 51, 156-162. doi: https://doi.org/10.1016/j.foodcont.2014.11.022.
484. Manetta, A.C., Giammarco, M., Di Giuseppe, L., Fusaro, I., Gramenzi, A., Formigoni, A., Vignola, G., Lambertini, L., 2009: Distribution of aflatoxin M_1 during Grana Padano cheese production from naturally contaminated milk. Food Chemistry 113, 595-599. doi: https://doi.org/10.1016/j.foodchem.2008.07.091.
485. Maleki, F., Abdi, S., Davodian, E., Haghani, K., Bakhtiyari, S., 2015: Exposure of infants to aflatoxin M_1 from mother's breast milk in Ilam, western Iran. Osong Public Health Research Perspectives 6, 283-287. doi: https://doi.org/10.1016/j.phrp.2015.10.001.
486. Moreno Guillamont, E., Lino, C.M., Baeta, M.L., Pena, A.S., Silveira, M.I.N., Mañes Vinuesa, J., 2005: A comparative study of extraction apparatus in HPLC analysis of ochratoxin A in muscle. Analytical and Bioanalytical Chemistry 383, 570-575. doi: https://doi.org/10.1007/s00216-005-0051-4.
487. Mohammed, S., Munissi, J.J.E., Nyandoro, S.S., 2016: Aflatoxin M_1 in raw milk and aflatoxin B_1 in feed from household cows in Singida, Tanzania. Food Additives and Contaminants: Part B 9, 85-90. doi: https://doi.org/10.1080/19393210.2015.1137361.
488. Akrami Mohajeri, F., Ghalebi, S.R., Rezaeian, M., Gheisari, H.R., Azad, H.K., Zolfaghari, A., Fallah, A.A., 2013: Aflatoxin M_1 in white and Lighvan cheese marketed in Rafsanjan, Iran. Food Control 33, 525-527. doi: https://doi.org/10.1016/j.foodcont.2013.04.002.
489. Tolosa, J., Barba, F.J., Font, G., Ferrer, E., 2019: Mycotoxin incidence in some fish products: QuEChERS methodology and liquid chromatography linear ion trap tandem mass spectrometry approach. Molecules 24, 527. doi: https://doi.org/10.3390/molecules24030527.
490. Santili, A.B.N., De Camargo, A.C., De Syllos Rosa Nunes, R., Da Gloria, E.M., Machado, P.F., Cassoli, L.D., Dos Santos Dias, C.T., Calori-Domingues, M.A., 2015: Aflatoxin M_1 in raw milk from different regions of São Paulo state - Brazil. Food Additives and Contaminants: Part B 8, 207-214. doi: https://doi.org/10.1080/19393210.2015.1048538.
491. Muñoz, K., Blaszkewicz, M., Campos, V., Vega, M., Degen, G.H., 2014: Exposure of infants to ochratoxin A with breast milk. Archives of Toxicology 88, 837-846. doi: https://doi.org/10.1007/s00204-013-1168-4.
492. Panahi, P., Kasaee, S., Mokhtari, A., Sharifi, A., Jangjou, A., 2011: Assessment of aflatoxin M_1 contamination in raw milk by ELISA in Urmia, Iran. American-Eurasian Journal of Toxicological Sciences 3, 231-233.
493. Ozsunar, A., Gumus, T., Arici, M., Demirci, M., 2010: Occurrence of aflatoxin M_1 in raw milk in Trakya region, Turkey. Asian Journal of Chemistry 22, 1879-1884.
494. Oliveira, C.A.F., Sebastião, L.S., Fagundes, H., Rosim, R.E., Fernandes, A.M., 2008: Aflatoxins and cyclopiazonic acid in feed and milk from dairy farms in São Paulo, Brazil. Food Additives and Contaminants: Part B 1, 147-152. doi: https://doi.org/10.1080/02652030802382865.
495. Pattono, D., Grosso, A., Stocco, P.P., Pazzi, M., Zeppa, G., 2013: Survey of the presence of patulin and ochratoxin A in traditional semi-hard cheeses. Food Control 33, 54-57. doi: https://doi.org/10.1016/j.foodcont.2013.02.019.
496. Pathirana, U.P.D., Wimalasiri, K.M.S., Silva, K.F.S.T., Gunarathne, S.P., 2010: Investigation of farm gate cow milk for aflatoxin M_1. Tropical Agricultural Research 21, 119-125.
497. Pietri, A., Fortunati, P., Mulazzi, A., Bertuzzi, T., 2016: Enzyme-assisted extraction for the HPLC determination of aflatoxin M_1 in cheese. Food Chemistry 192, 235-241. doi: https://doi.org/10.1016/j.foodchem.2015.07.006.
498. Pozzo, L., Cavallarin, L., Nucera, D., Antoniazzi, S., Schiavone, A., 2010: A survey of ochratoxyin A contamination in feeds and sera from organic and standard swine farms in northwest Italy. Journal of the Science of Food and Agriculture 90, 1467-1472. doi: https://doi.org/10.1002/jsfa.3965.

499. Pleadin, J., Zadravec, M., Brnić, D., Perković, I., Škrivanko, M., Kovačević, D., 2017: Moulds and mycotoxins detected in the regional speciality fermented sausage "Slavonski Kulen" during a 1-year production period. Food Additives and Contaminants: Part A 34, 282-290. doi: https://doi.org/10.1080/19440049.2016.1266395.
500. Pleadin, J., Staver, M.M., Vahčić, N., Kovačević, D., Milone, S., Saftić, L., Scortichini, G., 2015: Survey of aflatoxin B_1 and ochratoxin A occurrence in traditional meat products coming from Croatian households and markets. Food Control 52, 71-77. doi: https://doi.org/10.1016/j.foodcont.2014.12.027.
501. Pleadin, J., Kovačević, D., Perši, N., 2015: Ochratoxin A contamination of the autochthonous dry-cured meat product "Slavonski Kulen" during a six-month production process. Food Control 57, 377-384. doi: https://doi.org/10.1016/j.foodcont.2015.05.013.
502. Rahimi, E., Shakerian, A., Jafariyan, M., Ebrahimi, M., Riahi, M., 2009: Occurrence of aflatoxin M_1 in raw, pasteurized and UHT milk commercialized in Esfahan and Shahr-e Kord, Iran. Food Security 1, 317-320. doi: https://doi.org/10.1007/s12571-009-0028-9.
503. Rafiei, H., Dehghan, P., Pakshir, K., Pour, M.C., Akbari, M., 2014: The concentration of aflatoxin M_1 in the mothers' milk in Khorrambid City, Fars, Iran. Advanced Biomedical Research 3, 152. doi: https://doi.org/10.4103/2277-9175.137859.
504. Rohani, F.G., Aminaee, M.M., Kianfar, M., 2011: Survey of aflatoxin M_1 in cow's milk for human consumption in Kerman Province of Iran. Food Additives and Contaminants: Part B 4, 191-194. doi: https://doi.org/10.1080/19393210.2011.599866.
505. Pleadin, J., Mihaljević, Ž., Barbir, T., Vulić, A., Kmetič, I., Zadravec, M., Brumen, V., Mitak, M., 2015: Natural incidence of zearalenone in Croatian pig feed, urine and meat in 2014. Food Additives and Contaminants: Part B 8, 277-283. doi: https://doi.org/10.1080/19393210.2015.1089946 (and personal communication).
506. Riahi-Zanjani, B., Balali-Mood, M., 2013: Aflatoxin M_1 contamination in commercial pasteurized milk from local markets in Fariman, Iran. Mycotoxin Research 29, 271-274. doi: https://doi.org/10.1007/s12550-013-0179-6.
507. Rezaei, M., Parviz, M., Es'haghi Gorji, M., Shariatifar, N., Hosseini, M.A., Habibi, S., 2014: Occurrence of aflatoxin M_1 in milk in Qom, Iran. Italian Journal of Food Science 26, 325-328.
508. Santini, A., Raiola, A., Ferrantelli, V., Giangrosso, G., Macaluso, A., Bognanno, M., Galvano, F., Ritieni, A., 2013: Aflatoxin M_1 in raw, UHT milk and dairy products in Sicily (Italy). Food Additives and Contaminants: Part B 6, 181-186. doi: https://doi.org/10.1080/19393210.2013.780186.
509. Sani, A.M., Nikpooyan, H., 2012: Determination of aflatoxin M_1 in milk by high-performance liquid chromatography in Mashhad (north east of Iran). Toxicology and Industrial Health 29, 334-338. doi: https://doi.org/10.1177/0748233711434954.
510. Sahin, H.Z., Celik, M., Kotay, S., Kabak, B., 2016: Aflatoxins in dairy cow feed, raw milk and milk products from Turkey. Food Additives and Contaminants: Part B 9, 152-158. doi: https://doi.org/10.1080/19393210.2016.1152599.
511. Sadia, A., Jabbar, M.J., Deng, Y., Hussain, E.A., Riffat, S., Naveed, S., Arif, M., 2012: A survey of aflatoxin M_1 in milk and sweets of Punjab, Pakistan. Food Control 26, 235-240. doi: https://doi.org/10.1016/j.foodcont.2012.01.055.
512. Saccà, E., Boscolo, D., Vallati, A., Ventura, W., Bigaran, F., Piasentier, E., 2009: Aflatoxin occurrence in milk and supplied concentrates of goat farms of north-eastern Italy. Journal of the Science of Food and Agriculture 89, 487-493. doi: https://doi.org/10.1002/jsfa.3478.
513. Rubert, J., León, N., Sáez, C., Martins, C.P.B., Godula, M., Yusà, V., Mañes, J., Soriano, J.M., Soler, C., 2014: Evaluation of mycotoxins and their metabolites in human breast milk using liquid chromatography coupled to high resolution mass spectrometry. Analytica Chimica Acta 820, 39-46. doi: https://doi.org/10.1016/j.aca.2014.02.009.
514. Sefidgar, S.A.A., Azizi, G., Khosravi, A.R., Roudbar-Mohammadi, S., 2008: Presence of aflatoxin M_1 in raw milk at cattle farms in Babol, Iran. Pakistan Journal of Biological Sciences 11, 484-486.
515. Škrbić, B., Živančev, J., Antić, I., Godula, M., 2014: Levels of aflatoxin M_1 in different types of milk collected in Serbia: assessment of human and animal exposure. Food Control 40, 113-119. doi: https://doi.org/10.1016/j.foodcont.2013.11.039.
516. Škrbić, B., Antić, I., Živančev, J., 2015: Presence of aflatoxin M_1 in white and hard cheese samples from Serbia. Food Control 50, 111-117. doi: https://doi.org/10.1016/j.foodcont.2014.08.031.
517. Sineque, A.R., Macuamule, C.L., Dos Anjos, F.R., 2017: Aflatoxin B_1 contamination in chicken livers and gizzards from industrial and small abattoirs, measured by ELISA techníque in Maputo, Mozambique. International Journal of Environmental Research and Public Health 14 (9), 951. doi: https://doi.org/10.3390/ijerph14090951.
518. Siddappa, V., Nanjegowda, D.K., Viswanath, P., 2012: Occurrence of aflatoxin M_1 in some samples of UHT, raw & pasteurized milk from Indian states of Karnataka and Tamilnadu. Food and Chemical Toxicology 50, 4158-4162. doi: https://doi.org/10.1016/j.fct.2012.08.034.
519. Stoev, S.D., Dutton, M.F., Njobeh, P.B., Mosonik, J.S., Steenkamp, P.A., 2010: Mycotoxic nephropathy in Bulgarian pigs and chickens: complex aetiology and similarity to Balkan Endemic Nephropathy. Food Additives and Contaminants 27, 72-88. doi: https://doi.org/10.1080/02652030903207227.
520. Tabari, M., Tabari, K., Tabari, O., 2012: Aflatoxin M_1 determination in yoghurt produced in Guilan province of Iran using immunoaffinity column and high-performance liquid chromatography. Toxicology and Industrial Health 29, 72-76. doi: https://doi.org/10.1177/0748233712446729.
521. Tabari, M., Tabari, K., Tabari, O., 2011: Occurrence of aflatoxin M_1 in pasteurized doogh commercialized in Tehran, Iran. Australian Journal of Basic and Applied Sciences 5, 1734-1737.
522. Sun, W., Han, Z., Aerts, J., Nie, D., Jin, M., Shi, W., Zhao, Z., De Saeger, S., Zhao, Y., Wu, A., 2015: A reliable liquid chromatography-tandem mass spectrometry method for simultaneous determination of multiple mycotoxins in fresh fish and dried seafood. Journal of Chromatography A 1387, 42-48. doi: https://doi.org/10.1016/j.chroma.2015.01.071.
523. Temamogullari, F., Kanici, A., 2014: Short communication: aflatoxin M_1 in dairy products sold in Şanlıurfa, Turkey. Journal of Dairy Science 97, 162-165. doi: https://doi.org/10.3168/jds.2012-6184.
524. Tavakoli, H.R., Riazipour, M., Kamkar, A., Shaldehi, H.R., Mozaffari Nejad, A.S., 2012: Occurrence of aflatoxin M_1 in white cheese samples from Tehran, Iran. Food Control 23, 293-295. doi: https://doi.org/10.1016/j.foodcont.2011.07.024.
525. Tavakoli, H.R., Kamkar, A., Riazipour, M., Mozaffari Nejad, A.S., Shaldehi, H.R., 2013: Assessment of aflatoxin M_1 levels by enzyme-linked immunosorbent assay in yoghurt consumed in Tehran, Iran. Asian Journal of Chemistry 25, 2836-2838.
526. Torović, L., 2015: Aflatoxin M_1 in processed milk and infant formulae and corresponding exposure of adult population in Serbia in 2013-2014. Food Additives and Contaminants: Part B 8, 235-244. doi: https://doi.org/10.1080/19393210.2015.1063094.
527. Tolosa, J., Font, G., Mañes, J., Ferrer, E., 2014: Natural occurrence of emerging *Fusarium* mycotoxins in feed and fish from aquaculture. Journal of Agricultural and Food Chemistry 62, 12462-12470. doi: https://doi.org/10.1021/jf5036838.

528. Tolosa, J., Font, G., Mañes, J., Ferrer, E., 2013: Natural occurrence of *Fusarium* mycotoxins in aquaculture fish food. Revista de Toxicologia 30, 193-197.
529. Trombete, F.M., De Castro, I.M., Da Silva Teixeira, A., Saldanha, T., Fraga, M.E., 2014: Aflatoxin M_1 contamination in grated Parmesan cheese marketed in Rio de Janeiro - Brazil. Brazilian Archives of Biology and Technology 57, 269-273. doi: https://doi.org/10.1590/S1516-89132013005000015.
530. Vagef, R., Mahmoudi, R., 2013: Occurrence of aflatoxin M_1 in raw and pasteurized milk produced in west region of Iran (during summer and winter). International Food Research Journal 20, 1421-1425.
531. Wang, Y., Liu, X., Xiao, C., Wang, Z., Wang, J., Xiao, H., Cui, L., Xiang, Q., Yue, T., 2012: HPLC determination of aflatoxin M_1 in liquid milk and milk powder using solid phase extraction on OASIS HLB. Food Control 28, 131-134. doi: https://doi.org/10.1016/j.foodcont.2012.04.037.
532. Yosef, T.A., Al-Julaifi, M.Z., Hussein, Y.A., Al-Shokair, S.S., Al-Amer, A.S., 2014: Occurrence of aflatoxin M_1 in raw camel milk in El-Ahsa governorate, Saudi Arabia. Nature and Science 12, 1-7.
533. Xiong, J.L., Wang, Y.M., Ma, M.R., Liu, J.X., 2013: Seasonal variation of aflatoxin M_1 in raw milk from the Yangtze river delta region of China. Food Control 34, 703-706. doi: https://doi.org/10.1016/j.foodcont.2013.06.024.
534. Zhang, K., Wong, J.W., Hayward, D.G., Vaclavikova, M., Liao, C.-D., Trucksess, M.W., 2013: Determination of mycotoxins in milk-based products and infant formula using stable isotope dilution assay and liquid chromatography tandem mass spectrometry. Journal of Agricultural and Food Chemistry 61, 6265-6273. doi: https://doi.org/10.1021/jf4018838.
535. Zhu, R., Zhao, Z., Wang, J., Bai, B., Wu, A., Yan, L., Song, S., 2015: A simple sample pretreatment method for multi-mycotoxin determination in eggs by liquid chromatography tandem mass spectrometry. Journal of Chromatography A 1417, 1-7. doi: https://doi.org/10.1016/j.chroma.2015.09.028.
536. Sun, D., Qiu, N., Zhou, S., Lyu, B., Zhang, S., Li, J., Zhao, Y., Wu, Y., 2019: Development of sensitive and reliable UPLC-MS/MS methods for food analysis of emerging mycotoxins in China total diet study. Toxins (Basel) 11, 166. doi: https://doi.org/10.3390/toxins11030166.
537. Usleber, E., Dade, M., Schneider, E., Dietrich, R., Bauer, J., Märtlbauer, E., 2008: Enzyme immunoassay for mycophenolic acid in milk and cheese. Journal of Agricultural and Food Chemistry 56, 6857-6862. doi: https://doi.org/10.1021/jf801063w.
538. Cao, X., Li, X., Li, J., Niu, Y., Shi, L., Fang, Z., Zhang, T., Ding, H., 2018: Quantitative determination of carcinogenic mycotoxins in human and animal biological matrices and animal-derived foods using multi-mycotoxin and analyte-specific high performance liquid chromatography-tandem mass spectrometric methods. Journal of Chromatography B 1073, 191-200. doi: https://doi.org/10.1016/j.jchromb.2017.10.006.
539. Elaridi, J., Bassil, M., Abi Kharma, J., Daou, F., Hassan, H.F., 2017: Analysis of aflatoxin M_1 in breast milk and its association with nutritional and socioeconomic status of lactating mothers in Lebanon. Journal of Food Protection 80, 1737-1741. doi: https://doi.org/10.4315/0362-028XJFP-17-083.
540. Kunter, I., Hürer, N., Gülcan, H.O., Öztürk, B., Doğan, I., Şahin, G., 2017: Assessment of aflatoxin M_1 and heavy metal levels in mothers breast milk in Famagusta, Cyprus. Biological Trace Element Research 175, 42-49. doi: https://doi.org/10.1007/s12011-016-0750-z.
541. Tomerak, R.H., Shaban, H.H., Khalafallah, O.A., El Shazly, M.N., 2011: Assessment of exposure of Egyptian infants to aflatoxin M_1 through breast milk. Journal of the Egyptian Public Health Association 86, 51-55. doi: https://doi.org/10.1097/01.EXP.0000399138.90797.40.
542. Christofidou, M., Kafouris, D., Christodoulou, M., Stefani, D., Christoforou, E., Nafti, G., Christou, E., Aletrari, M., Ioannou-Kakouri, E., 2015: Occurrence, surveillance, and control of mycotoxins in food in Cyprus for the years 2004-2013. Food and Agricultural Immunology 26, 880-895. doi: https://doi.org/10.1080/09540105.2015.1039499.
543. Sartori, A.V., Swensson de Mattos, J., de Moraes, M.H.P., da Nóbrega, A.W., 2015: Determination of aflatoxins M_1, M_2, B_1, B_2, G_1, and G_2 and ochratoxin A in UHT and powdered milk by modified QuEChERS method and ultra-high-performance liquid chromatography tandem mass spectrometry. Food Analytical Methods 8, 2321-2330. doi: https://doi.org/10.1007/s12161-015-0128-4.
544. Mwanza, M., Abdel-Hadi, A., Ali, A.M., Egbuta, M., 2015: Evaluation of analytical assays efficiency to detect aflatoxin M_1 in milk from selected areas in Egypt and South Africa. Journal of Dairy Science 98, 6660-6667. doi: https://doi.org/10.3168/jds.2014-9220.
545. Abdallah, M.F., Girgin, G., Baydar, T., 2019: Mycotoxin detection in maize, commercial feed, and raw dairy milk samples from Assiut City, Egypt. Veterinary Sciences 6, 57. doi: https://doi.org/10.3390/vetsci6020057.
546. Keller, L.A.M., Aronovich, M., Keller, K.M., Castagna, A.A., Cavaglieri, L.R., da Rocha Rosa, C.A., 2016: Incidence of mycotoxins (AFB_1 and AFM_1) in feeds and dairy farms from Rio de Janeiro State, Brazil. Veterinary Medicine - Open Journal 1, 29-35. doi: https://doi.org/10.17140/VMOJ-1-106.
547. Xiong, J., Xiong, L., Zhou, H., Liu, Y., Wu, L., 2018: Occurrence of aflatoxin B_1 in dairy cow feedstuff and aflatoxin M_1 in UHT and pasteurized milk in central China. Food Control 92, 386-390. doi: https://doi.org/10.1016/j.foodcont.2018.05.022.
548. Rama, A., Montesissa, C., Lucatello, L., Galina, G., Benetti, C., Bajraktari, D., 2016: A study on the occurrence of aflatoxin M_1 in milk consumed in Kosovo during 2009-2010. Food Control 62, 52-55. doi: https://doi.org/10.1016/j.foodcont.2015.10.019.
549. De Paepe, E., Wauters, J., Van Der Borght, M., Claes, J., Huysman, S., Croubels, S., Vanhaecke, L., 2019: Ultra-high-performance liquid chromatography coupled to quadrupole orbitrap high-resolution mass spectrometry for multi-residue screening of pesticides, (veterinary) drugs and mycotoxins in edible insects. Food Chemistry 293, 187-196. doi: https://doi.org/10.1016/j.foodchem.2019.04.082.
550. Gizachew, D., Szonyi, B., Tegegne, A., Hanson, J., Grace, D., 2016: Aflatoxin contamination of milk and dairy feeds in the Greater Addis Ababa milk shed, Ethiopia. Food Control 59, 773-779. doi: https://doi.org/10.1016/j.foodcont.2015.06.060.
551. Molina, A., Chavarría, G., Alfaro-Cascante, M., Leiva, A., Granados-Chinchilla, F., 2019: Mycotoxins at the start of the food chain in Costa Rica: analysis of six *Fusarium* toxins and ochratoxin A between 2013 and 2017 in animal feed and aflatoxin M_1 in dairy products. Toxins (Basel) 11, 312. doi: https://doi.org/10.3390/toxins11060312.
552. Gonçalves, L., Dalla Rosa, A., Gonzales, S.L., Feltes, M.M.C., Badiale-Furlong, E., Dors, G.C., 2017: Incidence of aflatoxin M_1 in fresh milk from small farms. Food Science and Technology (Campinas) 37 (Special Issue), 11-15. doi: https://doi.org/10.1590/1678-457X.06317.
553. Shuib, N.S., Makahleh, A., Salhimi, S.M., Saad, B., 2017: Natural occurrence of aflatoxin M_1 in fresh cow milk and human milk in Penang, Malaysia. Food Control 73, 966-970. doi: https://doi.org/10.1016/j.foodcont.2016.10.013.
554. Jawaid, S., Talpur, F.N., Nizamani, S.M., Afridi, H.I., 2015 : Contamination profile of aflatoxin M_1 residues in milk supply chain of Sindh, Pakistan. Toxicology Reports 2, 1418-1422. doi: https://doi.org/10.1016/j.toxrep.2015.10.011.

555. Michlig, N., Signorini, M., Gaggiotti, M., Chiericatti, C., Basílico, J.C., Repetti, M.R., Beldomenico, H.R., 2016: Risk factors associated with the presence of aflatoxin M_1 in raw bulk milk from Argentina. Food Control 64, 151-156. doi: https://doi.org/10.1016/j.foodcont.2015.12.025.

556. Bellio, A., Bianchi, D.M., Gramaglia, M., Loria, A., Nucera, D., Gallina, S., Gili, M., Decastelli, L., 2016: Aflatoxin M_1 in cow's milk: method validation for milk sampled in northern Italy. Toxins (Basel) 8, 57. doi: https://doi.org/10.3390/toxins8030057.

557. Virdis, S., Scarano, C., Spanu, V., Murittu, G., Spanu, C., Ibba, I., De Santis, E.P.L., 2014: A survey on aflatoxin M_1 content in sheep and goat milk produced in Sardina region, Italy (2005-2013). Italian Journal of Food Safety 3, 206-209. doi: https://doi.org/10.4081/ijfs.2014.4517.

558. Armorini, S., Altafini, A., Zaghini, A., Roncada, P., 2016: Occurrence of aflatoxin M_1 in conventional and organic milk offered for sale in Italy. Mycotoxin Research 32, 237-246. doi: https://doi.org/10.1007/s12550-016-0256-8.

559. Schirone, M., Visciano P., Olivastri, A.M.A., Tofalo, R., Perpetuini, G., Suzzi, G., 2015: A one-year survey on aflatoxin M_1 in raw milk. Italian Journal of Food Science 27, 143-148.

560. Camarillo, E.H., Ramirez-Martinez, A., Carvajal-Moreno, M., Vargas-Ortíz, M., Wesolek, N., Jimenes, G.D.C.R., Alvarado, M.A.G., Roudot, A.-C., Cervantes, M.A.S., Robles-Olvera, V.J., 2018: Assessment of aflatoxin M_1 and M_2 exposure risk through Oaxaca cheese consumption in south-eastern Mexico. International Journal of Environmental Health Research 28, 202-213. doi: https://doi.org/10.1080/09603123.2018.1453054.

561. Yoon, B.R., Hong, S.-Y., Cho, S.M., Lee, K.R., Kim, M., Chung, S.H., 2016: Aflatoxin M_1 levels in dairy products from South Korea determined by high performance liquid chromatography with fluorescence detection. Journal of Food and Nutrition Research 55, 171-180.

562. Marimón Sibaja, K.V., Gonçalves, K.D.M., De Oliveira Garcia, S., Feltrin, A.C.P., Nogueira, W.V., Badiale-Furlong, E., Garda-Buffon, J., 2019: Aflatoxin M_1 and B_1 in Colombian milk powder and estimated risk exposure. Food Additives and Contaminants: Part B 12, 97-104. doi: https://doi.org/10.1080/19393210.2019.1567611.

563. Asghar, M.A., Ahmed, A., Asghar, M.A., 2018: Aflatoxin M_1 in fresh milk collected from local markets of Karachi, Pakistan. Food Additives and Contaminants: Part B 11, 167-174. doi: https://doi.org/10.1080/19393210.2018.1446459.

564. Beltrán, E., Ibáñez, M., Sancho, J.V., Cortés, M.A., Yusà, V., Hernández, F., 2011: UHPLC-MS/MS highly sensitive determination of aflatoxins, the aflatoxin metabolite M_1 and ochratoxin A in baby food and milk. Food Chemistry 126, 737-744. doi: https://doi.org/10.1016/j.foodchem.2010.11.056.

565. Kuboka, M.M., Imungi, J.K., Njue, L., Mutua, F., Grace, D., Lindahl, J.F., 2019: Occurrence of aflatoxin M_1 in raw milk traded in peri-urban Nairobi, and the effect of boiling and fermentation. Infection Ecology & Epidemiology 9, 1625703. doi: https://doi.org/10.1080/20008686.2019.1625703.

Alphabetical Bibliography

Abd Alla, E.A.M., Metwally, M.M., Mehriz, A.M., Abu Sree, Y.H., 1996: Sterigmatocystin: incidence, fate and production by *Aspergillus versicolor* in Ras cheese. Nahrung 40, 310-313 (147).

Abdallah, M.F., Girgin, G., Baydar, T., 2019: Mycotoxin detection in maize, commercial feed, and raw dairy milk samples from Assiut City, Egypt. Veterinary Sciences 6, 57. doi: https://doi.org/10.3390/vetsci6020057 (545).

Abd-Elghany, S.M., Sallam, K.I., 2015: Rapid determination of total aflatoxins and ochratoxins A in meat products by immuno-affinity fluorimetry. Food Chemistry 179, 253-256. doi: https://doi.org/10.1016/j.foodchem.2015.01.140 (392).

Abdulkadar, A.H.W., Al-Ali, A.A., Al-Kildi, A.M., Al-Jedah, J.H., 2004: Mycotoxins in food products available in Qatar. Food Control 15, 543-548. doi: https://doi.org/10.1016/j.foodcont.2003.08.008 (76).

Abdulrazzaq, Y.M., Osman, N., Yousif, Z.M., Al-Falahi, S., 2003: Aflatoxin M_1 in breast-milk of UAE women. Annals of Tropical Paediatrics 23, 173-179. doi: https://doi.org/10.1179/027249303225007671 (334).

Adejumo, O., Atanda, O., Raiola, A., Somorin, Y., Bandyopadhyay, R., Ritieni, A., 2013: Correlation between aflatoxin M_1 content of breast milk, dietary exposure to aflatoxin B_1 and socioeconomic stastus of lactating mothers in Ogun State, Nigeria. Food and Chemical Toxicology 56, 171-177. doi: https://doi.org/10.1016/j.fct.2013.02.027 (391).

Afshar, P., Shokrzadeh, M., Kalhori, S., Babaee, Z., Saeedi Saravi, S.S., 2013: Occurrence of ochratoxin A and aflatoxin M_1 in human breast milk in Sari, Iran. Food Control 31, 525-529. doi: https://doi.org/10.1016/j.foodcont.2012.12.009 (390).

Aidoo, K.E., Mohamed, S.M., Candlish, A.A., Tester, R.F., Elgerbi, A.M., 2011: Occurrence of fungi and mycotoxins in some commercial baby foods in North Africa. Food Nutrition Sciences 2, 751-758. doi: https://doi.org/10.4236/fns.2011.27103 (394).

Akkaya, L., Birdane, Y.O., Oguz, H., Cemek, M., 2006: Occurrence of aflatoxin M_1 in yogurt samples from Afyonkarahisar, Turkey. Bulletin of the Veterinary Institute in Pulawy 50, 517-519 (387).

Akrami Mohajeri, F., Ghalebi, S.R., Rezaeian, M., Gheisari, H.R., Azad, H.K., Zolfaghari, A., Fallah, A.A., 2013: Aflatoxin M_1 in white and Lighvan cheese marketed in Rafsanjan, Iran. Food Control 33, 525-527. doi: https://doi.org/10.1016/j.foodcont.2013.04.002 (488).

Alborzi, S., Pourabbas, B., Rashidi, M., Astaneh, B., 2006: Aflatoxin M_1 contamination in pasteurized milk in Shiraz (south of Iran). Food Control 17, 582-584. doi: https://doi.org/10.1016/j.foodcont.2005.03.009 (210).

Ali, M.A.I., El Zubeir, I.E.M., Fadel Elseed, A.M.A., 2014: Aflatoxin M_1 in raw and imported powdered milk sold in Khartoum state, Sudan. Food Additives and Contaminants 7, 208-212. doi: https://doi.org/10.1080/19393210.2014.887149 (393).

Ali, N., 2000: Aflatoxins in Malaysian food. Mycotoxins 50, 31-35 (80).

Alonso, V.A., Monge, M.P., Larriestra, A., Dalcero, A.M., Cavaglieri, L.R., Chiachiera, S.M., 2010: Naturally occurring aflatoxin M_1 in raw bulk milk from farm cooling tanks in Argentina. Food Additives and Contaminants: Part A 27, 373-379. doi: https://doi.org/10.1080/19440040903403362 (211).

Altafani, A., Fedrizzi, G., Roncada, P., 2018: Occurrence of ochratoxin A in typical salami produced in different regions of Italy. Mycotoxin Research 22, 77. doi: https://doi.org/10.1007/s12550-018-0338-x (423).

Alvito, P.C., Sizoo, E.A., Almeida, C.M.M., Van Egmond, H.P., 2010: Occurrence of aflatoxins and ochratoxin A in baby foods in Portugal. Food Analytical Methods 3, 22-30. doi: https://doi.org/10.1007/s12161-008-9064-x (90).

Amer, A.A., Ibrahim, M.A.E., 2010: Determination of aflatoxin M_1 in raw milk and traditional cheeses retailed in Egyptian markets. Journal of Toxicology and Environmental Health Sciences 2, 50-53 (396).

Amirkhizi, B., Arefhosseini, S.R., Ansarin, M., Nemati, M., 2015: Aflatoxin B_1 in eggs and chicken livers by dispersive liquid-liquid microextraction and HPLC. Food Additives and Contaminants: Part B 8, 245-249. doi: https://doi.org/10.1080/19393210.2015.1067649 (397).

Amra, H.A., 1998: Survey of aflatoxin M_1 in Egyptian raw milk by enzyme-linked immunosorbent assay. Revue de Médecine Vétérinaire 149, 695 (261).

Anderson, P.H., Wells, G.A.H., Jackman, R., Morgan, M.R.A., 1984: Ochratoxicosis and ochratoxin A residues in adult pig`s kidneys—a pilot survey. In: Moss, M.O., Frank, M. (eds): Proceedings of the 5th Meeting of Mycotoxins in Animal and Human Health. Surrey University Press, Guildford, UK, pp. 23-29 (169).

Andrade, P.D., Gomes Da Silva, J.L., Caldas, E.D., 2013: Simultaneous analysis of aflatoxins B_1, B_2, G_1, G_2, M_1 and ochratoxin A in breast milk by high-performance liquid chromatography/fluorescence after liquid-liquid extraction with low temperature purification (LLE-LTP). Journal of Chromatography A 1304, 61-68. doi: https://doi.org/10.1016/j.chroma.2013.06.049 (395).

Anfossi, L., Baggiani, C., Giovannoli, C., D'Arco, G., Passini, C., Giraudi, C., 2012: Occurrence of aflatoxin M_1 in Italian cheese: results of a survey conducted in 2010 and correlation with manufacturing, production season, milking animals, and maturation of cheese. Food Control 25, 125-130. doi: https://doi.org/10.1016/j.foodcont.2011.10.027 (400).

Apostolou, E., El-Nezami, H.S., Ahokas, J.T., Donohue, D.C., 1998: The evaluation of ochratoxin A in breast milk in Victoria (Australia). Revue de Médecine Véterinaire 149, 709 (343).

Ardic, M., Karakaya, Y., Atasever, M., Adiguzel, G., 2009: Aflatoxin M_1 levels of Turkish white brined cheese. Food Control 20, 196-199. doi: https://doi.org/10.1016/j.foodcont.2008.04.003 (151).

Armorini, S., Altafini, A., Zaghini, A., Roncada, P., 2016: Ochratoxin A in artisan salami in Veneto (Italy). Food Additives and Contaminants: Part B 9, 9-14. doi: https://doi.org/10.1080/19393210.2015.1098735 (399).

Armorini, S., Altafini, A., Zaghini, A., Roncada, P., 2016: Occurrence of aflatoxin M_1 in conventional and organic milk offered for sale in Italy. Mycotoxin Research 32, 237-246. doi: https://doi.org/10.1007/s12550-016-0256-8 (558).

Asam, S., Rychlik, M., 2013: Potential health hazards due to the occurrence of the mycotoxin tenuazonic acid in infant food. European Food Research and Technology 236, 491-497 doi: https://doi.org/10.1007/s00217-012-1901-x (108).

Asghar, M.A., Ahmed, A., Asghar, M.A., 2018: Aflatoxin M_1 in fresh milk collected from local markets of Karachi, Pakistan. Food Additives and Contaminants: Part B 11, 167-174. doi: https://doi.org/10.1080/19393210.2018.1446459 (563).

Aslam, N., Tipu, M.Y., Ishaq, M., Cowling, A., McGill, D., Warriach, H.M., Wynn, P., 2016: Higher levels of aflatoxin M_1 contamination and poorer composition of milk supplied by informal milk marketing chains in Pakistan. Toxins (Basel) 8, 347. doi: https://doi.org/10.3390/toxins8120347 (398) (and personal communication).

Assem, E., Mohamad, A., Oula, E.A., 2011: A survey on the occurrence of aflatoxin M_1 in raw and processed milk samples marketed in Lebanon. Food Control 22, 1856-1858. doi: https://doi.org/10.1016/j.foodcont.2011.04.026 (404).

Atasever, M., Yildirim, Y., Atasever, M., Tastekin, A., 2014: Assessment of aflatoxin M_1 in maternal breast milk in eastern Turkey. Food and Chemical Toxicology 66, 147-149. doi: https://doi.org/10.1016/j.fct.2014.01.037 (405).

Atasever, M.A., Atasever, M., Özturan, K., 2011: Aflatoxin M_1 levels in retail yogurt and ayran in Erzurum in Turkey. Turkish Journal of Veterinary and Animal Sciences 35, 59-62. doi: https://doi.org/10.3906/vet-0906-96 (389).

Awashti, V., Bahman, S., Thakur, L.K., Singh, S.K., Dua, A., Ganguly, S., 2012: Contaminants in milk and impact of heating: an assessment study. Indian Journal of Public Health 56, 95-99. doi: https://doi.org/10.4103/0019-557X.96985 (264).

Ayar, A., Sert, D., Çon, A.I., 2007: A study on the occurrence of aflatoxin in raw milk due to feeds. Journal of Food Safety 27, 199-207 (289).

Aycicek, H., Aksoy, A., Saygi, S., 2005: Determination of aflatoxin levels in some dairy and food products which consumed in Ankara, Turkey. Food Control 16, 263-266. doi: https://doi.org/10.1016/j.foodcont.2004.03.004 (11).

Aycicek, H., Yarsan, E., Sarimehmetoglu, B., Cakmak, O., 2002: Aflatoxin M_1 in white cheese and butter consumed in Istanbul, Turkey. Veterinary and Human Toxicology 44, 295-296 (10).

Aygun, O., Essiz, D., Durmaz, H., Yarsan, E., Altinas, L., 2009: Aflatoxin M_1 levels in Surk samples, a traditional Turkish cheese from southern Turkey. Bulletin of Environmental Contamination and Toxicology 83, 164-167. doi: https://doi.org/10.1007/s00128-009-9765-x (148).

Aziz, N.H., Youssef, Y.A., 1991: Occurrence of aflatoxins and aflatoxin-producing moulds in fresh and processed meat in Egypt. Food Additives and Contaminants 8, 321-331 (2).

Bahrami, R., Shahbazi, Y., Nikousefat, Z., 2016: Aflatoxin M_1 in milk and traditional dairy products from west part of Iran: occurrence and seasonal variation with an emphasis on risk assessment of human exposure. Food Control 62, 250-256. doi: https://doi.org/10.1016/j.foodcont.2015.10.039 (328).

Bakirci, I., 2001: A study on the occurrence of aflatoxin M_1 in milk and milk products produced in Van Province of Turkey. Food Control 12, 47-51 (258).

Barbieri, G., Bergamini, C., Ori, E., Resca, P., 1994: Aflatoxin M_1 in Parmesan cheese: HPLC determination. Journal of Food Science 59, 1313, 1331 (146).

Barjesteh, M.H., Azizi, G., Noshfar, E., 2010: Occurrence of aflatoxin M_1 in pasteurized and local yogurt in Mazandaran province (northern Iran) using ELISA. Global Veterinaria 4, 459-462 (388).

Barrios, M.J., Gualda, M.J., Cabanas, J.M., Medina, L.M., Jordano, R., 1996: Occurrence of aflatoxin M_1 in cheeses from the south of Spain. Journal of Food Protection 59, 898-900 (34).

Bartoš, J., Matyáš, Z., 1981: Research on aflatoxin M_1 in fresh milk. Veterinární Medicína 26, 419-423 (254).

Bartoš, J., Matyáš, Z., 1982: Examination of cheeses for sterigmatocystine. Veterinární Medicína 27, 747-752 (112).

Baydar, T., Erkekoglu, P., Sipahi, H., Sahin, G., 2007: Aflatoxin B_1, M_1 and ochratoxin A levels in infant formulae and baby foods marketed in Ankara, Turkey. Journal of Food and Drug Analysis 15, 89-92 (89).

Bellio, A., Bianchi, D.M., Gramaglia, M., Loria, A., Nucera, D., Gallina, S., Gili, M., Decastelli, L., 2016: Aflatoxin M_1 in cow's milk: method validation for milk sampled in northern Italy. Toxins (Basel) 8, 57. doi: https://doi.org/10.3390/toxins8030057 (556).

Beltrán, E., Ibáñez, M., Sancho, J.V., Cortés, M.A., Yusà, V., Hernández, F., 2011: UHPLC-MS/MS highly sensitive determination of aflatoxins, the aflatoxin metabolite M_1 and ochratoxin A in baby food and milk. Food Chemistry 126, 737-744. doi: https://doi.org/10.1016/j.foodchem.2010.11.056 (564).

Bento, H., Fernandes, A.M., Barbosa, M., 1989: Determination of the aflatoxin M_1 in pasteurized milk and in milk powder. Revista Portuguesa de Ciencias Veterinarias 84, 161-171 (292).

Biancardi, A., Piro, R., Galaverna, G., Dall'Asta, C., 2013: A simple and reliable liquid chromatography-tandem mass spectrometry method for determination of ochratoxin A in hard cheese. International Journal of Food Sciences and Nutrition 64, 632-640. doi: https://doi.org/10.3109/09637486.2013.763911 (474).

Biasucci, G., Calabrese, G., Di Giuseppe, R., Carrara, G., Colombo, F., Mandelli, B., Maj, M., Bertuzzi, T., Pietri, A., Rossi, F., 2011: The presence of ochratoxin A in cord serum and in human milk and its correspondence with maternal dietary habits. European Journal of Nutrition 50, 211-218. doi: https://doi.org/10.1007/s00394-010-0130-y (346).

Bijl, J.P., Van Peteghem, C.H., Dekeyser, D.A., 1987: Fluorimetric determination of aflatoxin M_1 in cheese. Journal of the Association of Official Analytical Chemists 70, 472-475 (110).

Bintvihok, A., Davitiyananda, D., Kositcharoenkul, S., Panichkriangkrai, W., Jamratchai, O., 1998: Residues of aflatoxins and their metabolites in chicken tissues in Thailand. The Journal of Toxicological Sciences 23 (Suppl II), 389 (189).

Blanco, J.L., Domínguez, L., Gómez-Lucía, E., Garayzabal, J.F.F., García, J.A., Suárez, G., 1988: Presence of aflatoxin M_1 in commercial ultra-high-temperature-treated milk. Applied and Environmental Microbiology 54, 1622-1623 (355).

Boccia, A., Micco, C., Miraglia, M., Scioli, M., 1986: A study on milk contamination by aflatoxin M_1 in a restricted area in central Italy. Microbiologie - Aliments - Nutrition 4, 293-298 (266).

Bognanno, M., La Fauci, L., Ritieni, A., Tafuri, A., De Lorenzo, A., Micari, P., Di Renzo, L., Ciappellano, S., Sarullo, V., Galvano, F., 2006: Survey of the occurrence of aflatoxin M_1 in ovine milk by HPLC and its confirmation by MS. Molecular Nutrition & Food Research 50, 300-305. doi: https://doi.org/10.1002/mnfr.200500224 (353).

Boudra, H., Barnouin, J., Dragacci, S., Morgavi, D.P., 2007: Aflatoxin M_1 and ochratoxin A in raw bulk milk from French dairy herds. Journal of Dairy Science 90, 3197-3201. doi: https://doi.org/10.3168/jds.2006-565 (212).

Boutrif, E., Jemmali, M., Campbell, A.D., Pohland, A.E., 1977: Aflatoxin in Tunisian foods and foodstuffs. Annales de la Nutrition et de l'Alimentation 31, 431-434 (26).

Bozzo, G., Ceci, E., Bonerba, E., di Pinto, A., Tantillo, G., de Giglio, E., 2012: Occurrence of ochratoxin A in the wild boar (*Sus scrofa*): chemical and histological analysis. Toxins (Basel) 4, 1440-1450. doi: https://doi.org/10.3390/toxins4121440 (408).

Braun, D., Ezekiel, C.N., Abia, W.A., Wisgrill, L., Degen, G.H., Turner, P.C., Marko, D., Warth, B., 2018: Monitoring early life mycotoxin exposures via LC-MS/MS breast milk analysis. Analytical Chemistry 90, 14569-14577. doi: https://doi.org/10.1021/acs.analchem.8b04576 (403).

Breitholtz-Emanuelsson, A., Olsen, M., Oskarsson, A., Palminger, I., Hult, K., 1993: Ochratoxin A in cow's milk and in human milk with corresponding human blood samples. Journal of the Association of Official Analytical Chemists International 76, 842-846 (315).

Büchmann, N.B., Hald, B., 1985: Analysis, occurrence and control of ochratoxin A residues in Danish pig kidneys. Food Additives and Contaminants 2, 193-199 (172).

Bulatao-Jayme, J., Almero, E.M., Castro Ma, C.A., Jardeleza Ma, T.R., Salamat, L.A., 1982: A case-control dietary study of primary liver cancer risk from aflatoxin exposure. International Journal of Epidemiology 11, 112-119 (70).

Burdaspal, P.A., Legarda, T.M., Pinilla, I., 1983: Note. Occurrence of aflatoxin M_1 contamination in milk. Revista de Agroquímica y Tecnologia de Alímentos 23, 287-290 (200).

Camarillo, E.H., Ramirez-Martinez, A., Carvajal-Moreno, M., Vargas-Ortíz, M., Wesolek, N., Jimenes, G.D.C.R., Alvarado, M.A.G., Roudot, A.-C., Cervantes, M.A.S., Robles-Olvera, V.J., 2018: Assessment of aflatoxin M_1 and M_2 exposure risk through Oaxaca cheese consumption in south-eastern Mexico. International Journal of Environmental Health Research 28, 202-213. doi: https://doi.org/10.1080/09603123.2018.1453054 (560).

Cano-Sancho, G., Marin, S., Ramos, A.J., Peris-Vicente, J., Sanchis, V., 2010: Occurrence of aflatoxin M_1 and exposure assessment in Catalonia (Spain). Revista Iberoamericana de Micología 37, 130-135. doi: https://doi.org/10.1016/j.riam.2010.05.003 (216).

Cantú-Cornelio, F., Aguilar-Toalá, J.E., de León-Rodríguez, C.I., Esparza-Romero, J., Vallejo-Cordoba, B., Gonzáles-Córdova, A.F., García, H.S., Hernández-Mendoza, A., 2016: Occurrence and factors associated with the presence of aflatoxin M_1 in breast milk samples of nursing mothers in central Mexico. Food Control 62, 16-22. doi: https://doi.org/10.1016/j.foodcont.2015.10.004 (409).

Cao, X., Li, X., Li, J., Niu, Y., Shi, L., Fang, Z., Zhang, T., Ding, H., 2018: Quantitative determination of carcinogenic mycotoxins in human and animal biological matrices and animal-derived foods using multi-mycotoxin and analyte-specific high performance liquid chromatography-tandem mass spectrometric methods. Journal of Chromatography B 1073, 191-200. doi: https://doi.org/10.1016/j.jchromb.2017.10.006 (538).

Capei, R., Neri, P., 2002: Occurrence of aflatoxin M_1 in milk and yoghurt offered for sale in Florence (Italy). Annali di Igiene 14, 313-319 (217).

Carvajal, M., Bolaños, A., Rojo, F., Méndez, I., 2003: Aflatoxin M_1 in pasteurized and ultrapasteurized milk with different fat content in Mexico. Journal of Food Protection 66, 1885-1892 (281).

Carvajal, M., Rojo, F., Méndez, I., Bolaños, A., 2003: Aflatoxin B_1 and its interconverting metabolite aflatoxicol in milk: the situation in Mexico. Food Additives and Contaminants 20, 1077-1086. doi: https://doi.org/10.1080/02652030310001594478 (231).

Cathey, C.G., Huang, Z.G., Sarr, A.B., Clement, B.A., Phillips, T.D., 1994: Development and evaluation of a minicolumn assay for the detection of aflatoxin M_1 in milk. Journal of Dairy Science 77, 1223-1231 (242).

Ceci, E., Bozzo, G., Bonerba, E., Di Pinto, A., Tantillo, M.G., 2007: Ochratoxin A detection by HPLC in target tissues of swine and cytological and histological analysis. Food Chemistry 105, 364-368. doi: https://doi.org/10.1016/j.foodchem.2006.12.019 (204).

Çelik, T.H., Sarımehmetoğlu, B., Küplülü, Ö., 2005: Aflatoxin M_1 contamination in pasteurised milk. Veterinarski Arhiv 75, 57-65 (282).

Cervino, C., Asam, S., Knopp, D., Rychlik, M., Niessner, R., 2008: Use of isotope-labeled aflatoxins for LC-MS/MS stable isotope dilution analysis of foods. Journal of Agricultural and Food Chemistry 56, 1873-1879. doi: https://doi.org/10.1021/jf073231z (49).

Chavarría, G., Granados-Chinchilla, F., Alfaro-Cascante, M., Molina, A., 2015: Detection of aflatoxin M_1 in milk, cheese and sour cream samples from Costa Rica using enzyme-assisted extraction and HPLC. Food Additives and Contaminants: Part B 8, 128-135. doi: https://doi.org/10.1080/19393210.2015.1015176 (411).

Chen, Y.-C., Liao, C.-D., Lin, H.-Y., Chiueh, L.-C., Shih, D.Y.-C., 2013: Survey of aflatoxin contamination in peanut products in Taiwan from 1997 to 2011. Journal of Food and Drug Analysis 21, 247-252. doi: https://doi.org/10.1016/j.jfda.2013.07.001 (82).

Chiavaro, E., Cacchioli, C., Berni, E., Spotti, E., 2005: Immunoaffinity clean-up and direct fluorescence measurement of aflatoxins B_1 and M_1 in pig liver: comparison with high-performance liquid chromatography determination. Food Additives and Contaminants 22, 1154-1161. doi: https://doi.org/10.1080/02652030500307115 (410).

Chiavaro, E., Lepiani, A., Colla, F., Bettoni, P., Pari, E., Spotti, E., 2002: Ochratoxin A determination in ham by immunoaffinity clean-up and a quick fluorometric method. Food Additives and Contaminants 19, 575-581. doi: https://doi.org/10.1080/02652030210123869 (165).

Choudhary, P.L., Sharma, R.S., Borkhatriya, V.N., Murthi, T.N., Wadodkar, U.R., 1997: Survey on the levels of aflatoxin M_1 in raw and market milk in and around Anand town. Indian Journal of Dairy Science 50, 156-158 (198).

Christofidou, M., Kafouris, D., Christodoulou, M., Stefani, D., Christoforou, E., Nafti, G., Christou, E., Aletrari, M., Ioannou-Kakouri, E., 2015: Occurrence, surveillance, and control of mycotoxins in food in Cyprus for the years 2004-2013. Food and Agricultural Immunology 26, 880-895. doi: https://doi.org/10.1080/09540105.2015.1039499 (542).

Chun, H.S., Kim, H.J., Ok, H.E., Hwang, J.-B., Chung, D.-H., 2007: Determination of aflatoxin levels in nuts and their products consumed in South Korea. Food Chemistry 102, 385-391. doi: https://doi.org/10.1016/j.foodchem.2006.05.031 (58).

Chung, S. W.-C., Kwong, K.P., Tang, A.S.P., Yeung, S.T.K., 2009: Ochratoxin A levels in foodstuffs marketed in Hong Kong. Journal of Food Composition and Analysis. 22, 756-761. doi: https://doi.org/10.1016/j.jfca.2009.02.014 (214).

Colak, H., Hampikyan, H., Ulusoy, B., Ergun, O., 2006: Comparison of a competitive ELISA with an HPLC method for the determination of aflatoxin M_1 in Turkish White, Kasar and Tulum cheeses. European Food Research and Technology 223, 719-723. doi: https://doi.org/10.1007/s00217-006-0258-4 (142).

Commission of the European Communities, 1997. Reports on tasks for scientific cooperation. Report of experts participating in task 3.2.2: assessment of dietary intake of ochratoxin A by the population of EU member states. Report EUR 17523 EN (192).

Commission of the European Communities, January 2002. Reports on tasks for scientific cooperation. Report of experts participating in task 3.2.7: assessment of dietary intake of ochratoxin A by the population of EU member states (215).

Copetti, M.V., Iamanaka, B.T., Nester, M.A., Efraim, P., Taniwaki, M.H., 2013: Occurrence of ochratoxin A in cocoa by-products and determination of its reduction during chocolate manufacture. Food Chemistry 136, 100-104. doi: https://doi.org/10.1016/j.foodchem.2012.07.093 (15).

Copetti, M.V., Iamanaka, B.T., Pereira, J.L., Lemes, D.P., Nakano, F., Taniwaki, M.H., 2012: Determination of aflatoxins in by-products of industrial processing of cocoa beans. Food Additives and Contaminants: Part A 29, 972-978. doi: https://doi.org/10.1080/19440049.2012.660657 (14).

Corbett, W.T., Brownie, C.F., Hagler, S.B., Hagler, W.M., Jr., 1988: An epidemiological investigation associating aflatoxin M_1 with milk production in dairy cattle. Veterinary and Human Toxicology 30, 5-8 (284).

Corbion, B., Fremy, J.M., 1978: Recherche des aflatoxines B_1 et M_1 dans les fromages de type "Camembert". Le Lait 58, 133-140 (127).

Coulter, J.B.S., Lamplugh, S.M., Suliman, G.I., Omer, M.I.A., Hendrickse, R.G., 1984: Aflatoxins in human breast milk. Annals of Tropical Paediatrics 4, 61-66 (336).

Cressey, P., Jones, S., 2010: Mycotoxin surveillance programme 2009-2010. Aflatoxins in nuts and nut products. ESR Client Report FW 10036. ESR, Christchurch, New Zealand (85).

Cressey, P., Pearson, A., 2014: Risk Profile: Mycotoxins in the New Zealand Food Supply. ESR Client Report FW 14005. ESR, Christchurch, New Zealand (66).

Cressey, P., Thomson, B., 2006: Risk Profile: Mycotoxins in the New Zealand Food Supply. ESR Client Report FW 0617. ESR, Christchurch, New Zealand (84).

Curtui, V.G., Gareis, M., Usleber, E., Märtlbauer, E., 2001: Survey of Romanian slaughtered pigs for the occurrence of mycotoxins ochratoxins A and B, and zearalenone. Food Additives and Contaminants 18, 730-738. doi: https://doi.org/10.1080/02652030110035101 (173).

Dall'Asta, C., De Dea Lindner, J., Galaverna, G., Dossena, A., Neviani, E., Marchelli, R., 2008: The occurrence of ochratoxin A in blue cheese. Food Chemistry 106, 729-734. doi: https://doi.org/10.1016/j.foodchem.2007.06.049 (121).

Dashti, B., Al-Hamli, S., Alomirah, H., Al-Zenki, S., Abbas, A.B., Sawaya, W., 2009: Levels of aflatoxin M_1 in milk, cheese consumed in Kuwait and occurrence of total aflatoxin in local and imported animal feed. Food Control 20, 686-690. doi: https://doi.org/10.1016/j.foodcont.2009.01.001 (41).

De Girolamo, A., Pereboom-De Fauw, D., Sizoo, E., Van Egmond, H.P., Gambacorta, L., Bouten, K., Stroka, J., Visconti, A., Solfrizzo, M., 2010: Determination of fumonisins B_1 and B_2 in maize-based baby food products by HPLC with fluorimetric detection after immunoaffinity column clean-up. World Mycotoxin Journal 3, 135-146. doi: https://doi.org/10.3920/WMJ2010.1213 (94).

De Oliveira, C.P., De Fátima Ferreira Soares, N., Veloso De Oliveira, T., Baffa Júnior, J.C., Azevedo De Silva, W., 2013: Aflatoxin M_1 occurrence in ultra high temperature (UHT) treated fluid milk from Minas Gerais/Brazil. Food Control 30, 90-92. doi: https://doi.org/10.1016/j.foodcont.2012.07.026 (414).

De Paepe, E., Wauters, J., Van Der Borght, M., Claes, J., Huysman, S., Croubels, S., Vanhaecke, L., 2019: Ultra-high-performance liquid chromatography coupled to quadrupole orbitrap high-resolution mass spectrometry for multi-residue screening of pesticides, (veterinary) drugs and mycotoxins in edible insects. Food Chemistry 293, 187-196. doi: https://doi.org/10.1016/j.foodchem.2019.04.082 (549).

De Sylos, C.M., Rodriguez-Amaya, D.B., Carvalho, P.R.N., 1996: Occurrence of aflatoxin M_1 in milk and dairy products commercialized in Campinas, Brazil. Food Additives and Contaminants 13, 169-172 (199).

Dehghan, P., Pakshir, K., Rafiei, H., Chadeganipour, M., Akbari, M., 2014: Prevalence of ochratoxin A in human milk in the Khorrambid town, Fars province, south of Iran. Jundishapur Journal of Microbiology 7 (7), e11220. doi: https://doi.org/10.5812/jjm.11220 (426).

Diaz, G.J., Espitia, E., 2006: Occurrence of aflatoxin M_1 in retail milk samples from Bogotá, Colombia. Food Additives and Contaminants 23, 811-815. doi: https://doi.org/10.1080/02652030600681617 (267).

Diaz, G.J., Sánchez, M.P., 2015: Determination of aflatoxin M_1 in breast milk as a biomarker of maternal and infant exposure in Colombia. Food Additives and Contaminants: Part A 32, 1192-1198. doi: https://doi.org/10.1080/19440049.2015.1049563 (413).

Díaz, S., Domínguez, L., Prieta, J., Blanco, J.L., Moreno, M.A., 1995: Application of a diphasic membrane procedure for surveying occurrence of aflatoxin M_1 in commercial milk. Journal of Agricultural and Food Chemistry 43, 2678-2680 (194).

Dieber, F., Köfer, J., 1999: Ochratoxin A-Nachweis im Serum steirischer Schlachtschweine. Deutsche Lebensmittel-Rundschau 95, 327-329 (382).

Dimitrieska-Stojković, E., Stojanovska-Dimzoska, B., Ilievska, G., Uzunov, R., Stojković, G., Hajrulai-Musliu, Z., Jankuloski, D., 2016: Assessment of aflatoxin contamination in raw milk and feed in Macedonia during 2013. Food Control 59, 201-206. doi: https://doi.org/10.1016/j.foodcont.2015.05.019 (412).

Dogan, M., Liman, B.C., Sagdic, O., 2006: Total aflatoxin levels of instant foods in Turkey. Archiv für Lebensmittelhygiene 57, 55-58 (95).

Domagala, J., Kisza, J., Blüthgen, A., Heeschen, W., 1997: Contamination of milk with aflatoxin M_1 in Poland. Milchwissenschaft 52, 631-633 (262).

Dostal, A., Jakusova, L., Cajdova, J., Hudeckova, H., 2008: Results of the first studies of occurrence of ochratoxin A in human milk in Slovakia. Bratislavslec Lekarske Listy 109, 276-278 (349).

Dragacci, S., Fremy, J.-M., 1993: Occurrence of aflatoxin M_1 in milk. Fifteen years of sanitary control. Sciences des Aliments 13, 711-722 (237).

Dragacci, S., Gleizes, E., Fremy, J.M., Candlish, A.A.G., 1995: Use of immunoaffinity chromatography as a purification step for the determination of aflatoxin M_1 in cheeses. Food Additives and Contaminants 12, 59-65 (129).

Dragacci, S., Grosso, F., Bire, R., Fremy, J.M., Coulon, S., 1999: A French monitoring programme for determining ochratoxin A occurrence in pig kidneys. Natural Toxins 7, 167-173 (170).

Duarte, S.C., Almeida, A.M., Teixeira, A.S., Pereira, A.L., Falcão, A.C., Pena, A., Lino, C.M., 2013: Aflatoxin M_1 in marketed milk in Portugal: assessment of human and animal exposure. Food Control 30, 411-417. doi: https://doi.org/10.1016/j.foodcont.2012.08.002 (428).

Duarte, S.C., Lino, C.M., Pena, A., 2013: Novel IAC-LC-ESI-MS^2 analytical set-up for ochratoxin A determination in pork. Food Chemistry 138, 1055-1061. doi: https://doi.org/10.1016/j.foodchem.2012.11.071 (364).

Dutton, M.F., Mwanza, M., De Kock, S., Khilosia, L.D., 2012: Mycotoxins in South African foods: a case study on aflatoxin M_1 in milk. Mycotoxin Research 28, 17-23. doi: https://doi.org/10.1007/s12550-011-0112-9 (427).

El Khoury A., Atoui, A., Yaghi, J., 2011: Analysis of aflatoxin M_1 in milk and yogurt and AFM_1 reduction by lactic acid bacteria used in Lebanese industry. Food Control 22, 1695-1699. doi: https://doi.org/10.1016/j.foodcont.2011.04.001 (432).

El Marnissi, B., Belkhou, R., Morgavi, D.P., Bennani, L., Boudra, H., 2012: Occurrence of aflatoxin M_1 in raw milk collected from traditional dairies in Morocco. Food and Chemical Toxicology 50, 2819-2821. doi: https://doi.org/10.1016/j.fct.2012.05.031 (312).

Elaridi, J., Bassil, M., Abi Kharma, J., Daou, F., Hassan, H.F., 2017: Analysis of aflatoxin M_1 in breast milk and its association with nutritional and socioeconomic status of lactating mothers in Lebanon. Journal of Food Protection 80, 1737-1741. doi: https://doi.org/10.4315/0362-028XJFP-17-083 (539).

Elgerbi, A.M., Aidoo, K.E., Candlish, A.A.G., Tester, R.F., 2004: Occurrence of aflatoxin M_1 in randomly selected North African milk and cheese samples. Food Additives and Contaminants 21, 592-597. doi: https://doi.org/10.1080/02652030410001687690 (150).

El-Gohary, A.H., 1995: Study on aflatoxins in some foodstuffs with special reference to public health hazard in Egypt. Asian-Australasian Journal of Animal Science 8, 571-575 (107).

El-Hoshy, S.M., 1999: Occurrence of zearalenone in milk, meat and their products with emphasis on influence of heat treatments on its level. Archiv für Lebensmittelhygiene 50, 140-143 (141).

Elkak, A., El Atat, O., Habib, J., Abbas, M., 2012: Occurrence of aflatoxin M_1 in cheese processed and marketed in Lebanon. Food Control 25, 140-143. doi: https://doi.org/10.1016/j.foodcont.2011.10.033 (433).

Elling, F., Hald, B., Jacobsen, Chr., Krogh, P., 1975: Spontaneous toxic nephropathy in poultry associated with ochratoxin A. Acta Pathologica et Microbiologica Scandinavica Section A 83, 739-741 (187).

Elmali, M., Yapar, K., Kart, A., Yaman, H., 2008: Aflatoxin M_1 levels in milk powder consumed in Turkey. Journal of Animal and Veterinary Advances 7, 643-645 (379).

El-Nezami, H.S., Nicoletti, G., Neal, G.E., Donohue, D.C., Ahokas, J.T., 1995: Aflatoxin M_1 in human breast milk samples from Victoria, Australia and Thailand. Food and Chemical Toxicology 33, 173-179 (325).

El-Sawi, N.M., El-Maghraby, O.M.O., Mohran, H.S., Abo-Gharbia, M.A., 1994: Abnormal contamination of cottage cheese in Egypt. Journal of Applied Animal Research 6, 81-90 (134).

El-Sayed Abd Alla, A.M., 1996: Natural occurrence of ochratoxin A and citrinin in food stuffs in Egypt. Mycotoxin Research 12, 41-44 (98).

El-Sayed Abd Alla, A.M., Neamat-Allah, A.A., Soher, E.A., 2000: Situation of mycotoxins in milk, dairy products and human milk in Egypt. Mycotoxin Research 16, 91-100 (138).

El-Sayed Abd Alla, A.M., Soher, E.A., Neamat-Allah, A.A., 2002: Human exposure to mycotoxins in Egypt. Mycotoxin Research 18, 23-30 (335).

Elshafie, S.Z.B., ElMubarak, A., El-Nagerabi, S.A.F., Elshafie, A.E., 2011: Aflatoxin B_1 contamination of traditionally processed peanuts butter for human consumption in Sudan. Mycopathologia 171, 435-439. doi: https://doi.org/10.1007/s11046-010-9378-2 (57)

El-Tras, W.F., El-Kady, N.N., Tayel, A.A., 2011: Infants exposure to aflatoxin M_1 as a novel foodborne zoonosis. Food and Chemical Toxicology 49, 2816-2819. doi: https://doi.org/10.1016/j.fct.2011.08.008 (340).

Elzupir, A.O., Abas, A.R.A., Hemmat Fadul, M., Modwi, A.K., Ali, N.M.I., Jadian, A.F.F., Ahmed, N.A.A., Adam, S.Y.A., Ahmed, N.A.M., Khairy A.A.A., Khalil, E.A.G., 2012: Aflatoxin M_1 in breast milk of nursing Sudanese mothers. Mycotoxin Research 28, 131-134. doi: https://doi.org/10.1007/s12550-012-0127-x (287).

Elzupir, A.O., Elhussein, A.M., 2010: Determination of aflatoxin M_1 in dairy cattle milk in Khartoum State, Sudan. Food Control 21, 945-946. doi: https://doi.org/10.1016/j.foodcont.2009.11.013 (285).

Elzupir, A.O., Makawi, S.Z.A., Elhussein, A.M., 2009: Determination of aflatoxins and ochratoxin A in dairy cattle feed and milk in Wad Medani, Sudan. Journal of Animal and Veterinary Advances 8, 2508-2511 (286).

Elzupir, A.O., Salih, A.O.A., Suliman, S.A., Adam, A.A., Elhussein, A.M., 2011: Aflatoxins in peanut butter in Khartoum State, Sudan. Mycotoxin Research 27, 183-186. doi: https://doi.org/10.1007/s12550-011-0094-7 (67).

Engel, G., 2000: Ochratoxin A in sweets, oil seeds and dairy products. Archiv für Lebensmittelhygiene 51, 98-101 (100).

Engel, G., von Milczewski, K.E., Prokopek, D., Teuber, M., 1982: Strain-specific synthesis of mycophenolic acid by *Penicillium roqueforti* in blue-veined cheese. Applied and Environmental Microbiology 43, 1034-1040 (434).

Er, B., Demirhan, B., Yentür, G., 2014: Short communication: investigation of aflatoxin M_1 levels in infant follow-on milks and infant formulas sold in the markets of Ankara, Turkey. Journal of Dairy Science 97, 3328-3331. doi: https://doi.org/10.3168/jds.2013-7831 (201).

Ertas, N., Gonulalan, Z., Yildirim, Y., Karadal, F., 2011: A survey of concentration of aflatoxin M_1 in dairy products marketed in Turkey. Food Control 22, 1956-1959. doi: https://doi.org/10.1016/j.foodcont.2011.05.009 (144).

Fallah, A.A., 2010: Aflatoxin M_1 contamination in dairy products marketed in Iran during winter and summer. Food Control 21, 1478-1481. doi: https://doi.org/10.1016/j.foodcont.2010.04.017 (435).

Fallah, A.A., 2010: Assessment of aflatoxin M_1 contamination in pasteurized and UHT milk marketed in central part of Iran. Food and Chemical Toxicology 48, 988-991. doi: https://doi.org/10.1016/j.fct.2010.01.014 (288).

Fallah, A.A., Barani, A., Nasiri, Z., 2015: Aflatoxin M_1 in raw milk in Qazvin province, Iran: a seasonal study. Food Additives and Contaminants: Part B 8, 195-198. doi: https://doi.org/10.1080/19393210.2015.1046193 (436).

Fallah, A.A., Fazlollahi, R., Emami, A., 2016: Seasonal study of aflatoxin M_1 contamination in milk of four dairy species in Yazd, Iran. Food Control 68, 77-82. doi: https://doi.org/10.1016/j.foodcont.2016.03.018 (437).

Fallah, A.A., Jafari, T., Fallah, A., Rahnama, M., 2009: Determination of aflatoxin M_1 levels in Iranian white and cream cheese. Food and Chemical Toxicology 47, 1872-1875. doi: https://doi.org/10.1016/j.fct.2009.04.042 (136).

Fallah, A.A., Rahnama, M., Jafari, T., Saei-Dehkordi, S.S., 2011: Seasonal variation of aflatoxin M_1 contamination in industrial and traditional Iranian dairy products. Food Control 22, 1653-1656. doi: https://doi.org/10.1016/j.foodcont.2011.03.024 (440).

Farah Nadira, A., Rosita, J., Norhaizan, M.E., Mohd Redzwan, S., 2017: Screening of aflatoxin M_1 occurrence in selected milk and dairy products in Terengganu, Malaysia. Food Control 73, 209-214. doi: https://doi.org/10.1016/j.foodcont.2016.08.004 (163).

Ferguson-Foos, J., Warren, J.D., 1984: Improved cleanup for liquid chromatographic analysis and fluorescence detection of aflatoxins M_1 and M_2 in fluid milk products. Journal of the Association of Official Analytical Chemists 67, 1111-1114 (314).

Finoli, C., Rondinini, G., 1989: Evaluation of infant formula contamination in Italy. Food Chemistry 32, 1-8 (369).

Finoli, C., Vecchio, A., 1997: Aflatoxin M_1 in goat dairy products. Microbiologie – Aliments – Nutrition 15, 47-52 (36).

Finoli, C., Vecchio, A., 2003: Occurrence of aflatoxins in feedstuff, sheep milk and dairy products in Western Sicily. Italian Journal of Animal Science 2, 191-196 (37).

Finoli, C., Vecchio, A., Galli, A., Dragoni, I., 2001: Roquefortine C occurrence in blue cheese. Journal of Food Protection 64, 246-251 (124).

Fontaine, K., Passeró, E., Vallone, L., Hymery, N., Coton, M., Jany, J.-L., Mounier, J., Coton, E., 2015: Occurrence of roquefortine C, mycophenolic acid and aflatoxin M_1 mycotoxins in blue-veined cheeses. Food Control 47, 634-640. doi: https://doi.org/10.1016/j.foodcont.2014.07.046 (439).

Food Standards Agency, 2001: Survey of milk for mycotoxins. Food-Survey-Information-Sheet No. 17/01 (208).

Food Standards Agency, 2002: Survey of nuts, nut products and dried tree fruits for mycotoxins. Food-Survey-Information-Sheet No. 21/02 (9).

Francis, O.J., Jr., Lipinski, L.J., Gaul, J.A., Campbell, A.D., 1982: High pressure liquid chromatographic determination of aflatoxins in peanut butter using a silica gel-packed flowcell for fluorescence detection. Journal of the Association of Official Analytical Chemists 65, 672-676 (60).

Fremy, J.-M., Boursier, B., 1981: Rapid determination of aflatoxin M_1 in dairy products by reversed-phase high-performance liquid chromatography. Journal of Chromatography 219, 156-161 (377).

Fremy, J.M., Chu, F.S., 1984: Direct enzyme-linked immunosorbent assay for determining aflatoxin M_1 at picogram levels in dairy products. Journal of the Association of Official Analytical Chemists 67, 1098-1101 (373).

Frenich, A.G., Romero-Gonzáles, R., Gómez-Pérez, M.L., Vidal, J.L.M., 2011: Multi-mycotoxin analysis in eggs using a QuEChERS-based extraction procedure and ultra-high-pressure liquid chromatography coupled to triple quadrupole mass spectrometry. Journal of Chromatography A 1218, 4349-4356. doi: https://doi.org/10.1016/j.chroma.2011.05.005 (155).

Fritz, W., Engst, R., 1981: Survey of selected mycotoxins in food. Journal of Environmental Science and Health B16, 193-210 (22).

Fukal, L., 1990: A survey of cereals, cereal products, feedstuffs and porcine kidneys for ochratoxin A by radioimmunoassay. Food Additives and Contaminants 7, 253-258 (174).

Fukal, L., 1991: Spontaneous occurrence of ochratoxin A residues in Czechoslovak slaughter pigs determined by immunoassay. Deutsche Lebensmittel-Rundschau 87, 316-319 (5).

Fukal, L., Březina, P., Marek, M., 1990: Immunochemical monitoring of aflatoxin M_1 occurrence in milk produced in Czechoslovakia. Deutsche Lebensmittel-Rundschau 86, 289-291 (232).

Gagliardi, S.J., Cheatle, T.F., Mooney, R.L., Llewellyn, G.C., O'Rear, C.E., Llewellyn, G.C., 1991: The occurrence of aflatoxin in peanut butter from 1982 to 1989. Journal of Food Protection 54, 627-631 (79).

Gajek, O., 1983: Aflatoxins in protein food for animals and milk. Roczniki Państwowego Zakładu Higieny 33, 415-420 (233).

Gallo, P., Salzillo, A., Rossini, C., Urbani, V., Serpe, L., 2006: Aflatoxin M_1 determination in milk: method validation and contamination levels in samples from southern Italy. Italian Journal of Food Science 18, 251-259 (225).

Galvano, F., Galofaro, V., De Angelis, A., Galvano, M., Bognanno, M., Galvano, G., 1998: Survey of the occurrence of aflatoxin M_1 in dairy products marketed in Italy. Journal of Food Protection 61, 738-741 (234).

Galvano, F., Galofaro, V., Ritieni, A., Bognanno, M., De Angelis, A., Galvano, G., 2001: Survey of the occurrence of aflatoxin M_1 in dairy products marketed in Italy: second year of observation. Food Additives and Contaminants 18, 644-646. doi: https://doi.org/10.1080/02652030110035381 (356).

Galvano, F., Pietri, A., Bertuzzi, T., Gagliardi, L., Ciotti, S., Luisi, S., Bognanno, M., La Fauci, L., Iacopino, A.M., Nigro, F., Li Volti, G., Vanella, L., Giammanco, G., Tina, G.L., Gazzolo, D., 2008: Maternal dietary habits and mycotoxin occurrence in human mature milk. Molecular Nutrition & Food Research 52, 496-501. doi: https://doi.org/10.1002/mnfr.200700266 (337).

García Londoño, V.A., Boasso A.C., De Paula, M.C.Z., Garcia, L.P., Scussel, V.M., Resnik, S., Pacín A., 2013: Aflatoxin M_1 survey on randomly collected milk powder commercialized in Argentina and Brazil. Food Control 34, 752-755. doi: https://doi.org/10.1016/j.foodcont.2013.06.030 (438).

Gareis, M., Märtlbauer, E., Bauer, J., Gedek, B., 1988: Determination of ochratoxin A in human milk. Zeitschrift für Lebensmittel-Untersuchung und –Forschung 186, 114-117 (344).

Gareis, M., Scheuer, R., 2000: Ochratoxin A in meat and meat products. Archiv für Lebensmittelhygiene 51, 102-104 (1).

Garmakhany, A.D., Zighamian, H., Sarhangpour R., Rasti, M., Aghajani, N., 2011: Occurrence of aflatoxin M_1 in raw and pasteurized milk in Esfahan province of Iran. Minerva Biotecnologica 23, 53-57 (422).

Garrido, N.S., Iha, M.H., Santos Ortolani, M.R., Duarte Fávaro, R.M., 2003: Occurrence of aflatoxins M_1 and M_2 in milk commercialized in Ribeirão Preto-SP, Brazil. Food Additives and Contaminants 20, 70-73. doi: https://doi.org/10.1080/0265203021000035371 (270).

Gazzotti, T., Lugoboni, B., Zironi, E., Barbarossa, A., Serraino, A., Pagliuca, G., 2009: Determination of fumonisin B_1 in bovine milk by LC-MS/MS. Food Control 20, 1171-1174. doi: https://doi.org/10.1016/j.foodcont.2009.02.009 (442).

Gelda, C.S., Luyt, L.J., 1977: Survey of total aflatoxin content in peanuts, peanut butter, and other foodstuffs. Annales de la Nutrition et de l'Alimentation 31, 477-483 (71).

Ghanem, I., Orfi, M., 2009: Aflatoxin M_1 in raw, pasteurized and powdered milk available in the Syrian market. Food Control 20, 603-605. doi: https://doi.org/10.1016/j.foodcont.2008.08.018 (293).

Ghazani, M.H.M., 2009: Aflatoxin M_1 contamination in pasteurized milk in Tabriz (northwest Iran). Food and Chemical Toxicology 47, 1624-1625. doi: https://doi.org/10.1016/j.fct.2009.04.011 (441).

Ghiasian, S.A., Maghsood, A.H., Neyestani, T.R., Mirhendi, S.H., 2007: Occurrence of aflatoxin M_1 in raw milk during the summer and winter seasons in Hamedan, Iran. Journal of Food Safety 27, 188-198 (294).

Ghidini, S., Zanardi, E., Battaglia, A., Varisco, G., Ferretti, E., Campanini, G., Chizzolini, R., 2005: Comparison of contaminant and residue levels in organic and conventional milk and meat products from northern Italy. Food Additives and Contaminants 22, 9-14. doi: https://doi.org/10.1080/02652030400027995 (295).

Gholampour Azizi, I., Khoushnevis, S.H., Hashemi, S.J., 2007: Aflatoxin M_1 level in pasteurized and sterilized milk of Babol city. Tehran University Medical Journal 65, Supplement 1, 20-24 (218).

Gilbert, J., Shepherd, M.J., 1985: A survey of aflatoxins in peanut butters, nuts and nut confectionery products by HPLC with fluorescence detection. Food Additives and Contaminants 2, 171-183 (18).

Gilbert, J., Shepherd, M.J., Wallwork, M.A., Knowles, M.E., 1984: A survey of the occurrence of aflatoxin M_1 in UK-produced milk for the period 1981-1983. Food Additives and Contaminants 1, 23-28 (235).

Gizachew, D., Szonyi, B., Tegegne, A., Hanson, J., Grace, D., 2016: Aflatoxin contamination of milk and dairy feeds in the Greater Addis Ababa milk shed, Ethiopia. Food Control 59, 773-779. doi: https://doi.org/10.1016/j.foodcont.2015.06.060 (550).

Godič Torkar, K., Vengušt, A., 2008: The presence of yeast, moulds and aflatoxin M_1 in raw milk and cheese in Slovenia. Food Control 19, 570-577. doi: https://doi.org/10.1016/j.foodcont.2007.06.008 (40).

Golge, O., 2014: A survey on the occurrence of aflatoxin M_1 in raw milk produced in Adana province of Turkey. Food Control 45, 150-155. doi: https://doi.org/10.1016/j.foodcont.2014.04.039 (444).

Goliński, P., Hult, K., Grabarkiewicz-Szczesna, J., Chelkowski, J., Kneblewski, P., Szebiotko, K., 1984: Mycotoxic porcine nephropathy and spontaneous occurrence of ochratoxin A residues in kidneys and blood of Polish swine. Applied and Environmental Microbiology 47, 1210-1212 (3).

Goliński, P., Hult, K., Grabarkiewicz-Szczęsna, J., Chelkowski, J., Szebiotko, K., 1985: Spontaneous occurrence of ochratoxin A residues in porcine kidney and serum samples in Poland. Applied and Environmental Microbiology 49, 1014-1015 (171).

Gómez-Arranz, E., Navarro-Blasco, I., 2010: Aflatoxin M_1 in Spanish infant formulae: occurrence and dietary intake regarding type, protein-base and physical state. Food Additives and Contaminants: Part B 3, 193-199. doi: https://doi.org/10.1080/19393210.2010.503353 (443).

Gonçalves, L., Dalla Rosa, A., Gonzales, S.L., Feltes, M.M.C., Badiale-Furlong, E., Dors, G.C., 2017: Incidence of aflatoxin M_1 in fresh milk from small farms. Food Science and Technology (Campinas) 37 (Special Issue), 11-15. doi: https://doi.org/10.1590/1678-457X.06317 (552).

Goto, T., Manabe, M., Matsuura, S., 1982: Analysis of aflatoxins in milk and milk products by high-performance liquid chromatography. Agricultural and Biological Chemistry 46, 801-802 (42).

Grajewski, J., Twarużek, M., Kosicki, R., 2012: High levels of ochratoxin A in blood serum and kidneys of wild boars *Sus scrofa* in Poland. Wildlife Biology 18, 272-279 doi: https://doi.org/10.2981/11-059 (and personal communication) (430).

Gresham, A., Done, S., Livesey, C., MacDonald, S., Chan, D., Sayers, R., Clark, C., Kemp, P., 2006: Survey of pigs' kidneys with lesions consistent with PMWS and PDNS and ochratoxicosis. Part 1: concentrations and prevalence of ochratoxin A. Veterinary Record 159, 737-742 (182).

Guan, D., Li, P., Zhang, Q., Zhang, W., Zhang, D., Jiang, J., 2011: An ultra-sensitive monoclonal antibody-based competitive enzyme immunoassay for aflatoxin M_1 in milk and infant milk products. Food Chemistry 125, 1359-1364. doi: https://doi.org/10.1016/j.foodchem.2010.10.006 (448).

Gul, O., Dervisoglu, M., 2014: Occurrence of aflatoxin M_1 in vacuum packed kashar cheeses in Turkey. International Journal of Food Properties 17, 273-282. doi: https://doi.org/10.1080/10942912.2011.631247 (348).

Gündinç, U., Filazi, A., 2009: Detection of aflatoxin M_1 concentrations in UHT milk consumed in Turkey markets by ELISA. Pakistan Journal of Biological Sciences 12, 653-656 (357).

Guo, Y., Yuan, Y., Yue, T., 2013: Aflatoxin M_1 in milk products in China and dietary risk assessment. Journal of Food Protection 76, 849-853. doi: https://doi.org/10.4315/0362-028X.JFP-12-419 (447).

Gürbay, A., Atasayar Sabuncuoğlu, S., Girgin, G., Şahin, G., Yiğit, Ş., Yurdakök, M., Tekinalp, G., 2010: Exposure of newborne to aflatoxin M_1 and B_1 from mothers' breast milk in Ankara, Turkey. Food and Chemical Toxicology 48, 314-319. doi: https://doi.org/10.1016/j.fct.2009.10.016 (323).

Gürbay, A., Aydın, S., Girgin, G., Engin, A.B., Şahin, G., 2006: Assessment of aflatoxin M_1 levels in milk in Ankara, Turkey. Food Control 17, 1-4. doi: https://doi.org/10.1016/j.foodcont.2004.07.008 (310).

Gürbay, A., Engin, A.B., Çağlayan, A., Şahin, G., 2006: Aflatoxin M_1 levels in commonly consumed cheese and yogurt samples in Ankara, Turkey. Ecology of Food and Nutrition 45, 449-459. doi: https://doi.org/10.1080/03670240600985274 (38).

Gürbay, A., Girgin, G., Atasayar Sabuncuoğlu, Şahin, G., Yurdakök, M., Yiğit, Ş., Tekinalp, G., 2010: Ochratoxin A: is it present in breast milk samples obtained from mothers from Ankara, Turkey? Journal of Applied Toxicology 30, 329-333. doi: https://doi.org/10.1002/jat.1499 (347).

Gürses, M., Erdoğan, A., Çetin, B., 2004: Occurrence of aflatoxin M_1 in some cheese types sold in Erzurum, Turkey. Turkish Journal of Veterinary and Animal Sciences 28, 527-530 (133).

Gutiérrez, R., Rosell, P., Vega, S., Pérez, J., Ramírez, A., Coronado, M., 2013: Self and foreign substances in organic and conventional milk produced in the eastern region of Mexico. Food and Nutrition Sciences 4, 586-593. doi: https://doi.org/10.4236/fns.2013.45076 (421).

Hampikyan, H., Bingol, E.B., Cetin, O., Colak, H., 2010: Determination of aflatoxin M_1 levels in Turkish white, kashar and tulum cheeses. Journal of Food, Agriculture and Environment 8, 13-15 (446).

Han, R.W., Zheng, N., Wang, J.Q., Zhen, Y.P., Xu, X.M., Li, S.L., 2013: Survey of aflatoxin in dairy cow feed and raw milk in China. Food Control 34, 35-39. doi: https://doi.org/10.1016/j.foodcont.2013.04.008 (449).

Haydar, M., Benelli, L., Brera, C., 1990: Occurrence of aflatoxins in Syrian foods and foodstuffs: a preliminary study. Food Chemistry 37, 261-268 (106).

Hendrickse, R.G., Coulter, J.B.S., Lamplugh, S.M., MacFarlane, S.B.J., Williams, T.E., Omer, M.I.A., Suliman, G.I., 1982: Aflatoxins and kwashiorkor: a study in Sudanese children. British Medical Journal 285, 843-846 (62).

Hernández-Martínez, R., Navarro-Blasco, I., 2010: Aflatoxin levels and exposure assessment of Spanish infant cereals. Food Additives and Contaminants: Part B 3, 275-288. doi: https://doi.org/10.1080/19393210.2010.531402 (96).

Herry, M.-P., Lemetayer, N., 1992: Aflatoxin B_1 contamination in oilseeds, dried fruits and spices. Microbiologie – Aliments – Nutrition 10, 261-266 (19).

Herzallah, S.M., 2009: Determination of aflatoxins in eggs, milk, meat and meat products using HPLC fluorescent and UV detectors. Food Chemistry 114, 1141-1146. doi: https://doi.org/10.1016/j.foodchem.2008.10.077 (157).

Heshmati, A., Milani, J.M., 2010: Contamination of UHT milk by aflatoxin M_1 in Iran. Food Control 21, 19-22. doi: https://doi.org/10.1016/j.foodcont.2009.03.013 (359).

Hisada, K., Terada, H., Yamamoto, K., Tsubouchi, H., Sakabe, Y., 1984: Reverse phase liquid chromatographic determination and confirmation of aflatoxin M_1 in cheese. Journal of the Association of Official Analytical Chemists 67, 601-606 (128).

Hisada, K., Yamamoto, K., Tsubouchi, H., Sakabe, Y., 1984: Natural occurrence of aflatoxin M_1 in imported and domestic cheese. (Studies on mycotoxins in foods. VI). Journal of Food Hygienic Society of Japan, Tokyo 25, 543-548 (120).

Holmberg, T., Breitholtz, A., Bengtsson, A., Hult, K., 1990: Ochratoxin A in swine blood in relation to moisture content in feeding barley at harvest. Acta Agriculturæ Scandinavica 40, 201-204 (6).

Honstead, J.P., Dreesen, D.W., Stubblefield, R.D., Shotwell, O.L., 1992: Aflatoxins in swine tissues during drought conditions: an epidemiologic study. Journal of Food Protection 55, 182-186 (190).

Huang, B., Han, Z., Cai, Z., Wu, Y., Ren, Y., 2010: Simultaneous determination of aflatoxin B_1, B_2, G_1, G_2, M_1 and M_2 in peanuts and their derivative products by ultra-high-performance liquid chromatography-tandem mass spectrometry. Analytica Chimica Acta 662, 62-68. doi: https://doi.org/10.1016/j.aca.2010.01.002 (65).

Huang, L.C., Zheng, N., Zheng, B.Q., Wen, F., Cheng, J.B., Han, R.W., Xu, X.M., Li, S.L., Wang, J.Q., 2014: Simultaneous determination of aflatoxin M_1, ochratoxin A, zearalenone and α-zearalenol by UHPLC-MS/MS. Food Chemistry 146, 242-249. doi: https://doi.org/10.1016/j.foodchem.2013.09.047 (456).

Hult, K., Hökby, E., Gatenbeck, S., Rutqvist, L., 1980: Ochratoxin A in blood from slaughter pigs in Sweden: use in evaluation of toxin content of consumed feed. Applied and Environmental Microbiology 39, 828-830. (455).

Hult, K., Rutqvist, L., Holmberg, T., Thafvelin, B., Gatenbeck, S., 1984: Ochratoxin A in blood of slaughter pigs. Nordisk Veterinaermedicin 36, 314-316 (7).

Hussain, I., Anwar, J., 2008: A study on contamination of aflatoxin M_1 in raw milk in the Punjab province of Pakistan. Food Control 19, 393-395. doi: https://doi.org/10.1016/j.foodcont.2007.04.019 (296).

Iacumin, L., Chiesa, L., Boscolo, D., Manzano, M., Cantoni, C., Orlic, S., Comi, G., 2009: Moulds and ochratoxin A on surfaces of artisanal and industrial dry sausages. Food Microbiology 26, 65-70. doi: https://doi.org/10.1016/j.fm.2008.07.006 (380).

Iha, M.H., Barbosa, C.B., Favaro, R.M.D., Trucksess, M.W., 2011: Chromatographic method for the determination of aflatoxin M_1 in cheese, yogurt, and dairy beverages. Journal of the Association of Official Analytical Chemists International 94, 1513-1518. doi: https://doi.org/10.5740/jaoac.int.10-432 (44).

Iha, M.H., Barbosa, C.B., Heck, A.R., Trucksess, M.W., 2014: Aflatoxin M_1 and ochratoxin A in human milk in Ribeirão Preto-SP, Brazil. Food Control 40, 310-313. doi: https://doi.org/10.1016/j.foodcont.2013.12.014 (454).

Iha, M.H., Barbosa, C.B., Okada, I.A., Trucksess, M.W., 2011: Occurrence of aflatoxin M_1 in dairy products in Brazil. Food Control 22, 1971-1974. doi: https://doi.org/10.1016/j.foodcont.2011.05.013 (453).

Iha, M.H., Barbosa, C.B., Okada, I.A., Trucksess, M.W., 2013: Aflatoxin M_1 in milk and distribution of aflatoxin M_1 during production and storage of yoghurt and cheese. Food Control 29, 1-6. doi: https://doi.org/10.1016/j.foodcont.2012.05.058 (452).

Ioannou-Kakouri, E., Aletrari, M., Christou, E., Hadjioannou-Ralli, A., Koliou, A., Akkelidou, D., 1999: Surveillance and control of aflatoxins B_1, B_2, G_1, G_2, and M_1 in foodstuffs in the Republic of Cyprus: 1992-1996. Journal of the Association of Official Analytical Chemists International 82, 883-892 (50).

Iqbal, S.Z., Asi, M.R., 2013: Assessment of aflatoxin M_1 in milk and milk products from Punjab, Pakistan. Food Control 30, 235-239. doi: https://doi.org/10.1016/j.foodcont.2012.06.026 (451).

Iqbal, S.Z., Asi, M.R., Ariño, A., 2011: Aflatoxin M_1 contamination in cow and buffalo milk samples from the North West Frontier Province (NWFP) and Punjab provinces of Pakistan. Food Additives and Contaminants: Part B 4, 282-288. doi: https://doi.org/10.1080/19393210.2011.637237 (450).

Iqbal, S.Z., Asi, M.R., Zuber, M., Akram, N., Batool, N., 2013: Aflatoxins contamination in peanut and peanut products commercially available in retail markets of Punjab, Pakistan. Food Control 32, 83-86. doi: https://doi.org/10.1016/j.foodcont.2012.11.024 (83).

Iqbal, S.Z., Nisar, S., Asi, M.R., Jinap, S., 2014: Natural incidence of aflatoxins, ochratoxin A and zearalenone in chicken meat and eggs. Food Control 43, 98-103. doi: https://doi.org/10.1016/j.foodcont.2014.02.046 (460).

Ishikawa, A.T., Takabayashi-Yamashita, C.R., Ono, E.Y.S., Bagatin, A.K., Rigobello, F.F., Kawamura, O., Hirooka, E.Y., Itano, E.N., 2016: Exposure assessment of infants to aflatoxin M_1 through consumption of breast milk and infant powdered milk in Brazil. Toxins (Basel) 8, 246. doi: https://doi.org/10.3390/toxins8090246 (161).

Ismail, A., Riaz, M., Levin, R.E., Akhtar, S., Gong, Y.Y., Hameed, A., 2016: Seasonal prevalence level of aflatoxin M_1 and its estimated daily intake in Pakistan. Food Control 60, 461-465. doi: https://doi.org/10.1016/j.foodcont.2015.08.025 (402).

Ismail, M.A., Zaky, Z.M., 1999: Evaluation of the mycological status of luncheon meat with special reference to aflatoxigenic moulds and aflatoxin residues. Mycopathologia 146, 147-154 (191).

Jafari, T., Fallah, A.A., Kheiri, S., Fadaei, A., Amini, S.A., 2017: Aflatoxin M_1 in human breast milk in Shahrekord, Iran and association with dietary factors. Food Additives and Contaminants: Part B 10, 128-136. doi: https://doi.org/10.1080/19393210.2017.1282545 (459).

Jafarian-Dehkordi, A., Pourradi, N., 2013: Aflatoxin M_1 contamination of human breast milk in Isfahan, Iran. Advanced Biomedical Research 2, 86. doi: https://doi.org/10.4103/2277-9175.122503 (420).

Jager, A.V., Tedesco, M.P., Souto, P.C.M.C., Oliveira, C.A.F., 2013: Assessment of aflatoxin intake in São Paulo, Brazil. Food Control 33, 87-92. doi: https://doi.org/10.1016/j.foodcont.2013.02.016 (458).

Jang, M.-R., Lee, C.-H., Cho, S.-H., Park, J.-S., Kwon, E.-Y., Lee, E.-J., Kim, S.-H., Kim, D.-B., 2007: A survey of total aflatoxins in food using high performance liquid chromatography-fluorescence detector (HPLC-FLD) and liquid chromatography tandem mass spectrometry (LC-MS/MS). Korean Journal of Food Science and Technology 39, 488-493 (20).

Jarvis, B., 1983: Mould and mycotoxins in mouldy cheeses. Microbiologie – Aliments – Nutrition 1, 187-191 (115).

Jawaid, S., Talpur, F.N., Nizamani, S.M., Afridi, H.I., 2015: Contamination profile of aflatoxin M_1 residues in milk supply chain of Sindh, Pakistan. Toxicology Reports 2, 1418-1422. doi: https://doi.org/10.1016/j.toxrep.2015.10.011 (554).

Jestoi, M., Rokka, M., Järvenpää, E., Peltonen, K., 2009: Determination of *Fusarium* mycotoxins beauvericin and enniatins (A, A_1, B, B_1) in eggs of laying hens using liquid chromatography-tandem mass spectrometry (LC-MS/MS). Food Chemistry 115, 1120-1127. doi: https://doi.org/10.1016/j.foodchem.2008.12.105 (158).

Jia, W., Chu, X., Ling, Y., Huang, J., Chang, J., 2014: Multi-mycotoxin analysis in dairy products by liquid chromatography coupled to quadrupole orbitrap mass spectrometry. Journal of Chromatography A 1345, 107-114. doi: https://doi.org/10.1016/j.chroma.2014.04.021 (457).

Jiménez, A.M., López De Cerain, A., Gonzalez-Peñas, E., Bello, J., 2001: Determination of ochratoxin A in pig liver-derived pâtés by high-performance liquid chromatography. Food Additives and Contaminants 18, 559-563. doi: https://doi.org/10.1080/02652030010025392 (362).

Jonsyn, F.E., Lahai, G.P., 1992: Mycotoxic flora and mycotoxins in smoke-dried fish from Sierra Leone. Nahrung 36, 485-489 (159).

Jonsyn, F.E., Maxwell, S.M., Hendrickse, R.G., 1995: Ochratoxin A and aflatoxins in breast milk samples from Sierra Leone. Mycopathologia 131, 121-126 (321).

Jørgensen, K., 1998: Survey of pork, poultry, coffee, beer and pulses for ochratoxin A. Food Additives and Contaminants 15, 550-554 (102).

Jørgensen, K., Petersen, A., 2002: Content of ochratoxin A in paired kidney and meat samples from healthy Danish slaughter pigs. Food Additives and Contaminants 19, 562-567. doi: https://doi.org/10.1080/02652030110113807 (175).

Josefsson, E., 1979: Study of ochratoxin A in pig kidneys. Vår Föda 31, 415-420 (183).

Kabak, B., 2012: Aflatoxin M_1 and ochratoxin A in baby formulae in Turkey: occurrence and safety evaluation. Food Control 26, 182-187. doi: https://doi.org/10.1016/j.foodcont.2012.01.032 (93).

Kachapulula, P.W., Akello, J., Bandyopadhyay, R., Cotty, P.J., 2018: Aflatoxin contamination of dried insects and fish in Zambia. Journal of Food Protection 81, 1508-1518. doi: https://doi.org/10.4315/0362-028X.JFP-17-527 (338).

Kamali, A., Mehni, S., Kamali, M., Taheri Sarvtin, M., 2017: Detection of ochratoxin A in human breast milk in Jiroft city, south of Iran. Current Medical Mycology 3, 1-4. doi: https://doi.org/10.29252/cmm.3.3.1 (416).

Kamber, U., 2005: Aflatoxin M_1 contamination of some commercial Turkish cheeses from markets in Kars, Turkey. Fresenius Environmental Bulletin 14, 1046-1049 (130).

Kamkar, A., 2005: A study on the occurrence of aflatoxin M_1 in raw milk produced in Sarab city of Iran. Food Control 16, 593-599. doi: https://doi.org/10.1016/j.foodcont.2004.06.021 (298).

Kamkar, A., 2006: A study on the occurrence of aflatoxin M_1 in Iranian Feta cheese. Food Control 17, 768-775. doi:https://doi.org/10.1016/j.foodcont.2005.04.018 (139).

Kamkar, A., 2008: The study of aflatoxin M_1 in UHT milk samples by ELISA. Journal of Veterinary Research 63, 7-12 (360).

Kamkar, A., Yazdankhah, S., Nafchi, A.M., Nejad, A.S.M., 2014: Aflatoxin M_1 in raw cow milk and buffalo milk in Shush city of Iran. Food Additives and Contaminants: Part B 7, 21-24. doi: https://doi.org/10.1080/19393210.2013.830277 (465).

Kaniou-Grigoriadu, I., Eleftheriadou, A., Mouratidou, T., Katikou, P., 2005: Determination of aflatoxin M_1 in ewe's milk samples and the produced curd and Feta cheese. Food Control 16, 257-261. doi: https://doi.org/10.1016/j.foodcont.2004.03.003 (318).

Kanungo, L., Bhand, S., 2014: A survey of aflatoxin M_1 in some commercial milk samples and infant formula milk in Goa, India. Food and Agricultural Immunology 25, 467-476. doi: https://doi.org/10.1080/09540105.2013.849051 (464).

Kara, R., Ince, S., 2014: Aflatoxin M_1 in buffalo and cow milk in Afyonkarahisar, Turkey. Food Additives and Contaminants: Part B 7, 7-10 doi: https://doi.org/10.1080/19393210.2013.825646 (463).

Karaioannoglou, P., Mantis, A., Koufidis, D., Koidis, P., Triantafillou, J., 1989: Occurrence of aflatoxin M_1 in raw and pasteurized milk and in Feta and Telme cheese samples. Milchwissenschaft 44, 746-748 (238).

Kart, A., Elmali, M., Yapar, K., Yaman, H., 2009: Occurrence of AFM_1 determined by ELISA in UHT (sterilized) and raw milk samples produced in Turkey. Asian Journal of Chemistry 21, 2047-2051 (462).

Karwowska, W., Pierzynowska, J., Janicki, A., Waskiewicz-Robak, B., Przybylska, A., 2004: Qualitative and quantitative analysis of filamentous fungi in air, food and ochratoxin A in human milk. Polish Journal of Food and Nutrition Sciences 13/54, SI 2, 41-44 (429).

Kav, K., Col, R., Tekinsen, K.K., 2011: Detection of aflatoxin M_1 levels by ELISA in white-brined Urfa cheese consumed in Turkey. Food Control 22, 1883-1886 doi: https://doi.org/10.1016/j.foodcont.2011.04.030 (461).

Kawamura, O., Sato, S., Nagura, M., Kishimoto, S., Ueno, I., Sato, S., Uda, T., Ito, Y., Ueno, Y., 1990: Enzyme-linked immunosorbent assay for detection and survey of ochratoxin A in livestock sera and mixed feeds. Food & Agricultural Immunology 2, 135-143 (431).

Kawamura, O., Wang, D.-S., Liang, Y.-X., Hasegawa, A., Saga, C., Visconti, A., Ueno, Y., 1994: Further survey of aflatoxin M_1 in milk powders by ELISA. Food & Agricultural Immunology 6, 465-467 (371).

Keller, L.A.M., Aronovich, M., Keller, K.M., Castagna, A.A., Cavaglieri, L.R., da Rocha Rosa, C.A., 2016: Incidence of mycotoxins (AFB_1 and AFM_1) in feeds and dairy farms from Rio de Janeiro State, Brazil. Veterinary Medicine - Open Journal 1, 29-35. doi: https://doi.org/10.17140/VMOJ-1-106 (546).

Keskin, Y., Başkaya, R., Karsli, S., Yurdun, T., Özyaral, O., 2009: Detection of aflatoxin M_1 in human breast milk and raw cow's milk in Istanbul, Turkey. Journal of Food Protection 72, 885-889 (299).

Khan, M.Z., Hameed, M.R., Hussain, T., Khan, A., Javed, I., Ahmad, I., Hussain, A., Saleemi, M.K., Islam, N.U., 2013: Aflatoxin residues in tissues of healthy and sick broiler birds at market age in Pakistan: a one year study. Pakistan Veterinary Journal 33, 423-427 (467).

Khoshnevis, S.H., Azizi, I.G., Shateri, S., Mousavizadeh, M., 2012: Determination of the aflatoxin M_1 in ice cream in Babol City (northern, Iran). Global Veterinaria 8, 205-208 (419).

Kiermeier, F., 1973: Aflatoxin residues in fluid milk. Pure and Applied Microbiology 35, 271-273 (239).

Kiermeier, F., Weiß, G., Behringer, G., Miller, M., 1977: On the presence and the content of aflatoxin M_1 in commercial cheese samples. Zeitschrift für Lebensmittel-Untersuchung und –Forschung 163, 268-271 (466).

Kiermeier, F., Weiss, G., Behringer, G., Miller, M., Ranfft, K., 1977: On the presence and the content of aflatoxin M_1 in milk shipped to a dairy plant. Zeitschrift für Lebensmittel-Untersuchung und –Forschung 163, 171-174 (240).

Kiliç Altun, S., Gürbüz, S., Ayağ, E., 2017: Aflatoxin M_1 in human breast milk in southeastern Turkey. Mycotoxin Research 33, 103-107. doi: https://doi.org/10.1007/s12550-016-0268-4 (470).

Kim, E.K., Shon, D. H., Ryu, D., Park, J.W., Hwang, H.J., Kim, Y. B., 2000: Occurrence of aflatoxin M_1 in Korean dairy products determined by ELISA and HPLC. Food Additives and Contaminants 17, 59-64 (241).

Kim, H.J., Lee, J.E., Kwak, B.-M., Ahn, J.-H., Jeong, S.-H., 2010: Occurrence of aflatoxin M_1 in raw milk from South Korea winter seasons using an immunoaffinity column and high performance liquid chromatography. Journal of Food Safety 30, 804-813 (469).

Kocasari, F.S., Tasci, F., Mor, F., 2012: Survey of aflatoxin M_1 in milk and dairy products consumed in Burdur, Turkey. International Journal of Dairy Technology 65, 365-371. doi: https://doi.org/10.1111/j.1471-0307.2012.00841.x (418).

Koirala, P., Kumar, S., Yadav, B.K., Premarajan, K.C., 2005: Occurrence of aflatoxin in some of the food and feed in Nepal. Indian Journal of Medical Sciences 59, 331-336 (23).

Kokkonen, M., Jestoi, M., Rizzo, A., 2005: Determination of selected mycotoxins in mould cheeses with liquid chromatography coupled to tandem with mass spectrometry. Food Additives and Contaminants 22, 449-456. doi: https://doi.org/10.1080/02652030500089861 (123).

Kos, J., Levic, J., Duragic, O., Kokic, B., Miladinovic, I., 2014: Occurrence and estimation of aflatoxin M_1 exposure in milk in Serbia. Food Control 38, 41-46. doi: https://doi.org/10.1016/j.foodcont.2013.09.060 (and personal communication) (468).

Kotowski, K., Grabarkiewicz-Szczesna, J., Waskiewicz, A., Kostecki, M., Golinski, P., 2000: Ochratoxin A in porcine blood and in consumed feed samples. Mycotoxin Research 16, 66-72 (8).

Kotowski, K., Kostecki, M., Grabarkiewicz-Szczęsna, J., Golinski, P., 1993: Ochratoxin A residue in kidney and blood of pigs. Medycyna Weterynaryjna 49, 554-556 (4).

Kovács, F., Sándor, G., Ványi, A., Domány, S., Zomborszky-Kovács, M., 1995: Detection of ochratoxin A in human blood and colostrum. Acta Veterinaria Hungarica 43, 393-400 (350).

Krogh, P., 1977: Ochratoxin A residues in tissues of slaughter pigs with nephropathy. Nordic Veterinary Medicine A 29, 402-405 (176).

Krogh, P., Hald, B., Plĕstina, R., Ćeović, S., 1977: Balkan (Endemic) Nephropathy and foodborne ochratoxin A: preliminary results of a survey of foodstuffs. Acta Pathologica Microbiologica Scandinavica Section B 85, 238-240 (363).

Kuboka, M.M., Imungi, J.K., Njue, L., Mutua, F., Grace, D., Lindahl, J.F., 2019: Occurrence of aflatoxin M_1 in raw milk traded in peri-urban Nairobi, and the effect of boiling and fermentation. Infection Ecology & Epidemiology 9, 1625703. doi: https://doi.org/10.1080/20008686.2019.1625703 (565).

Kumagai, S., Nakajima, M., Tabata, S., Ishikuro, E., Tanaka, T., Norizuki, H., Itoh, Y., Aoyama, K., Fujita, K., Kai, S., Sato, T., Saito, S., Yoshiike, N., Sugita-Konishi, Y., 2008: Aflatoxin and ochratoxin A contamination of retail foods and intake of these mycotoxins in Japan. Food Additives and Contaminants: Part A 25, 1101-1106. doi: https://doi.org/10.1080/02652030802226187 (47).

Kunter, I., Hürer, N., Gülcan, H.O., Öztürk, B., Doğan, I., Şahin, G., 2017: Assessment of aflatoxin M_1 and heavy metal levels in mothers breast milk in Famagusta, Cyprus. Biological Trace Element Research 175, 42-49. doi: https://doi.org/10.1007/s12011-016-0750-z (540).

Lafont, P., Siriwardana, M.G., Combemale, I., Lafont, J., 1979: Mycophenolic acid in marketed cheeses. Food and Cosmetics Toxicology 17, 147-149 (118).

Lafont, P., Siriwardana, M.G., DeBoer, E., 1990: Contamination of dairy products by fungal metabolites. Journal of Environmental Pathology, Toxicology and Oncology 10, 99-102 (114).

Laleh rokhi, M., Darsanaki, R., Mohammadi, M., Kolavani, M.H., Issazadeh, K., Aliabadi, M.A., 2013: Determination of aflatoxin M_1 levels in raw milk samples in Gilan, Iran. Advanced Studies in Biology 5, 151-156 (471).

Lamplugh, S.M., Hendrickse, R.G., Apeagyei, F., Mwammut, D.D., 1988: Aflatoxins in breast milk, neonatal cord blood, and serum of pregnant women. British Medical Journal 296, 968 (472).

Langseth, W., Nymoen, U., Bergsjø, B., 1993: Ochratoxin A in plasma of Norwegian swine determined by an HPLC column-switching method. Natural Toxins 1, 216-221 (384).

Le Bars, J., 1979: Cyclopiazonic acid production by *Penicillium camemberti* Thom and natural occurrence of this mycotoxin in cheese. Applied and Environmental Microbiology 38, 1052-1055 (116).

Lee, J.E., Kwak, B.-M., Ahn, J.-H., Jeon, T.-H., 2009: Occurrence of aflatoxin M_1 in raw milk in South Korea using an immunoaffinity column and liquid chromatography. Food Control 20, 136-138. doi: https://doi.org/10.1016/j.foodcont.2008.03.002 (475).

Leong, Y.-H., Ismail, N., Latif, A.A., Ahmad, R., 2010: Aflatoxin occurrence in nuts and commercial nutty products in Malaysia. Food Control 21, 334-338. doi: https://doi.org/10.1016/j.foodcont.2009.06.002 (63).

Li, F.-Q., Li, Y.-W., Wang, Y.-R., Luo, X.-Y., 2009: Natural occurrence of aflatoxins in Chinese peanut butter and sesame paste. Journal of Agricultural and Food Chemistry 57, 3519-3524. doi: https://doi.org/10.1021/jf804055n (61).

Liman, B.C., Şeybeck, N., 2001: Quantitative analysis of aflatoxin M_1 in raw milk contaminated at low levels. Toxicology Letters, 123, 42 (480).

Lin, L.-C., Liu, F.-M., Fu, Y.-M., Shih, Y.-C., 2004: Survey of aflatoxin M_1 contamination of dairy products in Taiwan. Journal of Food and Drug Analysis 12, 154-160 (291).

Lindahl, J.F., Kagera, I.N., Grace, D., 2018: Aflatoxin M_1 levels in different marketed milk products in Nairobi, Kenya. Mycotoxin Research 34, 1-7. doi: https://doi.org/10.1007/s12550-018-0323-4 (228).

Lohiya, G., Nichols, L., Hsieh, D., Lohiya, S., Nguyen, H., 1987: Aflatoxin content of foods served to a population with a high incidence of hepatocellular carcinoma. Hepatology 7, 750-752 (45).

López, C.E., Ramos, L.L., Ramadán, S.S., Bulacio, L.C., 2003: Presence of aflatoxin M_1 in milk for human consumption in Argentina. Food Control 14, 31-34 (268).

López, P., De Rijk, T., Sprong, R.C., Mengelers, M.J.B., Castenmiller, J.J.M., Alewijn, M. 2016: A mycotoxin-dedicated total diet study in the Netherlands in 2013: part II - occurrence. World Mycotoxin Journal 9, 89-108. doi: https://doi.org/10.3920/WMJ2015.1906 (275).

Losito, I., Monaci, L., Aresta, A., Zambonin, C.G., 2002: LC-ion trap electrospray MS-MS for the determination of cyclopiazonic acid in milk samples. Analyst 127, 499-502. doi: https://doi.org/10.1039/b200394p (223).

Losito, I., Monaci, L., Palmisano, F., Tantillo, G., 2004: Determination of ochratoxin A in meat products by high-performance liquid chromatography coupled to electrospray ionisation sequential mass spectrometry. Rapid Communications in Mass Spectrometry 18, 1965-1971. doi: https://doi.org/10.1002/rcm.1577 (185).

Luci, G., Intorre, L., Ferruzzi, G., Mani, D., Giuliotti, L., Pretti, C., Tognetti, R., Bertini, S., Meucci, V., 2018: Determination of ochratoxin A in tissues of wild boar (*Sus scrofa* L.) by enzymatic digestion (ED) coupled to high-performance liquid chromatography with a fluorescence detector (HPLC-FLD). Mycotoxin Research 34, 1-8. doi: https://doi.org/10.1007/s12550-017-0292-z (478).

Maaroufi, K., Achour, A., Betbeder, A.M., Hammami, M., Ellouz, F., Creppy, E.E., Bacha, H., 1995: Foodstuffs and human blood contamination by the mycotoxin ochratoxin A: correlation with chronic interstitial nephropathy in Tunisia. Archives of Toxicology 69, 552-558 (105).

MAFF - UK, 1995: Survey of aflatoxin M_1 in retail milk and milk products. Food-Surveillance-Information-Sheet No. 64 (131).

MAFF - UK, 1996: Aflatoxin surveillance of retail and imported nuts, nut products, dried figs and fig products. Food-Surveillance-Information-Sheet No. 81 (74).

MAFF - UK, 1996: Survey of aflatoxin M_1 in farm gate milk. Food-Surveillance-Information-Sheet No. 78 (207).

MAFF - UK, 1997: Survey of aflatoxins and ochratoxin A in cereals and retail products. Food-Surveillance-Information-Sheet No. 130 (361).

Magoha, H., De Meulenaer, B., Kimanya, M., Hipolite, D., Lachat, C., Kolsteren, P., 2014: Fumonisin B_1 contamination in breast milk and its exposure in infants under 6 months of age in Rombo, northern Tanzania. Food and Chemical Toxicology 74, 112-116. doi: https://doi.org/10.1016/j.fct.2014.09.008 (477).

Magoha, H., Kimanya, M., De Meulenaer, B., Roberfroid, D., Lachat, C., Kolsteren, P., 2014: Association between aflatoxin M_1 exposure through breast milk and growth impairment in infants from northern Tanzania. World Mycotoxin Journal 7, 277-284. doi: https://doi.org/10.3920/WMJ2014.1705 (479).

Mahdavi, R., Nikniaz, L., Arefhosseini, S.R., Vahed Jabbari, M., 2010: Determination of aflatoxin M_1 in breast milk samples in Tabriz-Iran. Maternal and Child Health Journal 14, 141-145. doi: https://doi.org/10.1007/s10995-008-0439-9 (341).

Mahmoudi, R., Norian, R., 2015: Aflatoxin B_1 and M_1 contamination in cow feeds and milk from Iran. Food and Agricultural Immunology 26, 131-137. doi: https://doi.org/10.1080/09540105.2013.876977 (476).

Majerus, P., Otteneder, H., Hower, C., 1989: Beitrag zum Vorkommen von Ochratoxin A in Schweineblutserum. Deutsche Lebensmittel-Rundschau 85, 307-313 (383).

Maleki, F., Abdi, S., Davodian, E., Haghani, K., Bakhtiyari, S., 2015: Exposure of infants to aflatoxin M_1 from mother's breast milk in Ilam, western Iran. Osong Public Health Research Perspectives 6, 283-287. doi: https://doi.org/10.1016/j.phrp.2015.10.001 (485).

Manetta, A.C., Giammarco, M., Di Giuseppe, L., Fusaro, I., Gramenzi, A., Formigoni, A., Vignola, G., Lambertini, L., 2009: Distribution of aflatoxin M_1 during Grana Padano cheese production from naturally contaminated milk. Food Chemistry 113, 595-599. doi: https://doi.org/10.1016/j.foodchem.2008.07.091 (484).

Mao, J., Lei, S., Liu, Y., Xiao, D., Fu, C., Zhong, L., Ouyang, H., 2015: Quantification of aflatoxin M_1 in raw milk by a core-shell column on a conventional HPLC with large volume injection and step gradient elution. Food Control 51, 156-162. doi: https://doi.org/10.1016/j.foodcont.2014.11.022 (483).

Maqbool, U., Anwar-Ul-Haq, Ahmad, M., 2009: ELISA determination of aflatoxin M_1 in milk and dairy products in Pakistan. Toxicological and Environmental Chemistry 91, 241-249. doi: https://doi.org/10.1080/02772240802144562 (482).

Maragos, C.M., Richard, J.L., 1994: Quantitation and stability of fumonisins B_1 and B_2 in milk. Journal of the Association of Official Analytical Chemists International 77, 1162-1167 (317).

Mardani, M., Rezapour, S., Rezapour, P., 2011: Survey of aflatoxins in Kashkineh: a traditional Iranian food. Iranian Journal of Microbiology 3, 147-151 (97).

Marimón Sibaja, K.V., Gonçalves, K.D.M., De Oliveira Garcia, S., Feltrin, A.C.P., Nogueira, W.V., Badiale-Furlong, E., Garda-Buffon, J., 2019: Aflatoxin M_1 and B_1 in Colombian milk powder and estimated risk exposure. Food Additives and Contaminants: Part B 12, 97-104. doi: https://doi.org/10.1080/19393210.2019.1567611 (562).

Markaki, P., Melissari, E., 1997: Occurrence of aflatoxin M_1 in commercial pasteurized milk determined with ELISA and HPLC. Food Additives and Contaminants 14, 451-456 (245).

Markov, K., Pleadin, J., Bevardi, M., Vahčić, N., Sokolić-Mihalak, D., Frece, J., 2013: Natural occurrence of aflatoxin B_1, ochratoxin and citrinin in Croatian fermented meat products. Food Control 34, 312-317. doi: https://doi.org/10.1016/j.foodcont.2013.05.002 (481).

Martins, H.M., Guerra, M.M., Bernardo, F., 2005: A six year survey (1999-2004) of the occurrence of aflatoxin M_1 in dairy products produced in Portugal. Mycotoxin Research 21, 192-195 (219).

Martins, M.L., Martins, H.M., 2000: Aflatoxin M_1 in raw and ultra high temperature-treated milk commercialized in Portugal. Food Additives and Contaminants 17, 871-874 (259).

Martins, M.L., Martins, H.M., 2004: Aflatoxin M_1 in yoghurts in Portugal. International Journal of Food Microbiology 91, 315-317. doi: https://doi.org/10.1016/S0168-1605(02)00363-X (386).

Massart, F., Micillo, F., Rivezzi, G., Perrone, L., Baggiani, A., Miccoli, M., Meucci, V., 2016: Zearalenone screening of human breast milk from the Naples area. Toxicological and Environmental Chemistry 98, 128-136. doi: https://doi.org/10.1080/02772248.2015.1101112 (274).

Matumba, L., Monjerezi, M., Biswick, T., Mwatseteza, J., Makumba, W., Kamangira, D., Mtukuso, A., 2014: A survey of the incidence and level of aflatoxin contamination in a range of locally and imported processed foods on Malawian retail market. Food Control 39, 87-91. doi: https://doi.org/10.1016/j.foodcont.2013.09.068 (68).

Matumba, L., Van Poucke, C., Monjerezi, M., Njumbe Ediage, E., De Saeger, S., 2015: Concentrating aflatoxins on the domestic market through groundnut export: a focus on Malawian groundnut value and supply chain. Food Control 51, 236-239. doi: https://doi.org/10.1016/j.foodcont.2014.11.035 (86).

Maxwell, S.M., Apeagyei, F., De Vries, H.R., Mwanmut, D.D., Hendrickse, R.G., 1989: Aflatoxins in breast milk, neonatal cord blood and sera of pregnant women. Journal of Toxicology - Toxin Reviews 8, 19-29 (320).

Meerarani, S., Ramadass, P., Padmanaban, V.D., Nachimuthu, K., 1997: Incidence of aflatoxin M_1 in milk samples around Chennai (Madras) City. Journal of Food Science and Technology 34, 506-508 (265).

Merla, C., Andreoli, G., Garino, C., Vicari, N., Tosi, G., Guglielminetti, M.L., Moretti, A., Biancardi, A., Alorio, M., Fabbi, M., 2018: Monitoring of ochratoxin A and ochratoxin-producing fungi in traditional salami manufactured in northern Italy. Mycotoxin Research 34, 107-116. doi: https://doi.org/10.1007/s12550-017-0305-y (and personal communication) (473).

Meucci, V., Costa, E., Razzuoli, E., Mengozzi, G., Soldani, G., 2005: Occurrence of ochratoxin A in blood of Italian slaughtered pigs. Toxicology Letters, 158S, S116 (415).

Meucci, V., Razzuoli, E., Soldani, G., Massart, F., 2010: Mycotoxin detection in infant formula milks in Italy. Food Additives and Contaminants: Part A 27, 64-71. doi: https://doi.org/10.1080/02652030903207201 (354).

Meucci, V., Soldani, G., Razzuoli, E., Saggeses, G., Massart, F., 2011: Mycoestrogen pollution of Italian infant food. Journal of Pediatrics. 159, 278-283. doi: https://doi.org/10.1016/j.peds.2011.01.028 (164).

Michlig, N., Signorini, M., Gaggiotti, M., Chiericatti, C., Basílico, J.C., Repetti, M.R., Beldomenico, H.R., 2016: Risk factors associated with the presence of aflatoxin M_1 in raw bulk milk from Argentina. Food Control 64, 151-156. doi: https://doi.org/10.1016/j.foodcont.2015.12.025 (555).

Milićević, D., Jurić, V., Stefanović, S., Jovanović, M., Janković, S., 2008: Survey of slaughtered pigs for occurrence of ochratoxin A and porcine nephropathy in Serbia. International Journal of Molecular Sciences 9, 2169-2183. doi: https://doi.org/10.3390/ijms9112169 (184).

Milićević, D., Jurić, V., Vuković, D.Z., Mandić, M.M., Baltić, T.M., 2009: Residue of ochratoxin A in swine tissues - risk assessment. Archive of Oncology 17, 56-60. doi: https://doi.org/10.2298/AOO0904056M (367).

Milićević, D., Stefanović, S., Janković, S., Radičević, T., 2012: Risk analysis and exposure assessment of ochratoxin A in Serbia. Veterinary World 5, 412-416. doi: https://doi.org/10.5455/vetworld.2012.412-416 (366).

Mitchell, N.J., Chen, C., Palumbo, J.D., Bianchini, A., Cappozzo, J., Stratton, J., Ryu, D., Wu, F., 2017: A risk assessment of dietary ochratoxin A in the United States. Food and Chemical Toxicology 100, 265-273. doi: https://doi.org/10.1016/j.fct.2016.12.037 (276).

Mohammadi, H., Shokrzadeh, M., Aliabadi, Z., Riahi-Zamjani, B., 2016: Occurrence of aflatoxin M_1 in commercial pasteurized milk samples in Sari, Mazandaran province, Iran. Mycotoxin Research 32, 85-87. doi: https://doi.org/10.1007/s12550-016-0243-0 (425).

Mohammed, S., Munissi, J.J.E., Nyandoro, S.S., 2016: Aflatoxin M_1 in raw milk and aflatoxin B_1 in feed from household cows in Singida, Tanzania. Food Additives and Contaminants: Part B 9, 85-90. doi: https://doi.org/10.1080/19393210.2015.1137361 (487).

Molina, A., Chavarría, G., Alfaro-Cascante, M., Leiva, A., Granados-Chinchilla, F., 2019: Mycotoxins at the start of the food chain in Costa Rica: analysis of six *Fusarium* toxins and ochratoxin A between 2013 and 2017 in animal feed and aflatoxin M_1 in dairy products. Toxins (Basel) 11, 312. doi: https://doi.org/10.3390/toxins11060312 (551).

Möller, T., Andersson, S., 1983: Aflatoxin M_1 in Swedish milk. Vår Föda 35, 461-465 (253).

Monaci, L., Palmisano, F., Matrella, R., Tantillo, G., 2005: Determination of ochratoxin A at part-per-trillion level in Italian salami by immunoaffinity clean-up high-performance liquid chromatography with fluorescence detection. Journal of Chromatography A 1090, 184-187. doi: https://doi.org/10.1016/j.chroma.2005.07.020 (381).

Monaci, L., Tantillo, G., Palmisano, F., 2004: Determination of ochratoxin A in pig tissues by liquid-liquid extraction and clean-up and high-performance liquid chromatography. Analytical and Bioanalytical Chemistry 378, 1777-1782. doi: https://doi.org/10.1007/s00216-004-2497-1 (186).

Montagna, M.T., Napoli, C., De Giglio, O., Iatta, R., Barbuti, G., 2008: Occurrence of aflatoxin M_1 in dairy products in southern Italy. International Journal of Molecular Sciences 9, 2614-2621. doi: https://doi.org/10.3390/ijms9122614 (43).

Moreno Guillamont, E., Lino, C.M., Baeta, M.L., Pena, A.S., Silveira, M.I.N., Mañes Vinuesa, J., 2005: A comparative study of extraction apparatus in HPLC analysis of ochratoxin A in muscle. Analytical and Bioanalytical Chemistry 383, 570-575. doi: https://doi.org/10.1007/s00216-005-0051-4 (486).

Morgan, M.R.A., Kang, A.S., Chan, H.W.-S., 1986: Aflatoxin determination in peanut butter by enzyme-linked immunosorbent assay. Journal of the Science of Food and Agriculture 37, 908-914 (75).

Morgan, M.R.A., McNerney, R., Chan, H.W.-S., Anderson, P.H., 1986: Ochratoxin A in pig kidney determined by enzyme-linked immunosorbent assay (ELISA). Journal of the Science of Food and Agriculture 37, 475-480 (177).

Mortimer, D.N., Shepherd, M.J., Gilbert, J., Morgan, M.R.A., 1987: A survey of the occurrence of aflatoxin B_1 in peanut butters by enzyme-linked immunosorbent assay. Food Additives and Contaminants 5, 127-132 (52).

Motawee, M.M., Bauer, J., McMahon, D.J., 2009: Survey of aflatoxin M_1 in cow, goat, buffalo and camel milks in Ismailia-Egypt. Bulletin of Environmental Contamination and Toxicology 83, 766-769. doi: https://doi.org/10.1007/s00128-009-9840-3 (226).

Mottaghianpour, E., Nazari, F., Reza Mehrasbi, M., Hosseini, M.-J, 2017: Occurrence of aflatoxin B_1 in baby foods marketed in Iran. Journal of the Science of Food and Agriculture. 97, 2690-2694. doi: https://doi.org/10.1002/jsfa.8092 (279).

Mounjouenpou, P., Mbang, J.A.A., Guyot, B., Guiraud, J.-P., 2012: Traditional procedures of cocoa processing and occurrence of ochratoxin - A in the derived products. Journal of Chemical and Pharmaceutical Research 4, 1332-1339 (17).

Movassagh, M.H., 2011: Presence of aflatoxin M_1 in UHT milk in Tabriz (northwest of Iran). Journal of Food Safety 31, 238-241. doi: https://doi.org/10.1111/j.1745-4565.2010.00291.x (368).

Movassagh Ghazani, M.H., 2009: Aflatoxin M_1 contamination in pasteurized milk in Tabriz (northwest of Iran). Food and Chemical Toxicology 47, 1624-1625. doi: https://doi.org/10.1016/j.fct.2009.04.011 (290).

Mühlemann, M., Lüthy, J., Hübner, P., 1997: Mycotoxin contamination of food in Ecuador. A: aflatoxins. Mitteilungen aus dem Gebiete der Lebensmitteluntersuchung und Hygiene 88, 474-496 (99).

Muñoz, K., Blaszkewicz, M., Campos, V., Vega, M., Degen, G.H., 2014: Exposure of infants to ochratoxin A with breast milk. Archives of Toxicology 88, 837-846. doi: https://doi.org/10.1007/s00204-013-1168-4 (491).

Muñoz, K., Campos, V., Blaszkewicz, M., Vega, M., Alvarez, A., Neira, J., Degen, G.H., 2010: Exposure of neonates to ochratoxin A: first biomonitoring results in human milk (colostrum) from Chile. Mycotoxin Research 26, 59-67. doi: https://doi.org/10.1007/s12550-009-0040-0 (352).

Mupunga, I., Lebelo, S.L., Mngqawa, P., Rheeder, J.P., Katerere, D.R., 2014: Natural occurrence of aflatoxins in peanuts and peanut butter from Bulawayo, Zimbabwe. Journal of Food Protection 77, 1814-1818. doi: https://doi.org/10.4315/0362-028X.JPF-14-129 (69).

Mutlu, A.G., Kursun, O., Kasimoglu, A., Dukel, M., 2010: Determination of aflatoxin M_1 levels and antibiotic residues in the traditional Turkish desserts and ice creams consumed in Burdur City Center. Journal of Animal and Veterinary Advances 9, 2035-2037 (154).

Mwanza, M., Abdel-Hadi, A., Ali, A.M., Egbuta, M., 2015: Evaluation of analytical assays efficiency to detect aflatoxin M_1 in milk from selected areas in Egypt and South Africa. Journal of Dairy Science 98, 6660-6667. doi: https://doi.org/10.3168/jds.2014-9220 (544).

Nakajima, M., Tabata, S., Akiyama, H., Itoh, Y., Tanaka, T., Sunagawa, H., Tyonan, T., Yoshizawa, T., Kumagai, S., 2004: Occurrence of aflatoxin M_1 in domestic milk in Japan during the winter season. Food Additives and Contaminants 21, 472-478. doi: https://doi.org/10.1080/026520304 10001677817 (269).

Nasir, M.S., Jolley, M.E., 2002: Development of a fluorescence polarization assay for the determination of aflatoxins in grains. Journal of Agricultural and Food Chemistry 50, 3116-3121. doi: https://doi.org/10.1021/jf011638c (78).

Navas, S.A., Sabino, M., Rodriguez-Amaya, D.B., 2005: Aflatoxin M_1 and ochratoxin A in a human milk bank in the city of São Paulo, Brazil. Food Additives and Contaminants 22, 457-462. doi: https://doi.org/10.1080/02652030500110550 (333).

Nemati, M., Mehran, M.A., Hamed, P.K., Masoud, A., 2010: A survey on the occurrence of aflatoxin M_1 in milk samples in Ardabil, Iran. Food Control 21, 1022-1024. doi: https://doi.org/10.1016/j.foodcont.2009.12.021 (309).

Neumann-Kleinpaul, A., Terplan, G., 1972: Zum Vorkommen von Aflatoxin M_1 in Trockenmilchprodukten. Archiv für Lebensmittelhygiene 6, 128-132 (374).

Nieuwenhof, F.F.J., Hoolwerf, J.D., Van Den Bedem, J.W., 1990: Evaluation of an enzyme immunoassay for the determination of aflatoxin M_1 in milk using antibody-coated polystyrene beads. Milchwissenschaft 45, 584-588 (246).

Nilüfer, D., Boyacioğlu, D., 2002: Comparative study of three different methods for the determination of aflatoxins in Tahini. Journal of Agricultural and Food Chemistry 50, 3375-3379. doi: https://doi.org/10.1021/jf020005a (88).

Noroozian, E., Lagerwerf, F., Lingeman, H., Brinkman, U.A.Th., Kerkhoff, M.A.T., 1999: Determination of roquefortine C in blue cheese using on-line column-switching liquid chromatography. Journal of Pharmaceutical and Biomedical Analysis 20, 611-619 (126).

Northolt, M.D., Van Egmond, H.P., Soentoro, P., Deijll, E., 1980: Fungal growth and the presence of sterigmatocystin in hard cheese. Journal of the Association of Official Analytical Chemists 63, 115-119 (113).

Noviandi, C.T., Razzazi, E., Agus, A., Böhm, J., Hulan, H.W., Wedhastri, S., Maryudhani, Y.B., Nuryono, Sardjono, Leibetseder, J., 2001: Natural occurrence of aflatoxin B_1 in some Indonesian food and feed products in Yogyakarta in year 1998-1999. Mycotoxin Research 17A, 174-178 (25).

Nuryono, N., Agus, A., Wedhastri, S., Maryudani, Y.B., Sigit Setyabudi, F.M.C., Böhm, J., Razzazi-Fazeli, E., 2009: A limited survey of aflatoxin M_1 in milk from Indonesia by ELISA. Food Control 20, 721-724. doi: https://doi.org/10.1016/j.foodcont.2008.09.005 (213).

Ok, H.E., Kim, H.J., Shim, W.B., Lee, H., Bae, D.-H., Chung, D.-H., Chun, H.S., 2007: Natural occurrence of aflatoxin B_1 in marketed foods and risk estimates of dietary exposure in Koreans. Journal of Food Protection 70, 2824-2828 (21).

Okumura, H., Okimoto, J., Kishimoto, S., Hasegawa, A., Kawamura, O., Nakajima, M., Miyabe, M., Ueno, Y., 1993: An improved indirect competitive ELISA for aflatoxin M_1 in milk powders using novel monoclonal antibodies. Food & Agricultural Immunology 5, 75-84 (378).

Oliveira, C.A., Rosmaninho, J., Rosim, R., 2006: Aflatoxin M_1 and cyclopiazonic acid in fluid milk traded in São Paulo, Brazil. Food Additives and Contaminants 23, 196-201. doi: https://doi.org/10.1080/02652030500398379 (263).

Oliveira, C.A.F., Ferraz, J.C.O., 2007: Occurrence of aflatoxin M_1 in pasteurised, UHT milk and milk powder from goat origin. Food Control 18, 375-378. doi: https://doi.org/10.1016/j.foodcont.2005.11.003 (319).

Oliveira, C.A.F., Germano, P.M.L., Bird, C., Pinto, C.A., 1997: Immunochemical assessment of aflatoxin M_1 in milk powder consumed by infants in São Paulo, Brazil. Food Additives and Contaminants 14, 7-10 (375).

Oliveira, C.A.F., Sebastião, L.S., Fagundes, H., Rosim, R.E., Fernandes, A.M., 2008: Aflatoxins and cyclopiazonic acid in feed and milk from dairy farms in São Paulo, Brazil. Food Additives and Contaminants: Part B 1, 147-152. doi: https://doi.org/10.1080/02652030802382865 (494).

Omar, S.S., 2012: Incidence of aflatoxin M_1 in human and animal milk in Jordan. Journal of Toxicology and Environmental Health Part A 75, 1404-1409. doi: https://doi.org/10.1080/15287394.2012.721174 (313).

Omar, S.S., 2016: Aflatoxin M_1 levels in raw milk, pasteurised milk and infant formula. Italian Journal of Food Safety 5:5788, 158-160. doi: https://doi.org/10.4081/ijfs.2016.5788 (278).

Ortiz, J., Jacxsens, L., Astudillo, G., Ballesteros, A., Donoso, S., Huybregts, L., De Meulenaer, B., 2018: Multiple mycotoxin exposure of infants and young children via breastfeeding and complementary/weaning foods consumption in Ecuadorian highlands. Food and Chemical Toxicology 118, 541-548. doi: https://doi.org/10.1016/j.fct.2018.06.008 (280).

Oruç, H.H., Sonal, S., 2002: Determination of aflatoxin M_1 levels in cheese and milk consumed in Bursa, Turkey. Veterinary and Human Toxicology 43, 292-293 (33).

Ostry, V., Malir, F., Dofkova, M., Skarkova, J., Pfohl-Leszkowicz, A., Ruprich, J., 2015: Ochratoxin A dietary exposure of ten population groups in the Czech Republic: comparison with data over the world. Toxins (Basel) 7 (9), 3608-3635. doi: https://doi.org/10.3390/toxins7093608 (101).

Oveisi, M.-R., Jannat, B., Sadeghi, N., Hajimahmoodi, M., Nikzad, A., 2007: Presence of aflatoxin M_1 in milk and infant milk products in Tehran, Iran. Food Control, 18, 1216-1218. doi: https://doi.org/10.1016/j.foodcont.2006.07.021 (92).

Ozsunar, A., Gumus, T., Arici, M., Demirci, M., 2010: Occurrence of aflatoxin M_1 in raw milk in Trakya region, Turkey. Asian Journal of Chemistry 22, 1879-1884 (493).

Palermo, D., Palermo, C., Rotunno, T., 2001: Survey of aflatoxin M_1 level in cow milk from Puglia Italy. Italian Journal of Food Science 13, 435-442 (247).

Panahi, P., Kasaee, S., Mokhtari, A., Sharifi, A., Jangjou, A., 2011: Assessment of aflatoxin M_1 contamination in raw milk by ELISA in Urmia, Iran. American-Eurasian Journal of Toxicological Sciences 3, 231-233 (492).

Patel, P.M., Netke, S.P., Gupta, B.S., Dabadghao, A.K., 1981: Note on the survey of consumer milk supplied to Jabalpur city for the incidence of aflatoxin M_1 and M_2. Indian Journal of Animal Science 51, 906 (220).

Pathirana, U.P.D., Wimalasiri, K.M.S., Silva, K.F.S.T., Gunarathne, S.P., 2010: Investigation of farm gate cow milk for aflatoxin M_1. Tropical Agricultural Research 21, 119-125 (496).

Pattono, D., Gallo, P.F., Civera, T., 2011: Detection and quantification of ochratoxin A in milk produced in organic farms. Food Chemistry 127, 374-377. doi: https://doi.org/10.1016/j.foodchem.2010.12.051 (224).

Pattono, D., Grosso, A., Stocco, P.P., Pazzi, M., Zeppa, G., 2013: Survey of the presence of patulin and ochratoxin A in traditional semi-hard cheeses. Food Control 33, 54-57. doi: https://doi.org/10.1016/j.foodcont.2013.02.019 (495).

Paul, R., Kalra, M.S., Singh, A., 1976: Incidence of aflatoxins in milk and milk products. Indian Journal of Dairy Science 29, 318-321 (27).

Pei, S.C., Zhang, Y.Y., Eremin, S.A., Lee, W.J., 2009: Detection of aflatoxin M_1 in milk products from China by ELISA using monoclonal antibodies. Food Control 20, 1080-1085. doi: https://doi.org/10.1016/j.foodcont.2009.02.004 (283).

Peng, K.-Y., Chen, C.-Y., 2009: Prevalence of aflatoxin M_1 in milk and its potential liver cancer risk in Taiwan. Journal of Food Protection 72, 1025-1029 (300).

Pettersson, H., Holmberg, T., Larsson, K., Kaspersson, A., 1989: Aflatoxins in acid-treated grain in Sweden and occurrence of aflatoxin M_1 in milk. Journal of the Science of Food and Agriculture 48, 411-420 (248).

Pfleger, R., Brandl, E., 1980: Aflatoxin residues in Austrian milk powder. Wiener Tierärztliche Monatsschrift 67, 101-106 (372).

Piątkowska, M., Sulyok, M., Pietruszka, K., Panasiuk, Ł., 2018: Pilot study for the presence of fungal metabolites in sheep milk from first spring milking. Journal of Veterinary Research 62, 167-172. doi: https://doi.org/10.2478/jvetres-2018-0026 (327).

Picinin, L.C.A., Oliveira Pinho Cerqueira, M.M., Azevedo Vargas, E., Quintão Lana, A.M., Toaldo, I.M., Bordignon-Luiz, M.T., 2013: Influence of climate conditions on aflatoxin M_1 contamination in raw milk from Minas Gerais State, Brazil. Food Control 31, 419-424. doi: https://doi.org/10.1016/j.foodcont.2012.10.024 (203).

Pietri, A., Bertuzzi, T., Bertuzzi, P., Piva, G., 1997: Aflatoxin M_1 occurrence in samples of Grana Padano cheese. Food Additives and Contaminants 14, 341-344 (140).

Pietri, A,. Bertuzzi, T., Gualla, A., Piva, G., 2006: Occurrence of ochratoxin A in raw ham muscles and in pork products from northern Italy. Italian Journal of Food Science 18, 99-106 (152).

Pietri, A., Bertuzzi, T., Moschini, M., Piva, G., 2003: Aflatoxin M_1 occurrence in milk samples destined for Parmigiano Reggiano cheese production. Italian Journal of Food Science 15, 301-306 (301).

Pietri, A., Fortunati, P., Mulazzi, A., Bertuzzi, T., 2016: Enzyme-assisted extraction for the HPLC determination of aflatoxin M_1 in cheese. Food Chemistry 192, 235-241. doi: https://doi.org/10.1016/j.foodchem.2015.07.006 (497).

Pietri, A., Gualla, A., Rastelli, S., Bertuzzi, T., 2011: Enzyme-assisted extraction for the HPLC determination of ochratoxin A in pork and dry-cured ham. Food Additives and Contaminants: Part A 28, 1717-1723. doi: https://doi.org/10.1080/19440049.2011.609490 (167).

Piñeiro, M., Dawson, R., Costarrica, M.L., 1996: Monitoring program for mycotoxin contamination in Uruguayan food and feeds. Natural Toxins 4, 242-245 (206).

Piva, G., Pietri, A., Galazzi, L., Curto, O., 1987: Aflatoxin M_1 occurrence in dairy products marketed in Italy. Food Additives and Contaminants 5, 133-139 (32).

Pleadin, J., Kovačević, D., Perši, N., 2015: Ochratoxin A contamination of the autochthonous dry-cured meat product "Slavonski Kulen" during a six-month production process. Food Control 57, 377-384. doi: https://doi.org/10.1016/j.foodcont.2015.05.013 (501).

Pleadin, J., Mihaljević, Ž., Barbir, T., Vulić, A., Kmetič, I., Zadravec, M., Brumen, V., Mitak, M., 2015: Natural incidence of zearalenone in Croatian pig feed, urine and meat in 2014. Food Additives and Contaminants: Part B 8, 277-283. doi: https://doi.org/10.1080/19393210.2015.1089946 (and personal communication) (505).

Pleadin, J., Staver, M.M., Vahčić, N., Kovačević, D., Milone, S., Saftić, L., Scortichini, G., 2015: Survey of aflatoxin B_1 and ochratoxin A occurrence in traditional meat products coming from Croatian households and markets. Food Control 52, 71-77. doi: https://doi.org/10.1016/j.foodcont.2014.12.027 (500).

Pleadin, J., Zadravec, M., Brnić, D., Perković, I., Škrivanko, M., Kovačević, D., 2017: Moulds and mycotoxins detected in the regional speciality fermented sausage "Slavonski Kulen" during a 1-year production period. Food Additives and Contaminants: Part A 34, 282-290. doi: https://doi.org/10.1080/19440049.2016.1266395 (499).

Polychronaki, N., Turner, P.C., Mykkänen, H., Gong, Y., Amra, H., Abdel-Wahhab, M., El-Nezami, H., 2006: Determinants of aflatoxin M_1 in breast milk in a selected group of Egyptian mothers. Food Additives and Contaminants 23, 700-708. doi: https://doi.org/10.1080/02652030600627222 (332).

Polychronaki, N., West, R.M., Turner, P.C., Amra, H., Abdel-Wahhab, M., Mykkänen, H., El-Nezami, H., 2007: A longitudinal assessment of aflatoxin M_1 excretion in breast milk of selected Egyptian mothers. Food and Chemical Toxicology 45, 1210-1215. doi: https://doi.org/10.1016/j.fct.2007.01.001 (329).

Polzhofer, K., 1977: Determination of aflatoxins in milk and milk products. Zeitschrift für Lebensmittel-Untersuchung und -Forschung 163, 175-177 (28).

Postupolski, J., Karlowski, K., Kubik, P., 2006: Ochratoxin A in maternal and foetal blood and in maternal milk. Roczniki Państwowego Zakładu Higieny 57, 23-30 (351).

Pozzo, L., Cavallarin, L:, Nucera, D., Antoniazzi, S., Schiavone, A., 2010: A survey of ochratoxyin A contamination in feeds and sera from organic and standard swine farms in northwest Italy. Journal of the Science of Food and Agriculture 90, 1467-1472. doi: https://doi.org/10.1002/jsfa.3965 (498).

Prado, G., Oliveira, M.S., Pereira, M.L., Abrantes, F.M., Santos, L.G., Veloso, T., 2000: Aflatoxin M_1 in samples of "Minas" cheese commercialized in the city of Belo Horizonte - Minas Gerais/Brazil. Ciência e Tecnologia de Alimentos 20, 398-400. doi: https://doi.org/10.1590/S0101-20612000000300020 (145).

Prado, G., Silva de Oliveira, M., Souza Lima, A., Aprigio Moreira, A.P., 2008: Occurrence of aflatoxin M_1 in Parmesan cheese consumed in Minas Gerais, Brazil. Ciência e Agrotecnologia 32, 1906-1911 (273).

Purchase, I.F.H., Vorster, L.J., 1968: Aflatoxin in commercial milk samples. South African Medical Journal 42, 219 (249).

Qian, G.-S., Ross, R.K., Yu, M.C., Yuan, J.-M., Gao, Y.-T., Henderson, B.E., Wogan, G.N., Groopman, J.D., 1994: A follow-up study of urinary markers of aflatoxin exposure and liver cancer risk in Shanghai, People's Republic of China. Cancer Epidemiology, Biomarkers & Prevention 3, 3-10 (46).

Rafiei, H., Dehghan, P., Pakshir, K., Pour, M.C., Akbari, M., 2014: The concentration of aflatoxin M_1 in the mothers' milk in Khorrambid City, Fars, Iran. Advanced Biomedical Research 3, 152. doi: https://doi.org/10.4103/2277-9175.137859 (503).

Rahimi, E., 2014: Survey of the occurrence of aflatoxin M_1 in dairy products marketed in Iran. Toxicology and Industrial Health 30, 750-754. doi: https://doi.org/10.1177/0748233712462476 (424).

Rahimi, E., Ameri, M., 2012: A survey of aflatoxin M_1 contamination in bulk milk samples from dairy bovine, ovine, and caprine herds in Iran. Bulletin of Environmental Contamination and Toxicology 89, 158-160. doi: https://doi.org/10.1007/s00128-012-0616-9 (311).

Rahimi, E., Bonyadian, M., Rafei, M., Kazemeini, H.R., 2010: Occurrence of aflatoxin M_1 in raw milk of five dairy species in Ahzav, Iran. Food and Chemical Toxicology 48, 129-131. doi: https://doi.org/10.1016/j.fct.2009.09.028 (227).

Rahimi, E., Karim, G., Shakerian, A., 2009: Occurrence of aflatoxin M_1 in traditional cheese consumed in Esfahan, Iran. World Mycotoxin Journal 2, 91-94. doi: https://doi.org/10.3920/WMJ2008.1082 (39).

Rahimi, E., Shakerian, A., Jafariyan, M., Ebrahimi, M., Riahi, M., 2009: Occurrence of aflatoxin M_1 in raw, pasteurized and UHT milk commercialized in Esfahan and Shahr-e Kord, Iran. Food Security 1, 317-320. doi: https://doi.org/10.1007/s12571-009-0028-9 (502).

Rajan, A., Ismail, P.K., Radhakrishnan, V., 1995: Survey of milk samples for aflatoxin M_1 in Thrissur, Kerala. Indian Journal of Dairy Science 48, 302-305 (250).

Ram, B.P., Hart, L.P., Cole, R.J., Pestka, J.J., 1986: Application of ELISA to retail survey of aflatoxin B_1 in peanut butter. Journal of Food Protection 49, 792-795 (53).

Rama, A., Latifi, F., Bajraktari, D., Ramadani, N., 2015: Assessment of aflatoxin M_1 levels in pasteurized and UHT milk consumed in Prishtina, Kosovo. Food Control 57, 351-354. doi: https://doi.org/10.1016/j.foodcont.2015.04.021 (401).

Rama, A., Montesissa, C., Lucatello, L., Galina, G., Benetti, C., Bajraktari, D., 2016: A study on the occurrence of aflatoxin M_1 in milk consumed in Kosovo during 2009-2010. Food Control 62, 52-55. doi: https://doi.org/10.1016/j.foodcont.2015.10.019 (548).

Ranfft, K., 1983: Aflatoxin M_1 in milkpowder - A fast method for determination and a stock-taking of occurrence. Milchwissenschaft 38, 328-329 (370).

Rastogi, S., Dwivedi, P.D., Khanna, S.K., Das, M., 2004: Detection of aflatoxin M_1 contamination in milk and infant milk products from Indian markets by ELISA. Food Control 15, 287-290. doi: https://doi.org/10.1016/S0956-7135(03)00078-1 (91).

Raza, R., 2006: Occurrence of aflatoxin M_1 in the milk marketed in the city of Karachi, Pakistan. Journal of the Chemical Society of Pakistan 28, 155-157 (229).

Razzazi-Fazeli, E., Noviandi, C.T., Porasuphatana, S., Agus, A., Böhm, J., 2004: A survey of aflatoxin B_1 and total aflatoxin contamination in baby food, peanut and corn products sold at retail in Indonesia analysed by ELISA and HPLC. Mycotoxin Research 20, 51-58 (64).

Redouane-Salah, S., Morgavi, D.P., Arhab, R., Messai, A., Boudra, H., 2015: Presence of aflatoxin M_1 in raw, reconstituted, and powdered milk samples collected in Algeria. Environmental Monitoring and Assessment 187: 375. doi: https://doi.org/10.1007/s10661-015-4627-y (406).

Rezaei, M., Parviz, M., Es'haghi Gorji, M., Shariatifar, N., Hosseini, M.A., Habibi, S., 2014: Occurrence of aflatoxin M_1 in milk in Qom, Iran. Italian Journal of Food Science 26, 325-328 (507).

Riahi-Zanjani, B., Balali-Mood, M., 2013: Aflatoxin M_1 contamination in commercial pasteurized milk from local markets in Fariman, Iran. Mycotoxin Research 29, 271-274. doi: https://doi.org/10.1007/s12550-013-0179-6 (506).

Richard, J.L., Arp, L.H., 1979: Natural occurrence of the mycotoxin penitrem A in moldy cream cheese. Mycopathologia 67, 107-109 (137).

Rodríguez, A., Rodríguez, M., Martín, A., Delgado, J., Córdoba, J.J., 2012: Presence of ochratoxin A on the surface of dry-cured Iberian ham after initial fungal growth in the drying stage. Meat Science 92, 728-734. doi: https://doi.org/10.1016/j.meatsci.2012.06.029 (168).

Rodríguez Velasco, M.L., Calonge Delso, M.M., Ortónez Escudero, D., 2003: ELISA and HPLC determination of the occurrence of aflatoxin M_1 in raw cow's milk. Food Additives and Contaminants 20, 276-280. doi: https://doi.org/10.1080/0265203021000045208 (272).

Rohani, F.G., Aminaee, M.M., Kianfar, M., 2011: Survey of aflatoxin M_1 in cow's milk for human consumption in Kerman Province of Iran. Food Additives and Contaminants: Part B 4, 191-194. doi: https://doi.org/10.1080/19393210.2011.599866 (504).

Rousseau, D.M., Candlish, A.A.G., Slegers, G.A., Van Peteghem, C.H., Stimson, W.H., Smith, J.E. 1987: Detection of ochratoxin A in porcine kidneys by a monoclonal antibody-based radioimmunoassay. Applied and Environmental Microbiology 53, 514-518 (181).

Rousseau, D.M., Slegers, G.A., Van Peteghem, C.H., 1986: Solid-phase radioimmunoassay of ochratoxin A in serum. Journal of Agricultural and Food Chemistry 34, 862-865 (385).

Rousseau, D.M., Van Peteghem, C.H., 1989: Spontaneous occurrence of ochratoxin A residues in porcine kidneys in Belgium. Bulletin of Environmental Contamination and Toxicology 42, 181-186 (178).

Roussi, V., Govaris, A., Varagouli, A., Botsoglou, N.A., 2002: Occurrence of aflatoxin M_1 in raw and market milk commercialized in Greece. Food Additives and Contaminants 19, 863-868. doi: https://doi.org/10.1080/02652030210146864 (195).

Ruangwises, N., Ruangwises, S., 2010: Aflatoxin M_1 contamination in raw milk within the central region of Thailand. Bulletin of Environmental Contamination and Toxicology 85, 195-198. doi: https://doi.org/10.1007/s00128-010-0056-3 (302).

Ruangwises, S., Ruangwises, N., 2009: Occurrence of aflatoxin M_1 in pasteurized milk of the school milk project in Thailand. Journal of Food Protection 72, 1761-1763 (303).

Rubert, J., León, N., Sáez, C., Martins, C.P.B., Godula, M., Yusà, V., Mañes, J., Soriano, J.M., Soler, C., 2014: Evaluation of mycotoxins and their metabolites in human breast milk using liquid chromatography coupled to high resolution mass spectrometry. Analytica Chimica Acta 820, 39-46. doi: https://doi.org/10.1016/j.aca.2014.02.009 (513).

Rubio, R., Licón, C.C., Berruga, M.I., Molina, M.P., Molina, A., 2011: Short communication: occurrence of aflatoxin M_1 in the Manchego cheese supply chain. Journal of Dairy Science 94, 2775-2778. doi: https://doi.org/10.3168/jds.2010-4017 (153).

Rutqvist, L., Björklund, N.-E., Hult, K., Gatenbeck, S., 1977: Spontaneous occurrence of ochratoxin residues in kidneys of fattening pigs. Zentralblatt für Veterinär Medizin A 24, 402-408 (179).

Saad, A.M., Abdelgadir, A.M., Moss, M.O., 1989: Aflatoxin in human and camel milk in Abu Dhabi, United Arab Emirates. Mycotoxin Research 5, 57-60 (230).

Saad, A.M., Abdelgadir, A.M., Moss, M.O., 1995: Exposure of infants to aflatoxin M_1 from mothers' breast milk in Abu Dhabi, UAE. Food Additives and Contaminants 12, 255-261 (331).

Sabino, M., Purchio, A., Milanez, T.V., 1996: Survey of aflatoxins B_1, M_1 and aflatoxicol in poultry and swine tissues from farms located in the states of Rio Grande do Sul and Santa Catarina, Brazil. Revista de Microbiologia 27, 189-191 (188).

Sabino, M., Purchio, A., Zorzetto, M.A.P., 1989: Variations in the levels of aflatoxin in cows milk consumed in the city of São Paulo, Brazil. Food Additives and Contaminants 6, 321-326 (251).

Saccà, E., Boscolo, D., Vallati, A., Ventura, W., Bigaran, F., Piasentier, E., 2009: Aflatoxin occurrence in milk and supplied concentrates of goat farms of north-eastern Italy. Journal of the Science of Food and Agriculture 89, 487-493. doi: https://doi.org/10.1002/jsfa.3478 (512).

Sadeghi, N., Oveisi, M.R., Jannat, B., Hajimahmoodi, M., Bonyani, H., Jannat, F., 2009: Incidence of aflatoxin M_1 in human breast milk in Tehran, Iran. Food Control 20, 75-78. doi: https://doi.org/10.1016/j.foodcont.2008.02.005 (339).

Sadia, A., Jabbar, M.J., Deng, Y., Hussain, E.A., Riffat, S., Naveed, S., Arif, M., 2012: A survey of aflatoxin M_1 in milk and sweets of Punjab, Pakistan. Food Control 26, 235-240. doi: https://doi.org/10.1016/j.foodcont.2012.01.055 (511).

Sahin, H.Z., Celik, M., Kotay, S., Kabak, B., 2016: Aflatoxins in dairy cow feed, raw milk and milk products from Turkey. Food Additives and Contaminants: Part B 9, 152-158. doi: https://doi.org/10.1080/19393210.2016.1152599 (510).

Sahindokuyucu Kocasari, F., 2014: Occurrence of aflatoxin M_1 in UHT milk and infant formula samples consumed in Burdur, Turkey. Environmental Monitoring and Assessment 186, 6363-6368. doi: https://doi.org/10.1007/s10661-014-3860-0 (202).

Saitanu, K., 1997: Incidence of aflatoxin M_1 in Thai milk products. Journal of Food Protection 60, 1010-1012 (196).

Saito, K., Nishijima, M., Yasuda, K., Kamimura, H., Ibe, A., Nagayama, T., Ushiyama, H., Naoi, Y., 1980: Natural occurrence of aflatoxins in commercial cheese. Journal of the Hygienic Society of Japan, Tokyo 21, 472-475 (29).

Salem, D.A., 2002: Natural occurrence of aflatoxins in feedstuffs and milk of dairy farms in Assiut Province, Egypt. Wiener Tierärztliche Monatsschrift 89, 86-91 (252).

Sani, A.M., Nikpooyan, H., 2012: Determination of aflatoxin M_1 in milk by high-performance liquid chromatography in Mashhad (north east of Iran). Toxicology and Industrial Health 29, 334-338. doi: https://doi.org/10.1177/0748233711434954 (509).

Sani, A.M., Nikpooyan, H., Moshiri, R., 2010: Aflatoxin M_1 contamination and antibiotic residue in milk in Khorasan province, Iran. Food and Chemical Toxicology 48, 2130-2132. doi: https://doi.org/10.1016/j.fct.2010.05.015 (304).

Santili, A.B.N., De Camargo, A.C., De Syllos Rosa Nunes, R., Da Gloria, E.M., Machado, P.F., Cassoli, L.D., Dos Santos Dias, C.T., Calori-Domingues, M.A., 2015: Aflatoxin M_1 in raw milk from different regions of São Paulo state - Brazil. Food Additives and Contaminants: Part B 8, 207-214. doi: https://doi.org/10.1080/19393210.2015.1048538 (490).

Santini, A., Raiola, A., Ferrantelli, V., Giangrosso, G., Macaluso, A., Bognanno, M., Galvano, F., Ritieni, A., 2013: Aflatoxin M_1 in raw, UHT milk and dairy products in Sicily (Italy). Food Additives and Contaminants: Part B 6, 181-186. doi: https://doi.org/10.1080/19393210.2013.780186 (508).

Sarımehmetoglu, B., Kuplulu, O., Celik, T.H., 2004: Detection of aflatoxin M_1 in cheese samples by ELISA. Food Control 15, 45-49. doi: https://doi.org/10.1016/S0956-7135(03)00006-9 (35).

Sartori, A.V., Swensson de Mattos, J., de Moraes, M.H.P., da Nóbrega, A.W., 2015: Determination of aflatoxins M_1, M_2, B_1, B_2, G_1, and G_2 and ochratoxin A in UHT and powdered milk by modified QuEChERS method and ultra-high-performance liquid chromatography tandem mass spectrometry. Food Analytical Methods 8, 2321-2330. doi: https://doi.org/10.1007/s12161-015-0128-4 (543).

Sassahara, M., Pontes Netto, D., Yanaka, E.K., 2005: Aflatoxin occurrence in foodstuff supplied to dairy cattle and aflatoxin M_1 in raw milk in the North of Paraná State. Food and Chemical Toxicology 43, 981-984. doi: https://doi.org/10.1016/j.fct.2005.02.003 (260).

Schirone, M., Visciano P., Olivastri, A.M.A., Tofalo, R., Perpetuini, G., Suzzi, G., 2015: A one-year survey on aflatoxin M_1 in raw milk. Italian Journal of Food Science 27, 143-148 (559).

Schuddeboom, L.J., 1983: Development of legislation concerning mycotoxins in dairy products in The Netherlands. Microbiologie – Aliments – Nutrition 1, 179-185 (244).

Schwartzbord, J.R., Brown, D.L., 2015: Aflatoxin contamination in Haitian peanut products and maize and the safety of oil processed from contaminated peanuts. Food Control 56, 114-118. doi: https://doi.org/10.1016/j.foodcont.2015.03.014 (81).

Scott, P.M., Kennedy, B.P.C., 1976: Analysis of blue cheese for roquefortine and other alkaloids from *Penicillium roqueforti*. Journal of Agricultural and Food Chemistry 24, 865-868 (122).

Sefidgar, S.A.A., Azizi, G., Khosravi, A.R., Roudbar-Mohammadi, S., 2008: Presence of aflatoxin M_1 in raw milk at cattle farms in Babol, Iran. Pakistan Journal of Biological Sciences 11, 484-486 (514).

Shank, R.C., Wogan, G.N., Gibson, J.B., Nondasuta, A., 1972: Dietary aflatoxins and human liver cancer. II. Aflatoxins in marketed foods and foodstuffs of Thailand and Hong Kong. Food Cosmetics and Toxicology 10, 61-69 (162).

Shon, D.-H., Lim, S.-H., Lee, Y.-W., 1996: Detection of aflatoxin in cow's milk by an enzyme-linked immunosorbent assay. Korean Journal of Applied Microbiology and Biotechnology 24, 630-635 (257).

Shuib, N.S., Makahleh, A., Salhimi, S.M., Saad, B., 2017: Natural occurrence of aflatoxin M_1 in fresh cow milk and human milk in Penang, Malaysia. Food Control 73, 966-970. doi: https://doi.org/10.1016/j.foodcont.2016.10.013 (553).

Shundo, L., Navas, S.A., Lamardo, L.C.A., Ruvieri, V., Sabino, M., 2009: Estimate of aflatoxin M_1 exposure in milk and occurrence in Brazil. Food Control 20, 655-657. doi: https://doi.org/10.1016/j.foodcont.2008.09.019 (305).

Shundo, L., Sabino, M., 2006: Aflatoxin M_1 in milk by immunoaffinity column cleanup with TLC/HPLC determination. Brazilian Journal of Microbiology 37, 164-167. doi: https://doi.org/10.1590/S1517-83822006000200013 (306).

Siame, B.A., Mpuchane, S.F., Gashe, B.A., Allotey, J., Teffera, G., 1998: Occurrence of aflatoxins, fumonisin B_1, and zearalenone in foods and feeds in Botswana. Journal of Food Protection 61, 1670-1673 (56).

Siddappa, V., Nanjegowda, D.K., Viswanath, P., 2012: Occurrence of aflatoxin M_1 in some samples of UHT, raw & pasteurized milk from Indian states of Karnataka and Tamilnadu. Food and Chemical Toxicology 50, 4158-4162. doi: https://doi.org/10.1016/j.fct.2012.08.034 (518).

Sineque, A.R., Macuamule, C.L., Dos Anjos, F.R., 2017: Aflatoxin B_1 contamination in chicken livers and gizzards from industrial and small abattoirs, measured by ELISA technique in Maputo, Mozambique. International Journal of Environmental Research and Public Health 14 (9), 951. doi: https://doi.org/10.3390/ijerph14090951 (517).

Sinha, A.K., Ranjan, K.S., 1991: A report of mycotoxin contamination in Bhutanese cheese. Journal of Food Science and Technology, India 28, 398-399 (119).

Skaug, M.A., 1999: Analysis of Norwegian milk and infant formulas for ochratoxin A. Food Additives and Contaminants 16, 75-78 (316).

Skaug, M.A., Helland, I., Solvoll, K., Saugstad, O.D., 2001: Presence of ochratoxin A in human milk in relation to dietary intake. Food Additives and Contaminants 18, 321-327. doi: https://doi.org/10.1080/02652030010021477 (345).

Skaug, M.A., Størmer, F.C., Saugstad, O.D., 1998: Ochratoxin A: a naturally occurring mycotoxin found in human milk samples from Norway. Acta Pædiatrica 87, 1275-1278 (342).

Škrbić, B., Antić, I., Živančev, J., 2015: Presence of aflatoxin M_1 in white and hard cheese samples from Serbia. Food Control 50, 111-117. doi: https://doi.org/10.1016/j.foodcont.2014.08.031 (516).

Škrbić, B., Živančev, J., Antić, I., Godula, M., 2014: Levels of aflatoxin M_1 in different types of milk collected in Serbia: assessment of human and animal exposure. Food Control 40, 113-119. doi: https://doi.org/10.1016/j.foodcont.2013.11.039 (515).

Sohrabi, N., Gharahkoli, H., 2016: A seasonal study for determination of aflatoxin M_1 level in dairy products in Iranshahr, Iran. Current Medical Mycology 2, 27-31. doi: https://doi.org/10.18869/acadpub.cmm.2.3.27 (417).

Souza, S.V.C., Vargas, E.A., Junqueira, R.G., 1999: Efficiency of an ELISA kit on the detection and quantification of aflatoxin M_1 in milk and it's occurrence in the state of Minas Gerais. Ciência e Tecnologia de Alimentos 19, 401-405. doi: https://doi.org/10.1590/S0101-20611999000300019 (221).

Srivastava, V.P., Bu-Abbas, A., Alaa-Basuny, W., Al-Johar, W., Al-Mufti, S., Siddiqui, M.K.J., 2001: Aflatoxin M_1 contamination in commercial samples of milk and dairy products in Kuwait. Food Additives and Contaminants 18, 993-997. doi: https://doi.org/10.1080/02652030110050357 (205).

Stoev, S.D., Dutton, M.F., Njobeh, P.B., Mosonik, J.S., Steenkamp, P.A., 2010: Mycotoxic nephropathy in Bulgarian pigs and chickens: complex aetiology and similarity to Balkan Endemic Nephropathy. Food Additives and Contaminants 27, 72-88. doi: https://doi.org/10.1080/02652030903207227 (519).

Stoloff, L., 1980: Aflatoxin M in perspective. Journal of Food Protection 43, 226-230 (255).

Stubblefield, R.D., 1979: The rapid determination of aflatoxin M_1 in dairy products. The Journal of the American Oil Chemist's Society 56, 800-802 (376).

Sun, D., Qiu, N., Zhou, S., Lyu, B., Zhang, S., Li, J., Zhao, Y., Wu, Y., 2019: Development of sensitive and reliable UPLC-MS/MS methods for food analysis of emerging mycotoxins in China total diet study. Toxins (Basel) 11, 166. doi: https://doi.org/10.3390/toxins11030166 (536).

Sun, W., Han, Z., Aerts, J., Nie, D., Jin, M., Shi, W., Zhao, Z., De Saeger, S., Zhao, Y., Wu, A., 2015: A reliable liquid chromatography-tandem mass spectrometry method for simultaneous determination of multiple mycotoxins in fresh fish and dried seafood. Journal of Chromatography A 1387, 42-48. doi: https://doi.org/10.1016/j.chroma.2015.01.071 (522).

Tabari, M., Tabari, K., Tabari, O., 2011: Occurrence of aflatoxin M_1 in pasteurized doogh commercialized in Tehran, Iran. Australian Journal of Basic and Applied Sciences 5, 1734-1737 (521).

Tabari, M., Tabari, K., Tabari, O., 2012: Aflatoxin M_1 determination in yoghurt produced in Guilan province of Iran using immunoaffinity column and high-performance liquid chromatography. Toxicology and Industrial Health 29, 72-76. doi: https://doi.org/10.1177/0748233712446729 (520).

Tabata, S., Kamimura, H., Tamura, Y., Yasuda, K., Ushiyama, H., Hashimoto, H., Nishijima, M., Nishima, T., 1987: Investigation of aflatoxins contamination in foods and foodstuffs. Journal of Food Hygienic Society of Japan 28, 395-401 (24).

Taguchi, S., Fukushima, S., Sumimoto, T., Yoshida, S., Nishimune, T., 1995: Aflatoxins in foods collected in Osaka, Japan, from 1988 to 1992. Journal of the Association of Official Analytical Chemists International 78, 325-327 (48).

Tajik, H., Rohani, S.M.R., Moradi, M., 2007: Detection of aflatoxin M_1 in raw and commercial pasteurized milk in Urmia, Iran. Pakistan Journal of Biological Sciences 10, 4103-4107 (307).

Tajkarimi, M., Aliabadi, F.S., Nejad, M.S., Pursoltani, H., Motallebi, A.A., Mahdavi, H., 2007: Seasonal study of aflatoxin M_1 contamination in milk in five regions in Iran. International Journal of Food Microbiology 116, 346-349. doi: https://doi.org/10.1016/j.ijfoodmicro.2007.02.008 (308).

Tam, J., Pantazopoulos, P., Scott, P.M., Moisey, J., Dabeka, R.W., Richard, I.D.K., 2011: Application of isotope dilution mass spectrometry: determination of ochratoxin A in the Canadian total diet study. Food Additives and Contaminants: Part A 28, 754-761. doi: https://doi.org/10.1080/19440049.2010.504750 (87).

Tavakoli, H.R., Kamkar, A., Riazipour, M., Mozaffari Nejad, A.S., Shaldehi, H.R., 2013: Assessment of aflatoxin M_1 levels by enzyme-linked immunosorbent assay in yoghurt consumed in Tehran, Iran. Asian Journal of Chemistry 25, 2836-2838 (525).

Tavakoli, H.R., Riazipour, M., Kamkar, A., Shaldehi, H.R., Mozaffari Nejad, A.S., 2012: Occurrence of aflatoxin M_1 in white cheese samples from Tehran, Iran. Food Control 23, 293-295. doi: https://doi.org/10.1016/j.foodcont.2011.07.024 (524).

Taveira, J.A., Midio, A.F., 2001: Incidence of aflatoxin M_1 in milk marketed in São Paulo, Brazil. Italian Journal of Food Science 13, 443-447 (197).

Tchana, A.N., Moundipa, P.F., Tchouanguep, F.M., 2010: Aflatoxin contamination in food and body fluids in relation to malnutrition and cancer status in Cameroon. International Journal of Environmental Research and Public Health 7, 178-188. doi: https://doi.org/10.3390/ijerph7010178 (156).

Tekinşen, K.K., Eken, H.S., 2008: Aflatoxin M_1 levels in UHT milk and kashar cheese consumed in Turkey. Food and Chemical Toxicology 46, 3287-3289. doi: https://doi.org/10.1016/j.fct.2008.07.014 (143).

Tekinşen, K.K., Tekinşen, O.C., 2005: Aflatoxin M_1 in white pickle and Van otlu (herb) cheeses consumed in southeastern Turkey. Food Control 16, 565-568. doi: https://doi.org/10.1016/j.foodcont.2004.02.006 (149).

Tekinşen, K.K., Uçar, G., 2008: Aflatoxin M_1 levels in butter and cream cheese consumed in Turkey. Food Control 19, 27-30. doi: https://doi.org/10.1016/j.foodcont.2007.01.003 (13).

Temamogullari, F., Kanici, A., 2014: Short communication: aflatoxin M_1 in dairy products sold in Şanlıurfa, Turkey. Journal of Dairy Science 97, 162-165 doi: https://doi.org/10.3168/jds.2012-6184 (523).

Tolosa, J., Barba, F.J., Font, G., Ferrer, E., 2019: Mycotoxin incidence in some fish products: QuEChERS methodology and liquid chromatography linear ion trap tandem mass spectrometry approach. Molecules 24, 527. doi: https://doi.org/10.3390/molecules24030527 (489).

Tolosa, J., Font, G., Mañes, J., Ferrer, E., 2013: Natural occurrence of *Fusarium* mycotoxins in aquaculture fish food. Revista de Toxicologia 30, 193-197 (528).

Tolosa, J., Font, G., Mañes, J., Ferrer, E., 2014: Natural occurrence of emerging *Fusarium* mycotoxins in feed and fish from aquaculture. Journal of Agricultural and Food Chemistry 62, 12462-12470. doi: https://doi.org/10.1021/jf5036838 (527).

Tomerak, R.H., Shaban, H.H., Khalafallah, O.A., El Shazly, M.N., 2011: Assessment of exposure of Egyptian infants to aflatoxin M_1 through breast milk. Journal of the Egyptian Public Health Association 86, 51-55. doi: https://doi.org/10.1097/01.EXP.0000399138.90797.40 (541).

Tonon, K.M., Savi, G.D., Scussel, V.M., 2018: Application of a LC-MS/MS method for multi-mycotoxin analysis in infant formula and milk-based products for young children commercialized in southern Brazil. Journal of Environmental Science and Health: Part B 2, 1-7. doi: https://doi.org/10.1080/03601234.2018.1474560 (407).

Torović, L., 2015: Aflatoxin M_1 in processed milk and infant formulae and corresponding exposure of adult population in Serbia in 2013-2014. Food Additives and Contaminants: Part B 8, 235-244. doi: https://doi.org/10.1080/19393210.2015.1063094 (526).

Toscani, T., Moseriti, A., Dossena, A., Dall'Astra, C., Simoncini, N., Virgili, R., 2007: Determination of ochratoxin A in dry-cured meat products by a HPLC-FLD quantitative method. Journal of Chromatography B 855, 242-248. doi: https://doi.org/10.1016/j.jchromb.2007.05.010 (166).

Trombete, F.M., De Castro, I.M., Da Silva Teixeira, A., Saldanha, T., Fraga, M.E., 2014: Aflatoxin M_1 contamination in grated Parmesan cheese marketed in Rio de Janeiro-Brazil. Brazilian Archives of Biology and Technology 57, 269-273. doi: https://doi.org/10.1590/S1516-89132013005000015 (529).

Trucksess, M.W., Page, S.W., 1986: Examination of imported cheeses for aflatoxin M_1. Journal of Food Protection 49, 632-633 (30).

Tung, T.C., Ling, K.H., 1968: Study on aflatoxin of foodstuffs in Taiwan. The Journal of Vitaminology 14, 48-52 (54).

Turconi, G., Guarcello, M., Liveri, C., Comizzoli, S., Maccarini, L., Castellazzi, A.M., Pietri, A., Piva, G., Roggi, C., 2004: Evaluation of xenobiotics in human milk and ingestion by the newborn. An epidemiological survey in Lombardy (northern Italy). European Journal of Nutrition 43, 191-197. doi: https://doi.org/10.1007/s00394-004-0458-2 (322).

Turcotte, A.-M., Scott, P.M., Tague, B., 2013: Analysis of cocoa products for ochratoxin A and aflatoxins. Mycotoxin Research 29, 193-201. doi: https://doi.org/10.1007/s12550-013-0167-x (16).

Tyllinen, M., Hintikka, E.-L., 1982: Occurrence of ochratoxin A in swine kidneys and feed in Finland. Nordisk Jordbruksforskning 64, 298-299 (180).

Ul Hassan, Z., Al Thani, R., Atia, F.A., Al Meer, S., Migheli, Q., Jaoua, S., 2018: Co-occurrence of mycotoxins in commercial formula milk and cereal-based baby food on the Qatar market. Food Additives and Contaminants: Part B 11, 191-197. doi: https://doi.org/10.1080/19393210.2018.1437785 (277).

Unusan, N., 2006: Occurrence of aflatoxin M_1 in UHT milk in Turkey. Food and Chemical Toxicology 44, 1897-1900. doi: https://doi.org/10.1016/j.fct.2006.06.010 (358).

Usleber, E., Dade, M., Schneider, E., Dietrich, R., Bauer, J., Märtlbauer, E., 2008: Enzyme immunoassay for mycophenolic acid in milk and cheese. Journal of Agricultural and Food Chemistry 56, 6857-6862. doi: https://doi.org/10.1021/jf801063w (537).

Vagef, R., Mahmoudi, R., 2013: Occurrence of aflatoxin M_1 in raw and pasteurized milk produced in west region of Iran (during summer and winter). International Food Research Journal 20, 1421-1425 (530).

Vahl, M., Jørgensen, K., 1998: Determination of aflatoxins in food using LC/MS/MS. Zeitschrift für Lebensmittel-Untersuchung und –Forschung 206, 243-245 (55).

Valitutti, F., De Santis, B., Trovato, C.M., Montuori, M., Gatti, S., Oliva, S., Brera, C., Catassi, C., 2018: Assessment of mycotoxin exposure in breast-feeding mothers with celiac disease. Nutrients 10, 336. doi: https://doi.org/10.3390/nu10030336 (160).

Var, I., Kabak, B., 2008: Detection of aflatoxin M_1 in milk and dairy products consumed in Adana, Turkey. International Journal of Dairy Technology 62, 15-18. doi: https://doi.org/10.1111/j.1471-0307.2008.00440.x (12).

Veršilovskis, A., Van Peteghem, C., De Saeger, S., 2009: Determination of sterigmatocystin in cheese by high-performance liquid chromatography-tandem mass spectrometry. Food Additives and Contaminants: Part A 26, 127-133. doi: https://doi.org/10.1080/02652030802342497 (111).

Veselý, D., Veselá, D., 1983: Determination of M_1 aflatoxin in milk and of its toxic effect on chick embryo. Veterinární Medicína 28, 57-61 (243).

Virdis, S., Corgiolu, G., Scarano, C., Pilo, A.L., De Santis, E.P.L., 2008: Occurrence of aflatoxin M_1 in tank bulk goat milk and ripened goat cheese. Food Control 19, 44-49. doi: https://doi.org/10.1016/j.foodcont.2007.02.001 (31).

Virdis, S., Scarano, C., Spanu, V., Murittu, G., Spanu, C., Ibba, I., De Santis, E.P.L., 2014: A survey on aflatoxin M_1 content in sheep and goat milk produced in Sardina region, Italy (2005-2013). Italian Journal of Food Safety 3, 206-209. doi: https://doi.org/10.4081/ijfs.2014.4517 (557).

Visconti, A., Bottalico, A., Solfrizzo, M., 1985: Aflatoxin M_1 in milk, in southern Italy. Mycotoxin Research 1, 71-75 (236).

Waliyar, F., Reddy, S.V., Subramaniam, K., Reddy, T.Y., Rama Devi, K., Craufurd, P.Q., Wheeler, T.R., 2003: Importance of mycotoxins in food and feed in India. Aspects of Applied Biology 68, 147-154 (256).

Wang, H., Zhou, X.-J., Liu, Y.-Q., Yang, H.-M., Guo, Q.-L., 2011: Simultaneous determination of chloramphenicol and aflatoxin M_1 residues in milk by triple quadrupole liquid chromatography-tandem mass spectrometry. Journal of Agricultural and Food Chemistry 59, 3532-3538. doi: https://doi.org/10.1021/jf2006062 (297).

Wang, L., Zhang, Q., Yan, Z., Tan, Y., Zhu, R., Yu, D., Yang, H., Wu, A., 2018: Occurrence and quantitative risk assessment of twelve mycotoxins in eggs and chicken tissues in China. Toxins (Basel) 10, 477. doi: https://doi.org/10.3390/toxins10110477 (326).

Wang, Y., Liu, X., Xiao, C., Wang, Z., Wang, J., Xiao, H., Cui, L., Xiang, Q., Yue, T., 2012: HPLC determination of aflatoxin M_1 in liquid milk and milk powder using solid phase extraction on OASIS HLB. Food Control 28, 131-134. doi: https://doi.org/10.1016/j.foodcont.2012.04.037 (531).

Ware, G.M., Thorpe, C.W., Pohland, A.E., 1980: Determination of roquefortine in blue cheese and blue cheese dressing by high pressure liquid chromatography with ultraviolet and electrochemical detectors. Journal of the Association of Official Analytical Chemists 63, 637-641 (125).

Wei, D.-L., Wei, R.-D., 1980: High pressure liquid chromatographic determination of aflatoxins in peanut and peanut products of Taiwan. Proceedings of the National Science Council, Republic of China 4, 152-155 (51).

Wild, C.P., Pionneau, F.A., Montesano, R., Mutiro, C.F., Chetsanga, C.J., 1987: Aflatoxin detected in human breast milk by immunoassay. International Journal of Cancer 40, 328-333 (330).

Winterlin, W., Hall, G., Hsieh, D.P.H., 1979: On-column chromatographic extraction of aflatoxin M_1 from milk and determination by reversed phase high performance liquid chromatography. Analytical Chemistry 51, 1873-1874 (222).

Wiśniewska-Dmytrow, H., Żmudzki, J., Burek, O., Pietruszka, K., 2013: Official control of ochratoxin A in food of animal origin in Poland between 2003 and 2012. Bulletin of the Veterinary Institute in Pulawy 57, 519-523. doi: https://doi.org/10.2478/bvip-2013-0090 (365).

Wood, G.E., 1989: Aflatoxins in domestic and imported foods and feeds. Journal of the Association of Official Analytical Chemists 72, 543-548 (72).

Xiong, J., Xiong, L., Zhou, H., Liu, Y., Wu, L., 2018: Occurrence of aflatoxin B_1 in dairy cow feedstuff and aflatoxin M_1 in UHT and pasteurized milk in central China. Food Control 92, 386-390. doi: https://doi.org/10.1016/j.foodcont.2018.05.022 (547).

Xiong, J.L., Wang, Y.M., Ma, M.R., Liu, J.X., 2013: Seasonal variation of aflatoxin M_1 in raw milk from the Yangtze river delta region of China. Food Control 34, 703-706. doi: https://doi.org/10.1016/j.foodcont.2013.06.024 (533).

Yang, L.-x., Liu, Y.-p., Miao, H., Dong, B., Yang, N.-j., Chang, F.-q., Sun, J.-b., 2011: Determination of aflatoxins in edible oil from markets in Hebei Province of China by liquid chromatography-tandem mass spectrometry. Food Additives and Contaminants: Part B 4, 244-247. doi: https://doi.org/10.1080/19393210.2011.632694 (109).

Yapar, K., Elmali, M., Kart, A., Yaman, H., 2008: Aflatoxin M_1 levels in different type of cheese products produced in Turkey. Medycyna Weterynaryjna 64, 53-55 (132).

Yaroglu, T., Oruc, H.H., Tayar, M., 2005: Aflatoxin M_1 levels in cheese samples from some provinces of Turkey. Food Control 16, 883-885. doi: https://doi.org/10.1016/j.foodcont.2004.08.001 (135).

Yentür, G., Er, B., Özkan, M.G., Öktem, A.B., 2006: Determination of aflatoxins in peanut butter and sesame samples using high-performance liquid chromatography method. European Food Research and Technology 224, 167-170. doi: https://doi.org/10.1007/s00217-006-0310-4 (59).

Yoon, B.R., Hong, S.-Y., Cho, S.M., Lee, K.R., Kim, M., Chung, S.H., 2016: Aflatoxin M_1 levels in dairy products from South Korea determined by high performance liquid chromatography with fluorescence detection. Journal of Food and Nutrition Research 55, 171-180 (561).

Yosef, T.A., Al-Julaifi, M.Z., Hussein, Y.A., Al-Shokair, S.S., Al-Amer, A.S., 2014: Occurrence of aflatoxin M_1 in raw camel milk in El-Ahsa governorate, Saudi Arabia. Nature and Science 12, 1-7 (532).

Younis, Y.M.H., Malik, K.M., 2003: TLC and HPLC assays of aflatoxin contamination in Sudanese peanuts and peanut products. Kuwait Journal of Science & Engineering 30, 79-93 (77).

Yurdun, T., Özmenteşe, N., 2001: Presence of aflatoxin M_1 in milk and dairy products in Turkey. Toxicology Letters 123 (Suppl), 39-40 (271).

Zambonin, C.G., Monaci, L., Aresta, A., 2001: Determination of cyclopiazonic acid in cheese samples using solid-phase microextraction and high performance liquid chromatography. Food Chemistry 75, 249-254 (117).

Zarba, A., Wild, C.P., Hall, A.J., Montesano, R., Hudson, G.J., Groopman, J.D., 1992: Aflatoxin M_1 in human breast milk from The Gambia, West Africa, quantified by combined monoclonal antibody immunoaffinity chromatography and HPLC. Carcinogenesis 13, 891-894 (324).

Zhang, K., Wong, J.W., Hayward, D.G., Vaclavikova, M., Liao, C.-D., Trucksess, M.W., 2013: Determination of mycotoxins in milk-based products and infant formula using stable isotope dilution assay and liquid chromatography tandem mass spectrometry. Journal of Agricultural and Food Chemistry 61, 6265-6273. doi: https://doi.org/10.1021/jf4018838 (534).

Zheng, N., Sun, P., Wang, J.Q., Zhen, Y.P., Han, R.W., Xu, X.M., 2013: Occurrence of aflatoxin M_1 in UHT milk and pasteurized milk in China market. Food Control 29, 198-201. doi: https://doi.org/10.1016/j.foodcont.2012.06.020 (193).

Zheng, N., Wang, J.-Q., Han, R.-W., Zhen, Y.P., Xu, X.-M., Sun, P., 2013: Survey of aflatoxin M_1 in raw milk in the five provinces of China. Food Additives and Contaminants: Part B 6, 110-115 doi: https://doi.org/10.1080/19393210.2012.763191 (445).

Zhou, J., Xu, J.-J., Cong, J.-M., Cai, Z.-X., Zhang, J.-S., Wang, J.-L., Ren, Y.-P., 2018: Optimization for quick, easy, cheap, effective, rugged and safe extraction of mycotoxins and veterinary drugs by response surface methodology for application to egg and milk. Journal of Chromatography A 1532, 20-29. doi: https://doi.org/10.1016/j.chroma.2017.11.050 (73).

Zhu, R., Zhao, Z., Wang, J., Bai, B., Wu, A., Yan, L., Song, S., 2015: A simple sample pretreatment method for multi-mycotoxin determination in eggs by liquid chromatography tandem mass spectrometry. Journal of Chromatography A 1417, 1-7. doi: https://doi.org/10.1016/j.chroma.2015.09.028 (535).

Zimmerli, B., Dick, R., 1995: Determination of ochratoxin A at the ppt level in human blood, serum, milk and some foodstuffs by high-performance liquid chromatography with enhanced fluorescence detection and immunoaffinity column cleanup: methodology and Swiss data. Journal of Chromatography B 666, 85-99 (104).

Zinedine, A., González-Osnaya, L., Soriano, J.M., Moltó, J.C., Idrissi, L., Mañes, J., 2007: Presence of aflatoxin M_1 in pasteurized milk from Morocco. International Journal of Food Microbiology 114, 25-29. doi: https://doi.org/10.1016/j.ijfoodmicro.2006.11.001 (209).

Zou, Z., He, Z., Li, H., Han, P., Tang, J., Xi, C., Li, Y., Zhang, L., Li, X., 2012: Development and application of a method for the analysis of two trichothecenes: deoxynivalenol and T-2 toxin in meat in China by HPLC-MS/MS. Meat Science 90, 613-617. doi: https://doi.org/10.1016/j.meatsci.2011.10.002 (103).